T0206113

# VECTORS, PURE AND APPLIED

A General Introduction to Linear Algebra

Many books on linear algebra focus purely on getting students through exams, but this text explains both the how and the why of linear algebra and enables students to begin thinking like mathematicians. The author demonstrates how different topics (geometry, abstract algebra, numerical analysis, physics) make use of vectors in different ways, and how these ways are connected, preparing students for further work in these areas.

The book is packed with hundreds of exercises ranging from the routine to the challenging. Sketch solutions of the easier exercises are available online.

T. W. KÖRNER is Professor of Fourier Analysis in the Department of Pure Mathematics and Mathematical Statistics at the University of Cambridge. His previous books include *Fourier Analysis* and *The Pleasures of Counting*.

# VECTORS, PURE AND APPLIED

## A General Introduction to Linear Algebra

T. W. KÖRNER

*Trinity Hall, Cambridge*

CAMBRIDGE
UNIVERSITY PRESS

# CAMBRIDGE
## UNIVERSITY PRESS

University Printing House, Cambridge CB2 8BS, United Kingdom

One Liberty Plaza, 20th Floor, New York, NY 10006, USA

477 Williamstown Road, Port Melbourne, VIC 3207, Australia

314-321, 3rd Floor, Plot 3, Splendor Forum, Jasola District Centre, New Delhi - 110025, India

79 Anson Road, #06-04/06, Singapore 079906

Cambridge University Press is part of the University of Cambridge.

It furthers the University's mission by disseminating knowledge in the pursuit of education, learning and research at the highest international levels of excellence.

www.cambridge.org
Information on this title: www.cambridge.org/9781107675223

First published 2013

*A catalogue record for this publication is available from the British Library*

*Library of Congress Cataloging in Publication data*
Körner, T. W. (Thomas William), 1946–
Vectors, pure and applied : a general introduction to linear algebra / T. W. Körner.
pages cm
Includes bibliographical references and index.
ISBN 978-1-107-03356-6 (hardback) – ISBN 978-1-107-67522-3 (paperback)
1. Vector algebra. 2. Algebras, Linear. I. Title.
QA200.K67 2013
516´.182 – dc23 2012036797

ISBN 978-1-107-03356-6 Hardback
ISBN 978-1-107-67522-3 Paperback

Additional resources for this publication at www.dpmms.cam.ac.uk/~twk/

In general the position as regards all such new calculi is this. – That one cannot accomplish by them anything that could not be accomplished without them. However, the advantage is, that, provided that such a calculus corresponds to the inmost nature of frequent needs, anyone who masters it thoroughly is able – without the unconscious inspiration which no one can command – to solve the associated problems, even to solve them mechanically in complicated cases in which, without such aid, even genius becomes powerless.... Such conceptions unite, as it were, into an organic whole, countless problems which otherwise would remain isolated and require for their separate solution more or less of inventive genius.

(Gauss *Werke*, Bd. 8, p. 298 (quoted by Moritz [24]))

For many purposes of physical reasoning, as distinguished from calculation, it is desirable to avoid explicitly introducing ... Cartesian coordinates, and to fix the mind at once on a point of space instead of its three coordinates, and on the magnitude and direction of a force instead of its three components. ... I am convinced that the introduction of the idea [of vectors] will be of great use to us in the study of all parts of our subject, and especially in electrodynamics where we have to deal with a number of physical quantities, the relations of which to each other can be expressed much more simply by [vectorial equations rather] than by the ordinary equations.

(Maxwell *A Treatise on Electricity and Magnetism* [21])

We [Halmos and Kaplansky] share a love of linear algebra. ... And we share a philosophy about linear algebra: we think basis-free, we write basis-free, but when the chips are down we close the office door and compute with matrices like fury.

(Kaplansky in *Paul Halmos: Celebrating Fifty Years of Mathematics* [17])

Marco Polo describes a bridge, stone by stone.

'But which is the stone that supports the bridge?' Kublai Khan asks.

'The bridge is not supported by one stone or another,' Marco answers, 'but by the line of the arch that they form.'

Kublai Khan remains silent, reflecting. Then he adds: 'Why do you speak to me of the stones? It is only the arch that matters to me.'

Polo answers: 'Without stones there is no arch.'

(Calvino *Invisible Cities* (translated by William Weaver) [8])

# Contents

# Introduction

There exist several fine books on vectors which achieve concision by only looking at vectors from a single point of view, be it that of algebra, analysis, physics or numerical analysis (see, for example, [18], [19], [23] and [28]). This book is written in the belief that it is helpful for the future mathematician to see all these points of view. It is based on those parts of the first and second year Cambridge courses which deal with vectors (omitting the material on multidimensional calculus and analysis) and contains roughly 60 to 70 hours of lectured material.

The first part of the book contains first year material and the second part contains second year material. Thus concepts reappear in increasingly sophisticated forms. In the first part of the book, the inner product starts as a tool in two and three dimensional geometry and is then extended to $\mathbb{R}^n$ and later to $\mathbb{C}^n$. In the second part, it reappears as an object satisfying certain axioms. I expect my readers to read, or skip, rapidly through familiar material, only settling down to work when they reach new results. The index is provided mainly to help such readers who come upon an unfamiliar term which has been discussed earlier. Where the index gives a page number in a different font (like **389**, rather than 389) this refers to an exercise. Sometimes I discuss the relation between the subject of the book and topics from other parts of mathematics. If the reader has not met the topic (morphisms, normal distributions, partial derivatives or whatever), she should simply ignore the discussion.

Random browsers are informed that, in statements involving $\mathbb{F}$, they may take $\mathbb{F} = \mathbb{R}$ or $\mathbb{F} = \mathbb{C}$, that $z^*$ is the complex conjugate of $z$ and that 'self-adjoint' and 'Hermitian' are synonyms. If $T : A \to B$ is a function we sometimes write $T(a)$ and sometimes $Ta$.

There are two sorts of exercises. The first form part of the text and provide the reader with an opportunity to think about what has just been done. There are sketch solutions to most of these on my home page www.dpmms.cam.ac.uk/~twk/.

These exercises are intended to be straightforward. If the reader does not wish to attack them, she should simply read through them. If she does attack them, she should remember to state reasons for her answers, whether she is asked to or not. Some of the results are used later, but no harm should come to any reader who simply accepts my word that they are true.

The second type of exercise occurs at the end of each chapter. Some provide extra background, but most are intended to strengthen the reader's ability to use the results of the preceding chapter. If the reader finds all these exercises easy or all of them impossible, she is reading the wrong book. If the reader studies the entire book, there are many more exercises than she needs. If she only studies an individual chapter, she should find sufficiently many to test and reinforce her understanding.

My thanks go to several student readers and two anonymous referees for removing errors and improving the clarity of my exposition. It has been a pleasure to work with Cambridge University Press.

I dedicate this book to the Faculty Board of Mathematics of the University of Cambridge. My reasons for doing this follow in increasing order of importance.

(1)  No one else is likely to dedicate a book to it.
(2)  No other body could produce Minute 39 (a) of its meeting of 18th February 2010 in which it is laid down that a basis is not an *ordered* set but an *indexed* set.
(3)  This book is based on syllabuses approved by the Faculty Board and takes many of its exercises from Cambridge exams.
(4)  I need to thank the Faculty Board and everyone else concerned for nearly 50 years spent as student and teacher under its benign rule. Long may it flourish.

# Part I

Familiar vector spaces

# 1

# Gaussian elimination

## 1.1 Two hundred years of algebra

In this section we recapitulate two hundred or so years of mathematical thought.

Let us start with a familiar type of brain teaser.

**Example 1.1.1** *Sally and Martin go to* The Olde Tea Shoppe. *Sally buys three cream buns and two bottles of pop for thirteen shillings, whilst Martin buys two cream buns and four bottles of pop for fourteen shillings. How much does a cream bun cost and how much does a bottle of pop cost?*

*Solution.* If Sally had bought six cream buns and four bottles of pop, then she would have bought twice as much and it would have cost her twenty six shillings. Similarly, if Martin had bought six cream buns and twelve bottles of pop, then he would have bought three times as much and it would have cost him forty two shillings. In this new situation, Sally and Martin would have bought the same number of cream buns, but Martin would have bought eight more bottles of pop than Sally. Since Martin would have paid sixteen shillings more, it follows that eight bottles of pop cost sixteen shillings and one bottle costs two shillings.

In our original problem, Sally bought three cream buns and two bottles of pop, which, we now know, must have cost her four shillings, for thirteen shillings. Thus her three cream buns cost nine shillings and each cream bun cost three shillings. □

As the reader well knows, the reasoning may be shortened by writing $x$ for the cost of one bun and $y$ for the cost of one bottle of pop. The information given may then be summarised in two equations

$$3x + 2y = 13$$
$$2x + 4y = 14.$$

In the solution just given, we multiplied the first equation by 2 and the second by 3 to obtain

$$6x + 4y = 26$$
$$6x + 12y = 42.$$

Subtracting the first equation from the second yields

$$8y = 16,$$

so $y = 2$ and substitution in either of the original equations gives $x = 3$.
    We can shorten the working still further. Starting, as before, with

$$3x + 2y = 13$$
$$2x + 4y = 14,$$

we retain the first equation and replace the second equation by the result of subtracting $2/3$ times the first equation from the second to obtain

$$3x + 2y = 13$$
$$\frac{8}{3}y = \frac{16}{3}.$$

The second equation yields $y = 2$ and substitution in the first equation gives $x = 3$.
    It is clear that we can now solve any number of problems involving Sally and Martin buying sheep and goats or yaks and xylophones. The general problem involves solving

$$ax + by = \alpha$$
$$cx + dy = \beta.$$

Provided that $a \neq 0$, we retain the first equation and replace the second equation by the result of subtracting $c/a$ times the first equation from the second to obtain

$$ax + by = \alpha$$
$$\left(d - \frac{cb}{a}\right)y = \beta - \frac{c\alpha}{a}.$$

Provided that $d - (cb)/a \neq 0$, we can compute $y$ from the second equation and obtain $x$ by substituting the known value of $y$ in the first equation.
    If $d - (cb)/a = 0$, then our equations become

$$ax + by = \alpha$$
$$0 = \beta - \frac{c\alpha}{a}.$$

There are two possibilities. Either $\beta - (c\alpha)/a \neq 0$, our second equation is inconsistent and the initial problem is insoluble, or $\beta - (c\alpha)/a = 0$, in which case the second equation says that $0 = 0$, and all we know is that

$$ax + by = \alpha$$

so, whatever value of $y$ we choose, setting $x = (\alpha - by)/a$ will give us a possible solution.

There is a second way of looking at this case. If $d - (cb)/a = 0$, then our original equations were

$$ax + by = \alpha$$

$$cx + \frac{cb}{a}y = \beta,$$

that is to say

$$ax + by = \alpha$$

$$c(ax + by) = a\beta$$

so, unless $c\alpha = a\beta$, our equations are inconsistent and, if $c\alpha = a\beta$, the second equation gives no information which is not already in the first.

So far, we have not dealt with the case $a = 0$. If $b \neq 0$, we can interchange the roles of $x$ and $y$. If $c \neq 0$, we can interchange the roles of the two equations. If $d \neq 0$, we can interchange the roles of $x$ and $y$ and the roles of the two equations. Thus we only have a problem if $a = b = c = d = 0$ and our equations take the simple form

$$0 = \alpha$$

$$0 = \beta.$$

These equations are inconsistent unless $\alpha = \beta = 0$. If $\alpha = \beta = 0$, the equations impose no constraints on $x$ and $y$ which can take any value we want.

Now suppose that Sally, Betty and Martin buy cream buns, sausage rolls and bottles of pop. Our new problem requires us to find $x$, $y$ and $z$ when

$$ax + by + cz = \alpha$$

$$dx + ey + fz = \beta$$

$$gx + hy + kz = \gamma.$$

It is clear that we are rapidly running out of alphabet. A little thought suggests that it may be better to try and find $x_1, x_2, x_3$ when

$$a_{11}x_1 + a_{12}x_2 + a_{13}x_3 = y_1$$

$$a_{21}x_1 + a_{22}x_2 + a_{23}x_3 = y_2$$

$$a_{31}x_1 + a_{32}x_2 + a_{33}x_3 = y_3.$$

Provided that $a_{11} \neq 0$, we can subtract $a_{21}/a_{11}$ times the first equation from the second and $a_{31}/a_{11}$ times the first equation from the third to obtain

$$a_{11}x_1 + a_{12}x_2 + a_{13}x_3 = y_1$$

$$b_{22}x_2 + b_{23}x_3 = z_2$$

$$b_{32}x_2 + b_{33}x_3 = z_3,$$

where

$$b_{22} = a_{22} - \frac{a_{21}a_{12}}{a_{11}} = \frac{a_{11}a_{22} - a_{21}a_{12}}{a_{11}}$$

and, similarly,

$$b_{23} = \frac{a_{11}a_{23} - a_{21}a_{13}}{a_{11}}, \quad b_{32} = \frac{a_{11}a_{32} - a_{31}a_{12}}{a_{11}} \quad \text{and} \quad b_{33} = \frac{a_{11}a_{33} - a_{31}a_{13}}{a_{11}}$$

whilst

$$z_2 = \frac{a_{11}y_2 - a_{21}y_1}{a_{11}} \quad \text{and} \quad z_3 = \frac{a_{11}y_3 - a_{31}y_1}{a_{11}}.$$

If we can solve the smaller system of equations

$$b_{22}x_2 + b_{23}x_3 = z_2$$
$$b_{32}x_2 + b_{33}x_3 = z_3,$$

then, knowing $x_2$ and $x_3$, we can use the equation

$$x_1 = \frac{y_1 - a_{12}x_2 - a_{13}x_3}{a_{11}}$$

to find $x_1$. In effect, we have reduced the problem of solving '3 linear equations in 3 unknowns' to the problem of solving '2 linear equations in 2 unknowns'. Since we know how to solve the smaller problem, we know how to solve the larger.

**Exercise 1.1.2** *Use the method just suggested to solve the system*

$$x + y + z = 1$$
$$x + 2y + 3z = 2$$
$$x + 4y + 9z = 6.$$

So far, we have assumed that $a_{11} \neq 0$. A little thought shows that, if $a_{ij} \neq 0$ for some $1 \leq i, j \leq 3$, then all we need to do is reorder our equations so that the $i$th equation becomes the first equation and reorder our variables so that $x_j$ becomes our first variable. We can then reduce the problem to one involving fewer variables as before.

If it is not true that $a_{ij} \neq 0$ for some $1 \leq i, j \leq 3$, then it must be true that $a_{ij} = 0$ for all $1 \leq i, j \leq 3$ and our equations take the peculiar form

$$0 = y_1$$
$$0 = y_2$$
$$0 = y_3.$$

These equations are inconsistent unless $y_1 = y_2 = y_3 = 0$. If $y_1 = y_2 = y_3 = 0$, the equations impose no constraints on $x_1$, $x_2$ and $x_3$ which can take any value we want.

We can now write down the general problem when $n$ people choose from a menu with $n$ items. Our problem is to find $x_1, x_2, \ldots, x_n$ when

$$a_{11}x_1 + a_{12}x_2 + \cdots + a_{1n}x_n = y_1$$
$$a_{21}x_1 + a_{22}x_2 + \cdots + a_{2n}x_n = y_2$$
$$\vdots$$
$$a_{n1}x_1 + a_{n2}x_2 + \cdots + a_{nn}x_n = y_n.$$

We can condense our notation further by using the summation sign and writing our system of equations as

$$\sum_{j=1}^{n} a_{ij}x_j = y_i \qquad [1 \le i \le n]. \qquad\qquad \bigstar$$

We say that we have '$n$ linear equations in $n$ unknowns' and talk about the '$n \times n$ problem'.

Using the insight obtained by reducing the $3 \times 3$ problem to the $2 \times 2$ case, we see at once how to reduce the $n \times n$ problem to the $(n-1) \times (n-1)$ problem. (We suppose that $n \ge 2$.)

*Step 1.* If $a_{ij} = 0$ for all $i$ and $j$, then our equations have the form

$$0 = y_i \qquad [1 \le i \le n].$$

Our equations are inconsistent unless $y_1 = y_2 = \ldots = y_n = 0$. If $y_1 = y_2 = \ldots = y_n = 0$, the equations impose no constraints on $x_1, x_2, \ldots, x_n$ which can take any value we want.

*Step 2.* If the condition of Step 1 does not hold, we can arrange, by reordering the equations and the unknowns, if necessary, that $a_{11} \ne 0$. We now subtract $a_{i1}/a_{11}$ times the first equation from the $i$th equation $[2 \le i \le n]$ to obtain

$$\sum_{j=2}^{n} b_{ij}x_j = z_i \qquad [2 \le i \le n] \qquad\qquad \bigstar\bigstar$$

where

$$b_{ij} = \frac{a_{11}a_{ij} - a_{i1}a_{1j}}{a_{11}} \quad \text{and} \quad z_i = \frac{a_{11}y_i - a_{i1}y_1}{a_{11}}.$$

*Step 3.* If the new set of equations $\bigstar\bigstar$ has no solution, then our old set $\bigstar$ has no solution. If our new set of equations $\bigstar\bigstar$ has a solution $x_i = x_i'$ for $2 \le i \le n$, then our old set $\bigstar$ has the solution

$$x_1 = \frac{1}{a_{11}}\left(y_1 - \sum_{j=2}^{n} a_{1j}x_j'\right)$$
$$x_i = x_i' \qquad [2 \le i \le n].$$

Note that this means that if ★★ has exactly one solution, then ★ has exactly one solution, and if ★★ has infinitely many solutions, then ★ has infinitely many solutions. We have already remarked that if ★★ has no solutions, then ★ has no solutions.

Once we have reduced the problem of solving our $n \times n$ system to that of solving an $(n - 1) \times (n - 1)$ system, we can repeat the process and reduce the problem of solving the new $(n - 1) \times (n - 1)$ system to that of solving an $(n - 2) \times (n - 2)$ system and so on. After $n - 1$ steps we will be faced with the problem of solving a $1 \times 1$ system, that is to say, solving a system of the form

$$ax = b.$$

If $a \neq 0$, then this equation has exactly one solution. If $a = 0$ and $b \neq 0$, the equation has no solution. If $a = 0$ and $b = 0$, every value of $x$ is a solution and we have infinitely many solutions.

Putting the observations of the two previous paragraphs together, we get the following theorem.

**Theorem 1.1.3** *The system of simultaneous linear equations*

$$\sum_{j=1}^{n} a_{ij}x_j = y_i \qquad [1 \leq i \leq n]$$

*has 0, 1 or infinitely many solutions.*

We shall see several different proofs of this result (for example, Theorem 1.4.5), but the proof given here, although long, is instructive.

## 1.2 Computational matters

The method just described for solving 'simultaneous linear equations' is called *Gaussian elimination*. Those who rate mathematical ideas by difficulty may find the attribution unworthy, but those who rate mathematical ideas by utility are happy to honour Gauss in this way.

In the previous section we showed how to solve $n \times n$ systems of equations, but it is clear that the same idea can be used to solve systems of $m$ equations in $n$ unknowns.

**Exercise 1.2.1** *If $m, n \geq 2$, show how to reduce the problem of solving the system of equations*

$$\sum_{j=1}^{n} a_{ij}x_j = y_i \qquad [1 \leq i \leq m] \qquad\qquad ★$$

*to the problem of solving a system of equations*

$$\sum_{j=2}^{n} b_{ij}x_j = z_i \qquad [2 \leq i \leq m]. \qquad\qquad ★★$$

**Exercise 1.2.2** *By using the ideas of Exercise 1.2.1, show that, if $m, n \geq 2$ and we are given a system of equations*

$$\sum_{j=1}^{n} a_{ij}x_j = y_i \qquad [1 \leq i \leq m], \qquad\qquad ★$$

*then at least one of the following must be true.*

(i) ★ *has no solution.*

(ii) ★ *has infinitely many solutions.*

(iii) *There exists a system of equations*

$$\sum_{j=2}^{n} b_{ij}x_j = z_i \qquad [2 \leq i \leq m] \qquad\qquad ★★$$

*with the property that if ★★ has exactly one solution, then ★ has exactly one solution, if ★★ has infinitely many solutions, then ★ has infinitely many solutions, and if ★★ has no solutions, then ★ has no solutions.*

If we repeat Exercise 1.2.1 several times, one of two things will eventually occur. If $n \geq m$, we will arrive at a system of $n - m + 1$ equations in one unknown. If $m > n$, we will arrive at 1 equation in $m - n + 1$ unknowns.

**Exercise 1.2.3** (i) *If $r \geq 1$, show that the system of equations*

$$a_i x = y_i \qquad [1 \leq i \leq r]$$

*has exactly one solution, has no solution or has an infinity of solutions. Explain when each case arises.*

(ii) *If $s \geq 2$, show that the equation*

$$\sum_{j=1}^{s} a_j x_j = b$$

*has no solution or has an infinity of solutions. Explain when each case arises.*

Combining the results of Exercises 1.2.2 and 1.2.3, we obtain the following extension of Theorem 1.1.3

**Theorem 1.2.4** *The system of equations*

$$\sum_{j=1}^{n} a_{ij}x_j = y_i \qquad [1 \leq i \leq m]$$

*has 0, 1 or infinitely many solutions. If $m > n$, then the system cannot have a unique solution (and so will have 0 or infinitely many solutions).*

**Exercise 1.2.5** *Consider the system of equations*

$$x + y = 2$$
$$ax + by = 4$$
$$cx + dy = 8.$$

*(i) Write down non-zero values of a, b, c and d such that the system has no solution.*

*(ii) Write down non-zero values of a, b, c and d such that the system has exactly one solution.*

*(iii) Write down non-zero values of a, b, c and d such that the system has infinitely many solutions.*

*Give reasons in each case.*

**Exercise 1.2.6** *Consider the system of equations*

$$x + y + z = 2$$
$$x + y + az = 4.$$

*For which values of a does the system have no solutions? For which values of a does the system have infinitely many solutions? Give reasons.*

How long does it take for a properly programmed computer to solve a system of $n$ linear equations in $n$ unknowns by Gaussian elimination? The exact time depends on the details of the program and the structure of the machine. However, we can get get a pretty good idea of the answer by counting up the number of elementary operations (that is to say, additions, subtractions, multiplications and divisions) involved.

When we reduce the $n \times n$ case to the $(n-1) \times (n-1)$ case, we subtract a multiple of the first row from the $j$th row and this requires *roughly* $2n$ operations. Since we do this for $j = 2, 3, \ldots, n - 1$ we need roughly $(n-1) \times (2n) \approx 2n^2$ operations. Similarly, reducing the $(n-1) \times (n-1)$ case to the $(n-2) \times (n-2)$ case requires about $2(n-1)^2$ operations and so on. Thus the reduction from the $n \times n$ case to the $1 \times 1$ case requires about

$$2\left(n^2 + (n-1)^2 + \cdots + 2^2\right)$$

operations.

**Exercise 1.2.7** *(i) Show that there exist A and B with $A \geq B > 0$ such that*

$$An^3 \geq \sum_{r=1}^{n} r^2 \geq Bn^3.$$

*(ii) (Not necessary for our argument.) By comparing $\sum_{r=1}^{n} r^2$ and $\int_1^{n+1} x^2 dx$, or otherwise, show that*

$$\sum_{r=1}^{n} r^2 \approx \frac{n^3}{3}.$$

Thus the number of operations required to reduce the $n \times n$ case to the $1 \times 1$ case increases like some multiple of $n^3$. Once we have reduced our system to the $1 \times 1$ case, the number of operations required to work backwards and solve the complete system is less than some multiple of $n^2$.

**Exercise 1.2.8** *Give a rough estimate of the number of operations required to solve the triangular system of equations*

$$\sum_{j=r}^{n} b_{jr}x_r = z_r \qquad [1 \leq r \leq n].$$

*[The roughness of the estimates is left to the taste of the reader.]*

Thus the total number of operations required to solve our initial system by Gaussian elimination increases like some multiple of $n^3$.

The reader may have learnt another method of solving simultaneous equations using determinants called Cramer's rule. If not, she should omit the next two paragraph and wait for our discussion in Section 4.5. Cramer's rule requires us to evaluate an $n \times n$ determinant (as well as lots of other determinants). If we evaluate this determinant by the 'top row rule', we need to evaluate $n$ minors, that is to say, determinants of size $(n - 1) \times (n - 1)$. Each of these new determinants requires the evaluation of $n - 1$ determinants of size $(n - 2) \times (n - 2)$, and so on. We will need roughly

$$An \times (n - 1) \times (n - 2) \times \cdots \times 1 = An!$$

operations where $A \geq 1$. Since $n!$ increases much faster than $n^3$, Cramer's rule is obviously unsatisfactory for large $n$.

The fact that Cramer's rule is unsatisfactory for large $n$ does not, of course, mean that it is unsatisfactory for small $n$. If we have to do hand calculations when $n = 2$ or $n = 3$, then Cramer's rule is no harder than Gaussian elimination. However, I strongly advise the reader to use Gaussian elimination in these cases as well, in order to acquire insight into what is actually going on.

**Exercise 1.2.9** *(For devotees of Cramer's rule only.) Write down a system of 4 linear equations in 4 unknowns and solve it (a) using Cramer's rule and (b) using Gaussian elimination.*

I hope never to have to solve a system of 10 linear equations in 10 unknowns, but I think that I could solve such a system within a reasonable time using Gaussian elimination and a basic hand calculator.

**Exercise 1.2.10** *What sort of time do I appear to consider reasonable?*

It is clear that even a desktop computer can be programmed to find the solution of 200 linear equations in 200 unknowns by Gaussian elimination in a very short time.[1]

---

[1] The author can remember when problems of this size were on the boundary of the possible for the biggest computers of the epoch. The storing and retrieval of the $200 \times 200 = 40\,000$ coefficients represented a major problem.

However, since the number of operations required increases with the cube of the number of unknowns, if we multiply the number of unknowns by 10, the time taken increases by a factor of 1000. Very large problems require new ideas which take advantage of special features of the particular system to be solved. We discuss such a new idea in the second half of Section 15.1.

When we introduced Gaussian elimination, we needed to consider the possibility that $a_{11} = 0$, since we cannot divide by zero. In numerical work it is unwise to divide by numbers close to zero since, if $a$ is small and $b$ is only known approximately, dividing $b$ by $a$ multiplies the error in the approximation by a large number. For this reason, instead of simply rearranging so that $a_{11} \neq 0$ we might rearrange so that $|a_{11}| \geq |a_{ij}|$ for all $i$ and $j$. (This is called 'pivoting' or 'full pivoting'. Reordering rows so that the largest element in a particular column occurs first is called 'row pivoting', and 'column pivoting' is defined similarly. Full pivoting requires substantially more work than row pivoting, and row pivoting is usually sufficient in practice.)

## 1.3 Detached coefficients

If we think about how a computer handles the solution of a system of equations

$$\sum_{j=1}^{3} a_{ij}x_j = y_i \qquad [1 \leq i \leq 3],$$

we see that it essentially manipulates an array

$$(A|\mathbf{y}) = \begin{pmatrix} a_{11} & a_{12} & a_{13} & y_1 \\ a_{21} & a_{22} & a_{23} & y_2 \\ a_{31} & a_{32} & a_{33} & y_3 \end{pmatrix}.$$

Let us imitate our computer and solve

$$x + 2y + z = 1$$
$$2x + 2y + 3z = 6$$
$$3x - 2y + 2z = 9$$

by manipulating the array

$$\begin{pmatrix} 1 & 2 & 1 & 1 \\ 2 & 2 & 3 & 6 \\ 3 & -2 & 2 & 9 \end{pmatrix}.$$

Subtracting 2 times the first row from the second row and 3 times the first row from the third, we get

$$\begin{pmatrix} 1 & 2 & 1 & 1 \\ 0 & -2 & 1 & 4 \\ 0 & -8 & -1 & 6 \end{pmatrix}.$$

We now look at the new array and subtract 4 times the second row from the third to get

$$\begin{pmatrix} 1 & 2 & 1 & | & 1 \\ 0 & -2 & 1 & | & 4 \\ 0 & 0 & -5 & | & -10 \end{pmatrix}$$

corresponding to the system of equations

$$x + 2y + z = 1$$
$$-2y + z = 4$$
$$-5z = -10.$$

We can now read off the solution

$$z = \frac{-10}{-5} = 2$$
$$y = \frac{1}{-2}(4 - 1 \times 2) = -1$$
$$x = 1 - 2 \times (-1) - 1 \times 2 = 1.$$

The idea of doing the calculations using the coefficients alone goes by the rather pompous title of 'the method of detached coefficients'.

We call an $m \times n$ (that is to say, $m$ rows and $n$ columns) array

$$A = \begin{pmatrix} a_{11} & a_{12} & \cdots & a_{1n} \\ a_{21} & a_{22} & \cdots & a_{2n} \\ \vdots & \vdots & & \vdots \\ a_{m1} & a_{m2} & \cdots & a_{mn} \end{pmatrix}$$

an $m \times n$ matrix and write $A = (a_{ij})_{1 \le i \le m}^{1 \le j \le n}$ or just $A = (a_{ij})$. We can rephrase the work we have done so far as follows.

**Theorem 1.3.1** *Suppose that A is an $m \times n$ matrix. By interchanging columns, interchanging rows and subtracting multiples of one row from another, we can reduce A to an $m \times n$ matrix $B = (b_{ij})$ where $b_{ij} = 0$ for $1 \le j \le i - 1$.*

As the reader is probably well aware, we can do better.

**Theorem 1.3.2** *Suppose that A is an $m \times n$ matrix and $p = \min\{n, m\}$. By interchanging columns, interchanging rows and subtracting multiples of one row from another, we can reduce A to an $m \times n$ matrix $B = (b_{ij})$ such that $b_{ij} = 0$ whenever $i \ne j$ and $1 \le i, j \le r$ and whenever $r + 1 \le i$ (for some r with $0 \le r \le p$).*

In the unlikely event that the reader requires a proof, she should observe that Theorem 1.3.2 follows by repeated application of the following lemma.

**Lemma 1.3.3** *Suppose that A is an $m \times n$ matrix, $p = \min\{n, m\}$ and $1 \le q \le p$. Suppose further that $a_{ij} = 0$ whenever $i \ne j$ and $1 \le i, j \le q - 1$ and whenever $q \le i$ and*

$j \leq q - 1$. *By interchanging columns, interchanging rows and subtracting multiples of one row from another, we can reduce $A$ to an $m \times n$ matrix $B = (b_{ij})$ where $b_{ij} = 0$ whenever $i \neq j$ and $1 \leq i, j \leq q$ and whenever $q + 1 \leq i$ and $j \leq q$.*

*Proof* If $a_{ij} = 0$ whenever $q + 1 \leq i$, then just take $A = B$. Otherwise, by interchanging columns (taking care only to move $k$th columns with $q \leq k$) and interchanging rows (taking care only to move $k$th rows with $q \leq k$), we may suppose that $a_{qq} \neq 0$. Now subtract $a_{iq}/a_{qq}$ times the $q$th row from the $i$th row for each $q \leq i$. $\qquad\qquad\square$

Theorem 1.3.2 has an obvious twin.

**Theorem 1.3.4** *Suppose that $A$ is an $m \times n$ matrix and $p = \min\{n, m\}$. By interchanging rows, interchanging columns and subtracting multiples of one column from another, we can reduce $A$ to an $m \times n$ matrix $B = (b_{ij})$ where $b_{ij} = 0$ whenever $i \neq j$ and $1 \leq i, j \leq r$ and whenever $r + 1 \leq j$ (for some $r$ with $0 \leq r \leq p$).*

**Exercise 1.3.5** *Illustrate Theorem 1.3.2 and Theorem 1.3.4 by carrying out appropriate operations on*

$$\begin{pmatrix} 1 & -1 & 3 \\ 2 & 5 & 2 \end{pmatrix}.$$

Combining the twin theorems, we obtain the following result. (Note that, if $a > b$, there are no $c$ with $a \leq c \leq b$.)

**Theorem 1.3.6** *Suppose that $A$ is an $m \times n$ matrix and $p = \min\{n, m\}$. There exists an $r$ with $0 \leq r \leq p$ with the following property. By interchanging rows, interchanging columns, subtracting multiples of one row from another and subtracting multiples of one column from another, we can reduce $A$ to an $m \times n$ matrix $B = (b_{ij})$ with $b_{ij} = 0$ unless $i = j$ and $1 \leq i \leq r$.*

We can obtain several simple variations.

**Theorem 1.3.7** *Suppose that $A$ is an $m \times n$ matrix and $p = \min\{n, m\}$. Then there exists an $r$ with $0 \leq r \leq p$ with the following property. By interchanging columns, interchanging rows and subtracting multiples of one row from another, and multiplying rows by non-zero numbers, we can reduce $A$ to an $m \times n$ matrix $B = (b_{ij})$ with $b_{ii} = 1$ for $1 \leq i \leq r$ and $b_{ij} = 0$ whenever $i \neq j$ and $1 \leq i, j \leq r$ and whenever $r + 1 \leq i$.*

**Theorem 1.3.8** *Suppose that $A$ is an $m \times n$ matrix and $p = \min\{n, m\}$. There exists an $r$ with $0 \leq r \leq p$ with the following property. By interchanging rows, interchanging columns, subtracting multiples of one row from another and subtracting multiples of one column from another and multiplying rows by a non-zero number, we can reduce $A$ to an $m \times n$ matrix $B = (b_{ij})$ such that there exists an $r$ with $0 \leq r \leq p$ such that $b_{ii} = 1$ if $1 \leq i \leq r$ and $b_{ij} = 0$ otherwise.*

**Exercise 1.3.9** *(i) Illustrate Theorem 1.3.8 by carrying out appropriate operations on*

$$\begin{pmatrix} 1 & -1 & 3 \\ 2 & 5 & 2 \\ 4 & 3 & 8 \end{pmatrix}.$$

*(ii) Illustrate Theorem 1.3.8 by carrying out appropriate operations on*

$$\begin{pmatrix} 2 & 4 & 5 \\ 3 & 2 & 1 \\ 4 & 1 & 3 \end{pmatrix}.$$

*By carrying out the same operations, solve the system of equations*

$$2x + 4y + 5z = -3$$
$$3x + 2y + z = 2$$
$$4x + y + 3z = 1.$$

[*If you are confused by the statement or proofs of the various results in this section, concrete examples along the lines of this exercise are likely to be more helpful than worrying about the general case.*]

**Exercise 1.3.10** *We use the notation of Theorem 1.3.8. Let $m = 3$, $n = 4$. Find, with reasons, $3 \times 4$ matrices $A$, all of whose entries are non-zero, for which $r = 3$, for which $r = 2$ and for which $r = 1$. Is it possible to find a $3 \times 4$ matrix $A$, all of whose entries are non-zero, for which $r = 0$? Give reasons.*

## 1.4 Another fifty years

In the previous section, we treated the matrix $A$ as a passive actor. In this section, we give it an active role by declaring that the $m \times n$ matrix $A$ acts on the $n \times 1$ matrix (or *column vector*) $\mathbf{x}$ to produce the $m \times 1$ matrix (or column vector) $\mathbf{y}$. We write

$$A\mathbf{x} = \mathbf{y}$$

with

$$\sum_{j=1}^{n} a_{ij}x_j = y_i \qquad \text{for } 1 \le i \le m.$$

In other words,

$$\begin{pmatrix} a_{11} & a_{12} & \cdots & a_{1n} \\ a_{21} & a_{22} & \cdots & a_{2n} \\ \vdots & \vdots & & \vdots \\ \vdots & \vdots & & \vdots \\ a_{m1} & a_{m2} & \cdots & a_{mn} \end{pmatrix} \begin{pmatrix} x_1 \\ x_2 \\ \vdots \\ x_n \end{pmatrix} = \begin{pmatrix} y_1 \\ y_2 \\ \vdots \\ y_m \end{pmatrix}.$$

We write $\mathbb{R}^n$ for the space of column vectors

$$\mathbf{x} = \begin{pmatrix} x_1 \\ x_2 \\ \vdots \\ x_n \end{pmatrix}$$

with $n$ entries. Since column vectors take up a lot of space, we often adopt one of the two notations

$$\mathbf{x} = (x_1 \ x_2 \ \ldots \ x_n)^T$$

or $\mathbf{x} = (x_1, x_2, \ldots, x_n)^T$. We also write $\pi_i \mathbf{x} = x_i$. It seems reasonable to call $\mathbf{x}$ an *arithmetic vector*.

As the reader probably knows, we can add vectors and multiply them by real numbers (scalars).

**Definition 1.4.1** *If* $\mathbf{x}, \mathbf{y} \in \mathbb{R}^n$ *and* $\lambda \in \mathbb{R}$, *we write* $\mathbf{x} + \mathbf{y} = \mathbf{z}$ *and* $\lambda \mathbf{x} = \mathbf{w}$ *where*

$$z_i = x_i + y_i, \quad w_i = \lambda x_i.$$

We write $\mathbf{0} = (0, 0, \ldots, 0)^T$, $-\mathbf{x} = (-1)\mathbf{x}$ and $\mathbf{x} - \mathbf{y} = \mathbf{x} + (-\mathbf{y})$.

The next lemma shows the kind of arithmetic we can do with vectors.

**Lemma 1.4.2** *Suppose that* $\mathbf{x}, \mathbf{y}, \mathbf{z} \in \mathbb{R}^n$ *and* $\lambda, \mu \in \mathbb{R}$. *Then the following relations hold.*
*(i)* $(\mathbf{x} + \mathbf{y}) + \mathbf{z} = \mathbf{x} + (\mathbf{y} + \mathbf{z})$.
*(ii)* $\mathbf{x} + \mathbf{y} = \mathbf{y} + \mathbf{x}$.
*(iii)* $\mathbf{x} + \mathbf{0} = \mathbf{x}$.
*(iv)* $\lambda(\mathbf{x} + \mathbf{y}) = \lambda\mathbf{x} + \lambda\mathbf{y}$.
*(v)* $(\lambda + \mu)\mathbf{x} = \lambda\mathbf{x} + \mu\mathbf{x}$.
*(vi)* $(\lambda\mu)\mathbf{x} = \lambda(\mu\mathbf{x})$.
*(vii)* $1\mathbf{x} = \mathbf{x}$, $0\mathbf{x} = \mathbf{0}$.
*(viii)* $\mathbf{x} - \mathbf{x} = \mathbf{0}$.

*Proof* This is an exercise in proving the obvious. For example, to prove (iv), we observe that (if we take $\pi_i$ as above)

$$\begin{aligned}
\pi_i\big(\lambda(\mathbf{x} + \mathbf{y})\big) &= \lambda\pi_i(\mathbf{x} + \mathbf{y}) && \text{(by definition)} \\
&= \lambda(x_i + y_i) && \text{(by definition)} \\
&= \lambda x_i + \lambda y_i && \text{(by properties of the reals)} \\
&= \pi_i(\lambda\mathbf{x}) + \pi_i(\lambda\mathbf{y}) && \text{(by definition)} \\
&= \pi_i(\lambda\mathbf{x} + \lambda\mathbf{y}) && \text{(by definition).}
\end{aligned}$$

$\square$

**Exercise 1.4.3** *I think it is useful for a mathematician to be able to prove the obvious. If you agree with me, choose a few more of the statements in Lemma 1.4.2 and prove them. If you disagree, ignore both this exercise and the proof above.*

The next result is easy to prove, but is central to our understanding of matrices.

**Theorem 1.4.4** *If $A$ is an $m \times n$ matrix $\mathbf{x}, \mathbf{y} \in \mathbb{R}^n$ and $\lambda, \mu \in \mathbb{R}$, then*

$$A(\lambda \mathbf{x} + \mu \mathbf{y}) = \lambda A \mathbf{x} + \mu A \mathbf{y}.$$

We say that $A$ acts *linearly* on $\mathbb{R}^n$.

*Proof* Observe that

$$\sum_{j=1}^{n} a_{ij}(\lambda x_j + \mu y_j) = \sum_{j=1}^{n}(\lambda a_{ij}x_j + \mu a_{ij}y_j) = \lambda \sum_{j=1}^{n} a_{ij}x_j + \mu \sum_{j=1}^{n} a_{ij}y_j$$

as required. □

To see why this remark is useful, observe that it gives a simple proof of the first part of Theorem 1.2.4.

**Theorem 1.4.5** *The system of equations*

$$\sum_{j=1}^{n} a_{ij}x_j = b_i \qquad [1 \leq i \leq m]$$

*has 0, 1 or infinitely many solutions.*

*Proof* We need to show that, if the system of equations has two distinct solutions, then it has infinitely many solutions.

Suppose that $A\mathbf{y} = \mathbf{b}$ and $A\mathbf{z} = \mathbf{b}$. Then, since $A$ acts linearly,

$$A(\lambda \mathbf{y} + (1 - \lambda)\mathbf{z}) = \lambda A\mathbf{y} + (1 - \lambda)A\mathbf{z} = \lambda \mathbf{b} + (1 - \lambda)\mathbf{b} = \mathbf{b}.$$

Thus, if $\mathbf{y} \neq \mathbf{z}$, there are infinitely many solutions

$$\mathbf{x} = \lambda \mathbf{y} + (1 - \lambda)\mathbf{z} = \mathbf{z} + \lambda(\mathbf{y} - \mathbf{z})$$

to the equation $A\mathbf{x} = \mathbf{b}$. □

With a little extra work, we can gain some insight into the nature of the infinite set of solutions.

**Definition 1.4.6** *A non-empty subset of $\mathbb{R}^n$ is called a* subspace *of $\mathbb{R}^n$ if, whenever $\mathbf{x}, \mathbf{y} \in E$ and $\lambda, \mu \in \mathbb{R}$, it follows that $\lambda \mathbf{x} + \mu \mathbf{y} \in E$.*

**Theorem 1.4.7** *If $A$ is an $m \times n$ matrix, then the set $E$ of column vectors $\mathbf{x} \in \mathbb{R}^n$ with $A\mathbf{x} = \mathbf{0}$ is a subspace of $\mathbb{R}^n$.*

*If $A\mathbf{y} = \mathbf{b}$, then $A\mathbf{z} = \mathbf{b}$ if and only if $\mathbf{z} = \mathbf{y} + \mathbf{e}$ for some $\mathbf{e} \in E$.*

*Proof* If $\mathbf{x}$, $\mathbf{y} \in E$ and $\lambda$, $\mu \in \mathbb{R}$, then, by linearity,

$$A(\lambda\mathbf{x} + \mu\mathbf{y}) = \lambda A\mathbf{x} + \mu A\mathbf{y} = \lambda\mathbf{0} + \mu\mathbf{0} = \mathbf{0}$$

so $\lambda\mathbf{x} + \mu\mathbf{y} \in E$. The set $E$ is non-empty, since

$$A\mathbf{0} = \mathbf{0},$$

so $\mathbf{0} \in E$.

If $A\mathbf{y} = \mathbf{b}$ and $\mathbf{e} \in E$, then

$$A(\mathbf{y} + \mathbf{e}) = A\mathbf{y} + A\mathbf{e} = \mathbf{b} + \mathbf{0} = \mathbf{b}.$$

Conversely, if $A\mathbf{y} = \mathbf{b}$ and $A\mathbf{z} = \mathbf{b}$, let us write $\mathbf{e} = \mathbf{y} - \mathbf{z}$. Then $\mathbf{z} = \mathbf{y} + \mathbf{e}$ and $\mathbf{e} \in E$ since

$$A\mathbf{e} = A(\mathbf{y} - \mathbf{z}) = A\mathbf{y} - A\mathbf{z} = \mathbf{b} - \mathbf{b} = \mathbf{0}$$

as required.                                                                   □

We could refer to $\mathbf{z}$ as a *particular solution* of $A\mathbf{x} = \mathbf{b}$ and $E$ as the space of *complementary solutions*.[2]

## 1.5 Further exercises

**Exercise 1.5.1** We work in $\mathbb{R}$. Show that the system

$$ax + by + cz = 0$$
$$cx + ay + bz = 0$$
$$bx + cy + az = 0$$

has a unique solution if and only if $a + b + c \neq 0$ and $a$, $b$ and $c$ are not all equal.

**Exercise 1.5.2** A glass of lemonade, 3 sandwiches and 7 biscuits together cost 14 pence (this is a Victorian puzzle), a glass of lemonade, 4 sandwiches and 10 biscuits together cost 17 pence. Required, to find the cost (1) of a glass of lemonade, a sandwich and a biscuit and (2) the cost of 2 glasses of lemonade, 3 sandwiches and 5 biscuits. (Knot 7 from *A Tangled Tale* by Lewis Carroll [10].)

**Exercise 1.5.3** (i) Find all the solutions of the following system of equations involving real numbers:

$$xy = 6$$
$$yz = 288$$
$$zx = 3.$$

---

[2] Later, we shall refer to $E$ as a *null-space* or *kernel*.

(ii) Explain how you would solve (or show that there were no solutions for) the system of equations

$$x^a y^b = u$$
$$x^c y^d = v$$

where $a$, $b$, $c$, $d$, $u$, $v \in \mathbb{R}$ are fixed with $u$, $v > 0$ and $x$ and $y$ are strictly positive real numbers to be found.

**Exercise 1.5.4** (You need to know about modular arithmetic for this question.) Use Gaussian elimination to solve the following system of integer *congruences* modulo 7:

$$x + y + z + w \equiv 6$$
$$x + 2y + 3z + 4w \equiv 6$$
$$x + 4y + 2z + 2w \equiv 0$$
$$x + y + 6z + w \equiv 2.$$

**Exercise 1.5.5** The set $S$ comprises all the triples $(x, y, z)$ of real numbers which satisfy the equations

$$x + y - z = b^2$$
$$x + ay + z = b$$
$$x + a^2 y - z = b^3.$$

Determine for which pairs of values of $a$ and $b$ (if any) the set $S$ (i) is empty, (ii) contains precisely one element, (iii) is finite but contains more than one element, or (iv) is infinite.

**Exercise 1.5.6** We work in $\mathbb{R}$. Use Gaussian elimination to determine the values of $a$, $b$, $c$ and $d$ for which the system of equations

$$x + ay + a^2 z + a^3 w = 0$$
$$x + by + b^2 z + b^3 w = 0$$
$$x + cy + c^2 z + c^3 w = 0$$
$$x + dy + d^2 z + d^3 w = 0$$

in $\mathbb{R}$ has a unique solution in $x$, $y$, $z$ and $w$.

# 2

# A little geometry

## 2.1 Geometric vectors

In the previous chapter we introduced column vectors and showed how they could be used to study simultaneous linear equations in many unknowns. In this chapter we show how they can be used to study some aspects of geometry.

There is an initial trivial, but potentially confusing, problem. Different branches of mathematics develop differently and evolve their own notation. This creates difficulties when we try to unify them. When studying simultaneous equations it is natural to use *column vectors* (so that a vector is an $n \times 1$ matrix), but centuries of tradition, not to mention ease of printing, mean that in elementary geometry we tend to use *row vectors* (so that a vector is a $1 \times n$ matrix). Since one cannot flock by oneself, I advise the reader to stick to the normal usage in each subject. Where the two usages conflict, I recommend using column vectors.

Let us start by looking at geometry in the plane. As the reader knows, we can use a Cartesian coordinate system in which each point of the plane corresponds to a unique ordered pair $(x_1, x_2)$ of real numbers. It is natural to consider $\mathbf{x} = (x_1, x_2) \in \mathbb{R}^2$ as a *row vector* and use the same kind of definitions of addition and (scalar) multiplication as we did for column vectors so that

$$\lambda(x_1, x_2) + \mu(y_1, y_2) = (\lambda x_1 + \mu y_1, \lambda x_2 + \mu y_2).$$

In school we may be taught that a straight line is the locus of all points in the $(x, y)$ plane given by

$$ax + by = c$$

where $a$ and $b$ are not both zero.

In the next couple of lemmas we shall show that this translates into the vectorial statement that a straight line joining distinct points $\mathbf{u}, \mathbf{v} \in \mathbb{R}^2$ is the set

$$\{\lambda \mathbf{u} + (1 - \lambda)\mathbf{v} : \lambda \in \mathbb{R}\}.$$

**Lemma 2.1.1** (*i*) *If* $\mathbf{u}$, $\mathbf{v} \in \mathbb{R}^2$ *and* $\mathbf{u} \neq \mathbf{v}$, *then we can write*

$$\{\lambda\mathbf{u} + (1 - \lambda)\mathbf{v} : \lambda \in \mathbb{R}\} = \{\mathbf{v} + \lambda\mathbf{w} : \lambda \in \mathbb{R}\}$$

*for some* $\mathbf{w} \neq \mathbf{0}$.

(*ii*) *If* $\mathbf{w} \neq \mathbf{0}$, *then*

$$\{\mathbf{v} + \lambda\mathbf{w} : \lambda \in \mathbb{R}\} = \{\lambda\mathbf{u} + (1 - \lambda)\mathbf{v} : \lambda \in \mathbb{R}\}$$

*for some* $\mathbf{u} \neq \mathbf{v}$.

*Proof* (i) Observe that

$$\lambda\mathbf{u} + (1 - \lambda)\mathbf{v} = \mathbf{v} + \lambda(\mathbf{u} - \mathbf{v})$$

and set $\mathbf{w} = (\mathbf{u} - \mathbf{v})$.

(ii) Reversing the argument of (i), we observe that

$$\mathbf{v} + \lambda\mathbf{w} = \lambda(\mathbf{w} + \mathbf{v}) + (1 - \lambda)\mathbf{v}$$

and set $\mathbf{u} = \mathbf{w} + \mathbf{v}$.  □

Naturally we think of

$$\{\mathbf{v} + \lambda\mathbf{w} : \lambda \in \mathbb{R}\}$$

as a line through $\mathbf{v}$ 'parallel to the vector $\mathbf{w}$' and

$$\{\lambda\mathbf{u} + (1 - \lambda)\mathbf{v} : \lambda \in \mathbb{R}\}$$

as a line through $\mathbf{u}$ and $\mathbf{v}$. The next lemma links these ideas with the school definition.

**Lemma 2.1.2** *If* $\mathbf{v} \in \mathbb{R}^2$ *and* $\mathbf{w} \neq \mathbf{0}$, *then*

$$\{\mathbf{v} + \lambda\mathbf{w} : \lambda \in \mathbb{R}\} = \{\mathbf{x} \in \mathbb{R}^2 : w_2 x_1 - w_1 x_2 = w_2 v_1 - w_1 v_2\}$$
$$= \{\mathbf{x} \in \mathbb{R}^2 : a_1 x_1 + a_2 x_2 = c\}$$

*where* $a_1 = w_2$, $a_2 = -w_1$ *and* $c = w_2 v_1 - v_2 w_2$.

*Proof* If $\mathbf{x} = \mathbf{v} + \lambda\mathbf{w}$, then

$$x_1 = v_1 + \lambda w_1, \quad x_2 = v_2 + \lambda w_2$$

and

$$w_2 x_1 - w_1 x_2 = w_2(v_1 + \lambda w_1) - w_1(v_2 + \lambda w_2) = w_2 v_1 - v_2 w_1.$$

To reverse the argument, observe that, since $\mathbf{w} \neq \mathbf{0}$, at least one of $w_1$ and $w_2$ must be non-zero. Without loss of generality, suppose that $w_1 \neq 0$. Then, if $w_2 x_1 - w_1 x_2 = w_2 v_1 - v_2 w_1$ and we set $\lambda = (x_1 - v_1)/w_1$, we obtain

$$v_1 + \lambda w_1 = x_1$$

and

$$v_2 + \lambda w_2 = \frac{w_1 v_2 + x_1 w_2 - v_1 w_2}{w_1}$$

$$= \frac{w_1 v_2 + (w_2 v_1 - w_1 v_2 + w_1 x_2) - w_2 v_1}{w_1}$$

$$= \frac{w_1 x_2}{w_1} = x_2$$

so $\mathbf{x} = \mathbf{v} + \lambda \mathbf{w}$. □

**Exercise 2.1.3** *If $a$, $b$, $c \in \mathbb{R}$ and $a$ and $b$ are not both zero, find distinct vectors $\mathbf{u}$ and $\mathbf{v}$ such that*

$$\{(x, y) : ax + by = c\}$$

*represents the line through $\mathbf{u}$ and $\mathbf{v}$.*

The following very simple exercise will be used later. The reader is free to check that she can prove the obvious or merely check that she understands why the results are true.

**Exercise 2.1.4** *(i) If $\mathbf{w} \neq \mathbf{0}$, then either*

$$\{\mathbf{v} + \lambda \mathbf{w} : \lambda \in \mathbb{R}\} \cap \{\mathbf{v}' + \mu \mathbf{w} : \mu \in \mathbb{R}\} = \emptyset$$

*or*

$$\{\mathbf{v} + \lambda \mathbf{w} : \lambda \in \mathbb{R}\} = \{\mathbf{v}' + \mu \mathbf{w} : \mu \in \mathbb{R}\}.$$

*(ii) If $\mathbf{u} \neq \mathbf{u}'$, $\mathbf{v} \neq \mathbf{v}'$ and there exist non-zero $\tau$, $\sigma \in \mathbb{R}$ such that*

$$\tau(\mathbf{u} - \mathbf{u}') = \sigma(\mathbf{v} - \mathbf{v}'),$$

*then either the lines joining $\mathbf{u}$ to $\mathbf{u}'$ and $\mathbf{v}$ to $\mathbf{v}'$ fail to meet or they are identical.*

*(iii) If $\mathbf{u}$, $\mathbf{u}'$, $\mathbf{u}'' \in \mathbb{R}^2$ satisfy an equation*

$$\mu\mathbf{u} + \mu'\mathbf{u}' + \mu''\mathbf{u}'' = \mathbf{0}$$

*with all the real numbers $\mu$, $\mu'$, $\mu''$ non-zero and $\mu + \mu' + \mu'' = 0$, then $\mathbf{u}$, $\mathbf{u}'$ and $\mathbf{u}''$ all lie on the same straight line.*

We use vectors to prove a famous theorem of Desargues. The result will not be used elsewhere, but is introduced to show the reader that vector methods can be used to prove interesting theorems.

**Example 2.1.5 [Desargues' theorem]** *Consider two triangles $ABC$ and $A'B'C'$ with distinct vertices such that lines $AA'$, $BB'$ and $CC'$ all intersect a some point $V$. If the lines $AB$ and $A'B'$ intersect at exactly one point $C''$, the lines $BC$ and $B'C'$ intersect at exactly one point $A''$ and the lines $CA$ and $CA'$ intersect at exactly one point $B''$, then $A''$, $B''$ and $C''$ lie on the same line.*

**Exercise 2.1.6** *Draw an example and check that (to within the accuracy of the drawing) the conclusion of Desargues' theorem holds. (You may need a large sheet of paper and some experiment to obtain a case in which A″, B″ and C″ all lie within your sheet.)*

**Exercise 2.1.7** *Try and prove the result without using vectors.*[1]

We translate Example 2.1.5 into vector notation.

**Example 2.1.8** *We work in $\mathbb{R}^2$. Suppose that* **a**, **a**′, **b**, **b**′, **c**, **c**′ *are distinct. Suppose further that* **v** *lies on the line through* **a** *and* **a**′, *on the line through* **b** *and* **b**′ *and on the line through* **c** *and* **c**′.

*If exactly one point* **a**″ *lies on the line through* **b** *and* **c** *and on the line through* **b**′ *and* **c**′ *and similar conditions hold for* **b**″ *and* **c**″, *then* **a**″, **b**″ *and* **c**″ *lie on the same line.*

We now prove the result.

*Proof* By hypothesis, we can find $\alpha$, $\beta$, $\gamma \in \mathbb{R}$, such that

$$\mathbf{v} = \alpha\mathbf{a} + (1 - \alpha)\mathbf{a}'$$
$$\mathbf{v} = \beta\mathbf{b} + (1 - \beta)\mathbf{b}'$$
$$\mathbf{v} = \gamma\mathbf{c} + (1 - \alpha)\mathbf{c}'.$$

Eliminating **v** between the first two equations, we get

$$\alpha\mathbf{a} - \beta\mathbf{b} = -(1 - \alpha)\mathbf{a}' + (1 - \beta)\mathbf{b}'. \qquad\qquad ★$$

We shall need to exclude the possibility $\alpha = \beta$. To do this, observe that, if $\alpha = \beta$, then

$$\alpha(\mathbf{a} - \mathbf{b}) = (1 - \alpha)(\mathbf{a}' - \mathbf{b}').$$

Since $\mathbf{a} \neq \mathbf{b}$ and $\mathbf{a}' \neq \mathbf{b}'$, we have $\alpha$, $1 - \alpha \neq 0$ and (see Exercise 2.1.4 (ii)) the straight lines joining **a** and **b** and **a**′ and **b**′ are either identical or do not intersect. Since the hypotheses of the theorem exclude both possibilities, $\alpha \neq \beta$.

We know that **c**″ is the unique point satisfying

$$\mathbf{c}'' = \lambda\mathbf{a} + (1 - \lambda)\mathbf{b}$$
$$\mathbf{c}'' = \lambda'\mathbf{a} + (1 - \lambda')\mathbf{b}$$

for some real $\lambda$, $\lambda'$, but we still have to find $\lambda$ and $\lambda'$. By inspection (see Exercise 2.1.9) we see that $\lambda = \alpha/(\alpha - \beta)$ and $\lambda' = (1 - \alpha)/(\beta - \alpha)$ do the trick and so

$$(\alpha - \beta)\mathbf{c}'' = \alpha\mathbf{a} - \beta\mathbf{b}.$$

Applying the same argument to **a**″ and **b**″, we see that

$$(\beta - \gamma)\mathbf{a}'' = \beta\mathbf{b} - \gamma\mathbf{c}$$
$$(\gamma - \alpha)\mathbf{b}'' = \gamma\mathbf{c} - \alpha\mathbf{a}$$
$$(\alpha - \beta)\mathbf{c}'' = \alpha\mathbf{a} - \beta\mathbf{b}.$$

---

[1] Desargues proved it long before vectors were thought of.

Adding our three equations we get

$$(\beta - \gamma)\mathbf{a}'' + (\gamma - \alpha)\mathbf{b}'' + (\alpha - \beta)\mathbf{c}'' = \mathbf{0}.$$

Thus (see Exercise 2.1.4 (ii)), since $\beta - \gamma$, $\gamma - \alpha$ and $\alpha - \beta$ are all non-zero, $\mathbf{a}''$, $\mathbf{b}''$ and $\mathbf{c}''$ all lie on the same straight line. ☐

**Exercise 2.1.9** *In the proof of Theorem 2.1.5 we needed to find $\lambda$ and $\lambda'$ so that*

$$\lambda\mathbf{a} + (1 - \lambda)\mathbf{b} = \lambda'\mathbf{a} + (1 - \lambda')\mathbf{b},$$

*knowing that*

$$\alpha\mathbf{a} - \beta\mathbf{b} = -(1 - \alpha)\mathbf{a}' + (1 - \beta)\mathbf{b}'. \qquad \bigstar$$

*The easiest way to use $\bigstar$ would be to have*

$$\frac{\lambda}{1 - \lambda} = -\frac{\alpha}{\beta}.$$

*Check that this corresponds to our choice of $\lambda$ in the proof. Obtain $\lambda'$ similarly. (Of course, we do not know that these choices will work and we now need to check that they give the desired result.)*

## 2.2 Higher dimensions

We now move from two to three dimensions. We use a Cartesian coordinate system in which each point of space corresponds to a unique ordered triple $(x_1, x_2, x_3)$ of real numbers. As before, we consider $\mathbf{x} = (x_1, x_2, x_3) \in \mathbb{R}^3$ as a *row vector* and use the same kind of definitions of addition and scalar multiplication as we did for column vectors so that

$$\lambda(x_1, x_2, x_3) + \mu(y_1, y_2, y_3) = (\lambda x_1 + \mu y_1, \lambda x_2 + \mu y_2, \lambda x_3 + \mu y_3).$$

If we say a straight line joining distinct points $\mathbf{u}$, $\mathbf{v} \in \mathbb{R}^3$ is the set

$$\{\lambda\mathbf{u} + (1 - \lambda)\mathbf{v} : \lambda \in \mathbb{R}\},$$

then a quick scan of our proof of Desargues' theorem shows it applies word for word to the new situation.

**Example 2.2.1 [Desargues' theorem in three dimensions]** *Consider two triangles $ABC$ and $A'B'C'$ with distinct vertices such that lines $AA'$, $BB'$ and $CC'$ all intersect at some point $V$. If the lines $AB$ and $A'B'$ intersect at exactly one point $C''$, the lines $BC$ and $B'C'$ intersect at exactly one point $A''$ and the lines $CA$ and $CA'$ intersect at exactly one point $B''$, then $A''$, $B''$ and $C''$ lie on the same line.*

(In fact, Desargues' theorem is most naturally thought about as a three dimensional theorem about the rules of perspective.)

The reader may, quite properly, object that I have not shown that the definition of a straight line given here corresponds to her definition of a straight line. However, I do not know her definition of a straight line. Under the circumstances, it makes sense for the author and reader to agree to accept the present definition for the time being. The reader is free to add the words 'in the vectorial sense' whenever we talk about straight lines.

There is now no reason to confine ourselves to the two or three dimensions of ordinary space. We can do 'vectorial geometry' in as many dimensions as we please. Our points will be row vectors

$$\mathbf{x} = (x_1, x_2, \ldots, x_n) \in \mathbb{R}^n$$

manipulated according to the rule

$$\lambda(x_1, x_2, \ldots, x_n) + \mu(y_1, y_2, \ldots, y_n) = (\lambda x_1 + \mu y_1, \lambda x_2 + \mu y_2, \ldots, \lambda x_n + \mu y_n)$$

and a straight line joining distinct points $\mathbf{u}, \mathbf{v} \in \mathbb{R}^n$ will be the set

$$\{\lambda \mathbf{u} + (1 - \lambda)\mathbf{v} : \lambda \in \mathbb{R}\}.$$

It seems reasonable to call $\mathbf{x}$ a *geometric* or *position* vector.

As before, Desargues' theorem and its proof pass over to the more general context.

**Example 2.2.2 [Desargues' theorem in $n$ dimensions]** *We work in $\mathbb{R}^n$. Suppose that $\mathbf{a}$, $\mathbf{a}'$, $\mathbf{b}$, $\mathbf{b}'$, $\mathbf{c}$, $\mathbf{c}'$ are distinct. Suppose further that $\mathbf{v}$ lies on the line through $\mathbf{a}$ and $\mathbf{a}'$, on the line through $\mathbf{b}$ and $\mathbf{b}'$ and on the line through $\mathbf{c}$ and $\mathbf{c}'$.*

*If exactly one point $\mathbf{a}''$ lies on the line through $\mathbf{b}$ and $\mathbf{c}$ and on the line through $\mathbf{b}'$ and $\mathbf{c}'$ and similar conditions hold for $\mathbf{b}''$ and $\mathbf{c}''$, then $\mathbf{a}''$, $\mathbf{b}''$ and $\mathbf{c}''$ lie on the same line.*

I introduced Desargues' theorem in order that the reader should not view vectorial geometry as an endless succession of trivialities dressed up in pompous phrases. My next example is simpler, but is much more important for future work. We start from a beautiful theorem of classical geometry.

**Theorem 2.2.3** *The lines joining the vertices of triangle to the mid-points of opposite sides meet at a point.*

**Exercise 2.2.4** *Draw an example and check that (to within the accuracy of the drawing) the conclusion of Theorem 2.2.3 holds.*

**Exercise 2.2.5** *Try to prove the result by trigonometry.*

In order to translate Theorem 2.2.3 into vectorial form, we need to decide what a mid-point is. We have not yet introduced the notion of distance, but we observe that, if we set

$$\mathbf{c}' = \frac{1}{2}(\mathbf{a} + \mathbf{b}),$$

then $\mathbf{c}'$ lies on the line joining $\mathbf{a}$ to $\mathbf{b}$ and

$$\mathbf{c}' - \mathbf{a} = \frac{1}{2}(\mathbf{b} - \mathbf{a}) = \mathbf{b} - \mathbf{c}',$$

so it is reasonable to call $\mathbf{c}'$ the mid-point of $\mathbf{a}$ and $\mathbf{b}$. (See also Exercise 2.3.12.) Theorem 2.2.3 can now be restated as follows.

**Theorem 2.2.6** *Suppose that* $\mathbf{a}$, $\mathbf{b}$, $\mathbf{c} \in \mathbb{R}^2$ *and*

$$\mathbf{a}' = \frac{1}{2}(\mathbf{b} + \mathbf{c}), \quad \mathbf{b}' = \frac{1}{2}(\mathbf{c} + \mathbf{a}), \quad \mathbf{c}' = \frac{1}{2}(\mathbf{a} + \mathbf{b}).$$

*Then the lines joining* $\mathbf{a}$ *to* $\mathbf{a}'$*,* $\mathbf{b}$ *to* $\mathbf{b}'$ *and* $\mathbf{c}$ *to* $\mathbf{c}'$ *intersect.*

*Proof* We seek a $\mathbf{d} \in \mathbb{R}^2$ and $\alpha$, $\beta$, $\gamma \in \mathbb{R}$ such that

$$\mathbf{d} = \alpha\mathbf{a} + (1 - \alpha)\mathbf{a}' = \alpha\mathbf{a} + \frac{1 - \alpha}{2}\mathbf{b} + \frac{1 - \alpha}{2}\mathbf{c}$$

$$\mathbf{d} = \beta\mathbf{b} + (1 - \beta)\mathbf{b}' = \frac{1 - \beta}{2}\mathbf{a} + \beta\mathbf{b} + \frac{1 - \beta}{2}\mathbf{c}$$

$$\mathbf{d} = \gamma\mathbf{c} + (1 - \gamma)\mathbf{c}' = \frac{1 - \gamma}{2}\mathbf{a} + \frac{1 - \gamma}{2}\mathbf{b} + \gamma\mathbf{c}.$$

By inspection, we see[2] that $\alpha$, $\beta$, $\gamma = 1/3$ and

$$\mathbf{d} = \frac{1}{3}(\mathbf{a} + \mathbf{b} + \mathbf{c})$$

satisfy the required conditions, so the result holds.                        □

As before, we see that there is no need to restrict ourselves to $\mathbb{R}^2$. The method of proof also suggests a more general theorem.

**Definition 2.2.7** *If the q points* $\mathbf{x}_1$, $\mathbf{x}_2$, $\ldots$, $\mathbf{x}_q \in \mathbb{R}^n$, *then their* centroid *is the point*

$$\frac{1}{q}(\mathbf{x}_1 + \mathbf{x}_2 + \cdots + \mathbf{x}_q).$$

**Exercise 2.2.8** *Suppose that* $\mathbf{x}_1$, $\mathbf{x}_2$, $\ldots$, $\mathbf{x}_q \in \mathbb{R}^n$. *Let* $\mathbf{y}_j$ *be the centroid of the* $q - 1$ *points* $\mathbf{x}_i$ *with* $1 \le i \le q$ *and* $i \ne j$. *Show that the q lines joining* $\mathbf{x}_j$ *to* $\mathbf{y}_j$ *for* $1 \le j \le q$ *all meet at the centroid of* $\mathbf{x}_1$, $\mathbf{x}_2$, $\ldots$, $\mathbf{x}_q$.

If $n = 3$ and $q = 4$, we recover a classical result on the geometry of the tetrahedron. We can generalise still further.

**Definition 2.2.9** *If each* $\mathbf{x}_j \in \mathbb{R}^n$ *is associated with a strictly positive real number* $m_j$ *for* $1 \le j \le q$, *then the* centre of mass[3] *of the system is the point*

$$\frac{1}{m_1 + m_2 + \cdots + m_q}(m_1\mathbf{x}_1 + m_2\mathbf{x}_2 + \cdots + m_q\mathbf{x}_q).$$

**Exercise 2.2.10** *(i) Suppose that* $\mathbf{x}_j \in \mathbb{R}^n$ *is associated with a strictly positive real number* $m_j$ *for* $1 \le j \le q$. *Let* $\mathbf{y}_j$ *be the centre of mass of the* $q - 1$ *points* $\mathbf{x}_i$ *with* $1 \le i \le q$ *and*

---

[2] That is we *guess* the correct answer and then *check* that our guess is correct.
[3] Traditionally called the *centre of gravity*. The change has been insisted on by the kind of person who uses 'Welsh rarebit' for 'Welsh rabbit' on the grounds that the dish contains no meat.

$i \neq j$. Then the $q$ lines joining $\mathbf{x}_j$ to $\mathbf{y}_j$ for $1 \leq j \leq q$ all meet at the centre of mass of $\mathbf{x}_1, \mathbf{x}_2, \ldots, \mathbf{x}_q$.

(ii) What mathematical (as opposed to physical) problem do we avoid by insisting that the $m_j$ are strictly positive?

## 2.3 Euclidean distance

So far, we have ignored the notion of distance. If we think of the distance between the points $\mathbf{x}$ and $\mathbf{y}$ in $\mathbb{R}^2$ or $\mathbb{R}^3$, it is natural to look at

$$\|\mathbf{x} - \mathbf{y}\| = \left( \sum_{j=1}^{n} (x_j - y_j)^2 \right)^{1/2} .$$

Let us make a formal definition.

**Definition 2.3.1** *If* $\mathbf{x} \in \mathbb{R}^n$, *we define the* norm (*or, more specifically, the* Euclidean norm) $\|\mathbf{x}\|$ *of* $\mathbf{x}$ *by*

$$\|\mathbf{x}\| = \left( \sum_{j=1}^{n} x_j^2 \right)^{1/2} ,$$

*where we take the positive square root.*

Some properties of the norm are easy to derive.

**Lemma 2.3.2** *Suppose that* $\mathbf{x} \in \mathbb{R}^n$ *and* $\lambda \in \mathbb{R}$. *The following results hold.*
(i) $\|\mathbf{x}\| \geq 0$.
(ii) $\|\mathbf{x}\| = 0$ *if and only if* $\mathbf{x} = \mathbf{0}$.
(iii) $\|\lambda\mathbf{x}\| = |\lambda| \|\mathbf{x}\|$.

**Exercise 2.3.3** *Prove Lemma 2.3.2.*

We would like to think of $\|\mathbf{x} - \mathbf{y}\|$ as the distance between $\mathbf{x}$ and $\mathbf{y}$, but we do not yet know that it has all the properties we expect of distance.

**Exercise 2.3.4** *Use school algebra to show that*

$$\|\mathbf{x} - \mathbf{y}\| + \|\mathbf{y} - \mathbf{z}\| \leq \|\mathbf{y} - \mathbf{z}\|. \qquad \bigstar$$

*(You may take* $n = 3$, *or even* $n = 2$ *if you wish.)*

The relation $\bigstar$ is called the triangle inequality. We shall prove it by an indirect approach which introduces many useful ideas.

We start by introducing the *inner product* or *dot product*.[4]

---

[4] For historical reasons it is also called the *scalar product*, but this can be confusing.

**Definition 2.3.5** *Suppose that* $\mathbf{x}$, $\mathbf{y} \in \mathbb{R}^n$. *We define the* inner product $\mathbf{x} \cdot \mathbf{y}$ *by*

$$\mathbf{x} \cdot \mathbf{y} = \frac{1}{4}(\|\mathbf{x} + \mathbf{y}\|^2 - \|\mathbf{x} - \mathbf{y}\|^2).$$

A quick calculation shows that our definition is equivalent to the more usual one.

**Lemma 2.3.6** *Suppose that* $\mathbf{x}$, $\mathbf{y} \in \mathbb{R}^n$. *Then*

$$\mathbf{x} \cdot \mathbf{y} = x_1 y_1 + x_2 y_2 + \cdots + x_n y_n.$$

*Proof* Left to the reader. □

In later chapters we shall use an alternative notation $\langle \mathbf{x}, \mathbf{y} \rangle$ for inner product. Other notations include $\mathbf{x}.\mathbf{y}$ and $(\mathbf{x}, \mathbf{y})$. The reader must be prepared to fall in with whatever notation is used.

Here are some key properties of the inner product.

**Lemma 2.3.7** *Suppose that* $\mathbf{x}$, $\mathbf{y}$, $\mathbf{w} \in \mathbb{R}^n$ *and* $\lambda$, $\mu \in \mathbb{R}$. *The following results hold.*
  (*i*) $\mathbf{x} \cdot \mathbf{x} \geq 0$.
  (*ii*) $\mathbf{x} \cdot \mathbf{x} = 0$ *if and only if* $\mathbf{x} = \mathbf{0}$.
  (*iii*) $\mathbf{x} \cdot \mathbf{y} = \mathbf{y} \cdot \mathbf{x}$.
  (*iv*) $\mathbf{x} \cdot (\mathbf{y} + \mathbf{w}) = \mathbf{x} \cdot \mathbf{y} + \mathbf{x} \cdot \mathbf{w}$.
  (*v*) $(\lambda \mathbf{x}) \cdot \mathbf{y} = \lambda(\mathbf{x} \cdot \mathbf{y})$.
  (*vi*) $\mathbf{x} \cdot \mathbf{x} = \|\mathbf{x}\|^2$.

*Proof* Simple verifications using Lemma 2.3.6. The details are left as an exercise for the reader. □

We also have the following trivial, but extremely useful, result.

**Lemma 2.3.8** *Suppose that* $\mathbf{a}$, $\mathbf{b} \in \mathbb{R}^n$. *If*

$$\mathbf{a} \cdot \mathbf{x} = \mathbf{b} \cdot \mathbf{x}$$

*for all* $\mathbf{x} \in \mathbb{R}^n$, *then* $\mathbf{a} = \mathbf{b}$.

*Proof* If the hypotheses hold, then

$$(\mathbf{a} - \mathbf{b}) \cdot \mathbf{x} = \mathbf{a} \cdot \mathbf{x} - \mathbf{b} \cdot \mathbf{x} = 0$$

for all $\mathbf{x} \in \mathbb{R}^n$. In particular, taking $\mathbf{x} = \mathbf{a} - \mathbf{b}$, we have

$$\|\mathbf{a} - \mathbf{b}\|^2 = (\mathbf{a} - \mathbf{b}) \cdot (\mathbf{a} - \mathbf{b}) = 0$$

and so $\mathbf{a} - \mathbf{b} = \mathbf{0}$. □

We now come to one of the most important inequalities in mathematics.[5]

---

[5] Since this is the view of both Gowers and Tao, I have no hesitation in making this assertion.

**Theorem 2.3.9 [The Cauchy–Schwarz inequality]** *If* $\mathbf{x}$, $\mathbf{y} \in \mathbb{R}^n$, *then*

$$|\mathbf{x} \cdot \mathbf{y}| \leq \|\mathbf{x}\| \|\mathbf{y}\|.$$

*Moreover* $|\mathbf{x} \cdot \mathbf{y}| = \|\mathbf{x}\| \|\mathbf{y}\|$ *if and only if we can find* $\lambda$, $\mu \in \mathbb{R}$ *not both zero such that* $\lambda \mathbf{x} = \mu \mathbf{y}$.

*Proof* (The clever proof given here is due to Schwarz.) If $\mathbf{x} = \mathbf{y} = \mathbf{0}$, then the theorem is trivial. Thus we may assume, without loss of generality, that $\mathbf{x} \neq \mathbf{0}$ and so $\|\mathbf{x}\| \neq 0$. If $\lambda$ is a real number, then, using the results of Lemma 2.3.7, we have

$$\begin{aligned}
0 &\leq \|\lambda \mathbf{x} + \mathbf{y}\|^2 \\
&= (\lambda \mathbf{x} + \mathbf{y}) \cdot (\lambda \mathbf{x} + \mathbf{y}) \\
&= (\lambda \mathbf{x}) \cdot (\lambda \mathbf{x}) + (\lambda \mathbf{x}) \cdot \mathbf{y} + \mathbf{y} \cdot (\lambda \mathbf{x}) + \mathbf{y} \cdot \mathbf{y} \\
&= \lambda^2 \mathbf{x} \cdot \mathbf{x} + 2\lambda \mathbf{x} \cdot \mathbf{y} + \mathbf{y} \cdot \mathbf{y} \\
&= \lambda^2 \|\mathbf{x}\|^2 + 2\lambda \mathbf{x} \cdot \mathbf{y} + \|\mathbf{y}\|^2 \\
&= \left( \lambda \|\mathbf{x}\| + \frac{\mathbf{x} \cdot \mathbf{y}}{\|\mathbf{x}\|} \right)^2 + \|\mathbf{y}\|^2 - \left( \frac{\mathbf{x} \cdot \mathbf{y}}{\|\mathbf{x}\|} \right)^2.
\end{aligned}$$

If we set

$$\lambda = -\frac{\mathbf{x} \cdot \mathbf{y}}{\|\mathbf{x}\|^2},$$

we obtain

$$0 \leq (\lambda \mathbf{x} + \mathbf{y}) \cdot (\lambda \mathbf{x} + \mathbf{y}) = \|\mathbf{y}\|^2 - \left( \frac{\mathbf{x} \cdot \mathbf{y}}{\|\mathbf{x}\|} \right)^2.$$

Thus

$$\|\mathbf{y}\|^2 - \left( \frac{\mathbf{x} \cdot \mathbf{y}}{\|\mathbf{x}\|} \right)^2 \geq 0 \qquad\qquad \bigstar$$

with equality only if

$$0 = \|\lambda \mathbf{x} + \mathbf{y}\|$$

so only if

$$\lambda \mathbf{x} + \mathbf{y} = \mathbf{0}.$$

Rearranging the terms in $\bigstar$ we obtain

$$(\mathbf{x} \cdot \mathbf{y})^2 \leq \|\mathbf{x}\|^2 \|\mathbf{y}\|^2$$

and so

$$|\mathbf{x} \cdot \mathbf{y}| \leq \|\mathbf{x}\| \|\mathbf{y}\|$$

with equality only if

$$\lambda \mathbf{x} + \mathbf{y} = \mathbf{0}.$$

We observe that, if $\lambda'$, $\mu' \in \mathbb{R}$ are not both zero and $\lambda' \mathbf{x} = \mu' \mathbf{y}$, then $|\mathbf{x} \cdot \mathbf{y}| = \|\mathbf{x}\| \|\mathbf{y}\|$, so the proof is complete. $\qquad\qquad\qquad\qquad\qquad\qquad\qquad\qquad\qquad\qquad\qquad\qquad\quad\square$

We can now prove the triangle inequality.

**Theorem 2.3.10** *Suppose that* $\mathbf{x}$, $\mathbf{y} \in \mathbb{R}^n$. *Then*

$$\|\mathbf{x}\| + \|\mathbf{y}\| \geq \|\mathbf{x} + \mathbf{y}\|$$

*with equality if and only if there exist* $\lambda$, $\mu \geq 0$ *not both zero such that* $\lambda \mathbf{x} = \mu \mathbf{y}$.

*Proof* Observe that

$$
\begin{aligned}
(\|\mathbf{x}\| + \|\mathbf{y}\|)^2 - \|\mathbf{x} + \mathbf{y}\|^2 &= (\|\mathbf{x}\| + \|\mathbf{y}\|)^2 + (\mathbf{x} + \mathbf{y}) \cdot (\mathbf{x} + \mathbf{y}) \\
&= (\|\mathbf{x}\|^2 + 2\|\mathbf{x}\| \|\mathbf{y}\| + \|\mathbf{y}\|^2) - (\|\mathbf{x}\|^2 + 2\mathbf{x} \cdot \mathbf{y} + \|\mathbf{y}\|^2) \\
&= 2(\|\mathbf{x}\| \|\mathbf{y}\| - \mathbf{x} \cdot \mathbf{y}) \\
&\geq 2(\|\mathbf{x}\| \|\mathbf{y}\| - |\mathbf{x} \cdot \mathbf{y}|) \geq 0
\end{aligned}
$$

with equality if and only if

$$\mathbf{x} \cdot \mathbf{y} \geq 0 \quad \text{and} \quad \|\mathbf{x}\| \|\mathbf{y}\| = |\mathbf{x} \cdot \mathbf{y}|.$$

Rearranging and taking positive square roots, we see that

$$\|\mathbf{x}\| + \|\mathbf{y}\| \geq \|\mathbf{x} + \mathbf{y}\|.$$

Since $(\lambda \mathbf{x}) \cdot (\mu \mathbf{x}) = \lambda \mu \|\mathbf{x}\|^2$, it is easy to check that we have equality if and only if there exist $\lambda$, $\mu \geq 0$ not both zero such that $\lambda \mathbf{x} = \mu \mathbf{y}$. $\qquad\qquad\qquad\square$

**Exercise 2.3.11** *Suppose that* $\mathbf{a}$, $\mathbf{b}$, $\mathbf{c} \in \mathbb{R}^n$. *Show that*

$$\|\mathbf{a} - \mathbf{b}\| + \|\mathbf{b} - \mathbf{c}\| \geq \|\mathbf{a} - \mathbf{c}\|.$$

*When does equality occur?*

*Deduce an inequality involving the length of the sides of a triangle* $ABC$.

**Exercise 2.3.12** (*This completes some unfinished business.*) *Suppose that* $\mathbf{a} \neq \mathbf{b}$. *Show that there exists a unique point* $\mathbf{x}$ *on the line joining* $\mathbf{a}$ *and* $\mathbf{b}$ *such that* $\|\mathbf{x} - \mathbf{a}\| = \|\mathbf{x} - \mathbf{b}\|$ *and that this point is given by* $\mathbf{x} = \frac{1}{2}(\mathbf{a} + \mathbf{b})$.

If we confine ourselves to two dimensions, we can give another characterisation of the inner product.

**Exercise 2.3.13** (*i*) *We work in the plane. If* $ABC$ *is a triangle with the angle between* $AB$ *and* $BC$ *equal to* $\theta$, *show, by elementary trigonometry, that*

$$|BC|^2 + |AB|^2 - 2|AB| \times |BC| \cos \theta = |AC|^2$$

*where* $|XY|$ *is the length of the side* $XY$.

(*ii*) *If* $\mathbf{a}$, $\mathbf{c} \in \mathbb{R}^n$ *show that*

$$\|\mathbf{a}\|^2 + \|\mathbf{c}\|^2 - 2\mathbf{a} \cdot \mathbf{c} = \|\mathbf{a} - \mathbf{c}\|^2.$$

(*iii*) *Returning to the plane* $\mathbb{R}^2$, *suppose that* $\mathbf{a}$, $\mathbf{c} \in \mathbb{R}^2$, *the point A is at* $\mathbf{a}$, *the point C is at* $\mathbf{c}$ *and* $\theta$ *is the angle between the line joining A to the origin and the line joining C to the origin. Show that*

$$\mathbf{a} \cdot \mathbf{c} = \|\mathbf{a}\| \|\mathbf{c}\| \cos \theta.$$

Because of this result, the inner product is sometimes defined in some such way as 'the product of the length of the two vectors times the cosine of the angle between them'. This is fine, so long as we confine ourselves to $\mathbb{R}^2$ and $\mathbb{R}^3$, but, once we consider vectors in $\mathbb{R}^4$ or more general spaces, this places a great strain on our geometrical intuition

We therefore turn the definition around as follows. Suppose that $\mathbf{a}$, $\mathbf{c} \in \mathbb{R}^n$ and $\mathbf{a}$, $\mathbf{c} \neq \mathbf{0}$. By the Cauchy–Schwarz inequality,

$$-1 \leq \frac{\mathbf{a} \cdot \mathbf{c}}{\|\mathbf{a}\| \|\mathbf{c}\|} \leq 1.$$

By the properties of the cosine function, there is a unique $\theta$ with $0 \leq \theta \leq \pi$ and

$$\cos \theta = \frac{\mathbf{a} \cdot \mathbf{c}}{\|\mathbf{a}\| \|\mathbf{c}\|}.$$

We *define* $\theta$ to be the angle between $\mathbf{a}$ and $\mathbf{c}$.

**Exercise 2.3.14** *Suppose that* $\mathbf{u}$, $\mathbf{v} \in \mathbb{R}^n$ *and* $\mathbf{u}$, $\mathbf{v} \neq \mathbf{0}$. *Show that, if we adopt the definition just given, and if the angle between* $\mathbf{u}$ *and* $\mathbf{v}$ *is* $\theta$, *then the angle between* $\mathbf{v}$ *and* $\mathbf{u}$ *is* $\theta$ *and the angle between* $\mathbf{u}$ *and* $-\mathbf{v}$ *is* $\pi - \theta$.

Most of the time, we shall be interested in a special case.

**Definition 2.3.15** *Suppose that* $\mathbf{u}$, $\mathbf{v} \in \mathbb{R}^n$. *We say that* $\mathbf{u}$ *and* $\mathbf{v}$ *are* orthogonal (*or* perpendicular) *if* $\mathbf{u} \cdot \mathbf{v} = 0$.

**Exercise 2.3.16** *Consider the* $2^n$ *vertices of a cube in* $\mathbb{R}^n$ *and the* $2^{n-1}$ *diagonals. (Part of the exercise is to decide what this means.) Show that no two diagonals can be perpendicular if n is odd.*

*For* $n = 4$, *what is the greatest number of mutually perpendicular diagonals and why? List all possible angles between the diagonals.*

Note that our definition of orthogonality means that the zero vector $\mathbf{0}$ is perpendicular to every vector. The symbolism $\mathbf{u} \perp \mathbf{v}$ is sometimes used to mean that $\mathbf{u}$ and $\mathbf{v}$ are perpendicular.

**Exercise 2.3.17** *We work in* $\mathbb{R}^3$. *Are the following statements always true or sometimes false? In each case give a proof or a counterexample.*

(*i*) *If* $\mathbf{u} \perp \mathbf{v}$, *then* $\mathbf{v} \perp \mathbf{u}$.

(*ii*) *If* $\mathbf{u} \perp \mathbf{v}$ *and* $\mathbf{v} \perp \mathbf{w}$, *then* $\mathbf{u} \perp \mathbf{w}$.

(*iii*) *If* $\mathbf{u} \perp \mathbf{u}$, *then* $\mathbf{u} = \mathbf{0}$.

**Exercise 2.3.18 [Pythagoras extended]** (*i*) *If* $\mathbf{u}, \mathbf{v} \in \mathbb{R}^2$ *and* $\mathbf{u} \perp \mathbf{v}$, *show that*

$$\|\mathbf{u}\|^2 + \|\mathbf{v}\|^2 = \|\mathbf{u} + \mathbf{v}\|^2.$$

*Why does this correspond to Pythagoras' theorem?*

(*ii*) *If* $\mathbf{u}, \mathbf{v}, \mathbf{w} \in \mathbb{R}^3$, $\mathbf{u} \perp \mathbf{v}$, $\mathbf{v} \perp \mathbf{w}$ *and* $\mathbf{w} \perp \mathbf{u}$ (*that is to say, the vectors are mutually perpendicular*), *show that*

$$\|\mathbf{u}\|^2 + \|\mathbf{v}\|^2 + \|\mathbf{w}\|^2 = \|\mathbf{u} + \mathbf{v} + \mathbf{w}\|^2.$$

(*iii*) *State and prove a corresponding result in* $\mathbb{R}^4$.

## 2.4 Geometry, plane and solid

We now look at some familiar geometric objects in a vectorial context. Our object is not to prove rigorous theorems, but to develop intuition.

For example, we shall say that a parallelogram in $\mathbb{R}^2$ is a figure with vertices $\mathbf{c}$, $\mathbf{c} + \mathbf{a}$, $\mathbf{c} + \mathbf{b}$, $\mathbf{c} + \mathbf{a} + \mathbf{b}$ and rely on the reader to convince herself that this corresponds to her pre-existing idea of a parallelogram.

**Exercise 2.4.1 [The parallelogram law]** *If* $\mathbf{a}, \mathbf{b} \in \mathbb{R}^n$, *show that*

$$\|\mathbf{a} + \mathbf{b}\|^2 + \|\mathbf{a} - \mathbf{b}\|^2 = 2(\|\mathbf{a}\|^2 + \|\mathbf{b}\|^2).$$

*If* $n = 2$, *interpret the equality in terms of the lengths of the sides and diagonals of a parallelogram.*

**Exercise 2.4.2** (*i*) *Prove that the diagonals of a parallelogram bisect each other.*

(*ii*) *Prove that the line joining one vertex of a parallelogram to the mid-point of an opposite side trisects the diagonal and is trisected by it.*

Here is a well known theorem of classical geometry.

**Example 2.4.3** *Consider a triangle* $ABC$. *The* altitude *through a vertex is the line through that vertex perpendicular to the opposite side. We assert that the three altitudes meet at a point.*

**Exercise 2.4.4** *Draw an example and check that* (*to within the accuracy of the drawing*) *the conclusion of Example 2.4.3 holds.*

*Proof of Example 2.4.3* If we translate the statement of Example 2.4.3 into vector notation, it asserts that, if $\mathbf{x}$ satisfies the equations

$$(\mathbf{x} - \mathbf{a}) \cdot (\mathbf{b} - \mathbf{c}) = 0$$

$$(\mathbf{x} - \mathbf{b}) \cdot (\mathbf{c} - \mathbf{a}) = 0,$$

then it satisfies the equation

$$(\mathbf{x} - \mathbf{c}) \cdot (\mathbf{a} - \mathbf{b}) = 0.$$

But adding the first two equations gives the third, so we are done. □

We consider the rather less interesting three dimensional case in Exercise 2.5.6 (ii). It turns out that the four altitudes of a tetrahedron only meet if the sides of the tetrahedron satisfy certain conditions. Exercises 2.5.7 and 2.5.8 give two other classical results which are readily proved by similar means to those used for Example 2.4.3.

We have already looked at the equation

$$ax + by = c$$

(where $a$ and $b$ are not both zero) for a straight line in $\mathbb{R}^2$. The inner product enables us to look at the equation in a different way. If $\mathbf{a} = (a, b)$, then $\mathbf{a}$ is a non-zero vector and our equation becomes

$$\mathbf{a} \cdot \mathbf{x} = c$$

where $\mathbf{x} = (x, y)$.

This equation is usually written in a different way.

**Definition 2.4.5** *We say that* $\mathbf{u} \in \mathbb{R}^n$ *is a* unit vector *if* $\|\mathbf{u}\| = 1$.

If we take

$$\mathbf{n} = \frac{1}{\|\mathbf{a}\|}\mathbf{a} \quad \text{and} \quad p = \frac{c}{\|\mathbf{a}\|},$$

we obtain the equation for a line in $\mathbb{R}^2$ as

$$\mathbf{n} \cdot \mathbf{x} = p,$$

where $\mathbf{n}$ is a unit vector.

**Exercise 2.4.6** *We work in* $\mathbb{R}^2$.

(*i*) *If* $\mathbf{u} = (u, v)$ *is a unit vector, show that there are exactly two unit vectors* $\mathbf{n} = (n, m)$ *and* $\mathbf{n}' = (n', m')$ *perpendicular to* $\mathbf{u}$. *Write them down explicitly.*

(*ii*) *Given a straight line written in the form*

$$\mathbf{x} = \mathbf{a} + t\mathbf{c}$$

(*where* $\mathbf{c} \neq \mathbf{0}$ *and* $t$ *ranges freely over* $\mathbb{R}$), *find a unit vector* $\mathbf{n}$ *and* $p$ *so that the line is described by* $\mathbf{n} \cdot \mathbf{x} = p$.

(*iii*) *Given a straight line written in the form*

$$\mathbf{n} \cdot \mathbf{x} = p$$

(*where* $\mathbf{n}$ *is a unit vector*), *find* $\mathbf{a}$ *and* $\mathbf{c} \neq \mathbf{0}$ *so that the line is described by*

$$\mathbf{x} = \mathbf{a} + t\mathbf{c}$$

*where* $t$ *ranges freely over* $\mathbb{R}$.

If we work in $\mathbb{R}^3$, it is easy to convince ourselves that the equation

$$\mathbf{n} \cdot \mathbf{x} = p,$$

where $\mathbf{n}$ is a unit vector, defines the plane perpendicular to $\mathbf{n}$ passing through the point $p\mathbf{n}$. (If the reader is worried by the informality of our arguments, she should note that we look at orthogonality in much greater depth and with due attention to rigour in Chapter 7.)

**Example 2.4.7** *Let $l$ be a straight line and $\pi$ a plane in $\mathbb{R}^3$. One of three things can occur.*
  *(i) $l$ and $\pi$ do not intersect.*
  *(ii) $l$ and $\pi$ intersect at a point.*
  *(iii) $l$ lies inside $\pi$ (so $l$ and $\pi$ intersect in a line).*

*Proof* Let $\pi$ have equation $\mathbf{n} \cdot \mathbf{x} = p$ (where $\mathbf{n}$ is a unit vector) and let $l$ be described by $\mathbf{x} = \mathbf{a} + t\mathbf{c}$ (with $\mathbf{c} \neq \mathbf{0}$) where $t$ ranges freely over $\mathbb{R}$.

Then the points of intersection (if any) are given by $\mathbf{x} = \mathbf{a} + s\mathbf{c}$ where

$$p = \mathbf{n} \cdot \mathbf{x} = \mathbf{n} \cdot (\mathbf{a} + s\mathbf{c}) = \mathbf{n} \cdot \mathbf{a} + s\mathbf{n} \cdot \mathbf{c}$$

that is to say, by

$$s\mathbf{n} \cdot \mathbf{c} = p - \mathbf{n} \cdot \mathbf{a}. \qquad\qquad ★$$

If $\mathbf{n} \cdot \mathbf{c} \neq 0$, then ★ has a unique solution in $s$ and we have case (ii). If $\mathbf{n} \cdot \mathbf{c} = 0$ (that is to say, if $\mathbf{n} \perp \mathbf{c}$), then one of two things may happen. If $p \neq \mathbf{n} \cdot \mathbf{a}$, then ★ has no solution and we have case (i). If $p = \mathbf{n} \cdot \mathbf{a}$, then every value of $s$ satisfies ★ and we have case (iii). □

In cases (i) and (iii), I would be inclined to say that $l$ is *parallel* to $\pi$.

**Example 2.4.8** *Let $\pi$ and $\pi'$ be planes in $\mathbb{R}^3$. One of three things can occur.*
  *(i) $\pi$ and $\pi'$ do not intersect.*
  *(ii) $\pi$ and $\pi'$ intersect in a line.*
  *(iii) $\pi$ and $\pi'$ coincide.*

*Proof* Let $\pi$ be given by $\mathbf{n} \cdot \mathbf{x} = p$ and $\pi'$ be given by $\mathbf{n}' \cdot \mathbf{x} = p'$.

If there exist real numbers $\mu$ and $\nu$ not both zero such that $\mu\mathbf{n} = \nu\mathbf{n}'$, then, in fact, $\mathbf{n}' = \pm\mathbf{n}$ and we can write $\pi'$ as

$$\mathbf{n} \cdot \mathbf{x} = q.$$

If $p \neq q$, then the pair of equations

$$\mathbf{n} \cdot \mathbf{x} = p, \quad \mathbf{n} \cdot \mathbf{x} = q$$

have no solutions and we have case (i). If $p = q$ the two equations are the same and we have case (ii).

If there do not exist real numbers $\mu$ and $\nu$ not both zero such that $\mu\mathbf{n} = \nu\mathbf{n}'$ (after Section 5.4, we will be able to replace this cumbrous phrase with the statement '$\mathbf{n}$ and $\mathbf{n}'$

are linearly independent'), then the points $\mathbf{x} = (x_1, x_2, x_3)$ of intersection of the two planes are given by a pair of equations

$$n_1 x_1 + n_2 x_2 + n_3 x_3 = p$$
$$n_1' x_1 + n_2' x_2 + n_3' x_3 = p'$$

and there do not exist exist real numbers $\mu$ and $\nu$ not both zero with

$$\mu(n_1, n_2, n_3) = \nu(n_1', n_2', n_3').$$

Applying Gaussian elimination, we have (possibly after relabelling the coordinates)

$$x_1 + c x_3 = a$$
$$x_2 + d x_3 = b$$

so

$$(x_1, x_2, x_3) = (a, b, 0) + x_3(-c, -d, 1)$$

that is to say

$$\mathbf{x} = \mathbf{a} + t\mathbf{c}$$

where $t$ is a freely chosen real number, $\mathbf{a} = (a, b, 0)$ and $\mathbf{c} = (-c, -d, 1) \neq \mathbf{0}$. We thus have case (ii). $\qquad\square$

**Exercise 2.4.9** *We work in $\mathbb{R}^3$. If*

$$\mathbf{n}_1 = (1, 0, 0), \quad \mathbf{n}_2 = (0, 1, 0), \quad \mathbf{n}_3 = (2^{-1/2}, 2^{-1/2}, 0),$$

$p_1 = p_2 = 0$, $p_3 = 2^{1/2}$ *and three planes $\pi_j$ are given by the equations*

$$\mathbf{n}_j \cdot \mathbf{x} = p_j,$$

*show that each pair of planes meet in a line, but that no point belongs to all three planes.*
 *Give similar example of planes $\pi_j$ obeying the following conditions.*
 *(i) No two planes meet.*
 *(ii) The planes $\pi_1$ and $\pi_2$ meet in a line and the planes $\pi_1$ and $\pi_3$ meet in a line, but $\pi_2$ and $\pi_3$ do not meet.*

Since a circle in $\mathbb{R}^2$ consists of all points equidistant from a given point, it is easy to write down the following equation for a circle

$$\|\mathbf{x} - \mathbf{a}\| = r.$$

We say that the circle has centre $\mathbf{a}$ and radius $r$. We demand $r > 0$.

**Exercise 2.4.10** *Describe the set*

$$\{\mathbf{x} \in \mathbb{R}^2 : \|\mathbf{x} - \mathbf{a}\| = r\}$$

*in the case $r = 0$. What is the set if $r < 0$?*

In exactly the same way, we say that a sphere in $\mathbb{R}^3$ of centre $\mathbf{a}$ and radius $r > 0$ is given by

$$\|\mathbf{x} - \mathbf{a}\| = r.$$

**Exercise 2.4.11 [Inversion]** (*i*) *We work in* $\mathbb{R}^2$ *and consider the map* $\mathbf{y} : \mathbb{R}^2 \setminus \{\mathbf{0}\} \to \mathbb{R}^2 \setminus \{\mathbf{0}\}$ *given by*

$$\mathbf{y}(\mathbf{x}) = \frac{1}{\|\mathbf{x}\|^2}\mathbf{x}.$$

(*We leave* $\mathbf{y}$ *undefined at* $\mathbf{0}$.) *Show that* $\mathbf{y}\big(\mathbf{y}(\mathbf{x})\big) = \mathbf{x}$.

(*ii*) *Suppose that* $\mathbf{a} \in \mathbb{R}^2$, $r > 0$ *and* $\|\mathbf{a}\| \neq r$. *Show that* $\mathbf{y}$ *takes the circle of radius* $r$ *and centre* $\mathbf{a}$ *to another circle. What are the radius and centre of the new circle?*

(*iii*) *Suppose that* $\mathbf{a} \in \mathbb{R}^2$, $r > 0$ *and* $\|\mathbf{a}\| = r$. *Show that* $\mathbf{y}$ *takes the circle of radius* $r$ *and centre* $\mathbf{a}$ (*omitting the point* $\mathbf{0}$) *to a line to be specified.*

(*iv*) *Generalise parts* (*i*), (*ii*) *and* (*iii*) *to* $\mathbb{R}^3$.

[*We refer to the transformation* $\mathbf{y}$ *as an* inversion. *We give a very pretty application in Exercise 2.5.14.*]

**Exercise 2.4.12 [Ptolemy's inequality]** *Let* $\mathbf{x}$, $\mathbf{y} \in \mathbb{R}^n \setminus \{\mathbf{0}\}$. *By squaring both sides of the equation, or otherwise, show that*

$$\left\| \frac{\mathbf{x}}{\|\mathbf{x}\|^2} - \frac{\mathbf{y}}{\|\mathbf{y}\|^2} \right\| = \frac{1}{\|\mathbf{x}\|\|\mathbf{y}\|}\|\mathbf{x} - \mathbf{y}\|.$$

*Hence, or otherwise, show that, if* $\mathbf{x}$, $\mathbf{y}$, $\mathbf{z} \in \mathbb{R}^n$,

$$\|\mathbf{z}\|\|\mathbf{x} - \mathbf{y}\| \leq \|\mathbf{y}\|\|\mathbf{z} - \mathbf{x}\| + \|\mathbf{x}\|\|\mathbf{y} - \mathbf{z}\|.$$

*If* $\mathbf{x}$, $\mathbf{y}$, $\mathbf{z} \neq \mathbf{0}$, *show that we have equality if and only if the points* $\|\mathbf{x}\|^{-2}\mathbf{x}$, $\|\mathbf{y}\|^{-2}\mathbf{y}$, $\|\mathbf{z}\|^{-2}\mathbf{z}$ *lie on a straight line.*

*Deduce that, if* $ABCD$ *is a quadrilateral in the plane,*

$$|AB||CD| + |BC||DA| \geq |AC||BD|. \qquad \bigstar$$

*Use Exercise 2.4.11 to show that* $\bigstar$ *becomes an equality if and only if* $A$, $B$, $C$ *and* $D$ *lie on a circle or straight line.*

*The statement that, if* $ABCD$ *is a 'cyclic quadrilateral', then*

$$|AB||CD| + |BC||DA| = |AC||BD|,$$

*is known as Ptolemy's theorem after the great Greek astronomer. Ptolemy and his predecessors used the theorem to produce what were, in effect, trigonometric tables.*

## 2.5 Further exercises

**Exercise 2.5.1** We work in $\mathbb{R}^3$. Let $A$, $B$, $C$ be strictly positive constants and $\mathbf{w}$ a fixed vector. Determine the vector $\mathbf{x}$ of smallest magnitude (i.e. with $\|\mathbf{x}\|$ as small as possible) which satisfies the simultaneous equations

$$A\mathbf{x} - B\mathbf{y} = 2\mathbf{w}$$
$$\mathbf{x} \cdot \mathbf{y} = C.$$

**Exercise 2.5.2** Describe geometrically the surfaces in $\mathbb{R}^3$ given by the following equations. Give brief reasons for your answers. (Formal proofs are not required.) We take $\mathbf{u}$ to be a fixed vector with $\|\mathbf{u}\| = 1$ and $\alpha$ and $\beta$ to be fixed real numbers with $1 > |\alpha| > 0$ and $\beta > 0$.

(i) $\mathbf{x} \cdot \mathbf{u} = \alpha \|\mathbf{x}\|$.

(ii) $\|\mathbf{x} - (\mathbf{x} \cdot \mathbf{u})\mathbf{u}\| = \beta$.

**Exercise 2.5.3** (i) By using the Cauchy–Schwarz inequality in $\mathbb{R}^3$, show that

$$x^2 + y^2 + z^2 \geq yz + zx + xy$$

for all real $x$, $y$, $z$.

(ii) By using the Cauchy–Schwarz inequality in $\mathbb{R}^4$ several times, show that only one choice of real numbers satisfies

$$3(x^2 + y^2 + z^2 + 4) - 2(yz + zx + xy) - 4(x + y + z) = 0$$

and find those numbers.

**Exercise 2.5.4** Let $\mathbf{a} \in \mathbb{R}^n$ be fixed. Suppose that vectors $\mathbf{x}$, $\mathbf{y} \in \mathbb{R}^n$ are related by the equation

$$\mathbf{x} + (\mathbf{x} \cdot \mathbf{y})\mathbf{y} = \mathbf{a}.$$

Show that

$$(\mathbf{x} \cdot \mathbf{y})^2 = \frac{\|\mathbf{a}\|^2 - \|\mathbf{x}\|^2}{2 + \|\mathbf{y}\|^2}$$

and deduce that

$$\|\mathbf{x}\|(1 + \|\mathbf{y}\|^2) \geq \|\mathbf{a}\| \geq \|\mathbf{x}\|.$$

Explain, with proof, the circumstances under which either of the two inequalities in the formula just given can be replaced by equalities, and describe the relation between $\mathbf{x}$, $\mathbf{y}$ and $\mathbf{a}$ in these circumstances.

**Exercise 2.5.5** We work in $\mathbb{R}^n$. Show that, if $\mathbf{w}$, $\mathbf{x}$, $\mathbf{y}$, $\mathbf{z} \in \mathbb{R}^n$, then

$$\sum_{j=1}^{n} |w_j x_j y_j z_j| \leq \left( \sum_{j=1}^{n} w_j^4 \sum_{j=1}^{n} x_j^4 \sum_{j=1}^{n} y_j^4 \sum_{j=1}^{n} z_j^4 \right)^{1/4}.$$

(We take the positive root.)

If we write $\|\mathbf{x}\|_4 = \left(\sum_{j=1}^{n} x_j^4\right)^{1/4}$, show that the following results hold for all $\mathbf{x}, \mathbf{y} \in \mathbb{R}^4$, $\lambda \in \mathbb{R}$.

(i) $\|\mathbf{x}\|_4 \geq 0$.

(ii) $\|\mathbf{x}\|_4 = 0 \Leftrightarrow \mathbf{x} = \mathbf{0}$.

(iii) $\|\lambda\mathbf{x}\|_4 = |\lambda|\|\mathbf{x}\|_4$.

(iv) $\|\mathbf{x} + \mathbf{y}\|_4 \leq \|\mathbf{x}\|_4 + \|\mathbf{y}\|_4$.

**Exercise 2.5.6** (i) We work in $\mathbb{R}^3$. Show that, if two pairs of opposite edges of a non-degenerate[6] tetrahedron $ABCD$ are perpendicular, then the third pair are also perpendicular to each other. Show also that, in this case, the sum of the lengths squared of the two opposite edges is the same for each pair.

(ii) The altitude through a vertex of a non-degenerate tetrahedron is the line through the vertex perpendicular to the opposite face. Translating into vectors, explain why $\mathbf{x}$ lies on the altitude through $\mathbf{a}$ if and only if

$$(\mathbf{x} - \mathbf{a}) \cdot (\mathbf{b} - \mathbf{c}) = 0 \quad \text{and} \quad (\mathbf{x} - \mathbf{a}) \cdot (\mathbf{c} - \mathbf{d}) = 0.$$

Show that the four altitudes of a non-degenerate tetrahedron meet only if each pair of opposite edges are perpendicular.

If each pair of opposite edges are perpendicular, show by observing that the altitude through $A$ lies in each of the planes formed by $A$ and the altitudes of the triangle $BCD$, or otherwise, that the four altitudes of the tetrahedron do indeed meet.

**Exercise 2.5.7** Consider a non-degenerate triangle in the plane with vertices $A, B, C$ given by the vectors $\mathbf{a}, \mathbf{b}, \mathbf{c}$.

Show that the equation

$$\mathbf{x} = \mathbf{a} + t\big(\|\mathbf{a} - \mathbf{c}\|(\mathbf{a} - \mathbf{b}) + \|\mathbf{a} - \mathbf{b}\|(\mathbf{a} - \mathbf{c})\big)$$

with $t \in \mathbb{R}$ defines a line which is the angle bisector at $A$ (i.e. passes through $A$ and makes equal angles with $AB$ and $AC$).

If the point $X$ lies on the angle bisector at $A$, the point $Y$ lies on $AB$ in such a way that $XY$ is perpendicular to $AB$ and the point $Z$ lies on $AC$, in such a way that $XZ$ is perpendicular to $AC$, show, using vector methods, that $XY$ and $XZ$ have equal length.

Show that the three angle bisectors at $A, B$ and $C$ meet at a point $Q$ given by

$$\mathbf{q} = \frac{\|\mathbf{a} - \mathbf{b}\|\mathbf{c} + \|\mathbf{b} - \mathbf{c}\|\mathbf{a} + \|\mathbf{c} - \mathbf{a}\|\mathbf{b}}{\|\mathbf{a} - \mathbf{b}\| + \|\mathbf{b} - \mathbf{c}\| + \|\mathbf{c} - \mathbf{a}\|}.$$

---

[6] *Non-degenerate* and *generic* are overworked and often deliberately vague adjectives used by mathematicians to mean 'avoiding special cases'. Thus a non-degenerate triangle has all three vertices distinct and the vertices of a non-degenerate tetrahedron do not lie in a plane. Even if you are only asked to consider non-degenerate cases, it is often instructive to think about what happens in the degenerate cases.

Show that, given three distinct lines $AB, BC, CA$, there is at least one circle which has those three lines as tangents.[7]

**Exercise 2.5.8** Consider a non-degenerate triangle in the plane with vertices $A, B, C$ given by the vectors $\mathbf{a}, \mathbf{b}, \mathbf{c}$.

Show that the points equidistant from $A$ and $B$ form a line $l_{AB}$ whose equation you should find in the form

$$\mathbf{n}_{AB} \cdot \mathbf{x} = p_{AB},$$

where the unit vector $\mathbf{n}_{AB}$ and the real number $p_{AB}$ are to be given explicitly. Show that $l_{AB}$ is perpendicular to $AB$. (The line $l_{AB}$ is called the perpendicular bisector of $AB$.)

Show that

$$\mathbf{n}_{AB} \cdot \mathbf{x} = p_{AB}, \quad \mathbf{n}_{BC} \cdot \mathbf{x} = p_{BC} \Rightarrow \mathbf{n}_{CA} \cdot \mathbf{x} = p_{CA}$$

and deduce that the three perpendicular bisectors meet in a point.

Deduce that, if three points $A, B$ and $C$ do not lie in a straight line, they lie on a circle.

**Exercise 2.5.9** (Not very hard.) Consider a non-degenerate tetrahedron. For each edge we can find a plane which which contains that edge and passes through the midpoint of the opposite edge. Show that the six planes all pass through a common point.

**Exercise 2.5.10 [The Monge point][8]** Consider a non-degenerate tetrahedron with vertices $A, B, C, D$ given by the vectors $\mathbf{a}, \mathbf{b}, \mathbf{c}, \mathbf{d}$. Use inner products to write down an equation for the so-called, 'midplane' $\pi_{AB,CD}$ which is perpendicular to $AB$ and passes through the mid-point of $CD$. Hence show that (with an obvious notation)

$$\mathbf{x} \in \pi_{AB,CD} \cap \pi_{BC,AD} \cap \pi_{AD,BC} \Rightarrow \mathbf{x} \in \pi_{AC,BD}.$$

Deduce that the six midplanes of a tetrahedron meet at a point.

**Exercise 2.5.11** Show that

$$\|\mathbf{x} - \mathbf{a}\|^2 \cos^2 \alpha = \big((\mathbf{x} - \mathbf{a}) \cdot \mathbf{n}\big)^2,$$

with $\|\mathbf{n}\| = 1$, is the equation of a right circular double cone in $\mathbb{R}^3$ whose vertex has position vector $\mathbf{a}$, axis of symmetry $\mathbf{n}$ and opening angle $\alpha$. Two such double cones, with vertices $\mathbf{a}_1$ and $\mathbf{a}_2$, have parallel axes and the same opening angle. Show that, if $\mathbf{b} = \mathbf{a}_1 - \mathbf{a}_2 \neq \mathbf{0}$, then the intersection of the cones lies in a plane with unit normal

$$\mathbf{N} = \frac{\mathbf{b} \cos^2 \alpha - \mathbf{n}(\mathbf{n} \cdot \mathbf{b})}{\sqrt{\|\mathbf{b}\|^2 \cos^4 \alpha + (\mathbf{n} \cdot \mathbf{b})^2 (1 - 2 \cos^2 \alpha)}}.$$

---

[7] Actually there are four. Can you spot what they are? Can you use the methods of this question to find them? The circle found in this question is called the 'incircle' and the other three are called the 'excircles'.

[8] Monge's ideas on three dimensional geometry were so useful to the French army that they were considered a state secret. I have been told that, when he was finally allowed to publish in 1795, the British War Office rushed out and bought two copies of his book for its library where they remained unopened for 150 years.

**Exercise 2.5.12** We work in $\mathbb{R}^3$. We fix $c \in \mathbb{R}$ and $\mathbf{a} \in \mathbb{R}^3$ with $c < \mathbf{a} \cdot \mathbf{a}$. Show that if

$$\mathbf{x} + \mathbf{y} = 2\mathbf{a}, \ \mathbf{x} \cdot \mathbf{y} = c,$$

then $\mathbf{x}$ lies on a sphere. You should find the centre and radius of the sphere explicitly.

**Exercise 2.5.13** We work in $\mathbb{R}^3$. Show that the equation of a sphere with centre $\mathbf{c}$ and radius $a$ is $F(\mathbf{r}) = 0$, where

$$F(\mathbf{r}) = \mathbf{r} \cdot \mathbf{r} - 2\mathbf{r} \cdot \mathbf{c} + k$$

and $k$ is a constant to be found explicitly.

Show that a line through $\mathbf{d}$ parallel to the unit vector $\mathbf{b}$ intersects the sphere in two distinct points $\mathbf{u}$ and $\mathbf{v}$ if and only if

$$F(\mathbf{d}) < \left(\mathbf{b} \cdot (\mathbf{d} - \mathbf{c})\right)^2$$

and that, if this is the case,

$$(\mathbf{u} - \mathbf{d}) \cdot (\mathbf{v} - \mathbf{d}) = F(\mathbf{d}).$$

If the line intersects the sphere at a single point, we call it a *tangent line*. Show that a tangent line passing through a point $\mathbf{w}$ on the sphere is perpendicular to the radius $\mathbf{w} - \mathbf{c}$ and, conversely, that every line passing through $\mathbf{w}$ and perpendicular to the radius $\mathbf{w} - \mathbf{c}$ is a tangent line. The tangent lines through $\mathbf{w}$ thus form a *tangent plane*.

Show that the condition for the plane $\mathbf{r} \cdot \mathbf{n} = p$ (where $\mathbf{n}$ is a unit vector) to be a tangent plane is that

$$(p - \mathbf{c} \cdot \mathbf{n})^2 = c^2 - k.$$

If two spheres given by

$$\mathbf{r} \cdot \mathbf{r} - 2\mathbf{r} \cdot \mathbf{c} + k = 0 \quad \text{and} \quad \mathbf{r} \cdot \mathbf{r} - 2\mathbf{r} \cdot \mathbf{c}' + k' = 0$$

cut each other at right angles, show that

$$2\mathbf{c} \cdot \mathbf{c}' = k + k'.$$

**Exercise 2.5.14 [Steiner's porism]** Suppose that a circle $\Gamma_0$ lies inside another circle $\Gamma_1$. We draw a circle $\Lambda_1$ touching both $\Gamma_0$ and $\Gamma_1$. We then draw a circle $\Lambda_2$ touching $\Gamma_0$, $\Gamma_1$ and $\Lambda_1$ (and lying outside $\Lambda_1$), a circle $\Lambda_3$ touching $\Gamma_0$, $\Gamma_1$ and $\Lambda_2$ (and lying outside $\Lambda_2$), ..., a circle $\Lambda_{j+1}$ touching $\Gamma_0$, $\Gamma_1$ and $\Lambda_j$ (and lying outside $\Lambda_j$) and so on. Eventually some $\Lambda_r$ will either cut $\Lambda_1$ in two distinct points or will touch it. If the second possibility occurs, we say that the circles $\Lambda_1$, $\Lambda_2$, ..., $\Lambda_r$ form a Steiner chain. There are excellent pictures of Steiner chains in Wikipedia and elsewhere on the web

Steiner's porism[9] asserts that if one choice of $\Lambda_1$ gives a Steiner chain, then all choices will give a Steiner chain.

---

[9] The Greeks used the word porism to denote a kind of corollary. However, later mathematicians decided that such a fine word should not be wasted and it now means a theorem which asserts that something always happens or never happens.

(i) Explain why Steiner's porism is true for concentric circles.

(ii) Give an example of two concentric circles for which a Steiner chain exists and another for which no Steiner chain exists.

(iii) Suppose that two circles lie in the $(x, y)$ plane with centres on the real axis. Suppose that the first circle cuts the $x$ axis at $a$ and $b$ and the second at $c$ and $d$. Suppose further that

$$0 < \frac{1}{a} < \frac{1}{c} < \frac{1}{d} < \frac{1}{b} \quad \text{and} \quad \frac{1}{a} + \frac{1}{b} < \frac{1}{c} + \frac{1}{d}.$$

By considering the behaviour of

$$f(x) = \left( \frac{1}{a-x} + \frac{1}{b-x} \right) - \left( \frac{1}{c-x} + \frac{1}{d-x} \right)$$

as $x$ increases from 0 towards $b$, or otherwise, show that there is an $x_0$ such that

$$\frac{1}{a - x_0} + \frac{1}{b - x_0} = \frac{1}{c - x_0} + \frac{1}{d - x_0}.$$

(iv) Using (iii), or otherwise, show that any two circles can be mapped to concentric circles by using translations, rotations and inversion (see Exercise 2.4.11).

(v) Deduce Steiner's porism.

# 3

# The algebra of square matrices

## 3.1 The summation convention

In our first chapter we showed how a system of $n$ linear equations in $n$ unknowns could be written compactly as

$$\sum_{j=1}^{n} a_{ij}x_j = b_i \qquad [1 \leq i \leq n].$$

In 1916, Einstein wrote to a friend (see, for example, [26])

I have made a great discovery in mathematics; I have suppressed the summation sign every time that the summation must be made over an index which occurs twice ...

Although Einstein wrote with his tongue in his cheek, the *Einstein summation convention* has proved very useful. When we use the summation convention with $i, j, \ldots$ running from 1 to $n$, then we must observe the following rules.

(1) Whenever the suffix $i$, say, occurs once in an expression forming part of an equation, then we have $n$ instances of the equation according as $i = 1, i = 2, \ldots$ or $i = n$.
(2) Whenever the suffix $i$, say, occurs twice in an expression forming part of an equation, then $i$ is a dummy variable, and we sum the expression over the values $1 \leq i \leq n$.
(3) The suffix $i$, say, will *never* occur more than twice in an expression forming part of an equation.

The summation convention appears baffling when you meet it first, but is easy when you get used to it. For the moment, whenever I use an expression involving the summation, I shall give the same expression using the older notation. Thus the system

$$a_{ij}x_j = b_i,$$

with the summation convention, corresponds to

$$\sum_{j=1}^{n} a_{ij}x_j = b_i \qquad [1 \leq i \leq n]$$

without it. In the same way, the equation

$$\mathbf{x} \cdot \mathbf{y} = \sum_{i=1}^{n} x_i y_i,$$

without the summation convention, corresponds to

$$\mathbf{x} \cdot \mathbf{y} = x_i y_i$$

with it.

Here is a proof of the parallelogram law (Exercise 2.4.1), using the summation convention.

$$
\begin{aligned}
\|\mathbf{a} + \mathbf{b}\|^2 + \|\mathbf{a} - \mathbf{b}\|^2 &= (a_i + b_i)(a_i + b_i) + (a_i - b_i)(a_i - b_i) \\
&= (a_i a_i + 2a_i b_i + b_i b_i) + (a_i a_i - 2a_i b_i + b_i b_i) \\
&= 2a_i a_i + 2b_i b_i \\
&= 2\|\mathbf{a}\|^2 + 2\|\mathbf{b}\|^2.
\end{aligned}
$$

Here is the same proof, not using the summation convention.

$$
\begin{aligned}
\|\mathbf{a} + \mathbf{b}\|^2 + \|\mathbf{a} - \mathbf{b}\|^2 &= \sum_{i=1}^{n}(a_i + b_i)(a_i + b_i) + \sum_{i=1}^{n}(a_i - b_i)(a_i - b_i) \\
&= \sum_{i=1}^{n}(a_i a_i + 2a_i b_i + b_i b_i) + \sum_{i=1}^{n}(a_i a_i - 2a_i b_i + b_i b_i) \\
&= 2\sum_{i=1}^{n} a_i a_i + 2\sum_{i=1}^{n} b_i b_i \\
&= 2\|\mathbf{a}\|^2 + 2\|\mathbf{b}\|^2.
\end{aligned}
$$

The reader should note that, unless I *explicitly* state that we are using the summation convention, we are not. If she is unhappy with any argument using the summation convention, she should first follow it with summation signs inserted and then remove the summation signs.

## 3.2 Multiplying matrices

Consider two $n \times n$ matrices $A = (a_{ij})$ and $B = (b_{ij})$. If

$$B\mathbf{x} = \mathbf{y} \quad \text{and} \quad A\mathbf{y} = \mathbf{z},$$

then

$$z_i = \sum_{k=1}^{n} a_{ik} y_k = \sum_{k=1}^{n} a_{ik} \left( \sum_{j=1}^{n} b_{kj} x_j \right)$$

$$= \sum_{k=1}^{n} \sum_{j=1}^{n} a_{ik} b_{kj} x_j = \sum_{j=1}^{n} \sum_{k=1}^{n} a_{ik} b_{kj} x_j$$

$$= \sum_{j=1}^{n} \left( \sum_{k=1}^{n} a_{ik} b_{kj} \right) x_j = \sum_{j=1}^{n} c_{ij} x_j$$

where

$$c_{ij} = \sum_{k=1}^{n} a_{ik} b_{kj} \qquad\qquad \bigstar$$

or, using the summation convention,

$$c_{ij} = a_{ik} b_{kj}.$$

**Exercise 3.2.1** *Write out the argument of the previous paragraph using the summation convention.*

It therefore makes sense to use the following definition.

**Definition 3.2.2** *If $A$ and $B$ are $n \times n$ matrices, then $AB = C$ where $C$ is the $n \times n$ matrix such that*

$$A(B\mathbf{x}) = C\mathbf{x}$$

*for all $\mathbf{x} \in \mathbb{R}^n$.*

The formula $\bigstar$ is complicated, but it is essential that the reader should get used to computations involving matrix multiplication.

It may be helpful to observe that, if we take

$$\mathbf{a}_i = (a_{i1}, a_{i2}, \ldots, a_{in})^T, \quad \mathbf{b}_j = (b_{1j}, b_{2j}, \ldots, b_{nj})^T$$

(so that $\mathbf{a}_i$ is the column vector obtained from the $i$th row of $A$ and $\mathbf{b}_j$ is the column vector corresponding to the $j$th column of $B$), then

$$c_{ij} = \mathbf{a}_i \cdot \mathbf{b}_j$$

and

$$AB = \begin{pmatrix} \mathbf{a}_1 \cdot \mathbf{b}_1 & \mathbf{a}_1 \cdot \mathbf{b}_2 & \mathbf{a}_1 \cdot \mathbf{b}_3 & \ldots & \mathbf{a}_1 \cdot \mathbf{b}_n \\ \mathbf{a}_2 \cdot \mathbf{b}_1 & \mathbf{a}_2 \cdot \mathbf{b}_2 & \mathbf{a}_2 \cdot \mathbf{b}_3 & \ldots & \mathbf{a}_2 \cdot \mathbf{b}_n \\ \vdots & \vdots & \vdots & & \vdots \\ \mathbf{a}_n \cdot \mathbf{b}_1 & \mathbf{a}_n \cdot \mathbf{b}_2 & \mathbf{a}_n \cdot \mathbf{b}_3 & \ldots & \mathbf{a}_n \cdot \mathbf{b}_n \end{pmatrix}.$$

Here is one way of explicitly multiplying two $3 \times 3$ matrices

$$X = \begin{pmatrix} a & b & c \\ d & e & f \\ g & h & i \end{pmatrix} \quad \text{and} \quad Y = \begin{pmatrix} \mathcal{A} & \mathcal{B} & \mathcal{C} \\ \mathcal{D} & \mathcal{E} & \mathcal{F} \\ \mathcal{G} & \mathcal{H} & \mathcal{I} \end{pmatrix}.$$

First write out three copies of $X$ next to each other as

$$\begin{pmatrix} a & b & c & a & b & c & a & b & c \\ d & e & f & d & e & f & d & e & f \\ g & h & i & g & h & i & g & h & i \end{pmatrix}.$$

Now fill in the 'row column multiplications' to get

$$XY = \begin{pmatrix} a\mathcal{A} + b\mathcal{D} + c\mathcal{G} & a\mathcal{B} + b\mathcal{E} + c\mathcal{H} & a\mathcal{C} + b\mathcal{F} + c\mathcal{I} \\ d\mathcal{A} + e\mathcal{D} + f\mathcal{G} & d\mathcal{B} + e\mathcal{E} + f\mathcal{H} & d\mathcal{C} + e\mathcal{F} + f\mathcal{I} \\ g\mathcal{A} + h\mathcal{D} + i\mathcal{G} & g\mathcal{B} + h\mathcal{E} + i\mathcal{H} & g\mathcal{C} + h\mathcal{F} + i\mathcal{I} \end{pmatrix}.$$

## 3.3 More algebra for square matrices

We can also add $n \times n$ matrices, though the result is less novel. We imitate the definition of multiplication.

Consider two $n \times n$ matrices $A = (a_{ij})$ and $B = (b_{ij})$. If

$$A\mathbf{x} = \mathbf{y} \quad \text{and} \quad B\mathbf{x} = \mathbf{z}$$

then

$$y_i + z_i = \sum_{j=1}^{n} a_{ij}x_j + \sum_{j=1}^{n} b_{ij}x_j = \sum_{j=1}^{n}(a_{ij}x_j + b_{ij}x_j) = \sum_{j=1}^{n}(a_{ij} + b_{ij})x_j = \sum_{j=1}^{n} c_{ij}x_j$$

where

$$c_{ij} = a_{ij} + b_{ij}.$$

It therefore makes sense to use the following definition.

**Definition 3.3.1** *If $A$ and $B$ are $n \times n$ matrices, then $A + B = C$ where $C$ is the $n \times n$ matrix such that*

$$A\mathbf{x} + B\mathbf{x} = C\mathbf{x}$$

*for all $\mathbf{x} \in \mathbb{R}^n$.*

In addition we can 'multiply a matrix $A$ by a scalar $\lambda$'.
Consider an $n \times n$ matrix $A = (a_{ij})$ and a $\lambda \in \mathbb{R}$. If

$$A\mathbf{x} = \mathbf{y},$$

then

$$\lambda y_i = \lambda \sum_{j=1}^{n} a_{ij} x_j = \sum_{j=1}^{n} \lambda(a_{ij} x_j) = \sum_{j=1}^{n} (\lambda a_{ij}) x_j = \sum_{j=1}^{n} c_{ij} x_j$$

where

$$c_{ij} = \lambda a_{ij}.$$

It therefore makes sense to use the following definition.

**Definition 3.3.2** *If $A$ is an $n \times n$ matrix and $\lambda \in \mathbb{R}$, then $\lambda A = C$ where $C$ is the $n \times n$ matrix such that*

$$\lambda(\mathbf{A}\mathbf{x}) = C\mathbf{x}$$

*for all $\mathbf{x} \in \mathbb{R}^n$.*

Once we have addition and the two kinds of multiplication, we can do quite a lot of algebra.

We have already met addition and scalar multiplication for column vectors with $n$ entries and for row vectors with $n$ entries. From the point of view of addition and scalar multiplication, an $n \times n$ matrix is simply another kind of vector with $n^2$ entries. We thus have an analogue of Lemma 1.4.2, proved in exactly the same way. (We write 0 for the $n \times n$ matrix all of whose entries are 0. We write $-A = (-1)A$ and $A - B = A + (-B)$.)

**Lemma 3.3.3** *Suppose that $A$, $B$ and $C$ are $n \times n$ matrices and $\lambda, \mu \in \mathbb{R}$. Then the following relations hold.*
  *(i) $(A + B) + C = A + (B + C)$.*
  *(ii) $A + B = B + A$.*
  *(iii) $A + 0 = A$.*
  *(iv) $\lambda(A + B) = \lambda A + \lambda B$.*
  *(v) $(\lambda + \mu)A = \lambda A + \mu A$.*
  *(vi) $(\lambda \mu)A = \lambda(\mu A)$.*
  *(vii) $1A = A$, $0A = 0$.*
  *(viii) $A - A = 0$.*

**Exercise 3.3.4** *Prove as many of the results of Lemma 3.3.3 as you feel you need to.*

We get new results when we consider matrix multiplication (that is to say, when we multiply matrices together). We first introduce a simple but important object.

**Definition 3.3.5** *Fix an integer $n \geq 1$. The Kronecker $\delta$ symbol associated with $n$ is defined by the rule*

$$\delta_{ij} = \begin{cases} 1 & \text{if } i = j \\ 0 & \text{if } i \neq j \end{cases}$$

for $1 \le i, j \le n$. The $n \times n$ identity matrix $I$ is given by $I = (\delta_{ij})$.

**Exercise 3.3.6** *Show that an $n \times n$ matrix $C$ satisfies the equation $C\mathbf{x} = \mathbf{x}$ for all $\mathbf{x} \in \mathbb{R}^n$ if and only if $C = I$.*

The following remark often comes in useful when we use the summation convention.

**Example 3.3.7** *If we use the summation convention,*

$$\delta_{ij} x_j = x_i \quad and \quad \delta_{ij} a_{jk} = a_{ik}.$$

*Proof* Observe that, if we do not use the summation convention,

$$\sum_{j=1}^{n} \delta_{ij} x_j = x_i \quad and \quad \sum_{j=1}^{n} \delta_{ij} a_{jk} = a_{ik},$$

so we get the required result when we do. $\qquad\square$

**Lemma 3.3.8** *Suppose that $A$, $B$ and $C$ are $n \times n$ matrices and $\lambda \in \mathbb{R}$. Then the following relations hold.*

(i) $(AB)C = A(BC)$.
(ii) $A(B + C) = AB + AC$.
(iii) $(B + C)A = BA + CA$.
(iv) $(\lambda A)B = A(\lambda B) = \lambda(AB)$.
(v) $AI = IA = A$.

*Proof* (i) We give three proofs.
*Proof by definition* Observe that, from Definition 3.2.2,

$$\big((AB)C\big)\mathbf{x} = (AB)(C\mathbf{x}) = A\big(B(C\mathbf{x})\big) = A\big((BC)\mathbf{x}\big) = \big(A(BC)\big)\mathbf{x}$$

for all $\mathbf{x} \in \mathbb{R}^n$ and so $(AB)C = A(BC)$.
*Proof by calculation* Observe that

$$\sum_{k=1}^{n} \left( \sum_{j=1}^{n} a_{ij} b_{jk} \right) c_{kj} = \sum_{k=1}^{n} \sum_{j=1}^{n} (a_{ij} b_{jk}) c_{kj} = \sum_{k=1}^{n} \sum_{j=1}^{n} a_{ij} (b_{jk} c_{kj})$$

$$= \sum_{j=1}^{n} \sum_{k=1}^{n} a_{ij} (b_{jk} c_{kj}) = \sum_{j=1}^{n} a_{ij} \left( \sum_{k=1}^{n} b_{jk} c_{kj} \right)$$

and so $(AB)C = A(BC)$.
*Proof by calculation using the summation convention* Observe that

$$(a_{ij} b_{jk}) c_{kj} = a_{ij} (b_{jk} c_{kj})$$

and so $(AB)C = A(BC)$.

Each of these proofs has its merits. The author thinks that the essence of what is going on is conveyed by the first proof and that the second proof shows the hidden machinery behind the short third proof.

(ii) Again we give three proofs.

*Proof by definition* Observe that, using our definitions and the fact that $A(\mathbf{u} + \mathbf{v}) = A\mathbf{u} + A\mathbf{v}$,

$$\big((A(B+C)\big)\mathbf{x} = A\big((B+C)\mathbf{x}\big) = A(B\mathbf{x}+C\mathbf{x})$$
$$= A(B\mathbf{x}) + A(C\mathbf{x}) = (AB)\mathbf{x} + (AC)\mathbf{x} = (AB+AC)\mathbf{x}$$

for all $\mathbf{x} \in \mathbb{R}^n$ and so $A(B+C) = AB + AC$.

*Proof by calculation* Observe that

$$\sum_{j=1}^{n} a_{ij}(b_{jk}+c_{jk}) = \sum_{j=1}^{n}(a_{ij}b_{jk}+a_{ij}c_{jk}) = \sum_{j=1}^{n} a_{ij}b_{jk} + \sum_{j=1}^{n} a_{ij}c_{jk}$$

and so $A(B+C) = AB + AC$.

*Proof by calculation using the summation convention* Observe that

$$a_{ij}(b_{jk}+c_{jk}) = a_{ij}b_{jk} + a_{ij}c_{jk}$$

and so $A(B+C) = AB + AC$.

We leave the remaining parts to the reader.                                    □

**Exercise 3.3.9** *Prove the remaining parts of Lemma 3.3.8 using each of the three methods of proof.*

**Exercise 3.3.10** *By considering a particular $n \times n$ matrix $A$, show that*

$$BA = A \text{ for all } A \Rightarrow B = I.$$

However, as the reader is probably already aware, the algebra of matrices differs in two unexpected ways from the kind of arithmetic with which we are familiar from school. The first is that we can no longer assume that $AB = BA$, even for $2 \times 2$ matrices. Observe that

$$\begin{pmatrix} 0 & 1 \\ 0 & 0 \end{pmatrix}\begin{pmatrix} 0 & 0 \\ 1 & 0 \end{pmatrix} = \begin{pmatrix} 0 \times 0 + 1 \times 1 & 0 \times 0 + 1 \times 0 \\ 0 \times 0 + 0 \times 1 & 0 \times 0 + 0 \times 0 \end{pmatrix} = \begin{pmatrix} 1 & 0 \\ 0 & 0 \end{pmatrix},$$

but

$$\begin{pmatrix} 0 & 0 \\ 1 & 0 \end{pmatrix}\begin{pmatrix} 0 & 1 \\ 0 & 0 \end{pmatrix} = \begin{pmatrix} 0 \times 0 + 0 \times 1 & 0 \times 1 + 0 \times 0 \\ 1 \times 0 + 0 \times 0 & 1 \times 1 + 0 \times 0 \end{pmatrix} = \begin{pmatrix} 0 & 0 \\ 0 & 1 \end{pmatrix}.$$

The second is that, even when $A$ is a non-zero $2 \times 2$ matrix, there may not be a matrix $B$ with $BA = I$ or a matrix $C$ with $AC = I$. Observe that

$$\begin{pmatrix} 1 & 0 \\ 0 & 0 \end{pmatrix}\begin{pmatrix} a & b \\ c & d \end{pmatrix} = \begin{pmatrix} 1 \times a + 0 \times c & 1 \times b + 0 \times d \\ 0 \times a + 0 \times c & 0 \times b + 0 \times d \end{pmatrix} = \begin{pmatrix} a & b \\ 0 & 0 \end{pmatrix} \neq \begin{pmatrix} 1 & 0 \\ 0 & 1 \end{pmatrix}$$

and

$$\begin{pmatrix} a & b \\ c & d \end{pmatrix}\begin{pmatrix} 1 & 0 \\ 0 & 0 \end{pmatrix} = \begin{pmatrix} a \times 1 + b \times 0 & a \times 0 + b \times 0 \\ c \times 1 + d \times 0 & c \times 0 + d \times 0 \end{pmatrix} = \begin{pmatrix} a & 0 \\ c & 0 \end{pmatrix} \neq \begin{pmatrix} 1 & 0 \\ 0 & 1 \end{pmatrix}$$

for all values of $a$, $b$, $c$ and $d$.

In the next section we discuss this phenomenon in more detail, but there is a further remark to be made before finishing this section to the effect that it is sometimes possible to do some algebra on non-square matrices. We will discuss the deeper reasons why this is so in Section 11.1, but for the moment we just give some computational definitions.

**Definition 3.3.11** *If $A = (a_{ij})_{1\le i\le m}^{1\le j\le n}$, $B = (b_{ij})_{1\le i\le m}^{1\le j\le n}$, $C = (c_{jk})_{1\le j\le n}^{1\le k\le p}$ and $\lambda \in \mathbb{R}$, we set*

$$\lambda A = (\lambda a_{ij})_{1\le i\le m}^{1\le j\le n},$$

$$A + B = (a_{ij} + b_{ij})_{1\le i\le m}^{1\le j\le n},$$

$$BC = \left(\sum_{j=1}^{n} b_{ij}c_{jk}\right)_{1\le i\le m}^{1\le k\le p}.$$

**Exercise 3.3.12** *Obtain definitions of $\lambda A$, $A + B$ and $BC$ along the lines of Definitions 3.3.2, 3.3.1 and 3.2.2.*

The conscientious reader will do the next two exercises in detail. The less conscientious reader will just glance at them, happy in my assurance that, once the ideas of this book are understood, the results are 'obvious'. As usual, we write $-A = (-1)A$.

**Exercise 3.3.13** *Suppose that $A$, $B$ and $C$ are $m \times n$ matrices, $\underline{0}$ is the $m \times n$ matrix with all entries 0, and $\lambda, \mu \in \mathbb{R}$. Then the following relations hold.*
*(i) $(A + B) + C = A + (B + C)$.*
*(ii) $A + B = B + A$.*
*(iii) $A + 0 = A$.*
*(iv) $\lambda(A + B) = \lambda A + \lambda B$.*
*(v) $(\lambda + \mu)A = \lambda A + \mu A$.*
*(vi) $(\lambda\mu)A = \lambda(\mu A)$.*
*(vii) $0A = \underline{0}$.*
*(viii) $A - A = \underline{0}$.*

**Exercise 3.3.14** *Suppose that $A$ is an $m \times n$ matrix, $B$ is an $m \times n$ matrix, $C$ is an $n \times p$ matrix, $F$ is a $p \times q$ matrix and $G$ is a $k \times m$ matrix and $\lambda \in \mathbb{R}$. Show that the following relations hold.*
*(i) $(AC)F = A(CF)$.*
*(ii) $G(A + B) = GA + GB$.*
*(iii) $(A + B)C = AC + AC$.*
*(iv) $(\lambda A)C = A(\lambda C) = \lambda(AC)$.*

## 3.4 Decomposition into elementary matrices

We start with a general algebraic observation.

**Lemma 3.4.1** *Suppose that $A$, $B$ and $C$ are $n \times n$ matrices. If $BA = I$ and $AC = I$, then $B = C$.*

*Proof* We have $B = BI = B(AC) = (BA)C = IC = C$.  □

Note that, if $BA = I$ and $AC_1 = AC_2 = I$, then the lemma tells us that $C_1 = B = C_2$ and that, if $AC = I$ and $B_1A = B_2A = I$, then the lemma tells us that $B_1 = C = B_2$.

We can thus make the following definition.

**Definition 3.4.2** *If A and B are $n \times n$ matrices such that $AB = BA = I$, then we say that A is* invertible *with inverse $A^{-1} = B$.*

The following simple lemma is very useful.

**Lemma 3.4.3** *If U and V are $n \times n$ invertible matrices, then $UV$ is invertible and $(UV)^{-1} = V^{-1}U^{-1}$.*

*Proof* Note that

$$(UV)(V^{-1}U^{-1}) = U(VV^{-1})U^{-1} = UIU^{-1} = UU^{-1} = I$$

and, by a similar calculation, $(V^{-1}U^{-1})(UV) = I$.  □

The following lemma links the existence of an inverse with our earlier work on equations.

**Lemma 3.4.4** *If A is an $n \times n$ square matrix with an inverse, then the system of equations*

$$\sum_{j=1}^{n} a_{ij}x_j = y_i \qquad [1 \le i \le n]$$

*has a unique solution for each choice of the $y_i$.*

*Proof* If we set $\mathbf{x} = A^{-1}\mathbf{y}$, then

$$A\mathbf{x} = A(A^{-1}\mathbf{y}) = (AA^{-1})\mathbf{y} = I\mathbf{y} = \mathbf{y}$$

so $\sum_{j=1}^{n} a_{ij}x_j = y_i$ for all $1 \le i \le n$ and a solution exists.

Conversely, if $\sum_{j=1}^{n} a_{ij}x_j = y_i$ for all $1 \le i \le n$, then $A\mathbf{x} = \mathbf{y}$ and

$$\mathbf{x} = I\mathbf{x} = (A^{-1}A)\mathbf{x} = A^{-1}(A\mathbf{x}) = A^{-1}\mathbf{y}$$

so the $x_j$ are uniquely determined.  □

Later, in Lemma 3.4.14, we shall show that, if the system of equations is always soluble, whatever the choice of $\mathbf{y}$, then an inverse exists. If a matrix $A$ has an inverse we shall say that it is *invertible* or *non-singular*.

The reader should note that, at this stage, we have not excluded the possibility that there might be an $n \times n$ matrix $A$ with a left inverse but no right inverse (in other words, there exists a $B$ such that $BA = I$, but there does not exist a $C$ with $AC = I$) or with a right inverse but no left inverse. Later we shall prove Lemma 3.4.13 which shows that this possibility does not arise.

There are many ways to investigate the existence or non-existence of matrix inverses and we shall meet several in the course of this book. Our first investigation will use the notion of an elementary matrix.

We define two sorts of $n \times n$ elementary matrices. The first sort are matrices of the form

$$E(r, s, \lambda) = (\delta_{ij} + \lambda \delta_{ir} \delta_{sj})$$

where $1 \leq r, s \leq n$ and $r \neq s$. (The summation convention will not apply to $r$ and $s$.) Such matrices are sometimes called *shear matrices*.

**Exercise 3.4.5** *If A and B are shear matrices, does it follow that* $AB = BA$? *Give reasons.* [*Hint: Try* $2 \times 2$ *matrices.*]

The second sort form a subset of the collection of matrices

$$P(\sigma) = (\delta_{\sigma(i)j})$$

where $\sigma : \{1, 2, \ldots, n\} \to \{1, 2, \ldots, n\}$ is a bijection. These are sometimes called *permutation matrices*. (Recall that $\sigma$ can be thought of as a shuffle or *permutation* of the integers $1, 2, \ldots, n$ in which $i$ goes to $\sigma(i)$.) We shall call any $P(\sigma)$ in which $\sigma$ interchanges only two integers an elementary matrix. More specifically, we demand that there be an $r$ and $s$ with $1 \leq r < s \leq n$ such that

$$\sigma(r) = s$$
$$\sigma(s) = r$$
$$\sigma(i) = i \text{ otherwise.}$$

The shear matrix $E(r, s, \lambda)$ has 1s down the diagonal and all other entries 0 apart from the $s$th entry of the $r$th row which is $\lambda$. The permutation matrix $P(\sigma)$ has the $\sigma(i)$th entry of the $i$th row 1 and all other entries in the $i$th row 0.

**Lemma 3.4.6** (*i*) *If we pre-multiply (i.e. multiply on the left) an* $n \times n$ *matrix A by* $E(r, s, \lambda)$ *with* $r \neq s$, *we add* $\lambda$ *times the* $s$th *row to the* $r$th *row but leave it otherwise unchanged.*

(*ii*) *If we post-multiply (i.e. multiply on the right) an* $n \times n$ *matrix A by* $E(r, s, \lambda)$ *with* $r \neq s$, *we add* $\lambda$ *times the* $r$th *column to the* $s$th *column but leave it otherwise unchanged.*

(*iii*) *If we pre-multiply an* $n \times n$ *matrix A by* $P(\sigma)$, *the* $i$th *row is moved to the* $\sigma(i)$th *row.*

(*iv*) *If we post-multiply an* $n \times n$ *matrix A by* $P(\sigma)$, *the* $\sigma(j)$th *column is moved to the* $j$th *column.*

(*v*) $E(r, s, \lambda)$ *is invertible with inverse* $E(r, s, -\lambda)$.

(*vi*) $P(\sigma)$ *is invertible with inverse* $P(\sigma^{-1})$.

*Proof* (i) Using the summation convention for $i$ and $j$, but keeping $r$ and $s$ fixed,

$$(\delta_{ij} + \lambda \delta_{ir} \delta_{sj}) a_{jk} = \delta_{ij} a_{jk} + \lambda \delta_{ir} \delta_{sj} a_{jk} = a_{ik} + \lambda \delta_{ir} a_{sk}.$$

(ii) Exercise for the reader.

(iii) Using the summation convention,

$$\delta_{\sigma(i)j} a_{jk} = a_{\sigma(i)k}.$$

(iv) Exercise for the reader.

(v) Direct calculation or apply (i) and (ii).

(vi) Direct calculation or apply (iii) and (iv).  □

**Exercise 3.4.7**  *Let $r \neq s$. Show that $E(r, s, \lambda)E(r, s, \mu) = E(r, s, \lambda + \mu)$.*

We now need a slight variation on Theorem 1.3.2.

**Theorem 3.4.8**  *Given any $n \times n$ matrix $A$, we can reduce it to a diagonal matrix $D = (d_{ij})$ with $d_{ij} = 0$ if $i \neq j$ by successive operations involving adding multiples of one row to another or interchanging rows.*

*Proof*  This is easy to obtain directly, but we shall deduce it from Theorem 1.3.2. This tells us that we can reduce the $n \times n$ matrix $A$, to a *diagonal matrix* $\tilde{D} = (\tilde{d}_{ij})$ with $\tilde{d}_{ij} = 0$ if $i \neq j$ by interchanging columns, interchanging rows and subtracting multiples of one row from another.

If we go through this process, but omit all the steps involving interchanging columns, we will arrive at a matrix $B$ such that each row and each column contain at most one non-zero element. By interchanging rows, we can now transform $B$ to a diagonal matrix and we are done.  □

Using Lemma 3.4.6, we can interpret this result in terms of elementary matrices.

**Lemma 3.4.9**  *Given any $n \times n$ matrix $A$, we can find elementary matrices $F_1$, $F_2$, ..., $F_p$ together with a diagonal matrix $D$ such that*

$$F_p F_{p-1} \ldots F_1 A = D.$$

A simple modification now gives the central theorem of this section.

**Theorem 3.4.10**  *Given any $n \times n$ matrix $A$, we can find elementary matrices $L_1$, $L_2$, ..., $L_p$ together with a diagonal matrix $D$ such that*

$$A = L_1 L_2 \ldots L_p D.$$

*Proof*  By Lemma 3.4.9, we can find elementary matrices $F_r$ and a diagonal matrix $D$ such that

$$F_p F_{p-1} \ldots F_1 A = D.$$

Since elementary matrices are invertible and their inverses are elementary (see Lemma 3.4.6), we can take $L_r = F_r^{-1}$ and obtain

$$L_1 L_2 \ldots L_p D = F_1^{-1} F_2^{-1} \ldots F_p^{-1} F_p \ldots F_2 F_1 A = A$$

as required.  □

There is an obvious connection with the problem of deciding when there is an inverse matrix.

**Lemma 3.4.11** *Let $D = (d_{ij})$ be a diagonal matrix.*

*(i) If all the diagonal entries $d_{ii}$ of $D$ are non-zero, $D$ is invertible and the inverse $D^{-1} = E$ where $E = (e_{ij})$ is given by $e_{ii} = d_{ii}^{-1}$ and $e_{ij} = 0$ for $i \neq j$.*

*(ii) If some of the diagonal entries of $D$ are zero, then $BD \neq I$ and $DB \neq I$ for all $B$.*

*Proof* (i) If all the diagonal entries of $D$ are non-zero, then, taking $E$ as proposed, we have

$$DE = ED = I$$

by direct calculation.

(ii) If $d_{rr} = 0$ for some $r$, then, if $B = (b_{ij})$ is any $n \times n$ matrix, we have

$$\sum_{j=1}^{n} b_{rj} d_{jk} = \sum_{j=1}^{n} b_{rj} \times 0 = 0,$$

so $BD$ has all entries in the $r$th row equal to zero. Thus $BD \neq I$. Similarly, $DB$ has all entries in the $r$th column equal to zero and $DB \neq I$. $\qquad\square$

**Lemma 3.4.12** *Let $L_1$, $L_2$, ..., $L_p$ be elementary $n \times n$ matrices and let $D$ be an $n \times n$ diagonal matrix. Suppose that*

$$A = L_1 L_2 \ldots L_p D.$$

*(i) If all the diagonal entries $d_{ii}$ of $D$ are non-zero, then $A$ is invertible.*

*(ii) If some of the diagonal entries of $D$ are zero, then $A$ is not invertible.*

*Proof* Since elementary matrices are invertible (Lemma 3.4.6 (v) and (vi)) and the product of invertible matrices is invertible (Lemma 3.4.3), we have $A = LD$ where $L$ is invertible.

If all the diagonal entries $d_{ii}$ of $D$ are non-zero, then $D$ is invertible and so, by Lemma 3.4.3, $A = LD$ is invertible.

If $A$ is invertible, then we can find a $B$ with $BA = I$. It follows that $(BL)D = B(LD) = I$ and, by Lemma 3.4.11 (ii), none of the diagonal entries of $D$ can be zero. $\qquad\square$

As a corollary we obtain a result promised at the beginning of this section.

**Lemma 3.4.13** *If $A$ and $B$ are $n \times n$ matrices such that $AB = I$, then $A$ and $B$ are invertible with $A^{-1} = B$ and $B^{-1} = A$.*

*Proof* Combine the results of Theorem 3.4.10 with those of Lemma 3.4.12. $\qquad\square$

Later we shall see how a more abstract treatment gives a simpler and more transparent proof of this fact.

We are now in a position to provide the complementary result to Lemma 3.4.4.

**Lemma 3.4.14** *If $A$ is an $n \times n$ square matrix such that the system of equations*

$$\sum_{j=1}^{n} a_{ij} x_j = y_i \qquad [1 \leq i \leq n]$$

*has a unique solution for each choice of $y_i$, then $A$ has an inverse.*

*Proof* If we fix $k$, then our hypothesis tells us that the system of equations

$$\sum_{j=1}^{n} a_{ij}x_j = \delta_{ik} \qquad [1 \le i \le n]$$

has a solution. Thus, for each $k$ with $1 \le k \le n$, we can find $x_{jk}$ with $1 \le j \le n$ such that

$$\sum_{j=1}^{n} a_{ij}x_{jk} = \delta_{ik} \qquad [1 \le i \le n].$$

If we write $X = (x_{jk})$ we obtain $AX = I$ so $A$ is invertible. $\qquad\square$

## 3.5 Calculating the inverse

Mathematics is full of objects which are very useful for studying the theory of a particular topic, but very hard to calculate in practice. Before seeking the inverse of an $n \times n$ matrix, you should always ask the question 'Do I really need to calculate the inverse or is there some easier way of attaining my object?' If $n$ is large, you will need to use a computer and you will either need to know the kind of problems that arise in the numerical inversion of large matrices or need to consult someone who does.[1]

Since students are unhappy with objects they cannot compute, I will show in this section how to invert matrices 'by hand'. The contents of this section should not be taken too seriously.

Suppose that we want to find the inverse of the matrix

$$A = \begin{pmatrix} 1 & 2 & -1 \\ 5 & 0 & 3 \\ 1 & 1 & 0 \end{pmatrix}.$$

In other words, we want to find a matrix

$$X = \begin{pmatrix} x_1 & x_2 & x_3 \\ y_1 & y_2 & y_3 \\ z_1 & z_2 & z_3 \end{pmatrix}$$

such that

$$\begin{pmatrix} 1 & 0 & 0 \\ 0 & 1 & 0 \\ 0 & 0 & 1 \end{pmatrix} = I = AX = \begin{pmatrix} x_1 + 2y_1 - z_1 & x_2 + 2y_2 - z_2 & x_3 + 2y_3 - z_3 \\ 5x_1 + 3z_1 & 5x_2 + 3z_2 & 5x_3 + 3z_3 \\ x_1 + y_1 & x_2 + y_2 & x_3 + y_3 \end{pmatrix}.$$

---

[1] A wise mathematician will, in any case, consult an expert.

We need to solve the three sets of simultaneous equations

$$x_1 + 2y_1 - z_1 = 1 \qquad x_2 + 2y_2 - z_2 = 0 \qquad x_3 + 2y_3 - z_3 = 0$$
$$5x_1 + 3z_1 = 0 \qquad 5x_2 + 3z_2 = 1 \qquad 5x_3 + 3z_3 = 0$$
$$x_1 + y_1 = 0 \qquad x_2 + y_2 = 0 \qquad x_3 + y_3 = 1.$$

Subtracting the third row from the first row, subtracting 5 times the third row from the second row and then interchanging the third and first rows, we get

$$x_1 + y_1 = 0 \qquad x_2 + y_2 = 0 \qquad x_3 + y_3 = 1$$
$$-5y_1 + 3z_1 = 0 \qquad -5y_2 + 3z_2 = 1 \qquad -5y_3 + 3z_3 = -5$$
$$y_1 - z_1 = 1 \qquad y_2 - z_2 = 0 \qquad y_3 - z_3 = -1.$$

Subtracting the third row from the first row, adding 5 times the third row to the second row and then interchanging the second and third rows, we get

$$x_1 + z_1 = -1 \qquad x_2 + z_2 = 0 \qquad x_3 + z_3 = 2$$
$$y_1 - z_1 = 1 \qquad y_2 - z_2 = 0 \qquad y_3 - z_3 = -1$$
$$-2z_1 = 5 \qquad -2z_2 = 1 \qquad -2z_3 = -10.$$

Multiplying the third row by $-1/2$ and then adding the new third row to the second row and subtracting the new third row from the first row, we get

$$x_1 = 3/2 \qquad x_2 = 1/2 \qquad x_3 = -3$$
$$y_1 = -3/2 \qquad y_2 = -1/2 \qquad y_3 = 4$$
$$z_1 = -5/2 \qquad z_2 = -1/2 \qquad z_3 = 5.$$

We can save time and ink by using the method of detached coefficients and setting the right-hand sides of our three sets of equations next to each other as follows

$$\begin{array}{ccc|ccc} 1 & 2 & -1 & 1 & 0 & 0 \\ 5 & 0 & 3 & 0 & 1 & 0 \\ 1 & 1 & -1 & 0 & 0 & 1 \end{array}$$

Subtracting the third row from the first row, subtracting 5 times the third row from the second row and then interchanging the third and first rows we get

$$\begin{array}{ccc|ccc} 1 & 1 & 0 & 0 & 0 & 1 \\ 0 & -5 & 3 & 0 & 1 & -5 \\ 0 & 1 & -1 & 1 & 0 & -1 \end{array}$$

Subtracting the third row from the first row, adding 5 times the third row to the second row and then interchanging the second and third rows, we get

$$\begin{array}{ccc|ccc} 1 & 0 & 1 & -1 & 0 & 2 \\ 0 & 1 & -1 & 1 & 0 & -1 \\ 0 & 0 & -2 & 5 & 1 & -10 \end{array}$$

Multiplying the third row by $-1/2$ and then adding the new third row to the second row and subtracting the new third row from the first row, we get

$$
\left[\begin{array}{ccc|ccc}
1 & 0 & 0 & 3/2 & 1/2 & -3 \\
0 & 1 & 0 & -3/2 & -1/2 & 4 \\
0 & 0 & 1 & -5/2 & -1/2 & 5
\end{array}\right]
$$

and looking at the right-hand side of our expression, we see that

$$
A^{-1} = \begin{pmatrix} 3/2 & 1/2 & -3 \\ -3/2 & -1/2 & 4 \\ -5/2 & -1/2 & 5 \end{pmatrix}.
$$

We thus have a recipe for finding the inverse of an $n \times n$ matrix $A$.

**Recipe** Write down the $n \times n$ matrix $I$ and call this matrix the second matrix. By a sequence of row operations of the following three types

(a)  interchange two rows,
(b)  add a multiple of one row to a different row,
(c)  multiply a row by a non-zero number,

reduce the matrix $A$ to the matrix $I$, whilst simultaneously applying exactly the same operations to the second matrix. At the end of the process the second matrix will take the form $A^{-1}$. If the systematic use of Gaussian elimination reduces $A$ to a diagonal matrix with some diagonal entries zero, then $A$ is not invertible.

**Exercise 3.5.1** *Let $L_1, L_2, \ldots, L_k$ be $n \times n$ elementary matrices. If $A$ is an $n \times n$ matrix with*

$$
L_k L_{k-1} \ldots L_1 A = I,
$$

*show that $A$ is invertible with*

$$
L_k L_{k-1} \ldots L_1 I = A^{-1}.
$$

*Explain why this justifies the use of the recipe described in the previous paragraph.*

**Exercise 3.5.2** *Explain why we do not need to use column interchange in reducing $A$ to $I$.*

Earlier we observed that the number of operations required to solve an $n \times n$ system of equations increases like $n^3$. The same argument shows that the number of operations required by our recipe also increases like $n^3$. If you are tempted to think that this is all you need to know about the matter, you should work through Exercise 3.6.1.

## 3.6 Further exercises

**Exercise 3.6.1** Compute the inverse of the following $3 \times 3$ matrix $A$ using the method of Section 3.5 (a) exactly, (b) rounding off each number in the calculation to three significant figures:

$$A = \begin{pmatrix} 1 & \frac{1}{2} & \frac{1}{3} \\ \frac{1}{2} & \frac{1}{3} & \frac{1}{4} \\ \frac{1}{3} & \frac{1}{4} & \frac{1}{5} \end{pmatrix}.$$

[Moral: 'Inverting a matrix on a computer' is not as straightforward as one might hope.]

**Exercise 3.6.2** The, almost trivial, result of this question is quite useful. Suppose that $A$, $B$, $C$, $D$, $E$, $F$, $G$ and $H$ are $n \times n$ matrices. Check that the following equation involving $2n \times 2n$ matrices holds:

$$\begin{pmatrix} A & B \\ C & D \end{pmatrix} \begin{pmatrix} E & F \\ G & H \end{pmatrix} = \begin{pmatrix} (AE + BG) & (AF + BH) \\ (CE + DG) & (CF + DH) \end{pmatrix}.$$

**Exercise 3.6.3 [Strassen–Winograd multiplication]** Check that the method of multiplying two $n \times n$ real matrices given in this book (and in general use) requires about $n^3$ multiplications of real numbers.

Continuing with the notation of the previous question, show that [2]

$$\begin{pmatrix} A & B \\ C & D \end{pmatrix} \begin{pmatrix} E & F \\ G & H \end{pmatrix} = \begin{pmatrix} (S_1 + S_4 - S_5 + S_7) & (S_3 + S_5) \\ (S_2 + S_4) & (S_1 - S_2 + S_3 + S_6) \end{pmatrix}$$

where

$$S_1 = (A + D)(E + H), \quad S_2 = (C + D)E, \quad S_3 = A(F - H), \quad S_4 = D(G - E),$$
$$S_5 = (A + B)H, \quad S_6 = (C - A)(E + F), \quad S_7 = (B - D)(G + H).$$

Conclude that we can find the result of multiplying two $2n \times 2n$ matrices by a method that only involves multiplying 7 pairs of $n \times n$ matrices. By repeating the argument, show that we can find the result of multiplying two $4n \times 4n$ matrices by a method that involves multiplying $49 = 7^2$ pairs of $n \times n$ matrices.

Show, by induction, that we can find the result of multiplying two $2^m \times 2^m$ matrices by a method that only involves multiplying $7^m$ pairs of $1 \times 1$ matrices, that is to say, $7^m$ multiplications of real numbers. If $n = 2^m$, show that we can find the result of multiplying two $n \times n$ matrices by using $n^{\log_2 7} \approx n^{2.8}$ multiplications of real numbers.[3]

On the whole, the complications involved in using the scheme sketched here are such that it is of little practical use, but it raises a fascinating and still open question. How small can $\beta$ be if, when $n$ is large, two $n \times n$ matrices can be multiplied using $n^\beta$ multiplications of real numbers? (For further details see [20], Volume 2.)

**Exercise 3.6.4** We work with $n \times n$ matrices.

(i) Show that a matrix $A$ commutes with every diagonal matrix $D$ (that is to say $AD = DA$) if and only if $A$ is diagonal. Characterise those matrices $A$ such that $AB = BA$ for every matrix $B$ and prove your statement.

---

[2] These formulae have been carefully copied down from elsewhere, but, *given the initial idea* that the result of this paragraph might be both true and useful, hard work and experiment could produce them (or one of their several variants).

[3] Because the ordinary method of adding two $n \times n$ matrices only involves $n^2$ additions, it turns out that the count of the total number of operations involved is dominated by the number of multiplications.

(ii) Characterise those matrices $A$ such that $AB = BA$ for every invertible matrix $B$ and prove your statement.

(iii) Suppose that the matrix $C$ has the property that $FE = C \Rightarrow EF = C$. By using your answer to part (ii), or otherwise, show that $C = \lambda I$ for some $\lambda \in \mathbb{R}$.

**Exercise 3.6.5 [Construction of $\mathbb{C}$ from $\mathbb{R}$]** Consider the space $M_2(\mathbb{R})$ of $2 \times 2$ matrices. If we take

$$I = \begin{pmatrix} 1 & 0 \\ 0 & 1 \end{pmatrix} \quad \text{and} \quad J = \begin{pmatrix} 0 & -1 \\ 1 & 0 \end{pmatrix},$$

show that (if $a, b, c, d \in \mathbb{R}$)

$$aI + bJ = cI + dJ \Rightarrow a = c, \ b = d$$
$$(aI + bJ) + (cI + dJ) = (a+c)I + (b+d)J$$
$$(aI + bJ)(cI + dJ) = (ac - bd)I + (ad + bc)J.$$

Why does this give a model for $\mathbb{C}$? (Answer this at the level you feel appropriate. We give a sequel to this question in Exercises 10.5.22 and 10.5.23.)

**Exercise 3.6.6** Let

$$A = \begin{pmatrix} 1 & 0 & 0 \\ 1 & -1 & 0 \\ 0 & 1 & 1 \end{pmatrix}.$$

Compute $A^2$ and $A^3$ and verify that

$$A^3 = A^2 + A - I.$$

Deduce that $A$ is invertible and calculate $A^{-1}$ explicitly.

Show that

$$A^{2n} = nA^2 - (n-1)I, \ A^{2n+1} = nA^2 + A - nI.$$

**Exercise 3.6.7** Consider the matrix

$$A = \begin{pmatrix} 0 & 1 \\ -1 & 0 \end{pmatrix}.$$

Compute $A^2$ and $A^3$.

If $M$ is an $n \times n$ matrix, we write

$$\exp M = I + \sum_{j=1}^{\infty} \frac{M^j}{j!},$$

where we look at convergence for each entry of the matrix. (The reader can proceed more or less formally, since we shall consider the matter in more depth in Exercise 15.5.19 and elsewhere.)

Show that $\exp tA = A \sin t - A^2 \cos t$ and identify the map $\mathbf{x} \mapsto (\exp tA)\mathbf{x}$. (If you cannot do so, make a note and return to this point after Section 7.3.)

**Exercise 3.6.8** Show that any $m \times n$ matrix $A$ can be reduced to a matrix $C = (c_{ij})$ with $c_{ii} = 1$ for all $1 \leq i \leq k$ (for some $k$ with $0 \leq k \leq \min\{n, m\}$) and $c_{ij} = 0$, otherwise, by a sequence of elementary row and column operations. Why can we write

$$C = PAQ,$$

where $P$ is an invertible $m \times m$ matrix and $Q$ is an invertible $n \times n$ matrix?

For which values of $n$ and $m$ (if any) can we find an example of an $m \times n$ matrix $A$ which cannot be reduced, by elementary row operations, to a matrix $C = (c_{ij})$ with $c_{ii} = 1$ for all $1 \leq i \leq k$ and $c_{ij} = 0$ otherwise? For which values of $n$ and $m$ (if any) can we find an example of an $m \times n$ matrix $A$ which cannot be reduced, by elementary row operations, to a matrix $C = (c_{ij})$ with $c_{ij} = 0$ if $i \neq j$ and $c_{ii} \in \{0, 1\}$? Give reasons for your answers.

**Exercise 3.6.9** (This exercise is intended to review concepts already familiar to the reader.) Recall that a function $f : A \rightarrow B$ is called *injective* if $f(a) = f(a') \Rightarrow a = a'$, and *surjective* if, given $b \in B$, we can find an $a \in A$ such that $f(a) = b$. If $f$ is both injective and surjective, we say that it is *bijective*.

(i) Give examples of $f_j : \mathbb{Z} \rightarrow \mathbb{Z}$ such that $f_1$ is neither injective nor surjective, $f_2$ is injective but not surjective, $f_3$ is surjective but not injective, $f_4$ is bijective.

(ii) Let $X$ be finite, but not empty. Either give an example or explain briefly why no such example exists of a function $g_j : X \rightarrow X$ such that $g_1$ is neither injective nor surjective, $g_2$ is injective but not surjective, $g_3$ is surjective but not injective, $g_4$ is bijective.

(iii) Consider $f : A \rightarrow B$. Show that, if there exists a $g : B \rightarrow A$ such that $(fg)(b) = b$ for all $b \in B$, then $f$ is surjective. Show that, if there exists an $h : B \rightarrow A$ such that $(hf)(a) = a$ for all $a \in A$, then $f$ is injective.

(iv) Consider $f : \mathbb{N} \rightarrow \mathbb{N}$ (where $\mathbb{N}$ denotes the positive integers). Show that, if $f$ is surjective, then there exists a $g : \mathbb{N} \rightarrow \mathbb{N}$ such that $(fg)(n) = n$ for every $n \in \mathbb{N}$. Give an example to show that $g$ need not be unique. Show that, if $f$ is injective, then there exists a $h : \mathbb{N} \rightarrow \mathbb{N}$ such that $(hf)(n) = n$ for every $n \in \mathbb{N}$. Give an example to show that $h$ need not be unique.

(v) Consider $f : A \rightarrow A$. Show that, if there exists a $g : A \rightarrow A$ such that $(fg)(a) = a$ for all $a \in A$ and there exists an $h : A \rightarrow A$ such that $(hf)(a) = a$ for all $a \in A$, then $h = g$.

(vi) Consider $f : A \rightarrow B$. Show that $f$ is bijective if and only if there exists a $g : B \rightarrow A$ such that $(fg)(b) = b$ for all $b \in B$ and such that $(gf)(a) = a$ for all $a \in A$. Show that, if $g$ exists, it is unique. We write $f^{-1} = g$.

# 4

# The secret life of determinants

## 4.1 The area of a parallelogram

Let us investigate the behaviour of the area $\mathcal{D}(\mathbf{a}, \mathbf{b})$ of the parallelogram with vertices $\mathbf{0}$, $\mathbf{a}$, $\mathbf{a} + \mathbf{b}$ and $\mathbf{b}$.

Our first observation is that it appears that

$$\mathcal{D}(\mathbf{a}, \mathbf{b} + \lambda\mathbf{a}) = \mathcal{D}(\mathbf{a}, \mathbf{b}).$$

To see this, examine Figure 4.1 showing the parallelogram $OAXB$ having vertices $O$ at $\mathbf{0}$, $A$ at $\mathbf{a}$, $X$ at $\mathbf{a} + \mathbf{b}$ and $B$ at $\mathbf{b}$ together with the parallelogram $OAX'B'$ having vertices $O$ at $\mathbf{0}$, $A$ at $\mathbf{a}$, $X'$ at $(1 + \lambda)\mathbf{a} + \mathbf{b}$ and $B'$ at $\lambda\mathbf{a} + \mathbf{b}$. Looking at the diagram, we see that (using congruent triangles)

$$\text{area } OBB' = \text{area } AXX'$$

and so

$$\mathcal{D}(\mathbf{a}, \mathbf{b} + \lambda\mathbf{a}) = \text{area } OAX'B' = \text{area } OAXB + \text{area } AXX' - \text{area } AXX'$$
$$= \text{area } OAXB = \mathcal{D}(\mathbf{a}, \mathbf{b}).$$

A similar argument shows that

$$\mathcal{D}(\mathbf{a} + \lambda\mathbf{b}, \mathbf{b}) = \mathcal{D}(\mathbf{a}, \mathbf{b}).$$

Encouraged by this, we now seek to prove that

$$\mathcal{D}(\mathbf{a} + \mathbf{c}, \mathbf{b}) = \mathcal{D}(\mathbf{a}, \mathbf{b}) + \mathcal{D}(\mathbf{c}, \mathbf{b}) \qquad \bigstar$$

by referring to Figure 4.2. This shows the parallelogram $OAXB$ having vertices $O$ at $\mathbf{0}$, $A$ at $\mathbf{a}$, $X$ at $\mathbf{a} + \mathbf{b}$ and $B$ at $\mathbf{b}$ together with the parallelogram $APQX$ having vertices $A$ at $\mathbf{a}$, $P$ at $\mathbf{a} + \mathbf{c}$, $Q$ at $\mathbf{a} + \mathbf{b} + \mathbf{c}$ and $X$ at $\mathbf{a} + \mathbf{b}$. Looking at the diagram we see that (using congruent triangles)

$$\text{area } OAP = \text{area } BXQ$$

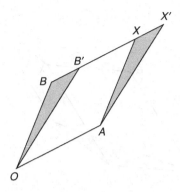

Figure 4.1 Shearing a parallelogram.

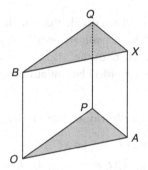

Figure 4.2 Adding two parallelograms.

and so

$$\mathcal{D}(\mathbf{a} + \mathbf{c}, \mathbf{b}) = \text{area } OPQB$$
$$= \text{area } OAXB + \text{area } APQX + \text{area } OAP - \text{area } BXQ$$
$$= \text{area } OAXB + \text{area } APQX = \mathcal{D}(\mathbf{a}, \mathbf{b}) + \mathcal{D}(\mathbf{c}, \mathbf{b}).$$

This seems fine until we ask what happens if we set $\mathbf{c} = -\mathbf{a}$ in ★ to obtain

$$\mathcal{D}(\mathbf{0}, \mathbf{b}) = \mathcal{D}(\mathbf{a} - \mathbf{a}, \mathbf{b}) = \mathcal{D}(\mathbf{a}, \mathbf{b}) + \mathcal{D}(-\mathbf{a}, \mathbf{b}).$$

Since we must surely take the area of the degenerate parallelogram[1] with vertices $\mathbf{0}, \mathbf{a}, \mathbf{a}, \mathbf{0}$
to be zero, we obtain

$$\mathcal{D}(-\mathbf{a}, \mathbf{b}) = -\mathcal{D}(\mathbf{a}, \mathbf{b})$$

and we are forced to consider negative areas.

Once we have seen one tiger in a forest, who knows what other tigers may lurk. If, instead of using the configuration in Figure 4.2, we use the configuration in Figure 4.3, the computation by which we 'proved' ★ looks more than a little suspect.

---

[1] There is a long mathematical tradition of insulting special cases.

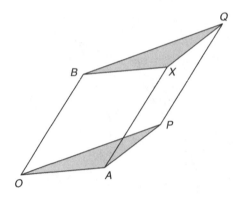

Figure 4.3 Adding two other parallelograms.

This difficulty can be resolved if we decide that simple polygons like triangles $ABC$ and parallelograms $ABCD$ will have 'ordinary area' if their boundary is described anti-clockwise (that is the points $A$, $B$, $C$ of the triangle and the points $A$, $B$, $C$, $D$ of the parallelogram are in anti-clockwise order) but 'minus ordinary area' if their boundary is described clockwise.

**Exercise 4.1.1**   (*i*) *Explain informally why, with this convention,*

$$\text{area } ABC = \text{area } BCA = \text{area } CAB$$
$$= -\text{area } ACB = -\text{area } BAC = -\text{area } CBA.$$

(*ii*) *Convince yourself that, with this convention, the argument for* ★ *continues to hold for Figure 4.3.*

We note that the rules we have given so far imply

$$\mathcal{D}(\mathbf{a}, \mathbf{b}) + \mathcal{D}(\mathbf{b}, \mathbf{a}) = \mathcal{D}(\mathbf{a} + \mathbf{b}, \mathbf{b}) + \mathcal{D}(\mathbf{a} + \mathbf{b}, \mathbf{a}) = \mathcal{D}(\mathbf{a} + \mathbf{b}, \mathbf{a} + \mathbf{b})$$
$$= \mathcal{D}(\mathbf{a} + \mathbf{b} - (\mathbf{a} + \mathbf{b}), \mathbf{a} + \mathbf{b}) = \mathcal{D}(\mathbf{0}, \mathbf{a} + \mathbf{b}) = 0,$$

so that

$$\mathcal{D}(\mathbf{a}, \mathbf{b}) = -\mathcal{D}(\mathbf{b}, \mathbf{a}).$$

**Exercise 4.1.2**  *State the rules used in each step of the calculation just given. Why is the rule* $\mathcal{D}(\mathbf{a}, \mathbf{b}) = -\mathcal{D}(\mathbf{b}, \mathbf{a})$ *consistent with the rule deciding whether the area of a parallelogram is positive or negative?*

If the reader feels that the convention 'areas of figures described with anti-clockwise boundaries are positive, but areas of figures described with clockwise boundaries are negative' is absurd, she should reflect on the convention used for integrals that

$$\int_a^b f(x)\,dx = -\int_b^a f(x)\,dx.$$

Even if the reader is not convinced that negative areas are a good thing, let us agree for the moment to consider a notion of area for which

$$\mathcal{D}(\mathbf{a} + \mathbf{c}, \mathbf{b}) = \mathcal{D}(\mathbf{a}, \mathbf{b}) + \mathcal{D}(\mathbf{c}, \mathbf{b}) \qquad\qquad \bigstar$$

holds. We observe that if $p$ is a strictly positive integer

$$\mathcal{D}(p\mathbf{a}, \mathbf{b}) = \mathcal{D}(\underbrace{\mathbf{a} + \mathbf{a} + \cdots + \mathbf{a}}_{p}, \mathbf{b})$$

$$= \underbrace{\mathcal{D}(\mathbf{a}, \mathbf{b}) + \mathcal{D}(\mathbf{a}, \mathbf{b}) + \cdots + \mathcal{D}(\mathbf{a}, \mathbf{b})}_{p}$$

$$= p\mathcal{D}(\mathbf{a}, \mathbf{b}).$$

Thus, if $p$ and $q$ are strictly positive integers,

$$q\mathcal{D}(\tfrac{p}{q}\mathbf{a}, \mathbf{b}) = \mathcal{D}(p\mathbf{a}, \mathbf{b}) = p\mathcal{D}(\mathbf{a}, \mathbf{b})$$

and so

$$\mathcal{D}(\tfrac{p}{q}\mathbf{a}, \mathbf{b}) = \tfrac{p}{q}\mathcal{D}(\mathbf{a}, \mathbf{b}).$$

Using the rules $\mathcal{D}(-\mathbf{a}, \mathbf{b}) = -\mathcal{D}(\mathbf{a}, \mathbf{b})$ and $\mathcal{D}(\mathbf{0}, \mathbf{b}) = 0$, we thus have

$$\mathcal{D}(\lambda\mathbf{a}, \mathbf{b}) = \lambda\mathcal{D}(\mathbf{a}, \mathbf{b})$$

for all rational $\lambda$. Continuity considerations now lead us to the formula

$$\mathcal{D}(\lambda\mathbf{a}, \mathbf{b}) = \lambda\mathcal{D}(\mathbf{a}, \mathbf{b})$$

for all real $\lambda$. We also have

$$\mathcal{D}(\mathbf{a}, \lambda\mathbf{b}) = -\mathcal{D}(\lambda\mathbf{b}, \mathbf{a}) = -\lambda\mathcal{D}(\mathbf{b}, \mathbf{a}) = \lambda\mathcal{D}(\mathbf{a}, \mathbf{b}).$$

We now put all our rules together to calculate $\mathcal{D}(\mathbf{a}, \mathbf{b})$. (We use column vectors.) Suppose that $a_1, a_2, b_1, b_2 \neq 0$. Then

$$\mathcal{D}\left(\begin{pmatrix} a_1 \\ a_2 \end{pmatrix}, \begin{pmatrix} b_1 \\ b_2 \end{pmatrix}\right) = \mathcal{D}\left(\begin{pmatrix} a_1 \\ 0 \end{pmatrix}, \begin{pmatrix} b_1 \\ b_2 \end{pmatrix}\right) + \mathcal{D}\left(\begin{pmatrix} 0 \\ a_2 \end{pmatrix}, \begin{pmatrix} b_1 \\ b_2 \end{pmatrix}\right)$$

$$= \mathcal{D}\left(\begin{pmatrix} a_1 \\ 0 \end{pmatrix}, \begin{pmatrix} b_1 \\ b_2 \end{pmatrix} - \frac{b_1}{a_1}\begin{pmatrix} a_1 \\ 0 \end{pmatrix}\right) + \mathcal{D}\left(\begin{pmatrix} 0 \\ a_2 \end{pmatrix}, \begin{pmatrix} b_1 \\ b_2 \end{pmatrix} - \frac{b_2}{a_2}\begin{pmatrix} 0 \\ a_2 \end{pmatrix}\right)$$

$$= \mathcal{D}\left(\begin{pmatrix} a_1 \\ 0 \end{pmatrix}, \begin{pmatrix} 0 \\ b_2 \end{pmatrix}\right) + \mathcal{D}\left(\begin{pmatrix} 0 \\ a_2 \end{pmatrix}, \begin{pmatrix} b_1 \\ 0 \end{pmatrix}\right)$$

$$= \mathcal{D}\left(\begin{pmatrix} a_1 \\ 0 \end{pmatrix}, \begin{pmatrix} 0 \\ b_2 \end{pmatrix}\right) - \mathcal{D}\left(\begin{pmatrix} b_1 \\ 0 \end{pmatrix}, \begin{pmatrix} 0 \\ a_2 \end{pmatrix}\right)$$

$$= (a_1b_2 - a_2b_1)\mathcal{D}\left(\begin{pmatrix} 1 \\ 0 \end{pmatrix}, \begin{pmatrix} 0 \\ 1 \end{pmatrix}\right) = a_1b_2 - a_2b_1.$$

(Observe that we know that the area of a unit square is 1.)

**Exercise 4.1.3** (*i*) *Justify each step of the calculation just performed.*
(*ii*) *Check that it remains true that*

$$D\left(\begin{pmatrix} a_1 \\ a_2 \end{pmatrix}, \begin{pmatrix} b_1 \\ b_2 \end{pmatrix}\right) = a_1 b_2 - a_2 b_1$$

*for all choices of $a_j$ and $b_j$ including zero values.*

If

$$A = \begin{pmatrix} a_1 & b_1 \\ a_2 & b_2 \end{pmatrix},$$

we shall write

$$DA = D\left(\begin{pmatrix} a_1 \\ a_2 \end{pmatrix}, \begin{pmatrix} b_1 \\ b_2 \end{pmatrix}\right) = a_1 b_2 - a_2 b_1.$$

The mountain has laboured and brought forth a mouse. The reader may feel that five pages of discussion is a ridiculous amount to devote to the calculation of the area of a simple parallelogram, but we shall find still more to say in the next section.

## 4.2 Rescaling

We know that the area of a region is unaffected by translation. It follows that the area of a parallelogram with vertices $\mathbf{c}, \mathbf{x} + \mathbf{c}, \mathbf{x} + \mathbf{y} + \mathbf{c}, \mathbf{y} + \mathbf{c}$ is

$$D\left(\begin{pmatrix} x_1 \\ x_2 \end{pmatrix}, \begin{pmatrix} y_1 \\ y_2 \end{pmatrix}\right) = D\begin{pmatrix} x_1 & y_1 \\ x_2 & y_2 \end{pmatrix}.$$

We now consider a $2 \times 2$ matrix

$$A = \begin{pmatrix} a_1 & b_1 \\ a_2 & b_2 \end{pmatrix}$$

and the map $\mathbf{x} \mapsto A\mathbf{x}$. We write $\mathbf{e}_1 = (1, 0)^T$, $\mathbf{e}_2 = (0, 1)^T$ and observe that $A\mathbf{e}_1 = \mathbf{a}$, $A\mathbf{e}_2 = \mathbf{b}$. The square $\mathbf{c}, \delta\mathbf{e}_1 + \mathbf{c}, \delta\mathbf{e}_1 + \delta\mathbf{e}_2 + \mathbf{c}, \delta\mathbf{e}_2 + \mathbf{c}$ is therefore mapped to the parallelogram with vertices

$$A\mathbf{c}, \quad A(\delta\mathbf{e}_1 + \mathbf{c}) = \delta\mathbf{a} + A\mathbf{c},$$

$$A(\delta\mathbf{e}_1 + \delta\mathbf{e}_2 + \mathbf{c}) = \delta\mathbf{a} + \delta\mathbf{b} + A\mathbf{c}, \quad A(\delta\mathbf{e}_2 + \mathbf{c}) = \delta\mathbf{b} + A\mathbf{c}$$

which has area

$$D\begin{pmatrix} \delta a_1 & \delta b_1 \\ \delta a_2 & \delta b_2 \end{pmatrix} = \delta^2 DA.$$

Thus the map takes squares of side $\delta$ with sides parallel to the coordinates (which have area $\delta^2$) to parallelograms of area $\delta^2 DA$.

If we want to find the area of a simply shaped subset $\Gamma$ of the plane like a disc, one common technique is to use squared paper and count the number of squares lying within $\Gamma$. If we we use squares of side $\delta$, then the map $\mathbf{x} \mapsto A\mathbf{x}$ takes those squares to parallelograms of area $\delta^2 \mathcal{D}A$. If we have $N(\delta)$ squares within $\Gamma$ and $\mathbf{x} \mapsto A\mathbf{x}$ takes $\Gamma$ to $\Gamma'$, it is reasonable to suppose

$$\text{area } \Gamma \approx N(\delta)\delta^2 \quad \text{and} \quad \text{area } \Gamma' \approx N(\delta)\delta^2 \mathcal{D}A$$

and so

$$\text{area } \Gamma' \approx \text{area } \Gamma \times \mathcal{D}A.$$

Since we expect the approximation to get better and better as $\delta \to 0$, we conclude that

$$\text{area } \Gamma' = \Gamma \times \mathcal{D}A.$$

Thus $\mathcal{D}A$ is the *area scaling factor* for the map $\mathbf{x} \mapsto A\mathbf{x}$ under the transformation $\mathbf{x} \mapsto A\mathbf{x}$. When $\mathcal{D}A$ is negative, this tells us that the mapping $\mathbf{x} \mapsto A\mathbf{x}$ interchanges clockwise and anti-clockwise.

Let $A$ and $B$ be two $2 \times 2$ matrices. The discussion of the last paragraph tells us that $\mathcal{D}A$ is the scaling factor for area under the transformation $\mathbf{x} \mapsto A\mathbf{x}$ and $\mathcal{D}B$ is the scaling factor for area under the transformation $\mathbf{x} \mapsto B\mathbf{x}$. Thus the scaling factor $\mathcal{D}(AB)$ for the transformation $\mathbf{x} \mapsto AB\mathbf{x}$ which is obtained by first applying the transformation $\mathbf{x} \mapsto B\mathbf{x}$ and then the transformation $\mathbf{x} \mapsto A\mathbf{x}$ must be the product of the scaling factors $\mathcal{D}B$ and $\mathcal{D}A$. In other words, we must have

$$\mathcal{D}(AB) = \mathcal{D}B \times \mathcal{D}A = \mathcal{D}A \times \mathcal{D}B.$$

**Exercise 4.2.1** *The argument above is instructive, but not rigorous. Recall that*

$$\mathcal{D} \begin{pmatrix} a_{11} & a_{21} \\ a_{21} & a_{22} \end{pmatrix} = a_{11}a_{22} - a_{12}a_{21}.$$

*Check algebraically that, indeed,*

$$\mathcal{D}(AB) = \mathcal{D}A \times \mathcal{D}B.$$

By Theorem 3.4.10, we know that, given any $2 \times 2$ matrix $A$, we can find elementary matrices $L_1, L_2, \ldots, L_p$ together with a diagonal matrix $D$ such that

$$A = L_1 L_2 \ldots L_p D.$$

We now know that

$$\mathcal{D}A = \mathcal{D}L_1 \times \mathcal{D}L_2 \times \cdots \times \mathcal{D}L_p \times \mathcal{D}D.$$

In the last but one exercise of this section, you are asked to calculate $\mathcal{D}E$ for each of the matrices $E$ which appear in this formula.

**Exercise 4.2.2** *In this exercise you are asked to prove results algebraically in the manner of Exercise 4.1.3 and then think about them geometrically.*

(*i*) *Show that* $DI = 1$. *Why is this a natural result?*

(*ii*) *Show that, if*

$$E = \begin{pmatrix} 0 & 1 \\ 1 & 0 \end{pmatrix},$$

*then* $DE = -1$. *Show that* $E(x, y)^T = (y, x)^T$ (*informally,* $E$ *interchanges the x and y axes*). *By considering the effect of* $E$ *on points* $(\cos t, \sin t)^T$ *as t runs from* $0$ *to* $2\pi$, *or otherwise, convince yourself that the map* $\mathbf{x} \to E\mathbf{x}$ *does indeed 'convert anti-clockwise to clockwise'.*

(*iii*) *Show that, if*

$$E = \begin{pmatrix} 1 & \lambda \\ 0 & 1 \end{pmatrix} \quad or \quad E = \begin{pmatrix} 1 & 0 \\ \lambda & 1 \end{pmatrix},$$

*then* $DE = 1$ (*informally, shears leave area unaffected*).

(*iii*) *Show that if*

$$D \begin{pmatrix} a & 0 \\ 0 & b \end{pmatrix},$$

*then* $DE = ab$. *Why should we expect this?*

Our final exercise prepares the way for the next section.

**Exercise 4.2.3** (*No writing required.*) *Go through the chapter so far, working in* $\mathbb{R}^3$ *and looking at the volume* $D(\mathbf{a}, \mathbf{b}, \mathbf{c})$ *of the parallelepiped with vertex* $\mathbf{0}$ *and neighbouring vertices* $\mathbf{a}$, $\mathbf{b}$ *and* $\mathbf{c}$.

### 4.3 $3 \times 3$ determinants

By now the reader may be feeling annoyed and confused. What precisely are the rules obeyed by $D$ and can some be deduced from others? Even worse, can we be sure that they are not contradictory? What precisely have we proved and how rigorously have we proved it? Do we know enough about area and volume to be sure of our 'rescaling' arguments? What is this business about clockwise and anti-clockwise?

Faced with problems like these, mathematicians employ a strategy which delights them and annoys pedagogical experts. We start again from the beginning and develop the theory from a new definition which we pretend has unexpectedly dropped from the skies.

**Definition 4.3.1** (*i*) *We set* $\epsilon_{12} = -\epsilon_{21} = 1$, $\epsilon_{rr} = 0$ *for* $r = 1, 2$.

(*ii*) *We set*

$$\epsilon_{123} = \epsilon_{312} = \epsilon_{231} = 1$$
$$\epsilon_{321} = \epsilon_{213} = \epsilon_{132} = -1$$
$$\epsilon_{rst} = 0 \text{ otherwise} \quad [1 \leq r, s, t \leq 3].$$

(iii) *If A is the 2 × 2 matrix ($a_{ij}$), we define*

$$\det A = \sum_{i=1}^{2} \epsilon_{ij} a_{i1} a_{j2}.$$

(iv) *If A is the 3 × 3 matrix ($a_{ij}$), we define*

$$\det A = \sum_{i=1}^{3} \sum_{j=1}^{3} \sum_{j=1}^{3} \epsilon_{ijk} a_{i1} a_{j2} a_{k3}.$$

We call det $A$ the *determinant* of $A$.

**Exercise 4.3.2** *Secretly check that*

$$\mathcal{D}A = \det A$$

*(at least in the 2 × 2 case).*

**Exercise 4.3.3** *Check that*

$$\epsilon_{rst} = -\epsilon_{srt} = -\epsilon_{rts} = -\epsilon_{tsr}$$

*for $1 \leq r, s, t \leq 3$. (Thus interchanging two indices multiplies $\epsilon_{rst}$ by $-1$.)*

The symbol $\epsilon_{ijk}$ and its generalisations are sometimes called *Levi-Civita symbols.*[2]
In this section we develop the theory of determinants in the 3 × 3 case, leaving the easier 2 × 2 case to the reader. Note that, if we use the summation convention for $i$, $j$ and $k$, then the definition of the determinant takes the pleasing form

$$\det A = \epsilon_{ijk} a_{i1} a_{j2} a_{k3}.$$

**Lemma 4.3.4** *We consider the 3 × 3 matrix $A = (a_{ij})$.*
*(i) If $\tilde{A}$ is formed from A by interchanging two columns, then det $\tilde{A} = -\det A$.*
*(ii) If A has two columns the same, then det $A = 0$.*
*(iii) If $\tilde{A}$ is formed from A by adding a multiple of one column to another, then det $\tilde{A} =$* det $A$.
*(iv) If $\tilde{A}$ is formed from A by multiplying one column by $\lambda$, then det $\tilde{A} = \lambda \det A$.*
*(v) det $I = 1$.*

*Proof* (i) Suppose that we interchange the second and third columns. Then, using the summation convention,

$$\det \tilde{A} = \epsilon_{ijk} a_{i1} a_{j3} a_{k2} \qquad \text{(by the definition of det and } \tilde{A})$$
$$= \epsilon_{ijk} a_{i1} a_{k2} a_{j3}$$
$$= -\epsilon_{ikj} a_{i1} a_{k2} a_{j3} \qquad \text{(by Exercise 4.3.3)}$$
$$= -\epsilon_{ijk} a_{i1} a_{j2} a_{k3} \qquad \text{(since } j \text{ and } k \text{ are dummy variables)}$$
$$= -\det A.$$

---

[2] When asked what he liked best about Italy, Einstein replied 'Spaghetti and Levi-Civita'.

The other cases follow in the same way.

(ii) Let $\tilde{A}$ be the matrix formed by interchanging the identical columns. Then $A = \tilde{A}$, so

$$\det A = \det \tilde{A} = -\det A$$

and $\det A = 0$.

(iii) By (i), we need only consider the case when we add $\lambda$ times the second column of $A$ to the first. Then

$$\det \tilde{A} = \epsilon_{ijk}(a_{i1} + \lambda a_{i2})a_{j2}a_{k3} = \epsilon_{ijk}a_{i1}a_{j2}a_{k3} + \lambda\epsilon_{ijk}a_{i2}a_{j2}a_{k3} = \det A + 0 = \det A,$$

using (ii) to tell us that the determinant of a matrix with two identical columns is zero.

(iv) By (i), we need only consider the case when we multiply the first column of $A$ by $\lambda$. Then

$$\det \tilde{A} = \epsilon_{ijk}(\lambda a_{i1})a_{j2}a_{k3} = \lambda(\epsilon_{ijk}a_{i1}a_{j2}a_{k3}) = \lambda \det A.$$

(v) $\det I = \epsilon_{ijk}\delta_{i1}\delta_{j2}\delta_{k3} = \epsilon_{123} = 1.$ $\qquad\square$

We can combine the results of Lemma 4.3.4 with the results on post-multiplication by elementary matrices obtained in Lemma 3.4.6.

**Lemma 4.3.5** *Let $A$ be a $3 \times 3$ matrix.*

*(i)* $\det AE_{r,s,\lambda} = \det A$, $\det E_{r,s,\lambda} = 1$ *and* $\det AE_{r,s,\lambda} = \det A \det E_{r,s,\lambda}$.

*(ii) Suppose that $\sigma$ is a permutation which interchanges two integers $r$ and $s$ and leaves the rest unchanged. Then* $\det AP(\sigma) = -\det A$, $\det P(\sigma) = -1$ *and* $\det AP(\sigma) = \det A \det P(\sigma)$.

*(iii) Suppose that $D = (d_{ij})$ is a $3 \times 3$ diagonal matrix (that is to say, a $3 \times 3$ matrix with all non-diagonal entries zero). If $d_{rr} = d_r$, then* $\det AD = d_1 d_2 d_3 \det A$, $\det D = d_1 d_2 d_3$ *and* $\det AD = \det A \det D$.

*Proof* (i) By Lemma 3.4.6 (ii), $AE_{r,s,\lambda}$ is the matrix obtained from $A$ by adding $\lambda$ times the $r$th column to the $s$th column. Thus, by Lemma 4.3.4 (iii),

$$\det AE_{r,s,\lambda} = \det A.$$

By considering the special case $A = I$, we have $\det E_{r,s,\lambda} = 1$. Putting the two results together, we have $\det AE_{r,s,\lambda} = \det A \det E_{r,s,\lambda}$.

(ii) By Lemma 3.4.6 (iv), $AP(\sigma)$ is the matrix obtained from $A$ by interchanging the $r$th column with the $s$th column. Thus, by Lemma 4.3.4 (i),

$$\det AP(\sigma) = -\det A.$$

By considering the special case $A = I$, we have $\det P(\sigma) = -1$. Putting the two results together, we have $\det AP(\sigma) = \det A \det P(\sigma)$.

(iii) The summation convention is not suitable here, so we do not use it. By direct calculation, $AD = \tilde{A}$ where

$$\tilde{a}_{ij} = \sum_{k=1}^{3} a_{ik} d_{kj} = d_j a_{ij}.$$

Thus $AD$ is the result of multiplying the $j$th column of $A$ by $d_j$ for $1 \le j \le 3$. Applying Lemma 4.3.4 (iv) three times, we obtain

$$\det AD = d_1 d_2 d_3 \det A.$$

By considering the special case $A = I$, we have $\det D = d_1 d_2 d_3$. Putting the two results together, we have $\det AD = \det A \det D$. $\qquad\square$

We can now exploit Theorem 3.4.10 which tells us that, given any $3 \times 3$ matrix $A$, we can find elementary matrices $L_1, L_2, \ldots, L_p$ together with a diagonal matrix $D$ such that

$$A = L_1 L_2 \ldots L_p D.$$

**Theorem 4.3.6** *If $A$ and $B$ are $3 \times 3$ matrices then $\det BA = \det B \det A$.*

*Proof* We know that we can write $A$ in the form given in the paragraph above so

$$
\begin{aligned}
\det BA &= \det(BL_1 L_2 \ldots L_p D) \\
&= \det\left(BL_1 L_2 \ldots L_p\right) \det D \\
&= \det\left(BL_1 L_2 \ldots L_{p-1}\right) \det L_p \det D \\
&\;\;\vdots \\
&= \det B \det L_1 \det L_2 \ldots \det L_p \det D.
\end{aligned}
$$

Looking at the special case $B = I$, we see that

$$\det A = \det L_1 \det L_2 \ldots \det L_p \det D,$$

and so $\det BA = \det B \det A$. $\qquad\square$

We can also obtain an important test for the existence of a matrix inverse.

**Theorem 4.3.7** *If $A$ is a $3 \times 3$ matrix, then $A$ is invertible if and only if $\det A \ne 0$.*

*Proof* Write $A$ in the form

$$A = L_1 L_2 \ldots L_p D$$

with $L_1, L_2, \ldots, L_p$ elementary and $D$ diagonal. By Lemma 4.3.5, we know that, if $E$ is an elementary matrix, then $|\det E| = 1$. Thus

$$|\det A| = |\det L_1||\det L_2| \ldots |\det L_p||\det D| = |\det D|.$$

Lemma 3.4.12 tells us that $A$ is invertible if and only if all the diagonal entries of $D$ are non-zero. Since $\det D$ is the product of the diagonal entries of $D$, it follows that $A$ is invertible if and only if $\det D \neq 0$ and so if and only if $\det A \neq 0$. $\qquad\square$

The reader may already have met treatments of the determinant which use row manipulation rather than column manipulation. We now show that this comes to the same thing.

**Definition 4.3.8** *If* $A = (a_{ij})_{1 \leq i \leq n}^{1 \leq j \leq m}$ *is an* $n \times m$ *matrix we define the* transposed matrix *(or, more usually, the* matrix transpose*)* $A^T = C$ *to be the* $m \times n$ *matrix* $C = (c_{rs})_{1 \leq r \leq m}^{1 \leq s \leq n}$ *where* $c_{ij} = a_{ji}$.

**Lemma 4.3.9** *If* $A$ *and* $B$ *are two* $n \times n$ *matrices, then* $(AB)^T = B^T A^T$.

*Proof* Let $A = (a_{ij})$, $B = (b_{ij})$, $A^T = (\tilde{a}_{ij})$ and $B^T = (\tilde{b}_{ij})$. If we use the summation convention with $i$, $j$ and $k$ ranging over $1, 2, \ldots, n$, then

$$a_{jk} b_{ki} = \tilde{a}_{kj} \tilde{b}_{ik} = \tilde{b}_{ik} \tilde{a}_{kj}.$$

$\qquad\square$

**Exercise 4.3.10** *Suppose that* $A$ *is an* $n \times m$ *and* $B$ *an* $m \times p$ *matrix. Show that* $(AB)^T = B^T A^T$. *(Note that you cannot use the summation convention here.)*

**Lemma 4.3.11** *We use our standard notation for* $3 \times 3$ *elementary matrices.*
   *(i)* $E_{r,s,\lambda}^T = E_{s,r,\lambda}$.
   *(ii) If* $\sigma$ *is a permutation which interchanges two integers* $r$ *and* $s$ *and leaves the rest unchanged, then* $P(\sigma)^T = P(\sigma)$.
   *(iii) If* $D$ *is a diagonal matrix, then* $D^T = D$.
   *(iv) If* $E$ *is an elementary matrix or a diagonal matrix, then* $\det E^T = \det E$.
   *(v) If* $A$ *is any* $3 \times 3$ *matrix, then* $\det A^T = \det A$.

*Proof* Parts (i) to (iv) are immediate. Since we can find elementary matrices $L_1, L_2, \ldots, L_p$ together with a diagonal matrix $D$ such that

$$A = L_1 L_2 \ldots L_p D,$$

part (i) tells us that

$$\det A^T = \det D^T L_p^T L_{p-1}^T L_p^T = \det D^T \det L_p^T \det L_{p-1}^T \ldots \det L_1^T$$
$$= \det D \det L_p \det L_{p-1} \ldots \det L_1 = \det A$$

as required. $\qquad\square$

Since transposition interchanges rows and columns, Lemma 4.3.4 on columns gives us a corresponding lemma for operations on rows.

**Lemma 4.3.12** *We consider the* $3 \times 3$ *matrix* $A = (a_{ij})$.
   *(i) If* $\tilde{A}$ *is formed from* $A$ *by interchanging two rows, then* $\det \tilde{A} = -\det A$.
   *(ii) If* $A$ *has two rows the same, then* $\det A = 0$.

*(iii) If $\tilde{A}$ is formed from A by adding a multiple of one row to another, then $\det \tilde{A} = \det A$.*

*(iv) If $\tilde{A}$ is formed from A by multiplying one row by λ, then $\det \tilde{A} = \lambda \det A$.*

**Exercise 4.3.13** *Describe geometrically, as well as you can, the effect of the mappings from $\mathbb{R}^3$ to $\mathbb{R}^3$ given by $\mathbf{x} \mapsto E_{r,s,\lambda}\mathbf{x}$, $\mathbf{x} \mapsto D\mathbf{x}$ (where D is a diagonal matrix) and $\mathbf{x} \mapsto P(\sigma)\mathbf{x}$ (where σ interchanges two integers and leaves the third unchanged).*

*For each of the above maps $\mathbf{x} \mapsto M\mathbf{x}$, convince yourself that $\det M$ is the appropriate scaling factor for volume. (In the case of $P(\sigma)$ you will mutter something about right-handed sets of coordinates being taken to left-handed coordinates and your muttering does not have to carry conviction.[3])*

*Let A be any 3 × 3 matrix. By considering A as the product of diagonal and elementary matrices, conclude that $\det A$ is the appropriate scaling factor for volume under the transformation $\mathbf{x} \mapsto A\mathbf{x}$.*

*[The reader may feel that we should try to prove this* rigorously, *but a rigorous proof would require us to produce an exact statement of what we mean by area and volume. All this can be done, but requires more time and effort than one might think.]*

**Exercise 4.3.14** *We shall not need the idea, but, for completeness, we mention that, if $\lambda > 0$ and $M = \lambda I$, the map $\mathbf{x} \mapsto M\mathbf{x}$ is called a* dilation *(or* dilatation*) by a factor of λ. Describe the map $\mathbf{x} \mapsto M\mathbf{x}$ geometrically and state the associated scaling factor for volume.*

**Exercise 4.3.15** *(i) Use Lemma 4.3.12 to obtain the pretty and useful formula*

$$\epsilon_{rst}a_{ir}a_{js}a_{kt} = \epsilon_{ijk} \det A.$$

*(Here, $A = (a_{ij})$ is a 3 × 3 matrix and we use the summation convention.)*

*Use the fact that $\det A = \det A^T$ to show that*

$$\epsilon_{ijk}a_{ir}a_{js}a_{kt} = \epsilon_{rst} \det A.$$

*(ii) Use the formula of (i) to obtain an alternative proof of Theorem 4.3.6 which states that $\det AB = \det A \det B$.*

**Exercise 4.3.16** *Let $\epsilon_{ijkl}$ $[1 \leq i, j, k, l \leq 4]$ be an expression such that $\epsilon_{1234} = 1$ and interchanging any two of the suffices of $\epsilon_{ijkl}$ multiplies the expression by $-1$. If $A = (a_{ij})$ is a 4 × 4 matrix we set*

$$\det A = \epsilon_{ijkl}a_{i1}a_{j2}a_{k3}a_{l4}.$$

*Develop the theory of the determinant of 4 × 4 matrices along the lines of the preceding section.*

---

[3] We discuss this a bit more in Chapter 7, when we talk about $O(\mathbb{R}^n)$ and $SO(\mathbb{R}^n)$ and in Section 10.3, when we talk about the physical implications of handedness.

## 4.4 Determinants of $n \times n$ matrices

A very wide-awake reader might ask how we know that the expression $\epsilon_{ijkl}$ of Exercise 4.3.16 actually exists with the properties we have assigned to it.

**Exercise 4.4.1** *Does there exist a non-trivial expression*

$$\chi_{ijklm} \qquad [1 \le i, j, k, l, m \le 5]$$

*such that cycling three suffices multiplies the expression by* $-1$? *(More formally, if* $r, s, t$
*are distinct integers between 1 and 5, then moving the suffix in the* $r$*th place to the* $s$*th
place, the suffix in the* $s$*th place to the* $t$*th place and the suffix in the* $t$*th place to the* $r$*th
place multiplies the expression by* $-1$.)
[*We take a slightly deeper look in Exercise 4.6.2 which uses a little group theory.*]

In the case of $\epsilon_{ijkl}$, we could just write down the $4^4$ values of $\epsilon_{ijkl}$ corresponding to
the possible choices of $i$, $j$, $k$ and $l$ (or, more sensibly, the 24 non-zero values of $\epsilon_{ijkl}$
corresponding to the possible choices of $i$, $j$, $k$ and $l$ with the four integers unequal).
However, we cannot produce the general result in this way.

Instead we proceed as follows. We start with a couple of definitions.

**Definition 4.4.2** *We write $S_n$ for the collection of bijections*

$$\sigma : \{1, 2, \ldots, n\} \to \{1, 2, \ldots, n\}.$$

*If* $\tau, \sigma \in S_n$, *then we write* $(\tau\sigma)(r) = \tau\big(\sigma(r)\big)$.

Many of my readers will know $S_n$ by the name of 'the permutation group on
$\{1, 2, \ldots, n\}$'.

**Definition 4.4.3** *We define the signature function* $\zeta : S_n \to \{-1, 1\}$ *by*

$$\zeta(\sigma) = \frac{\prod_{1 \le r < s \le n} \big(\sigma(s) - \sigma(r)\big)}{\prod_{1 \le r < s \le n}(s - r)}.$$

Thus, if $n = 3$,

$$\zeta(\sigma) = \frac{\big(\sigma(2) - \sigma(1)\big)\big(\sigma(3) - \sigma(1)\big)\big(\sigma(3) - \sigma(2)\big)}{(2 - 1)(3 - 1)(3 - 2)}$$

and, if $\tau(1) = 2, \tau(2) = 3, \tau(3) = 1$,

$$\zeta(\tau) = \frac{(3 - 2)(1 - 2)(1 - 3)}{(2 - 1)(3 - 1)(3 - 2)} = 1.$$

**Exercise 4.4.4** *If* $n = 4$, *write out* $\zeta(\sigma)$ *in full.*
*Compute* $\zeta(\tau)$ *if* $\tau(1) = 2$, $\tau(2) = 3$, $\tau(3) = 1$, $\tau(4) = 4$. *Compute* $\zeta(\rho)$ *if* $\rho(1) = 2$,
$\rho(2) = 3$, $\rho(3) = 4$, $\rho(4) = 1$.

**Lemma 4.4.5** *Let $\zeta$ be the signature function for $S_n$.*
(i) $\zeta(\sigma) = \pm 1$ *for all* $\sigma \in S_n$.

(*ii*) *If* $\tau, \sigma \in S_n$, *then* $\zeta(\tau\sigma) = \zeta(\tau)\zeta(\sigma)$.

(*iii*) *If* $\rho(1) = 2$, $\rho(2) = 1$ *and* $\rho(j) = j$ *otherwise, then* $\zeta(\rho) = -1$.

(*iv*) *If* $\tau$ *interchanges* 1 *and* $i$ *with* $i \neq 1$ *and leaves the remaining integers fixed, then* $\zeta(\kappa) = -1$.

(*v*) *If* $\kappa$ *interchanges two distinct integers* $i$ *and* $j$ *and leaves the rest fixed then* $\zeta(\kappa) = -1$.

*Proof* (i) Observe that each *unordered pair* $\{r, s\}$ with $r \neq s$ and $1 \leq r, s \leq n$ occurs exactly once in the set

$$\Gamma = \{\{r, s\} : 1 \leq r < s \leq n\}$$

and exactly once in the set

$$\Gamma_\sigma = \{\{\sigma(r), \sigma(s)\} : 1 \leq r < s \leq n\}.$$

Thus

$$\left| \prod_{1 \leq r < s \leq n} (\sigma(s) - \sigma(r)) \right| = \prod_{1 \leq r < s \leq n} |\sigma(s) - \sigma(r)|$$

$$= \prod_{1 \leq r < s \leq n} |s - r| = \prod_{1 \leq r < s \leq n} (s - r) > 0$$

and so

$$|\zeta(\sigma)| = \frac{\left| \prod_{1 \leq r < s \leq n} (\sigma(s) - \sigma(r)) \right|}{\prod_{1 \leq r < s \leq n} (s - r)} = 1.$$

(ii) Again, using the fact that each *unordered pair* $\{r, s\}$ with $r \neq s$ and $1 \leq r, s \leq n$ occurs exactly once in the set $\Gamma_\sigma$, we have

$$\zeta(\tau\sigma) = \frac{\prod_{1 \leq r < s \leq n} (\tau\sigma(s) - \tau\sigma(r))}{\prod_{1 \leq r < s \leq n} (s - r)}$$

$$= \frac{\prod_{1 \leq r < s \leq n} (\tau\sigma(s) - \tau\sigma(r))}{\prod_{1 \leq r < s \leq n} (\sigma(s) - \sigma(r))} \cdot \frac{\prod_{1 \leq r < s \leq n} (\sigma(s) - \sigma(r))}{\prod_{1 \leq r < s \leq n} (s - r)} = \zeta(\tau)\zeta(\sigma).$$

(iii) For the given $\rho$,

$$\prod_{3 \leq r < s \leq n} (\rho(s) - \rho(r)) = \prod_{3 \leq r < s \leq n} (r - s), \quad \prod_{3 \leq s \leq n} (\rho(s) - \rho(1)) = \prod_{3 \leq s \leq n} (s - 2),$$

$$\prod_{3 \leq s \leq n} (\rho(s) - \rho(2)) = \prod_{3 \leq s \leq n} (s - 1).$$

Thus

$$\zeta(\rho) = \frac{\rho(2) - \rho(1)}{2 - 1} = \frac{1 - 2}{2 - 1} = -1.$$

(iv) If $i = 2$, then the result follows from part (iii). If not, let $\rho$ be as in part (iii) and let $\alpha \in S_n$ be the permutation which interchanges 2 and $i$ leaving the remaining integers unchanged. Then, by inspection,

$$\alpha\rho\alpha(r) = \tau(r)$$

for all $1 \le r \le n$ and so $\alpha\rho\alpha = \tau$. Parts (ii) and (i) now tell us that

$$\zeta(\tau) = \zeta(\alpha)\zeta(\rho)\zeta(\alpha) = \zeta(\alpha)^2\zeta(\rho) = -1.$$

(v) Use an argument like that of (iv). $\qquad\square$

**Exercise 4.4.6** (i) *We use the notation of the proof just concluded. Check that $\alpha\rho\alpha(r) = \tau(r)$ by considering the cases $r = 1$, $r = 2$, $r = i$ and $r \notin \{1, 2, i\}$ in turn.*
(ii) *Write out the proof of Lemma 4.4.5 (v).*

Exercise 4.6.1 shows that $\zeta$ is the unique function with the properties described in Lemma 4.4.5.

**Exercise 4.4.7** *Check that, if we write*

$$\epsilon_{\sigma(1)\sigma(2)\sigma(3)\sigma(4)} = \zeta(\sigma)$$

*for all $\sigma \in S_4$ and*

$$\epsilon_{rstu} = 0$$

*whenever $1 \le r, s, t, u \le 4$ and not all of the $r, s, t, u$ are distinct, then $\epsilon_{ijkl}$ satisfies all the conditions required in Exercise 4.3.16.*

If we wish to define the determinant of an $n \times n$ matrix $A = (a_{ij})$ we can define $\epsilon_{ijk...uv}$ (with $n$ suffices) in the obvious way and set

$$\det A = \epsilon_{ijk...uv}a_{i1}a_{j2}a_{k3}\dots a_{u\,n-1}a_{vn}$$

(where we use the summation convention with range $1, 2, \dots, n$). Alternatively, but entirely equivalently, we can set

$$\det A = \sum_{\sigma \in S_n} \zeta(\sigma)a_{\sigma(1)1}a_{\sigma(2)2}\dots a_{\sigma(n)n}.$$

All the results we established in the $3 \times 3$ case together with their proofs go though essentially unaltered to the $n \times n$ case.

**Exercise 4.4.8** *We use the notation just established. Show that if $\sigma \in S_n$ then $\zeta(\sigma) = \zeta(\sigma^{-1})$. Use the definition*

$$\det A = \sum_{\sigma \in S_n} \zeta(\sigma)a_{\sigma(1)1}a_{\sigma(2)2}\dots a_{\sigma(n)n}$$

*to show that $\det A^T = \det A$.*

The next exercise shows how to evaluate the determinant of a matrix that appears from time to time in various parts of algebra and, in particular, in this book. It also suggests how mathematicians might have arrived at the approach to the signature used in this section.

**Exercise 4.4.9** **[The Vandermonde determinant]**[4] (*i*) *Compute*

$$\det \begin{pmatrix} 1 & 1 \\ x & y \end{pmatrix}.$$

(*ii*) *Consider the function* $F : \mathbb{R}^3 \to \mathbb{R}$ *given by*

$$F(x, y, z) = \det \begin{pmatrix} 1 & 1 & 1 \\ x & y & z \\ x^2 & y^2 & z^2 \end{pmatrix}.$$

*Explain why* $F$ *is a multinomial of degree* 3. *By considering* $F(x, x, z)$, *show that* $F$ *has* $y - x$ *as a factor. Explain why* $F(x, y, z) = A(y - x)(z - y)(z - x)$ *for some constant* $A$. *By looking at the coefficient of* $yz^2$, *or otherwise, show that*

$$F(x, y, z) = (y - x)(z - y)(z - x).$$

(*iii*) *Consider the* $n \times n$ *matrix* $V$ *with* $v_{rs} = x_s^{r-1}$. *Show that, if we set*

$$F(x_1, x_2, \ldots, x_n) = \det V,$$

*then*

$$F(x_1, x_2, \ldots, x_n) = \prod_{i > j}(x_i - x_j).$$

(*iv*) *Suppose that* $\sigma \in S_n$, *all the* $x_r$ *are distinct, and we set*

$$\tilde{\zeta}_{\mathbf{x}}(\sigma) = \frac{F(x_{\sigma(1)}, x_{\sigma(2)}, \ldots, x_{\sigma(n)})}{F(x_1, x_2, \ldots, x_n)}.$$

*Show that* $\tilde{\zeta}_{\mathbf{x}}$ *is the signature function.*

The reader should be aware of an alternative notation for determinants illustrated in the equation

$$\begin{vmatrix} a_{11} & a_{12} & a_{13} \\ a_{21} & a_{22} & a_{23} \\ a_{31} & a_{32} & a_{33} \end{vmatrix} = \det \begin{pmatrix} a_{11} & a_{12} & a_{13} \\ a_{21} & a_{22} & a_{23} \\ a_{31} & a_{32} & a_{33} \end{pmatrix}.$$

## 4.5 Calculating determinants

Much of this section deals with how *not* to evaluate a determinant. We shall use the determinant of a $4 \times 4$ matrix $A = (a_{ij})$ as a typical example, but our main interest is in what happens if we compute the determinant of an $n \times n$ matrix when $n$ is large.

---

[4] Vandermonde was an important figure in the development of the idea of the determinant, but appears never to have considered the determinant which bears his name.

In 1963, at the age of eighteen, I had never heard of matrices, but could evaluate $3 \times 3$ Vandermonde determinants on sight. Just as a palaeontologist is said to be able to reconstruct a dinosaur from a single bone, so it might be possible to reconstruct the 1960s Cambridge mathematics entrance papers from this one fact.

It is obviously foolish to work directly from the definition

$$\det A = \sum_{i=1}^{4} \sum_{j=1}^{4} \sum_{k=1}^{4} \sum_{l=1}^{4} \epsilon_{ijkl} a_{i1} a_{j2} a_{k3} a_{l4}$$

since only $24 = 4!$ of the $4^4 = 256$ terms are non-zero. The alternative definition

$$\det A = \sum_{\sigma \in S_4} \zeta(\sigma) a_{\sigma(1)1} a_{\sigma(2)2} a_{\sigma(3)3} a_{\sigma(4)4}$$

requires us to compute $24 = 4!$ terms using $3 = 4 - 1$ multiplications for each term and then add them together. This is feasible, but the analogous method for an $n \times n$ matrix involves the computation of $n!$ terms and is thus impractical even for quite small $n$.

**Exercise 4.5.1** *Estimate the number of multiplications required for $n = 10$ and for $n = 20$.*

However, matters are rather different if we have an upper triangular or lower triangular matrix.

**Definition 4.5.2** *An $n \times n$ matrix $A = (a_{ij})$ is called* upper triangular *(or* right triangular*) if $a_{ij} = 0$ for $j < i$. An $n \times n$ matrix $A = (a_{ij})$ is called* lower triangular *(or* left triangular*) if $a_{ij} = 0$ for $j > i$.*

**Exercise 4.5.3** *Show that a matrix which is both upper triangular and lower triangular must be diagonal.*

**Lemma 4.5.4** *(i) If $A = (a_{ij})$ is an upper or lower triangular $n \times n$ matrix, and $\sigma \in S_n$, then $a_{\sigma(1)1} a_{\sigma(2)2} \ldots a_{\sigma(n)n} = 0$ unless $\sigma$ is the identity map (that is to say, $\sigma(i) = i$ for all $i$).*

*(ii) If $A = (a_{ij})$ is an upper or lower triangular $n \times n$ matrix, then*

$$\det A = a_{11} a_{22} \ldots a_{nn}.$$

*Proof* Immediate.                                                                        □

We thus have a reasonable method for computing the determinant of an $n \times n$ matrix. Use elementary row and column operations to reduce the matrix to upper or lower triangular form (keeping track of any scale change introduced) and then compute the product of the diagonal entries.

**Exercise 4.5.5** *(i) Suppose that $A$ is an $r \times r$ matrix and $B$ an $s \times s$ matrix. Let*

$$C = \begin{pmatrix} A & 0 \\ 0 & B \end{pmatrix}$$

*(in other words, $c_{ij} = a_{ij}$ if $1 \le i, j \le r$, $c_{ij} = b_{i-r, j-r}$ if $r + 1 \le i, j \le r + s$ and $c_{ij} = 0$ otherwise). By using the ideas of the previous paragraph, or otherwise (there are lots of ways of doing this exercise), show that*

$$\det C = \det A \det B.$$

(*ii*) *Find* $2 \times 2$ *matrices A, B, C and D such that* $\det A = \det B = \det C = \det D = 0$, *but*

$$\det \begin{pmatrix} A & B \\ C & D \end{pmatrix} = 1.$$

[*There is thus no general method for computing determinants 'by blocks' although special cases like that in part* (*i*) *can be very useful.*]

When working by hand, we may introduce various modifications as in the following typical calculation:

$$\det \begin{pmatrix} 2 & 4 & 6 \\ 3 & 1 & 2 \\ 5 & 2 & 3 \end{pmatrix} = 2 \det \begin{pmatrix} 1 & 2 & 3 \\ 3 & 1 & 2 \\ 5 & 2 & 3 \end{pmatrix} = 2 \det \begin{pmatrix} 1 & 2 & 3 \\ 0 & -5 & -7 \\ 0 & -8 & -12 \end{pmatrix}$$

$$= 2 \det \begin{pmatrix} -5 & -7 \\ -8 & -12 \end{pmatrix} = 8 \det \begin{pmatrix} 5 & 7 \\ 2 & 3 \end{pmatrix}$$

$$= 8 \det \begin{pmatrix} 5 & 2 \\ 2 & 1 \end{pmatrix} = 8 \det \begin{pmatrix} 1 & 0 \\ 2 & 1 \end{pmatrix} = 8.$$

**Exercise 4.5.6** *Justify each step.*

As the reader probably knows, there is another method for calculating $3 \times 3$ matrices called *row* expansion described in the next exercise.

**Exercise 4.5.7** (*i*) *Show by direct algebraic calculation (there are only* $3! = 6$ *expressions involved) that*

$$\det \begin{pmatrix} a_{11} & a_{12} & a_{13} \\ a_{21} & a_{22} & a_{23} \\ a_{31} & a_{32} & a_{33} \end{pmatrix}$$

$$= a_{11} \det \begin{pmatrix} a_{22} & a_{23} \\ a_{32} & a_{33} \end{pmatrix} - a_{12} \det \begin{pmatrix} a_{21} & a_{23} \\ a_{31} & a_{33} \end{pmatrix} + a_{31} \det \begin{pmatrix} a_{21} & a_{22} \\ a_{31} & a_{32} \end{pmatrix}.$$

(*ii*) *Use the result of* (*i*) *to compute*

$$\det \begin{pmatrix} 2 & 4 & 6 \\ 3 & 1 & 2 \\ 5 & 2 & 3 \end{pmatrix}.$$

Most people (including the author) use the method of row expansion (or mix row operations and row expansion) to evaluate $3 \times 3$ matrices but, as we remarked on page 11, row expansion of an $n \times n$ matrix also involves about $n!$ operations, so (in the absence of special features) it should not be used for numerical calculation when $n \geq 4$.

In the remainder of this section we discuss row and column expansion for $n \times n$ matrices and Cramer's rule. The material is not very important, but provides useful exercises for keen students.[5]

---

[5] Less keen students can skim the material and omit the final set of exercises.

**Exercise 4.5.8** **[Column expansion]** *If $A = (a_{ij})$ is a $4 \times 4$ matrix, let us write*

$$\mathcal{F}(A) = a_{11} \det \begin{pmatrix} a_{22} & a_{23} & a_{24} \\ a_{32} & a_{33} & a_{34} \\ a_{42} & a_{43} & a_{43} \end{pmatrix} - a_{21} \det \begin{pmatrix} a_{12} & a_{13} & a_{14} \\ a_{32} & a_{33} & a_{34} \\ a_{42} & a_{43} & a_{44} \end{pmatrix}$$

$$+ a_{31} \det \begin{pmatrix} a_{12} & a_{13} & a_{14} \\ a_{22} & a_{23} & a_{24} \\ a_{42} & a_{43} & a_{44} \end{pmatrix} - a_{41} \det \begin{pmatrix} a_{12} & a_{13} & a_{14} \\ a_{22} & a_{23} & a_{24} \\ a_{32} & a_{33} & a_{34} \end{pmatrix}.$$

*(i) Show that, if $\tilde{A}$ is the matrix formed from $A$ by interchanging the first row and the $j$th row (with $1 \neq j$), then*

$$\mathcal{F}(\tilde{A}) = -\mathcal{F}(A).$$

*(ii) By applying (i) three times, or otherwise, show that, if $\tilde{A}$ is the matrix formed from $A$ by interchanging the $i$th row and the $j$th row (with $i \neq j$), then*

$$\mathcal{F}(\tilde{A}) = -\mathcal{F}(A).$$

*Deduce that, if $A$ has two rows identical, $\mathcal{F}(A) = 0$.*

*(iii) Show that, if $\tilde{A}$ is the matrix formed from $A$ by multiplying the first row by $\lambda$,*

$$\mathcal{F}(\tilde{A}) = \lambda \mathcal{F}(A).$$

*Deduce that, if $\bar{A}$ is the matrix formed from $A$ by multiplying the $i$th row by $\lambda$,*

$$\mathcal{F}(\bar{A}) = \lambda \mathcal{F}(A).$$

*(iv) Show that, if $\tilde{A}$ is the matrix formed from $A$ by adding $\lambda$ times the $i$th row to the first row $[i \neq 1]$,*

$$\mathcal{F}(\tilde{A}) = \mathcal{F}(A).$$

*Deduce that, if $\bar{A}$ is the matrix formed from $A$ by adding $\lambda$ times the $i$th row to the $j$th row $[i \neq j]$,*

$$\mathcal{F}(\bar{A}) = \mathcal{F}(A).$$

*(v) Use Theorem 3.4.8 to show that*

$$\mathcal{F}(A) = \det A.$$

Since $\det A^T = \det A$, the proof of the validity of *column expansion* given as Exercise 4.5.8 immediately implies the validity of *row expansion* for a $4 \times 4$ matrix $A = (a_{ij})$:

$$\mathcal{D}(A) = a_{11} \det \begin{pmatrix} a_{22} & a_{23} & a_{24} \\ a_{32} & a_{33} & a_{34} \\ a_{42} & a_{43} & a_{44} \end{pmatrix} - a_{12} \det \begin{pmatrix} a_{21} & a_{23} & a_{24} \\ a_{31} & a_{33} & a_{34} \\ a_{41} & a_{43} & a_{44} \end{pmatrix}$$

$$+ a_{13} \det \begin{pmatrix} a_{21} & a_{22} & a_{24} \\ a_{31} & a_{32} & a_{34} \\ a_{41} & a_{42} & a_{44} \end{pmatrix} - a_{14} \det \begin{pmatrix} a_{21} & a_{22} & a_{23} \\ a_{31} & a_{32} & a_{33} \\ a_{41} & a_{42} & a_{43} \end{pmatrix}.$$

It is easy to generalise our results to $n \times n$ matrices.

**Definition 4.5.9** *Suppose $n \geq 2$. Let $A$ be an $n \times n$ matrix. If $M_{ij}$ is the $(n-1) \times (n-1)$ matrix formed by removing the $i$th row and $j$th column from $A$, we write*

$$A_{ij} = (-1)^{i+j} \det M_{ij}.$$

*The $A_{ij}$ are called the* cofactors *of $A$.*

**Exercise 4.5.10** *Suppose $n \geq 2$. Let $A$ be an $n \times n$ matrix $a_{ij}$ with cofactors $A_{ij}$.*

*(i) Check that the argument of Exercise 4.5.7 and the paragraph that follows applies in the general case and deduce that*

$$\sum_{j=1}^{n} a_{1j} A_{1j} = \det A.$$

*(ii) By considering the effect of interchanging rows, show that*

$$\sum_{j=1}^{n} a_{ij} A_{ij} = \det A$$

*for all $1 \leq i \leq n$. (We talk of 'expanding by the $i$th row'.)*

*(iii) By considering what happens when a matrix has two rows the same, show that*

$$\sum_{j=1}^{n} a_{ij} A_{kj} = 0$$

*whenever $i \neq k$, $1 \leq i, k \leq n$.*

*(iv) Summarise your results in the formula*

$$\sum_{j=1}^{n} a_{ij} A_{kj} = \delta_{kj} \det A. \qquad \bigstar$$

If we define the *adjugate matrix* Adj $A = B$ by taking $b_{ij} = A_{ji}$ (thus Adj $A$ is the transpose of the matrix of cofactors of $A$), then equation $\bigstar$ may be rewritten in a way that deserves to be stated as a theorem.

**Theorem 4.5.11** *If $n \geq 2$ and $A$ is an $n \times n$ matrix, then*

$$A \, \text{Adj} \, A = (\det A)I.$$

**Theorem 4.5.12** *Let $A$ be an $n \times n$ matrix. If $\det A \neq 0$, then $A$ is invertible with inverse*

$$A^{-1} = \frac{1}{\det A} \, \text{Adj} \, A.$$

*If $\det A = 0$, then $A$ is not invertible.*

*Proof* If $\det A \neq 0$, then we can apply $\bigstar$. If $A^{-1}$ exists, then

$$\det A \det A^{-1} = \det A A^{-1} = \det I = 1$$

so det $A \neq 0$.                                                                      □

Theorem 4.5.12 gives another proof of Theorem 4.3.7.

The next exercise merely emphasises part of the proof just given.

**Exercise 4.5.13** *If A is an n × n invertible matrix, show that* $\det A^{-1} = (\det A)^{-1}$.

Since the direct computation of Adj $A$ involves finding $n^2$ determinants of $(n-1) \times$ $(n-1)$ matrices, this result is more important in theory than in practical computation.

We conclude with Cramer's rule. This topic was historically very important in popularising the idea of a determinants. (It was said that success in the entry to the major French engineering schools depended on mastering the rule.[6]) However, for the reasons already explained, it not often useful in a modern context.

**Exercise 4.5.14** **[Cramer's rule]** *Suppose that $n \geq 2$, A is an n × n matrix and* **b** *a column vector of length n. Write $B_i$ for the n × n matrix obtained by replacing the ith column of A by* **b**. *Show that*

$$\det B_j = \sum_{k=1}^{n} b_k A_{kj}.$$

*If A is invertible and* **x** *is the solution of show that*

$$x_j = \frac{\det B_j}{\det A}.$$

(*This is* Cramer's rule.)

**Exercise 4.5.15** *According to one Internet site, 'Cramer's Rule is a handy way to solve for just one of the variables without having to solve the whole system of equations.' Comment.*

**Exercise 4.5.16** *We can define the* permanent *of an n × n square matrix by*

$$\text{perm}(A) = \sum_{\sigma \in S_n} \prod_{i=1}^{n} a_{\sigma(i)i}.$$

(*Compare our standard formula* $\det A = \sum_{\sigma \in S_n} \zeta(\sigma) \prod_{i=1}^{n} a_{\sigma(i)i}$.)

(*i*) *Show that* perm $A^T$ = perm $A$.

(*ii*) *Is it true that* perm $A \neq 0$ *implies A invertible? Is is it true that A invertible implies* perm $A \neq 0$? *Give reasons.*
[*Hint: Look at the 2 × 2 case.*]

(*iii*) *Explain how to calculate* perm $A$ *by row expansion.*

(*iv*) *If* $|a_{ij}| \leq K$, *show that* $|\text{perm } A| \leq n!K^n$. *Give an example to show that this result is best possible whatever the value of n.*

---

[6] 'Cette méthode était tellement en faveur, que les examens aux écoles des services publics ne roulait pour ainsi dire que sur elle; on était admis ou réjeté suivant qu'on la possedait bien ou mal.' (Quoted in Chapter 1 of [25].)

(v) *If $|a_{ij}| \le K$, show that $|\det A| \le n!K^n$. Can you do better? (Please spend a few minutes on this, since, otherwise, when you see Hadamard's inequality in Exercise 7.6.13, you will say 'I could have thought of that!'* [7])

**Exercise 4.5.17** *We say that an $n \times n$ matrix $A$ with real entries is* antisymmetric *if $A = -A^T$. Which of the following result are true and which are false for $n \times n$ antisymmetric matrices $A$? Give reasons.*

(i) *If $n = 2$ and $A \neq 0$, then $\det A \neq 0$.*

(ii) *If $n$ is even and $A \neq 0$, then $\det A \neq 0$.*

(iii) *If $n$ is odd, then $\det A = 0$.*

## 4.6 Further exercises

**Exercise 4.6.1** Show that any permutation $\sigma \in S_n$ can be written as

$$\sigma = \tau_1 \tau_2 \dots \tau_p$$

where $\tau_j$ is a *transposition* (that is to say, $\tau_j$ is a permutation which interchanges two integers and leaves the rest fixed).

Suppose that $\tilde{\zeta} : S_n \to \mathbb{C}$ has the following two properties.

(A) If $\sigma, \tau \in S_n$, then $\tilde{\zeta}(\sigma \tau) = \tilde{\zeta}(\sigma)\tilde{\zeta}(\tau)$.

(B) If $\tau(1) = 2$, $\tau(2) = 1$ and $\tau(j) = j$ otherwise, then $\tilde{\zeta}(\tau) = -1$.

Show that $\tilde{\zeta}$ is the signature function.

Explain why we have shown that, if a permutation is obtained from an even number of transpositions, it cannot be obtained from an odd number of transpositions and, if a permutation is obtained from an odd number of transpositions, it cannot be obtained from an even number number of transpositions. In particular, the identity permutation cannot be expressed as the composition of an odd number of permutations.

[If we had proved this result without using the signature, we could have used it to *define* the signature $\zeta(\sigma)$ as $(-1)^r$ when we can obtain $\sigma$ as the product of $r$ transpositions.]

**Exercise 4.6.2** (This continues Exercise 4.6.1, but requires a tiny bit of knowledge of group theory.)

(i) Show that $\mathbb{C} \setminus \{0\}$ is a group under multiplication.

(ii) By using the result of the first paragraph of Exercise 4.6.1, or otherwise, show that, if $n \ge 2$, the only homomorphisms $\theta : S_n \to \mathbb{C}$ are the trivial homomorphism $\theta$ with $\theta\sigma = 1$ for all $\sigma \in S_n$ and the signature function $\zeta$.

**Exercise 4.6.3** If $A$ and $B$ are $n \times n$ matrices and $AB = 0$, show that either $A = 0$ or $B = 0$ or $\det A = \det B = 0$.

[This is *easy* with the tools of this chapter, but, when we have talked about the rank of mappings in the next chapter, the result will appear *obvious*.]

---

[7] The idea is in plain sight, but, if you do discover Hadamard's inequality independently, the author raises his hat to you.

**Exercise 4.6.4** Consider the equation

$$Ax = h$$

where $A$ is an $n \times n$ matrix with integral entries (that is to say, $a_{ij} \in \mathbb{Z}$) and $\mathbf{h}$ is a column vector with integral entries. Show that the solution vector $\mathbf{x}$ will have integral entries if $\det A$ divides each entry $h_i$ of $\mathbf{h}$. Give an example to show that this condition is not necessary.

Now suppose that $A$ and $\mathbf{h}$ have rational entries and $\det A = 0$. Is it true that the equation $Ax = h$ always has a solution vector with rational entries? Is it true that every solution vector must have rational entries? Is it true that, if there exists a solution vector, then there exists a solution vector with rational entries? Give proofs or counterexamples as appropriate.

**Exercise 4.6.5** Let

$$A = \begin{pmatrix} a & b \\ c & d \end{pmatrix}.$$

Show that $\det(tI - A) = t^2 + ut + v$, where $u$ and $v$ are to be found in terms of $a, b, c$ and $d$.

Show by direct calculation that $A^2 + uA + vI = 0$.

[In Example 6.4.4 we shall see an easier proof which also indicates why the result is true.]

**Exercise 4.6.6** If $f_r, g_r$ are once times differentiable functions from $\mathbb{R}$ to $\mathbb{R}$ and

$$H(x) = \det \begin{pmatrix} f_1(x) & g_1(x) \\ f_2(x) & g_2(x) \end{pmatrix},$$

show that

$$H'(x) = \det \begin{pmatrix} f_1(x) & g_1(x) \\ f_2'(x) & g_2'(x) \end{pmatrix} + \det \begin{pmatrix} f_1'(x) & g_1'(x) \\ f_2(x) & g_2(x) \end{pmatrix}.$$

If $f_r, g_r, h_r$ are once times differentiable functions from $\mathbb{R}$ to $\mathbb{R}$ and

$$F(x) = \det \begin{pmatrix} f_1(x) & g_1(x) & h_1(x) \\ f_2(x) & g_2(x) & h_2(x) \\ f_3(x) & g_3(x) & h_3(x) \end{pmatrix},$$

express $F'(x)$ as the sum of three similar determinants. If $f, g, h$ are five times differentiable and

$$G(x) = \det \begin{pmatrix} f(x) & g(x) & h(x) \\ f'(x) & g'(x) & h'(x) \\ f''(x) & g''(x) & h''(x) \end{pmatrix},$$

compute $G'(x)$ and $G''(x)$ as determinants.

**Exercise 4.6.7 [Cauchy's proof of L'Hôpital's rule]** (This question requires Rolle's theorem from analysis.)

Suppose that $u$, $v$, $w : [a, b] \to \mathbb{R}$ are continuous on $[a, b]$ and differentiable on $(a, b)$. Let

$$f(t) = \det \begin{pmatrix} u(a) & u(b) & u(t) \\ v(a) & v(b) & v(t) \\ w(a) & w(b) & w(t) \end{pmatrix}.$$

Verify that $f$ satisfies the conditions of Rolle's theorem and deduce that there exists a $c \in (a, b)$ such that

$$\det \begin{pmatrix} u(a) & u(b) & u'(c) \\ v(a) & v(b) & v'(c) \\ w(a) & w(b) & w'(c) \end{pmatrix} = 0.$$

By choosing $w$ appropriately, prove that, if $v'(t) \neq 0$ for all $t \in (a, b)$, there exists a $c \in (a, b)$ such that

$$\frac{u(b) - u(a)}{v(b) - v(a)} = \frac{u'(c)}{v'(c)}.$$

Suppose now that, in addition, $u(a)$, $v(a) = 0$, but $v(t) \neq 0$ for all $t \in (a, b)$. If $t \in (a, b)$, show that there exists a $c_t \in (a, t)$ such that $u(t)/v(t) = u'(c_t)/v'(c_t)$. If, in addition, $u'(x)/v'(x) \to l$ as $x \to a$ through values $x > a$, deduce that $u(t)/v(t) \to l$ as $t \to a$ through values $t > a$.

**Exercise 4.6.8** State a condition in terms of determinants for the two sets of three equations for $x_j$, $y_j$

$$a_{i1}x_1 + a_{i2}x_2 + a_{i3}x_3 = c_i, \qquad b_{i1}y_1 + b_{i2}y_2 + b_{i3}y_3 = x_i \qquad [i = 1, 2, 3]$$

to have a unique solution for the $x_i$ and for the $y_j$. State a condition in terms of determinants for the two sets of three equations to have a unique solution for the $y_j$. Give reasons in both cases.

Show that the equations

$$x_1 + 2x_2 = 2 \qquad\qquad 3y_1 + y_2 + 4y_3 = x_1$$
$$x_1 + x_2 + x_3 = 1 \qquad\qquad -y_1 + 2y_2 - 3y_3 = x_2$$
$$3x_2 - x_3 = k \qquad\qquad y_1 + 5y_2 - 2y_3 = x_3$$

are inconsistent if $k \neq 7$ and find the most general solution for the $x_j$ and $y_j$ if $k = 7$.

**Exercise 4.6.9** Let $a_1, a_2, a_3$ be real numbers and write $s_r = a_1^r + a_2^r + a_3^r$. If

$$S = \begin{pmatrix} s_0 & s_1 & s_2 \\ s_1 & s_2 & s_3 \\ s_2 & s_3 & s_4 \end{pmatrix},$$

show that $S = VV^T$ where $V$ is a suitable $3 \times 3$ Vandermonde matrix (see Exercise 4.4.9) and hence find $\det S$.

Generalise the result to $n \times n$ matrices.

**Exercise 4.6.10** This short question uses Exercise 3.6.2. Suppose that $C$ and $S$ are $n \times n$ matrices such that $CS = SC$ and $C^2 + S^2 = I$. If

$$Z = \begin{pmatrix} C & S \\ -S & C \end{pmatrix} \quad \text{and} \quad W = \begin{pmatrix} C & -S \\ S & C \end{pmatrix}$$

show that $\det Z = \det W$ and calculate $(\det Z)^2$.

**Exercise 4.6.11** Let $A$ and $B$ be $n \times n$ matrices. If

$$C = \begin{pmatrix} I & B \\ -A & 0 \end{pmatrix} \quad \text{and} \quad D = \begin{pmatrix} I & B \\ 0 & AB \end{pmatrix},$$

show that the $2n \times 2n$ matrix $C$ can be transformed into the $2n \times 2n$ matrix $D$ by row operations which you should specify. By considering the determinants of $C$ and $D$, obtain another proof that $\det AB = \det A \det B$.

**Exercise 4.6.12** If $A$ is an invertible $n \times n$ matrix, show that $\det(\mathrm{Adj}\ A) = (\det A)^{n-1}$ and that $\mathrm{Adj}(\mathrm{Adj}\ A) = (\det A)^{n-2} A$. What can you say if $A$ is not invertible?
[If you have problems with the last sentence, you should note that we take the matter up again in Exercise 5.7.10.]

**Exercise 4.6.13** Let $P(n)$ be the $n \times n$ matrix with $p_{ii}(n) = a$, $p_{ij}(n) = b$ for $i \neq j$. Find $\det P(n)$. In particular, find the determinant of the $n \times n$ matrix $A$ with diagonal entries $0$ and all other entries $1$. (In other words, $a_{ij} = 1$ if $i \neq j$, $a_{ij} = 0$ if $i = j$.)

**Exercise 4.6.14** Let $A(n)$ be the $n \times n$ matrix given by

$$A(n) = \begin{pmatrix} a & b & 0 & 0 & \dots & 0 & 0 & 0 \\ c & a & b & 0 & \dots & 0 & 0 & 0 \\ 0 & c & a & b & \dots & 0 & 0 & 0 \\ \vdots & \vdots & \vdots & \vdots & & \vdots & \vdots & \vdots \\ 0 & 0 & 0 & 0 & \dots & 0 & c & a \end{pmatrix}.$$

By considering row expansions, find a linear relation between $\det A(n)$, $\det A(n-1)$ and $\det A(n-2)$.

(i) Find $\det A(n)$ if $a = 1 + bc$. (Look carefully at any special cases that arise.)

(ii) If $a = 2 \cos \theta$ and $b = c = 1$, show that

$$\det A(n) = \frac{\sin(n+1)\theta}{\sin \theta}$$

if $\sin \theta \neq 0$. Find the values of $\det A(n)$ when $\sin \theta = 0$.

**Exercise 4.6.15** Let $A(n) = \big(a_{ij}(n)\big)$ be the $n \times n$ matrix given by

$$a_{ij}(n) = \begin{cases} i & \text{if } i \leq j, \\ j & \text{otherwise.} \end{cases}$$

Find $\det A(n)$.

**Exercise 4.6.16** Prove that

$$\det \begin{pmatrix} 1+x_1 & x_2 & x_3 & \cdots & x_n \\ x_1 & 1+x_2 & x_3 & \cdots & x_n \\ \vdots & \vdots & \vdots & \ddots & \vdots \\ x_1 & x_2 & x_3 & \cdots & 1+x_n \end{pmatrix} = 1 + x_1 + x_2 + \cdots + x_n.$$

Use this result to find $\det P(n)$ with $P(n)$ as in Exercise 4.6.13.

**Exercise 4.6.17** Show that if $A$ is an $n \times n$ matrix with all entries 1 or $-1$, then $\det A$ is a multiple of $2^{n-1}$.

If $B = (b_{ij})$ is the $n \times n$ matrix with $b_{ij} = 1$ if $1 \le i \le j \le n$ and $b_{ij} = -1$ otherwise, show that $\det B = 2^{n-1}$.

**Exercise 4.6.18** Let $M_2(\mathbb{R})$ be the set of all real $2 \times 2$ matrices. We write

$$I = \begin{pmatrix} 1 & 0 \\ 0 & 1 \end{pmatrix}, \quad J = \begin{pmatrix} 0 & 1 \\ 1 & 0 \end{pmatrix}, \quad K = \begin{pmatrix} 0 & 0 \\ 1 & 0 \end{pmatrix}, \quad L = \begin{pmatrix} 1 & 0 \\ 0 & 0 \end{pmatrix}.$$

Suppose that $D : M_2(\mathbb{R}) \to \mathbb{R}$ is a function such that $D(AB) = D(A)D(B)$ for all $A, B \in M_2(\mathbb{R})$ and $D(I) \ne D(J)$. Prove that $D$ has the following properties.

(i) $D(0) = 0$, $D(I) = 1$, $D(J) = -1$, $D(K) = D(L) = 0$.

(ii) If $B$ is obtained from $A$ by interchanging its rows or its columns, then $D(A) = -D(B)$.

(iii) If one row or one column of $A$ vanishes, then $D(A) = 0$.

(iv) $D(A) = 0$ if and only if $A$ is singular.

Give an example of such a $D$ which is not the determinant function.

**Exercise 4.6.19** Consider four distinct points $(x_j, y_j)$ in the plane. Let us write

$$A = \begin{pmatrix} x_1^2 + y_1^2 & x_1 & y_1 & 1 \\ x_2^2 + y_2^2 & x_2 & y_2 & 1 \\ x_3^2 + y_3^2 & x_3 & y_3 & 1 \\ x_4^2 + y_4^2 & x_4 & y_4 & 1 \end{pmatrix} \quad \text{and} \quad B = \begin{pmatrix} 1 & -2x_1 & -2y_1 & x_1^2 + y_1^2 \\ 1 & -2x_2 & -2y_2 & x_2^2 + y_2^2 \\ 1 & -2x_3 & -2y_3 & x_3^2 + y_3^2 \\ 1 & -2x_4 & -2y_4 & x_4^2 + y_4^2 \end{pmatrix}.$$

(i) Show that the equation

$$A(1, -2x_0, -2y_0, x_0^2 + y_0^2 - t)^T = (0, 0, 0, 0)^T$$

has a solution if and only if $\det A = 0$.

(ii) Hence show that the four equations $(x_j - x_0)^2 + (y_j - y_0)^2 = t$ $[1 \le j \le 4]$ are consistent if and only if $\det A = 0$.

(iii) Use the fact that there is exactly one circle or straight line through three distinct points in the plane to show that the four distinct points $(x_j, y_j)$ lie on the same circle or straight line if and only if $\det A = 0$.

(iv) By considering the equations

$$B(x_0^2 + y_0^2, x_0, y_0, 1)^T = (0, 0, 0, 0)^T,$$

show that the four distinct points $(x_j, y_j)$ lie on the same circle or straight line if and only if $\det B = 0$.

(v) By computing $AB^T$, show that the four distinct points $(x_j, y_j)$ lie on the same circle or straight line if and only if

$$\det \begin{pmatrix} 0 & d_{12} & d_{13} & d_{14} \\ d_{21} & 0 & d_{23} & d_{24} \\ d_{31} & d_{32} & 0 & d_{34} \\ d_{41} & d_{42} & d_{43} & 0 \end{pmatrix} = 0,$$

where $d_{ij} = (x_i - x_j)^2 + (y_i - y_j)^2$.

(vi) Write down the corresponding result in three dimensions and check in as much detail as you consider appropriate that the proof goes through in exactly the same manner. [The kind of determinant which appears in part (v) is known as a Cayley–Menger determinant.]

**Exercise 4.6.20** If $A = (a_{i,j})$ is an $n \times n$ matrix and $B = (b_{k,l})$ is an $m \times m$ matrix, we set $A \otimes B = C$ where $C = (c_{r,s})$ is the $nm \times nm$ matrix given by the rule

$$c_{m(i-1)+k, m(j-1)+l} = a_{i,j} b_{k,l}$$

for $1 \le i \le n, 1 \le j \le n, 1 \le k \le m, 1 \le l \le m$.

Show that $A \otimes B = (A \otimes I_m)(I_n \otimes B)$ where $I_p$ is the $p \times p$ identity matrix. Deduce that $\det(A \otimes B) = (\det A)^m (\det B)^n$.

# 5

# Abstract vector spaces

## 5.1 The space $\mathbb{C}^n$

So far, in this book, we have only considered vectors and matrices with real entries. However, as the reader may have already remarked, there is nothing in Chapter 1 on Gaussian elimination which will not work equally well when applied to $m$ linear equations with complex coefficients in $n$ complex unknowns. In particular, there is nothing to prevent us considering complex row and column vectors $(z_1, z_2, \ldots, z_n)$ and $(z_1, z_2, \ldots, z_n)^T$ with $z_j \in \mathbb{C}$ and complex $m \times n$ matrices $A = (a_{ij})$ with $a_{ij} \in \mathbb{C}$. (If we are going to make use of the complex number $i$, it may be better to use other suffices and talk about $A = (a_{rs})$.)

**Exercise 5.1.1** *Explain why we cannot replace $\mathbb{C}$ by $\mathbb{Z}$ in the discussion of the previous paragraph.*

However, this smooth process does not work for the geometry of Chapter 2. It is possible to develop complex geometry to mirror real geometry, but, whilst an ancient Greek mathematician would have no difficulty understanding the meaning of the theorems of Chapter 2 as they apply to the plane or three dimensional space, he or she[1] would find the complex analogues (when they exist) incomprehensible. Leaving aside the question of the meaning of theorems of complex geometry, the reader should note that the naive translation of the definition of inner product from real to complex vectors does not work very well. (We shall give an appropriate translation in Section 8.4.)

Continuing our survey, we see that Chapter 3 on the algebra of $n \times n$ matrices carries over word for word to the complex case. Something more interesting happens in Chapter 4. Here the *geometrical* arguments of the first two sections (and one or two similar observations elsewhere) are either meaningless or, at the least, carry no intuitive conviction when applied to the complex case but, once we define the determinant *algebraically*, the development proceeds identically in the real and complex cases.

**Exercise 5.1.2** *Check that, in the parts of the book where I claim this, there is, indeed, no difference between the real and complex cases.*

---

[1] Remember Hypatia.

## 5.2 Abstract vector spaces

We have already met several kinds of objects that 'behave like vectors'. What are the properties that we demand of a mathematical system in which the objects 'behave like vectors'? To deal with both real and complex vectors simultaneously we adopt the following convention.

**Convention 5.2.1** We shall write $\mathbb{F}$ to mean either $\mathbb{C}$ or $\mathbb{R}$.

Our definition of a vector space is obtained by recasting Lemma 1.4.2 as a *definition*.

**Definition 5.2.2** *A vector space $U$ over $\mathbb{F}$ is a set containing an element $\mathbf{0}$ equipped with addition and scalar multiplication with the properties given below.*[2]
*Suppose that $\mathbf{x}, \mathbf{y}, \mathbf{z} \in U$ and $\lambda, \mu \in \mathbb{F}$. Then the following relations hold.*
*(i)* $(\mathbf{x} + \mathbf{y}) + \mathbf{z} = \mathbf{x} + (\mathbf{y} + \mathbf{z})$.
*(ii)* $\mathbf{x} + \mathbf{y} = \mathbf{y} + \mathbf{x}$.
*(iii)* $\mathbf{x} + \mathbf{0} = \mathbf{x}$.
*(iv)* $\lambda(\mathbf{x} + \mathbf{y}) = \lambda\mathbf{x} + \lambda\mathbf{y}$.
*(v)* $(\lambda + \mu)\mathbf{x} = \lambda\mathbf{x} + \mu\mathbf{x}$.
*(vi)* $(\lambda\mu)\mathbf{x} = \lambda(\mu\mathbf{x})$.
*(vii)* $1\mathbf{x} = \mathbf{x}$ *and* $0\mathbf{x} = \mathbf{0}$.

It seems reasonable to call the elements of $U$ *abstract* vectors or just vectors.

As usual in these cases, there is a certain amount of fussing around establishing that the rules do everything we want. (Exercise 5.7.7 provides some more fussing for those who like that sort of thing.) We do this in the next lemma which the reader can more or less ignore.

**Lemma 5.2.3** *Let $U$ be the vector space of Definition 5.2.2.*
*(i) If $\mathbf{x} + \mathbf{0}' = \mathbf{x}$ for all $\mathbf{x} \in U$, then $\mathbf{0}' = \mathbf{0}$. (In other words, the zero vector is unique.)*
*(ii) $\mathbf{x} + (-1)\mathbf{x} = \mathbf{0}$ for all $\mathbf{x} \in U$.*
*(iii) If $\mathbf{x} + \mathbf{0}' = \mathbf{x}$ for some $\mathbf{x} \in U$, then $\mathbf{0}' = \mathbf{0}$.*

*Proof* (i) By the stated properties of $\mathbf{0}$ and $\mathbf{0}'$

$$\mathbf{0}' = \mathbf{0} + \mathbf{0}' = \mathbf{0}.$$

(ii) We have

$$\mathbf{x} + (-1)\mathbf{x} = 1\mathbf{x} + (-1)\mathbf{x} = \big(1 + (-1)\big)\mathbf{x} = 0\mathbf{x} = \mathbf{0}.$$

---

[2] If the reader feels this is insufficiently formal, she should replace the paragraph with the following mathematical boiler plate.
   *A vector space $(U, \mathbb{F}, +, .)$ is a set $U$ containing an element $\mathbf{0}$ together with maps $A : U^2 \to U$ and $M : \mathbb{F} \times U \to U$ such that, writing $\mathbf{x} + \mathbf{y} = A(\mathbf{x}, \mathbf{y})$ and $\lambda\mathbf{x} = M(\lambda, \mathbf{x})$ [$\mathbf{x}, \mathbf{y} \in U$, $\lambda \in \mathbb{F}$], the system has the properties given below.*

(iii) We have

$$\mathbf{0} = \mathbf{x} + (-1)\mathbf{x} = (\mathbf{x} + \mathbf{0}') + (-1)\mathbf{x}$$
$$= (\mathbf{0}' + \mathbf{x}) + (-1)\mathbf{x} = \mathbf{0}' + (\mathbf{x} + (-1)\mathbf{x})$$
$$= \mathbf{0}' + \mathbf{0} = \mathbf{0}'.$$

$\square$

We shall write

$$-\mathbf{x} = (-1)\mathbf{x}, \quad \mathbf{y} - \mathbf{x} = \mathbf{y} + (-\mathbf{x})$$

and so on. Since there is no ambiguity, we shall drop brackets and write

$$\mathbf{x} + \mathbf{y} + \mathbf{z} = \mathbf{x} + (\mathbf{y} + \mathbf{z}).$$

In abstract algebra, most systems give rise to subsystems, and abstract vector spaces are no exception.

**Definition 5.2.4** *If $V$ is a vector space over $\mathbb{F}$, we say that $U \subseteq V$ is a* subspace *of $V$ if the following three conditions hold.*
*(i) Whenever $\mathbf{x}, \mathbf{y} \in U$, we have $\mathbf{x} + \mathbf{y} \in U$.*
*(ii) Whenever $\mathbf{x} \in U$ and $\lambda \in \mathbb{F}$, we have $\lambda\mathbf{x} \in U$.*
*(iii) $\mathbf{0} \in U$.*

Condition (iii) is a convenient way of ensuring that $U$ is not empty.

**Lemma 5.2.5** *If $U$ is a subspace of a vector space $V$ over $\mathbb{F}$, then $U$ is itself a vector space over $\mathbb{F}$ (if we use the same operations).*

*Proof* Proof by inspection. $\square$

**Lemma 5.2.6** *Let $X$ be a non-empty set and $\mathbb{F}^X$ the collection of all functions $f : X \to \mathbb{F}$. If we define the* pointwise sum *$f + g$ of any $f, g \in \mathbb{F}^X$ by*

$$(f + g)(x) = f(x) + g(x)$$

*and pointwise* scalar multiple *$\lambda f$ of any $f \in \mathbb{F}^X$ and $\lambda \in \mathbb{F}$ by*

$$(\lambda f)(x) = \lambda f(x),$$

*then $\mathbb{F}^X$ is a vector space.*

*Proof* The checking is lengthy, but trivial. For example, if $\lambda, \mu \in \mathbb{F}$ and $f \in \mathbb{F}^X$, then

$$\big((\lambda + \mu)f\big)(x) = (\lambda + \mu)f(x) \qquad \text{(by definition)}$$
$$= \lambda f(x) + \mu f(x) \qquad \text{(by properties of } \mathbb{F})$$
$$= (\lambda f)(x) + (\mu f)(x) \qquad \text{(by definition)}$$
$$= (\lambda f + \mu f)(x) \qquad \text{(by definition)}$$

for all $x \in X$ and so, by the definition of equality of functions,

$$(\lambda + \mu)f = \lambda f + \mu f,$$

as required by condition (v) of Definition 5.2.2.                                    □

**Exercise 5.2.7** *Choose a couple of conditions from Definition 5.2.2 and verify that they hold for $\mathbb{F}^X$. (You should choose the conditions which you think are hardest to verify.)*

**Exercise 5.2.8** *If we take $X = \{1, 2, \ldots, n\}$ which, already known, vector space, do we obtain? (You should give an informal but reasonably convincing argument.)*

Lemmas 5.2.5 and 5.2.6 immediately reveal a large number of vector spaces.

**Example 5.2.9** *(i) The set $C([a, b])$ of all continuous functions $f : [a, b] \to \mathbb{R}$ is a vector space under pointwise addition and scalar multiplication.*

*(ii) The set $C^{\infty}(\mathbb{R})$ of all infinitely differentiable functions $f : \mathbb{R} \to \mathbb{R}$ is a vector space under pointwise addition and scalar multiplication.*

*(iii) The set $\mathcal{P}$ of all polynomials $P : \mathbb{R} \to \mathbb{R}$ is a vector space under pointwise addition and scalar multiplication.*

*(iv) The collection $c$ of all two sided sequences*

$$\mathbf{a} = (\ldots, a_{-2}, a_{-1}, a_0, a_1, a_2, \ldots)$$

*of complex numbers with $\mathbf{a} + \mathbf{b}$ the sequence with $j$th term $a_j + b_j$ and $\lambda\mathbf{a}$ the sequence with $j$th term $a_j + b_j$ is a vector space.*

*Proof* (i) Observe that $C([a, b])$ is a subspace of $\mathbb{R}^{[a,b]}$.

(ii) and (iii) Left to the reader.

(iv) Observe that $c = \mathbb{C}^{\mathbb{Z}}$.                                    □

In the next section we make use of the following improvement on Lemma 5.2.6.

**Lemma 5.2.10** *Let $X$ be a non-empty set and $V$ a vector space over $\mathbb{F}$. Write $L$ for the collection of all functions $f : X \to V$. If we define the* pointwise *sum $f + g$ of any $f, g \in L$ by*

$$(f + g)(x) = f(x) + g(x)$$

*and* pointwise *scalar multiple $\lambda f$ of any $f \in L$ and $\lambda \in \mathbb{F}$ by*

$$(\lambda f)(x) = \lambda f(x),$$

*then $L$ is a vector space.*

*Proof* Left as an exercise to the reader to do as much or as little of as she wishes.    □

In general, it is easiest to show that something is a vector space by showing that it is a subspace of some $\mathbb{F}^X$ or some $L$ of the type described in Lemma 5.2.10 and to show that something is not a vector space by showing that one of the conditions of Definition 5.2.2 fails.[3]

**Exercise 5.2.11** *Which of the following are vector spaces with pointwise addition and scalar multiplication? Give reasons.*

(i) *The set of all three times differentiable functions* $f : \mathbb{R} \to \mathbb{R}$.

(ii) *The set of continuous functions* $f : \mathbb{R} \to \mathbb{R}$ *with* $f(t) \geq 0$ *for all* $t \in \mathbb{R}$.

(iii) *The set of all polynomials* $P : \mathbb{R} \to \mathbb{R}$ *with* $P(1) = 1$.

(iv) *The set of all polynomials* $P : \mathbb{R} \to \mathbb{R}$ *with* $P'(1) = 0$.

(v) *The set of all polynomials* $P : \mathbb{R} \to \mathbb{R}$ *with* $\int_0^1 P(t)\, dt = 0$.

(vi) *The set of all continuous functions* $f : \mathbb{R} \to \mathbb{R}$ *with* $\int_{-1}^1 f(t)^3\, dt = 0$.

(vii) *The set of all polynomials of degree exactly* 3.

(viii) *The set of all polynomials of even degree.*

## 5.3 Linear maps

If the reader has followed any course in abstract algebra she will have met the notion of a morphism,[4] that is to say a mapping which preserves algebraic structure. A vector space morphism corresponds to the much older notion of a linear map.

**Definition 5.3.1** *Let* $U$ *and* $V$ *be vector spaces over* $\mathbb{F}$. *We say that a function* $T : U \to V$ *is a linear map if*

$$T(\lambda \mathbf{x} + \mu \mathbf{y}) = \lambda T\mathbf{x} + \mu T\mathbf{y}$$

*for all* $\mathbf{x}, \mathbf{y} \in U$ *and* $\lambda, \mu \in \mathbb{F}$.

**Exercise 5.3.2** *If* $T : U \to V$ *is a linear map, show that* $T(\mathbf{0}) = \mathbf{0}$.

Since the time of Newton, mathematicians have realised that the fact that a mapping is linear gives a very strong handle on that mapping. They have also discovered an ever wider collection of linear maps[5] in subjects ranging from celestial mechanics to quantum theory and from statistics to communication theory.

**Exercise 5.3.3** *Consider the vector space* $\mathcal{D}$ *of infinitely differentiable functions* $f : \mathbb{R} \to \mathbb{R}$. *Show that the following maps are linear.*

(i) $\delta : \mathcal{D} \to \mathbb{R}$ *given by* $\delta(f) = f(0)$.

(ii) $D : \mathcal{D} \to \mathcal{D}$ *given by* $(Df)(x) = f'(x)$.

(iii) $K : \mathcal{D} \to \mathcal{D}$ *where* $(Kf)(x) = (x^2 + 1)f(x)$.

(iv) $J : \mathcal{D} \to \mathcal{D}$ *where* $(Jf)(x) = \int_0^x f(t)\, dt$.

---

[3] Like all such statements, this is just an expression of opinion and carries no guarantee.

[4] If not, she should just ignore this sentence.

[5] Of course not everything is linear. To quote Swinnerton-Dyer, 'The great discovery of the 18th and 19th centuries was that nature is linear. The great discovery of the 20th century was that nature is not.' But linear problems remain a good place to start.

From this point of view, **we do not study vector spaces for their own sake but for the sake of the linear maps that they support.**

**Lemma 5.3.4** *If $U$ and $V$ are vector spaces over $\mathbb{F}$, then the set $\mathcal{L}(U, V)$ of linear maps $T : U \to V$ is a vector space under pointwise addition and scalar multiplication.*

*Proof* We know, by Lemma 5.2.10, that the collection $L$ of all functions $f : U \to V$ is a vector space under pointwise addition and scalar multiplication, so all we need do is show that $\mathcal{L}(U, V)$ is a subspace of $L$.

Observe first that the zero mapping $\underline{0} : U \to V$ given by $\underline{0}\mathbf{u} = \mathbf{0}$ is linear, since

$$\underline{0}(\lambda_1\mathbf{u}_1 + \lambda_2\mathbf{u}_2) = \mathbf{0} = \lambda_1\mathbf{0} + \lambda_2\mathbf{0} = \lambda_1\underline{0}\mathbf{u}_1 + \lambda_2\underline{0}\mathbf{u}_2.$$

Next observe that, if $S, T \in \mathcal{L}(U, V)$ and $\lambda \in \mathbb{F}$, then

$$
\begin{aligned}
(T &+ S)(\lambda_1\mathbf{u}_1 + \lambda_2\mathbf{u}_2) \\
&= T(\lambda_1\mathbf{u}_1 + \lambda_2\mathbf{u}_2) + S(\lambda_1\mathbf{u}_1 + \lambda_2\mathbf{u}_2) && \text{(by definition)} \\
&= (\lambda_1 T\mathbf{u}_1 + \lambda_2 T\mathbf{u}_2) + (\lambda_1 S\mathbf{u}_1 + \lambda_2 S\mathbf{u}_2) && \text{(since } S \text{ and } T \text{ are linear)} \\
&= \lambda_1(T\mathbf{u}_1 + S\mathbf{u}_1) + \lambda_2(T\mathbf{u}_2 + S\mathbf{u}_2) && \text{(collecting terms)} \\
&= \lambda_1(S + T)\mathbf{u}_1 + \lambda_2(S + T)\mathbf{u}_2
\end{aligned}
$$

and, by the same kind of argument (which the reader should write out),

$$(\lambda T)(\lambda_1\mathbf{u}_1 + \lambda_2\mathbf{u}_2) = \lambda_1(\lambda T)\mathbf{u}_1 + \lambda_2(\lambda T)\mathbf{u}_2$$

for all $\mathbf{u}_1, \mathbf{u}_2 \in U$ and $\lambda_1, \lambda_2 \in \mathbb{F}$. Thus $S + T$ and $\lambda T$ are linear and $\mathcal{L}(U, V)$ is, indeed, a subspace of $L$ as required. $\qquad\square$

From now on $\mathcal{L}(U, V)$ will denote the vector space of Lemma 5.3.4. In more advanced work the elements of $\mathcal{L}(U, V)$ are often called *linear operators* or just *operators*. We usually write the zero map as $0$ rather than $\underline{0}$.

Similar arguments establish the following simple, but basic, result.

**Lemma 5.3.5** *If $U$, $V$, and $W$ are vector spaces over $\mathbb{F}$ and $T \in \mathcal{L}(U, V)$, $S \in \mathcal{L}(V, W)$, then the composition $ST \in \mathcal{L}(U, W)$.*

*Proof* Left as an exercise for the reader. $\qquad\square$

**Lemma 5.3.6** *Suppose that $U$, $V$, and $W$ are vector spaces over $\mathbb{F}$, that $T, T_1, T_2 \in \mathcal{L}(U, V)$, $S, S_1, S_2 \in \mathcal{L}(V, W)$ and that $\lambda \in \mathbb{F}$. Then the following results hold.*
  *(i) $(S_1 + S_2)T = S_1 T + S_2 T$.*
  *(ii) $S(T_1 + T_2) = ST_1 + ST_2$.*
  *(iii) $(\lambda S)T = S(\lambda T) = \lambda(ST)$.*

*Proof* To prove part (i), observe that, by repeated use of our definitions,

$$
\begin{aligned}
\big((S_1 + S_2)T\big)\mathbf{u} &= (S_1 + S_2)(T\mathbf{u}) = S_1(T\mathbf{u}) + S_2(T\mathbf{u}) \\
&= (S_1 T)\mathbf{u} + (S_2 T)\mathbf{u} = (S_1 T + S_2 T)\mathbf{u}
\end{aligned}
$$

for all $\mathbf{u} \in U$ and so, by the definition of what it means for two functions to be equal, $(S_1 + S_2)T = S_1 T + S_2 T$.

The remaining parts are left as an exercise. $\qquad\square$

The proof of the next result requires a little more thought.

**Lemma 5.3.7** *If $U$ and $V$ are vector spaces over $\mathbb{F}$ and $T \in \mathcal{L}(U, V)$ is a bijection, then $T^{-1} \in \mathcal{L}(V, U)$.*

*Proof* The statement that $T$ is a bijection is equivalent to the statement that an inverse function $T^{-1} : V \to U$ exists. We have to show that $T^{-1}$ is linear. To see this, observe that

$$
\begin{aligned}
T\left(T^{-1}(\lambda \mathbf{x} + \mu \mathbf{y})\right) &= (\lambda \mathbf{x} + \mu \mathbf{y}) && \text{(definition)} \\
&= \lambda T T^{-1}\mathbf{x} + \mu T T^{-1}\mathbf{y} && \text{(definition)} \\
&= T\left(\lambda T^{-1}\mathbf{x} + \mu T^{-1}\mathbf{y}\right) && (T \text{ linear})
\end{aligned}
$$

so, applying $T^{-1}$ to both sides (or just noting that $T$ is bijective),

$$
T^{-1}(\lambda \mathbf{x} + \mu \mathbf{y}) = \lambda T^{-1}\mathbf{x} + \mu T^{-1}\mathbf{y}
$$

for all $\mathbf{x}, \mathbf{y} \in U$ and $\lambda, \mu \in \mathbb{F}$ as required. $\qquad\square$

Whenever we study abstract structures, we need the notion of isomorphism.

**Definition 5.3.8** *We say that two vector spaces $U$ and $V$ over $\mathbb{F}$ are* isomorphic *if there exists a linear map $T : U \to V$ which is bijective. We write $U \cong V$ to mean that $U$ and $V$ are isomorphic.*

Since anything that happens in $U$ is exactly mirrored (via the map $\mathbf{u} \mapsto T\mathbf{u}$) by what happens in $V$ and anything that happens in $V$ is exactly mirrored (via the map $\mathbf{v} \mapsto T^{-1}\mathbf{v}$) by what happens in $U$, isomorphic vector spaces may be considered as identical *from the point of view of abstract vector space theory.*

**Definition 5.3.9** *If $U$ is a vector space over $\mathbb{F}$ then the* identity map *$\iota : U \to U$ is defined by $\iota\mathbf{x} = \mathbf{x}$ for all $\mathbf{x} \in U$.*

(The Greek letter $\iota$ is written as $i$ without the dot and pronounced iota.)

**Exercise 5.3.10** *Let $U$ be a vector space over $\mathbb{F}$.*

(i) *Show that the identity map $\iota \in \mathcal{L}(U, U)$.*

(ii) *If $\alpha \in \mathcal{L}(U, U)$ is a bijection, show that $\alpha^{-1}\alpha = \alpha^{-1}\alpha = \iota$. If $\alpha, \beta \in \mathcal{L}(U, U)$ and $\alpha\beta = \beta\alpha = \iota$, show that $\alpha$ is a bijection and $\alpha^{-1} = \beta$.*

(iii) *Let $\mathcal{D}$, $D$ and $J$ be as in Exercise 5.3.3 and take $U = \mathcal{D}$. Explain why $DJ = \iota$, but show that $JD \neq \iota$. Show that $J$ is injective, but not surjective and $D$ is surjective, but not injective.*

[*This is possible because $\mathcal{D}$ is infinite dimensional. See Exercise 5.5.5.*]

(iv) *If $\alpha, \beta \in \mathcal{L}(U, U)$ are invertible, show that $\alpha\beta$ is invertible and $(\alpha\beta)^{-1} = \beta^{-1}\alpha^{-1}$.*

The following remark is often helpful.

**Exercise 5.3.11** *If $U$ and $V$ are vector spaces over $\mathbb{F}$, then a linear map $T : U \to V$ is injective if and only if $T\mathbf{u} = \mathbf{0}$ implies $\mathbf{u} = \mathbf{0}$.*

**Definition 5.3.12** *If $U$ is a vector space over $\mathbb{F}$, we write $GL(U)$ or $GL(U, \mathbb{F})$ for the collection of bijective linear maps $\alpha : U \to U$.*

The reader may well be familiar with the definition of group. If not the brief discussion that follows provides all that is necessary for the purposes of this book. (A few exercises such as Exercise 6.8.19 go a little further.)

**Definition 5.3.13** *A group is a set $G$ equipped with a multiplication $\times$ having the following properties.*
  *(i) If $a$, $b \in G$, then $a \times b \in G$.*
  *(ii) If $a$, $b$, $c \in G$, then $a \times (b \times c) = (a \times b) \times c$.*
  *(iii) There exists an element $e \in G$ such that $e \times a = a \times e = a$.*
  *(iv) If $a \in G$, then we can find an $a^{-1} \in G$ such that $a \times a^{-1} = a^{-1} \times a = e$.*

**Exercise 5.3.14** *If $U$ is a vector space over $\mathbb{F}$ show that $GL(U)$ with multiplication defined by composition satisfies the axioms for a group.*[6]
  *Show that $GL(\mathbb{R}^2)$ is not Abelian (that is to say, show that there exist $\alpha$, $\beta \in GL(\mathbb{R}^2)$ with $\alpha\beta \neq \beta\alpha$).*

We shall make very little use of the group concept, but we shall come across several *subgroups* of $GL(U)$ which turn out to be useful in physics.

**Definition 5.3.15** *A subset $H$ of $GL(U)$ is called a* matrix group *if it is a subgroup of $GL(U)$, that is to say, the following conditions hold.*
  *(i) $\iota \in H$.*
  *(ii) If $\alpha \in H$, then $\alpha^{-1} \in H$.*
  *(iii) If $\alpha$, $\beta \in H$, then $\alpha\beta \in H$.*

If $H$ and $K$ are subgroups of $GL(U)$ and $K \subseteq H$, we say that $K$ is a subgroup of $H$.

**Exercise 5.3.16** *Let $U$ be a vector space over $\mathbb{F}$. State, with reasons, which of the following statements are always true.*
  *(i) If $\mathbf{a} \in U$, then the set of $\alpha \in GL(U)$ with $\alpha\mathbf{a} = \mathbf{a}$ is a subgroup of $GL(U)$.*
  *(ii) If $\mathbf{a}$, $\mathbf{b} \in U$, then the set of $\alpha \in GL(U)$ with $\alpha\mathbf{a} = \mathbf{b}$ is a subgroup of $GL(U)$.*
  *(iii) The set of $\alpha \in GL(U)$ with $\alpha^2 = \iota$ is a subgroup of $GL(U)$.*
*[Exercises 6.8.40 and 6.8.41 give more practice in these ideas.]*

---

[6] $GL(U)$ is called the general linear group.

### 5.4 Dimension

Just as our work on determinants may have struck the reader as excessively computational, so this section on dimension may strike the reader as excessively abstract. It is possible to derive a useful theory of vector spaces without determinants and it is just about possible to deal with vectors in $\mathbb{R}^n$ without a clear definition of dimension, but in both cases we have to imitate the participants of an elegantly dressed party determined to ignore the presence of very large elephants.

It is certainly impossible to do advanced work without knowing the contents of this section, so I suggest that the reader bites the bullet and gets on with it.

**Definition 5.4.1** *Let $U$ be a vector space over $\mathbb{F}$.*

*(i) We say that the vectors $\mathbf{f}_1$, $\mathbf{f}_2$, $\ldots$, $\mathbf{f}_n \in U$ span $U$ if, given any $\mathbf{u} \in U$, we can find $\lambda_1$, $\lambda_2$, $\ldots$, $\lambda_n \in \mathbb{F}$ such that*

$$\sum_{j=1}^{n} \lambda_j \mathbf{f}_j = \mathbf{u}.$$

*(ii) We say that the vectors $\mathbf{e}_1$, $\mathbf{e}_2$, $\ldots$, $\mathbf{e}_n \in U$ are* linearly independent *if, whenever $\lambda_1$, $\lambda_2$, $\ldots$, $\lambda_n \in \mathbb{F}$ and*

$$\sum_{j=1}^{n} \lambda_j \mathbf{e}_j = \mathbf{0},$$

*it follows that $\lambda_1 = \lambda_2 = \ldots = \lambda_n = 0$.*

*(iii) If the vectors $\mathbf{e}_1$, $\mathbf{e}_2$, $\ldots$, $\mathbf{e}_n \in U$ span $U$ and are linearly independent, we say that they form a* basis *for $U$.*

The reason for our definition of a basis is given by the next lemma.

**Lemma 5.4.2** *Let $U$ be a vector space over $\mathbb{F}$. The vectors $\mathbf{e}_1$, $\mathbf{e}_2$, $\ldots$, $\mathbf{e}_n \in U$ form a basis for $U$ if and only if each $\mathbf{x} \in U$ can be written* uniquely *in the form*

$$\mathbf{x} = \sum_{j=1}^{n} x_j \mathbf{e}_j$$

*with $x_j \in \mathbb{F}$.*

*Proof* We first prove the if part. Since $\mathbf{e}_1$, $\mathbf{e}_2$, $\ldots$, $\mathbf{e}_n$ span $U$, we can certainly write

$$\mathbf{x} = \sum_{j=1}^{n} x_j \mathbf{e}_j$$

with $x_j \in \mathbb{F}$. We need to show that the expression is unique.

To this end, suppose that

$$\mathbf{x} = \sum_{j=1}^{n} x_j \mathbf{e}_j \quad \text{and} \quad \mathbf{x} = \sum_{j=1}^{n} x'_j \mathbf{e}_j$$

with $x_j$, $x_j' \in \mathbb{F}$. Then

$$\mathbf{0} = \mathbf{x} - \mathbf{x} = \sum_{j=1}^{n} x_j \mathbf{e}_j - \sum_{j=1}^{n} x_j' \mathbf{e}_j = \sum_{j=1}^{n} (x_j - x_j') \mathbf{e}_j,$$

so, since $\mathbf{e}_1$, $\mathbf{e}_2$, $\ldots$, $\mathbf{e}_n$ are linearly independent, $x_j - x_j' = 0$ and $x_j = x_j'$ for all $j$.

The only if part is even simpler. The vectors $\mathbf{e}_1$, $\mathbf{e}_2$, $\ldots$, $\mathbf{e}_n$ automatically span $U$. To prove independence, observe that, if

$$\sum_{j=1}^{n} \lambda_j \mathbf{e}_j = \mathbf{0},$$

then

$$\sum_{j=1}^{n} \lambda_j \mathbf{e}_j = \sum_{j=1}^{n} 0 \mathbf{e}_j$$

so, by uniqueness, $\lambda_j = 0$ for all $j$.                                        $\square$

The reader may think of the $x_j$ as the coordinates of $\mathbf{x}$ with respect to the basis $\mathbf{e}_1$, $\mathbf{e}_2$, $\ldots$, $\mathbf{e}_n$.

**Exercise 5.4.3** *Suppose that* $\mathbf{e}_1$, $\mathbf{e}_2$, $\ldots$, $\mathbf{e}_n$ *form a basis for a vector space* $U$ *over* $\mathbb{F}$. *If* $\mathbf{y} \in U$, *it follows by the previous result that there are unique* $a_j \in \mathbb{F}$ *such that*

$$\mathbf{y} = a_1 \mathbf{e}_1 + a_2 \mathbf{e}_2 + \cdots + a_n \mathbf{e}_n.$$

*Suppose that* $\mathbf{y} \neq \mathbf{0}$. *Show that* $\mathbf{e}_1 + \mathbf{y}$, $\mathbf{e}_2 + \mathbf{y}$, $\ldots$, $\mathbf{e}_n + \mathbf{y}$ *are linearly independent if and only if*

$$a_1 + a_2 + \cdots + a_n + 1 \neq 0.$$

**Lemma 5.4.4** *Let* $U$ *be a vector space over* $\mathbb{F}$ *which is non-trivial in the sense that it is not the space consisting of* $\mathbf{0}$ *alone.*

*(i) If the vectors* $\mathbf{e}_1$, $\mathbf{e}_2$, $\ldots$, $\mathbf{e}_n$ *span* $U$, *then either they form a basis or there exists one of these vectors such that, when it is removed, the remaining vectors will still span* $U$.

*(ii) If* $\mathbf{e}_1$ *spans* $U$, *then* $\mathbf{e}_1$ *forms a basis for* $U$.

*(iii) Any finite collection of vectors which span* $U$ *contains a basis for* $U$.

*(iv)* $U$ *has a basis if and only if it has a finite spanning set.*

*Proof* (i) If $\mathbf{e}_1$, $\mathbf{e}_2$, $\ldots$, $\mathbf{e}_n$ do not form a basis, then they are not linearly independent, so we can find $\lambda_1$, $\lambda_2$, $\ldots$, $\lambda_n \in \mathbb{F}$ not all zero such that

$$\sum_{j=1}^{n} \lambda_j \mathbf{e}_j = \mathbf{0}.$$

By renumbering, if necessary, we may suppose that $\lambda_n \neq 0$, so

$$\mathbf{e}_n = \sum_{j=1}^{n-1} \mu_j \mathbf{e}_j$$

with $\mu_j = -\lambda_j/\lambda_n$.

We claim that $\mathbf{e}_1, \mathbf{e}_2, \ldots, \mathbf{e}_{n-1}$ span $U$. For, if $\mathbf{u} \in U$, we know, by hypothesis, that there exist $\nu_j \in \mathbb{F}$ with

$$\mathbf{u} = \sum_{j=1}^{n} \nu_j \mathbf{e}_j$$

and so

$$\mathbf{u} = \nu_n \mathbf{e}_n + \sum_{j=1}^{n-1} \nu_j \mathbf{e}_j = \sum_{j=1}^{n-1} (\nu_j + \mu_j \nu_n) \mathbf{e}_j.$$

(ii) If $\lambda \mathbf{e}_1 = \mathbf{0}$ with $\lambda \neq 0$, then

$$\mathbf{e}_1 = \lambda^{-1}(\lambda \mathbf{e}_1) = \lambda^{-1}\mathbf{0} = \mathbf{0}$$

which is impossible.

(iii) Use (i) repeatedly.

(iv) We have proved the 'if' part. The only if part follows by the definition of a basis. $\square$

Consistency requires that we take the empty set $\emptyset$ to be the basis of the vector space $U = \{\mathbf{0}\}$.

Lemma 5.4.4 (i) is complemented by another simple result.

**Lemma 5.4.5** *Let $U$ be a vector space over $\mathbb{F}$. If the vectors $\mathbf{e}_1, \mathbf{e}_2, \ldots, \mathbf{e}_n$ are linearly independent, then either they form a basis or we may find a further $\mathbf{e}_{n+1} \in U$ so that $\mathbf{e}_1, \mathbf{e}_2, \ldots, \mathbf{e}_{n+1}$ are linearly independent.*

*Proof* If the vectors $\mathbf{e}_1, \mathbf{e}_2, \ldots, \mathbf{e}_n$ do not form a basis, then there exists an $\mathbf{e}_{n+1} \in U$ such that there do not exist $\mu_j \in \mathbb{F}$ with $\mathbf{e}_{n+1} = \sum_{j=1}^{n} \mu_j \mathbf{e}_j$.

If $\sum_{j=1}^{n+1} \lambda_j \mathbf{e}_j = \mathbf{0}$, then, if $\lambda_{n+1} \neq 0$, we have

$$\mathbf{e}_{n+1} = \sum_{j=1}^{n} (-\lambda_j/\lambda_{n+1}) \mathbf{e}_j$$

which is impossible by the previous paragraph. Thus $\lambda_{n+1} = 0$ and

$$\sum_{j=1}^{n} \lambda_j \mathbf{e}_j = \mathbf{0},$$

so, by hypothesis, $\lambda_1 = \lambda_2 = \ldots = \lambda_n = 0$. We have shown that $\mathbf{e}_1, \mathbf{e}_2, \ldots, \mathbf{e}_{n+1}$ are linearly independent. $\square$

We now come to a crucial result in vector space theory which states that, if a vector space has a finite basis, then all its bases contain the same number of elements. We use a kind of 'etherialised Gaussian elimination' called the Steinitz replacement lemma.

**Lemma 5.4.6 [The Steinitz replacement lemma]** *Let $U$ be a vector space over $\mathbb{F}$ and let $m \geq n - 1 \geq r \geq 0$. If $\mathbf{e}_1$, $\mathbf{e}_2$, ..., $\mathbf{e}_n$ are linearly independent and the vectors*

$$\mathbf{e}_1, \mathbf{e}_2, \ldots, \mathbf{e}_r, \mathbf{f}_{r+1}, \mathbf{f}_{r+2}, \ldots, \mathbf{f}_m$$

*span $U$ (if $r = 0$, this means that $\mathbf{f}_1$, $\mathbf{f}_2$, ..., $\mathbf{f}_m$ span $U$), then $m \geq r + 1$ and, after renumbering the $\mathbf{f}_j$ if necessary,*

$$\mathbf{e}_1, \mathbf{e}_2, \ldots, \mathbf{e}_r, \mathbf{e}_{r+1}, \mathbf{f}_{r+2}, \ldots, \mathbf{f}_m$$

*span the space.*[7]

*Proof* Since $\mathbf{e}_1$, $\mathbf{e}_2$, ..., $\mathbf{e}_r$, $\mathbf{f}_{r+1}$, $\mathbf{f}_{r+2}$, ..., $\mathbf{f}_m$ span $U$, it follows, in particular, that there exist $\lambda_j \in \mathbb{F}$ such that

$$\mathbf{e}_{r+1} = \sum_{j=1}^{r} \lambda_j \mathbf{e}_j + \sum_{j=r+1}^{n} \lambda_j \mathbf{f}_j.$$

If $\lambda_j = 0$ for $r + 1 \leq j \leq m$ (and so, in particular, if $m = r$), then

$$\sum_{j=1}^{r} \lambda_j \mathbf{e}_j + (-1)\mathbf{e}_{r+1} = \mathbf{0},$$

contradicting the hypothesis that the $\mathbf{e}_j$ are independent.

Thus $m \geq r + 1$ and, *after renumbering the $\mathbf{f}_j$ if necessary*, we may suppose that $\lambda_{r+1} \neq 0$ so, after algebraic rearrangement,

$$\mathbf{f}_{j+1} = \sum_{j=1}^{r+1} \mu_j \mathbf{e}_j + \sum_{j=r+2}^{n} \mu_j \mathbf{f}_j$$

where $\mu_j = -\lambda_j / \lambda_{r+1}$ for $j \neq r + 1$ and $\mu_{r+1} = 1/\lambda_{r+1}$.

We proceed to show that

$$\mathbf{e}_1, \mathbf{e}_2, \ldots, \mathbf{e}_r, \mathbf{e}_{r+1}, \mathbf{f}_{r+2}, \ldots, \mathbf{f}_m$$

span $U$. If $\mathbf{u} \in U$, then, by hypothesis, we can find $\nu_j \in \mathbb{F}$ such that

$$\mathbf{u} = \sum_{j=1}^{r} \nu_j \mathbf{e}_j + \sum_{j=r+1}^{n} \nu_j \mathbf{f}_j,$$

---

[7] In science fiction films, the inhabitants of some innocent village are replaced, one by one, by things from outer space. In the Steinitz replacement lemma, the original elements of the spanning collection are replaced, one by one, by elements from the linearly independent collection.

so

$$\mathbf{u} = \sum_{j=1}^{r} (\nu_j + \nu_{r+1}\mu_j)\mathbf{e}_j + \nu_{r+1}\mu_{r+1}\mathbf{e}_{r+1} + \sum_{j=r+2}^{n} (\nu_j + \nu_{r+1}\mu_j)\mathbf{f}_j$$

and we are done.                                                              □

The Steinitz replacement lemma has several important corollaries.

**Theorem 5.4.7** *Let U be a vector space over* $\mathbb{F}$.

(i) *If the vectors* $\mathbf{e}_1$, $\mathbf{e}_2$, ..., $\mathbf{e}_n$ *are linearly independent and the vectors* $\mathbf{f}_1$, $\mathbf{f}_2$, ..., $\mathbf{f}_m$ *span U, then* $m \geq n$.

(ii) *If U has a finite basis, then all bases have the same number of elements.*

(iii) *If U has a finite basis, then any subspace of U has a finite basis.*

(iv) *If U has a basis with n elements, then any collection of n vectors which span U will be a basis for U and any collection of n linearly independent vectors will be a basis for U.*

(v) *If U has a finite basis, then any collection of linearly independent vectors can be extended to form a basis.*

*Proof* (i) Suppose, if possible, that $m < n$. By applying the Steinitz replacement lemma $m$ times, we see that $\mathbf{e}_1$, $\mathbf{e}_2$, ..., $\mathbf{e}_m$ span $U$. Thus we can find $\lambda_1$, $\lambda_2$, ..., $\lambda_m$ such that

$$\mathbf{e}_n = \sum_{j=1}^{m} \lambda_j \mathbf{e}_j$$

and so

$$\sum_{j=1}^{m} \lambda_j \mathbf{e}_j + (-1)\mathbf{e}_n = \mathbf{0}$$

contradicting linear independence.

(ii) If the vectors $\mathbf{e}_1$, $\mathbf{e}_2$, ..., $\mathbf{e}_n$ and $\mathbf{f}_1$, $\mathbf{f}_2$, ..., $\mathbf{f}_m$ are both bases, then, since the $\mathbf{e}_j$ are linearly independent and the $\mathbf{f}_k$ span, part (i) tells us that $m \geq n$. Reversing the roles, we get $n \geq m$, so $n = m$.

(iii) Let $U$ have a basis with $n$ elements and let $V$ be a subspace. If we use Lemma 5.4.5 to find a sequence $\mathbf{e}_1$, $\mathbf{e}_2$, ... of linearly independent vectors in $V$, then, by part (i), the process must terminate after at most $n$ steps and Lemma 5.4.5 tells us that we will then have a basis for $V$.

(iv) By Lemma 5.4.4 (i), any collection of $n$ vectors which spanned $U$ and was not a basis would contain a spanning collection with $n - 1$ members, which is impossible by part (i). By Lemma 5.4.5 any collection of $n$ linearly independent vectors which was not a basis could be extended to a collection of $n + 1$ linearly independent vectors which is impossible by part (i).

(v) Suppose that $\mathbf{e}_1$, $\mathbf{e}_2$, ..., $\mathbf{e}_k$ are linearly independent and $\mathbf{f}_1$, $\mathbf{f}_2$, ..., $\mathbf{f}_n$ is a basis for $U$. Applying the Steinitz replacement lemma $k$ times we obtain *possibly after renumbering the* $\mathbf{f}_j$, a spanning set

$$\mathbf{e}_1, \ \mathbf{e}_2, \ \ldots, \ \mathbf{e}_k, \ \mathbf{f}_{k+1}, \ \mathbf{f}_{k+2}, \ \ldots, \ \mathbf{f}_n$$

which must be a basis by (iv). □

Theorem 5.4.7 enables us to introduce the notion of dimension.

**Definition 5.4.8** *If a vector space $U$ over $\mathbb{F}$ has no finite spanning set, then we say that $U$ is* infinite dimensional. *If $U$ is non-trivial and has a finite spanning set, we say that $U$ is* finite dimensional *with dimension the size of any basis of $U$. If $U = \{\mathbf{0}\}$, we say that $U$ has* zero dimension.

Theorem 5.4.7 immediately gives us the following result.[8]

**Theorem 5.4.9** *Any subspace of a finite dimensional vector space is itself finite dimensional. The dimension of a subspace cannot exceed the dimension of the original space.*

Here is a typical result on dimension. We write $\dim X$ for the dimension of $X$.

**Lemma 5.4.10** *Let $V$ and $W$ be subspaces of a vector space $U$ over $\mathbb{F}$.*
*(i) The sets $V \cap W$ and*

$$V + W = \{\mathbf{v} + \mathbf{w} : \mathbf{v} \in V, \ \mathbf{w} \in W\}$$

*are subspaces of $U$.*
*(ii) If $V$ and $W$ are finite dimensional, then so are $V \cap W$ and $V + W$. We have*

$$\dim(V \cap W) + \dim(V + W) = \dim V + \dim W.$$

*Proof* (i) Left as an easy, but recommended, exercise for the reader.

(ii) Since we are talking about dimension, we must introduce a basis. It is a good idea in such cases to 'find the basis of the smallest space available'. With this advice in mind, we observe that $V \cap W$ is a subspace of the finite dimensional space $V$ and so is finite dimensional with basis $\mathbf{e}_1, \mathbf{e}_2, \ldots, \mathbf{e}_k$ say. By Theorem 5.4.7 (iv), we can extend this to a basis $\mathbf{e}_1, \mathbf{e}_2, \ldots, \mathbf{e}_k, \mathbf{e}_{k+1}, \mathbf{e}_{k+2}, \ldots, \mathbf{e}_{k+l}$ of $V$ and to a basis $\mathbf{e}_1, \mathbf{e}_2, \ldots, \mathbf{e}_k, \mathbf{e}_{k+l+1}, \mathbf{e}_{k+l+2}, \ldots, \mathbf{e}_{k+l+m}$ of $W$. We claim that $\mathbf{e}_1, \mathbf{e}_2, \ldots, \mathbf{e}_k, \mathbf{e}_{k+1}, \mathbf{e}_{k+2}, \ldots, \mathbf{e}_{k+l}, \mathbf{e}_{k+l+1}, \mathbf{e}_{k+l+2}, \ldots, \mathbf{e}_{k+l+m}$ form a basis of $V + W$.

First we show that the purported basis spans $V + W$. If $\mathbf{u} \in V + W$, then we can find $\mathbf{v} \in V$ and $\mathbf{w} \in W$ such that $\mathbf{u} = \mathbf{v} + \mathbf{w}$. By the definition of a basis we can find $\lambda_1, \lambda_2, \ldots, \lambda_k, \lambda_{k+1}, \lambda_{k+2}, \ldots, \lambda_{k+l} \in \mathbb{F}$ and $\mu_1, \mu_2, \ldots, \mu_k, \mu_{k+l+1}, \mu_{k+l+2}, \ldots, \mu_{k+l+m} \in \mathbb{F}$ such that

$$\mathbf{v} = \sum_{j=1}^{k} \lambda_j \mathbf{e}_j + \sum_{j=k+1}^{k+l} \lambda_j \mathbf{e}_j \quad \text{and} \quad \mathbf{w} = \sum_{j=1}^{k} \mu_j \mathbf{e}_j + \sum_{j=k+l+1}^{k+l+m} \mu_j \mathbf{e}_j.$$

---

[8] The result may appear obvious, but I recall a lecture by a distinguished engineer in which he correctly predicted the failure of a US Navy project on the grounds that it required finding five linearly independent vectors in a space of dimension three.

It follows that

$$\mathbf{u} = \mathbf{v} + \mathbf{w} = \sum_{j=1}^{k}(\lambda_j + \mu_j)\mathbf{e}_j + \sum_{j=k+1}^{k+l}\lambda_j\mathbf{e}_j + \sum_{j=k+l+1}^{k+l+m}\mu_j\mathbf{e}_j$$

and we have a spanning set.

Next we show that the purported basis is linearly independent. To this end, suppose that $\lambda_j \in \mathbb{F}$ and

$$\sum_{j=1}^{k+l+m}\lambda_j\mathbf{e}_j = \mathbf{0}.$$

We then have

$$\sum_{j=k+l+1}^{k+l+m}\lambda_j\mathbf{e}_j = -\sum_{j=1}^{k+l}\lambda_j\mathbf{e}_j \in V$$

and, automatically,

$$\sum_{j=k+l+1}^{k+l+m}\lambda_j\mathbf{e}_j \in W.$$

Thus

$$\sum_{j=k+l+1}^{k+l+m}\lambda_j\mathbf{e}_j \in V \cap W$$

and so we can find $\mu_1, \mu_2, \ldots, \mu_k$ such that

$$\sum_{j=k+l+1}^{k+l+m}\lambda_j\mathbf{e}_j = \sum_{j=1}^{k}\mu_j\mathbf{e}_j.$$

We thus have

$$\sum_{j=1}^{k}\mu_j\mathbf{e}_j + \sum_{j=k+l+1}^{k+l+m}(-\lambda_j)\mathbf{e}_j = \mathbf{0}.$$

Since $\mathbf{e}_1, \mathbf{e}_2, \ldots, \mathbf{e}_k$ $\mathbf{e}_{k+l+1}, \mathbf{e}_{k+l+2}, \ldots, \mathbf{e}_{k+l+m}$ form a basis for $W$, they are independent and so

$$\mu_1 = \mu_2 = \ldots = \mu_k = -\lambda_{k+l+1} = -\lambda_{k+l+2} = \ldots = -\lambda_{k+l+m} = 0.$$

In particular, we have shown that

$$\lambda_{k+l+1} = \lambda_{k+l+2} = \ldots = \lambda_{k+l+m} = 0.$$

Exactly the same kind of argument shows that

$$\lambda_{k+1} = \lambda_{k+2} = \ldots = \lambda_{k+l} = 0.$$

We now know that

$$\sum_{j=1}^{k} \lambda_j \mathbf{e}_j = \mathbf{0}$$

so, since we are dealing with a basis for $V \cap W$, we have

$$\lambda_1 = \lambda_2 = \ldots = \lambda_k = 0.$$

We have proved that $\lambda_j = 0$ for $1 \le j \le k + l + m$ so we have linear independence and our purported basis is, indeed, a basis.

The dimensions of the various spaces involved can now be read off as follows.

$$\dim(V \cap W) = k, \quad \dim V = k + l, \quad \dim W = k + m, \quad \dim(V + W) = k + l + m.$$

Thus

$$\dim(V \cap W) + \dim(V + W) = k + (k + l + m) = 2k + l + m$$
$$= (k + l) + (k + m) = \dim V + \dim W$$

and we are done.                                                                    □

**Exercise 5.4.11**  *We work in $\mathbb{R}^4$. Let*

$$U = \{(x, y, z, w) : x + y - 2z + w = 0, \ -x + y + z - 3w = 0\},$$
$$V = \{(x, y, z, w) : x - 2y + z + 2w = 0, \ y + z - 3w = 0\}.$$

*Explain why $U$ and $V$ are subspaces of $\mathbb{R}^4$. Find a basis of $U \cap V$, extend it to a basis of $U$ and extend the resulting basis to a basis of $U + V$.*

**Exercise 5.4.12**  *(i) Let $V$ and $W$ be subspaces of a finite dimensional vector space $U$ over $\mathbb{F}$. By using Lemma 5.4.10, or otherwise, show that*

$$\min\{\dim U, \dim V + \dim W\} \ge \dim(V + W) \ge \max\{\dim V, \dim W\}.$$

*(ii) Suppose that $n$, $r$, $s$ and $t$ are positive integers with*

$$\min\{n, r + s\} \ge t \ge \max\{r, s\}.$$

*Show that any vector space $U$ over $\mathbb{F}$ of dimension $n$ contains subspaces $V$ and $W$ such that*

$$\dim V = r, \quad \dim W = s, \quad \dim(V + W) = t.$$

The next exercise should always be kept in mind when talking about 'standard bases'.

**Exercise 5.4.13**  *Show that*

$$E = \{\mathbf{x} \in \mathbb{R}^3 : x_1 + x_2 + x_3 = 0\}$$

*is a subspace of* $\mathbb{R}^3$. *Write down a basis for E and show that it is a basis. Do you think that everybody who does this exercise will choose the same basis? What does this tell us about the notion of a 'standard basis'?*

*[Exercise 5.5.19 gives other examples where there are no 'natural' bases.]*

The final exercise of this section introduces an idea which reappears throughout mathematics.

**Exercise 5.4.14** (*i*) *Suppose that U is a vector space over* $\mathbb{F}$. *Suppose that* $\mathcal{V}$ *is a non-empty collection of subspaces of U. Show that, if we write*

$$\bigcap_{V \in \mathcal{V}} V = \{\mathbf{e} : \mathbf{e} \in V \text{ for all } V \in \mathcal{V}\},$$

*then* $\bigcap_{V \in \mathcal{V}} V$ *is a subspace of U.*

(*ii*) *Suppose that U is a vector space over* $\mathbb{F}$ *and E is a non-empty subset of U. If we write* $\mathcal{V}$ *for the collection of all subspaces V of U with* $V \supseteq E$, *show that* $W = \bigcap_{V \in \mathcal{V}} V$ *is a subspace of U such that (a)* $E \subseteq W$ *and (b) whenever* $W'$ *is a subspace of U containing E we have* $W \subseteq W'$. (*In other words, W is the* smallest *subspace of U containing E.*)

(*iii*) *Continuing with the notation of (ii), show that if* $E = \{\mathbf{e}_j : 1 \le j \le n\}$, *then*

$$W = \left\{ \sum_{j=1}^{n} \lambda_j \mathbf{e}_j : \lambda_j \in \mathbb{F} \text{ for } 1 \le j \le n \right\}.$$

We call the set $W$ described in Exercise 5.4.14 the subspace of $U$ spanned by $E$ and write $W = \operatorname{span} E$.

## 5.5 Image and kernel

In this section we revisit the system of simultaneous linear equations

$$A\mathbf{x} = \mathbf{b}$$

using the idea of dimension.

**Definition 5.5.1** *If U and V are vector spaces over* $\mathbb{F}$ *and* $\alpha : U \to V$ *is linear, then the set*

$$\alpha(U) = \{\alpha\mathbf{u} : \mathbf{u} \in U\}$$

*is called the* image (*or* image space) *of* $\alpha$ *and the set*

$$\alpha^{-1}(\mathbf{0}) = \{\mathbf{u} \in U : \alpha\mathbf{u} = \mathbf{0}\}$$

*is the* kernel (*or* null-space) *of* $\alpha$. *We write* $\ker \alpha = \alpha^{-1}(\mathbf{0})$ *and* $\operatorname{im} \alpha = \alpha(U)$.

**Lemma 5.5.2** *Let U and V be vector spaces over* $\mathbb{F}$ *and let* $\alpha : U \to V$ *be a linear map.*
(*i*) $\alpha(U)$ *is a subspace of V.*
(*ii*) $\alpha^{-1}(\mathbf{0})$ *is a subspace of U.*

*Proof* It is left as a very strongly recommended exercise for the reader to check that the conditions of Definition 5.2.4 apply. □

**Exercise 5.5.3** *Let U and V be vector spaces over* $\mathbb{F}$ *and let* $\alpha : U \to V$ *be a linear map. If* $\mathbf{v} \in V$, *but* $\mathbf{v} \neq \mathbf{0}$, *is it* (a) *always true,* (b) *sometimes true and sometimes false or* (c) *always false that*

$$\alpha^{-1}(\mathbf{v}) = \{\mathbf{u} \in U \, : \, \alpha\mathbf{u} = \mathbf{v}\}$$

*is a subspace of U? Give reasons.*

The proof of the next theorem requires care, but the result is very useful.

**Theorem 5.5.4 [The rank-nullity theorem]** *Let U and V be vector spaces over* $\mathbb{F}$ *and let* $\alpha : U \to V$ *be a linear map. If U is finite dimensional, then* $\alpha(U)$ *and* $\alpha^{-1}(\mathbf{0})$ *are finite dimensional and*

$$\dim \alpha(U) + \dim \alpha^{-1}(\mathbf{0}) = \dim U.$$

Here dim $X$ means the dimension of $X$. We call $\dim \alpha(U)$ the *rank* of $\alpha$ and $\dim \alpha^{-1}(\mathbf{0})$ the *nullity* of $\alpha$. We do not need $V$ to be finite dimensional, but the reader will miss nothing if she only considers the case when $V$ is finite dimensional.

*Proof* I repeat the opening sentences of the proof of Lemma 5.4.10. Since we are talking about dimension, we must introduce a basis. It is a good idea in such cases to 'find the basis of the smallest space available'. With this advice in mind, we choose a basis $\mathbf{e}_1, \mathbf{e}_2, \ldots, \mathbf{e}_k$ for $\alpha^{-1}(\mathbf{0})$. (Since $\alpha^{-1}(\mathbf{0})$ is a subspace of a finite dimensional space, Theorem 5.4.9 tells us that it must itself be finite dimensional.) By Theorem 5.4.7 (iv), we can extend this to basis $\mathbf{e}_1, \mathbf{e}_2, \ldots, \mathbf{e}_n$ of $U$.

We claim that $\alpha\mathbf{e}_{k+1}, \alpha\mathbf{e}_{k+2}, \ldots, \alpha\mathbf{e}_n$ form a basis for $\alpha(U)$. The proof splits into two parts. First observe that, if $\mathbf{u} \in U$, then, by the definition of a basis, we can find $\lambda_1, \lambda_2, \ldots, \lambda_n \in \mathbb{F}$ such that

$$\mathbf{u} = \sum_{j=1}^{n} \lambda_j \mathbf{e}_j,$$

and so, using linearity and the fact that $\alpha\mathbf{e}_j = \mathbf{0}$ for $1 \leq j \leq k$,

$$\alpha\left(\sum_{j=1}^{n} \lambda_j \mathbf{e}_j\right) = \sum_{j=1}^{n} \lambda_j \alpha\mathbf{e}_j = \sum_{j=k+1}^{n} \lambda_j \alpha\mathbf{e}_j.$$

Thus $\alpha\mathbf{e}_{k+1}, \alpha\mathbf{e}_{k+2}, \ldots, \alpha\mathbf{e}_n$ span $\alpha(U)$.

To prove linear independence, we suppose that $\lambda_{k+1}, \lambda_{k+2}, \ldots, \lambda_n \in \mathbb{F}$ are such that

$$\sum_{j=k+1}^{n} \lambda_j \alpha\mathbf{e}_j = \mathbf{0}.$$

By linearity,

$$\alpha\left(\sum_{j=k+1}^{n} \lambda_j \mathbf{e}_j\right) = \mathbf{0},$$

and so

$$\sum_{j=k+1}^{n} \lambda_j \mathbf{e}_j \in \alpha^{-1}(\mathbf{0}).$$

Since $\mathbf{e}_1, \mathbf{e}_2, \ldots, \mathbf{e}_k$ form a basis for $\alpha^{-1}(\mathbf{0})$, we can find $\mu_1, \mu_2, \ldots, \mu_k \in \mathbb{F}$ such that

$$\sum_{j=k+1}^{n} \lambda_j \mathbf{e}_j = \sum_{j=1}^{k} \mu_j \mathbf{e}_j.$$

Setting $\lambda_j = -\mu_j$ for $1 \le j \le k$, we obtain

$$\sum_{j=1}^{n} \lambda_j \mathbf{e}_j = \mathbf{0}.$$

Since the $\mathbf{e}_j$ are independent, $\lambda_j = 0$ for all $1 \le j \le n$ and so, in particular, for all $k + 1 \le j \le n$. We have shown that $\alpha\mathbf{e}_{k+1}, \alpha\mathbf{e}_{k+2}, \ldots, \alpha\mathbf{e}_n$ are linearly independent and so, since we have already shown that they span $\alpha(U)$, it follows that they form a basis.

By the definition of dimension, we have

$$\dim U = n, \quad \dim \alpha^{-1}(\mathbf{0}) = k, \quad \dim \alpha(U) = n - k,$$

so

$$\dim \alpha(U) + \dim \alpha^{-1}(\mathbf{0}) = (n - k) + k = n = \dim U$$

and we are done. □

**Exercise 5.5.5** *Let $U$ be a finite dimensional space and let $\alpha \in \mathcal{L}(U, U)$. Show that the following statements are equivalent.*

    *(i) $\alpha$ is injective.*
    *(ii) $\alpha$ is surjective.*
    *(iii) $\alpha$ is bijective.*
    *(iv) $\alpha$ is invertible.*

*[Compare Exercise 5.3.10 (iii). You may also wish to consider how obvious our abstract treatment makes the result of Lemma 3.4.13.]*

**Exercise 5.5.6** *Let $U = V = \mathbb{R}^3$ and let $\alpha : U \to V$ be the linear map with*

$$\alpha\begin{pmatrix} 1 \\ 0 \\ 0 \end{pmatrix} = \begin{pmatrix} a \\ b \\ b \end{pmatrix}, \alpha\begin{pmatrix} 0 \\ 1 \\ 0 \end{pmatrix} = \begin{pmatrix} b \\ a \\ b \end{pmatrix} \text{ and } \alpha\begin{pmatrix} 0 \\ 0 \\ 1 \end{pmatrix} = \begin{pmatrix} b \\ b \\ a \end{pmatrix}.$$

*Find the values of a and b, if any, for which α has rank 3, 2, 1 and 0. In each case find a basis for $\alpha^{-1}(\mathbf{0})$, extend it to a basis for U and identify a basis for $\alpha(U)$.*

We can extract a little extra information from the proof of Theorem 5.5.4 which will come in handy later.

**Lemma 5.5.7** *Let U be a vector space of dimension n and V a vector space of dimension m over $\mathbb{F}$ and let $\alpha : U \to V$ be a linear map. We can find a q with $0 \le q \le \min\{n, m\}$, a basis $\mathbf{u}_1, \mathbf{u}_2, \ldots, \mathbf{u}_n$ for U and a basis $\mathbf{v}_1, \mathbf{v}_2, \ldots, \mathbf{v}_m$ for V such that*

$$\alpha \mathbf{u}_j = \begin{cases} \mathbf{v}_j & \text{for } 1 \le j \le q, \\ \mathbf{0} & \text{otherwise.} \end{cases}$$

*Proof* We use the notation of the proof of Theorem 5.5.4. Set $q = n - k$ and $\mathbf{u}_j = \mathbf{e}_{n+1-j}$. If we take

$$\mathbf{v}_j = \alpha \mathbf{u}_j \quad \text{for } 1 \le j \le q,$$

we know that $\mathbf{v}_1, \mathbf{v}_2, \ldots, \mathbf{v}_q$ are linearly independent and so can be extended to a basis $\mathbf{v}_1$, $\mathbf{v}_2, \ldots, \mathbf{v}_m$ for V. We have achieved the required result. $\qquad \square$

We use Theorem 5.5.4 in conjunction with a couple of very simple results.

**Lemma 5.5.8** *Let U and V be vector spaces over $\mathbb{F}$ and let $\alpha : U \to V$ be a linear map. Consider the equation*

$$\alpha \mathbf{u} = \mathbf{v}, \qquad\qquad\qquad \bigstar$$

*where $\mathbf{v}$ is a fixed element of V and $\mathbf{u} \in U$ is to be found.*
  *(i) $\bigstar$ has a solution if and only if $\mathbf{v} \in \alpha(U)$.*
  *(ii) If $\mathbf{u} = \mathbf{u}_0$ is a solution of $\bigstar$, then the solutions of $\bigstar$ are precisely those $\mathbf{u}$ with*

$$\mathbf{u} \in \mathbf{u}_0 + \alpha^{-1}(\mathbf{0}) = \{\mathbf{u}_0 + \mathbf{w} : \mathbf{w} \in \alpha^{-1}(\mathbf{0})\}.$$

*Proof* (i) This is a tautology.
  (ii) Left as an exercise. $\qquad\qquad\qquad\qquad\qquad\qquad\qquad\qquad\qquad\qquad \square$

Combining Theorem 5.5.4 and Lemma 5.5.8, gives the following result.

**Lemma 5.5.9** *Let U and V be vector spaces over $\mathbb{F}$ and let $\alpha : U \to V$ be a linear map. Suppose further that U is finite dimensional with dimension n. Then the set of solutions of*

$$\alpha \mathbf{u} = \mathbf{0}$$

*forms a vector subspace of U of dimension k, say.*
  *The set of $\mathbf{v} \in V$ such that*

$$\alpha \mathbf{u} = \mathbf{v} \qquad\qquad\qquad\qquad \bigstar$$

*has a solution, is a finite dimensional subspace of V with dimension $n - k$.*

*If* $\mathbf{u} = \mathbf{u}_0$ *is a solution of* ★, *then the set*

$$\{\mathbf{u} - \mathbf{u}_0 \ : \ \mathbf{u} \ \text{is a solution of} \ ★\}$$

*is a subspace of U with dimension k.*

*Proof* Immediate. □

It is easy to apply these results to a system of $m$ linear equations in $n$ unknowns. We introduce the notion of the column rank of a matrix.

**Definition 5.5.10** *Let A be an* $m \times n$ *matrix over* $\mathbb{F}$ *with columns the column vectors* $\mathbf{a}_1$, $\mathbf{a}_2, \ldots, \mathbf{a}_n$. *The* column rank[9] *of A is the dimension of the subspace of* $\mathbb{F}^m$ *spanned by* $\mathbf{a}_1$, $\mathbf{a}_2, \ldots, \mathbf{a}_n$.

**Theorem 5.5.11** *Let A be an* $m \times n$ *matrix over* $\mathbb{F}$ *with columns the column vectors* $\mathbf{a}_1$, $\mathbf{a}_2$, $\ldots, \mathbf{a}_n$ *and let* $\mathbf{b}$ *be a column vector with m entries. Consider the system of linear equations*

$$A\mathbf{x} = \mathbf{b}. \qquad\qquad ★$$

*(i)* ★ *has a solution if and only if*

$$\mathbf{b} \in \text{span}\{\mathbf{a}_1, \mathbf{a}_2, \ldots, \mathbf{a}_n\}$$

*(that is to say,* $\mathbf{b}$ *lies in the subspace spanned by* $\mathbf{a}_1, \mathbf{a}_2, \ldots, \mathbf{a}_n$*).*

*(ii)* ★ *has a solution if and only if*

$$\text{rank } A = \text{rank}(A|\mathbf{b}),$$

*where* $(A|\mathbf{b})$ *is the* $m \times (n+1)$ *matrix formed from A by adjoining* $\mathbf{b}$ *as the* $n + 1$*st column.*[10]

*(iii) If we write*

$$N = \{\mathbf{x} \in \mathbb{F}^n \ : \ A\mathbf{x} = \mathbf{0}\},$$

*then N is a subspace of* $\mathbb{F}^n$ *of dimension* $n - \text{rank } A$. *If* $\mathbf{x}_0$ *is a solution of* ★, *then the solutions of* ★ *are precisely the* $\mathbf{x} = \mathbf{x}_0 + \mathbf{u}$ *where* $\mathbf{u} \in N$.

*Proof* (i) We give the proof at greater length than is really necessary. Let $\alpha : \mathbb{F}^n \to \mathbb{F}^n$ be the linear map defined by

$$\alpha(\mathbf{x}) = A\mathbf{x}.$$

Then

$$\alpha(\mathbb{F}^n) = \{\alpha(\mathbf{x}) \ : \ \mathbf{x} \in \mathbb{F}^n\} = \{A\mathbf{x} \ : \ \mathbf{x} \in \mathbb{F}^n\}$$

$$= \left\{ \sum_{j=1}^{n} x_j \mathbf{a}_j \ : \ x_j \in \mathbb{F} \text{ for } 1 \leq j \leq n \right\}$$

$$= \text{span}\{\mathbf{a}_1, \mathbf{a}_2, \ldots, \mathbf{a}_n\},$$

---

[9] Often called simply the *rank*. Exercise 5.5.13 explains why we can drop the reference to columns.
[10] $(A|\mathbf{b})$ is sometimes called the *augmented matrix*.

and so

$$\bigstar \text{ has a solution} \Leftrightarrow \text{there is an } \mathbf{x} \text{ such that } \alpha(\mathbf{x}) = \mathbf{b}$$
$$\Leftrightarrow \mathbf{b} \in \alpha(\mathbb{F}^n)$$
$$\Leftrightarrow \mathbf{b} \in \text{span}\{\mathbf{a}_1, \mathbf{a}_2, \ldots, \mathbf{a}_n\},$$

as stated.

(ii) Observe that, using part (i),

$$\bigstar \text{ has a solution} \Leftrightarrow \mathbf{b} \in \text{span}\{\mathbf{a}_1, \mathbf{a}_2, \ldots, \mathbf{a}_n\}$$
$$\Leftrightarrow \text{span}\{\mathbf{a}_1, \mathbf{a}_2, \ldots, \mathbf{a}_m, \mathbf{b}\} = \text{span}\{\mathbf{a}_1, \mathbf{a}_2, \ldots, \mathbf{a}_n\}$$
$$\Leftrightarrow \dim \text{span}\{\mathbf{a}_1, \mathbf{a}_2, \ldots, \mathbf{a}_m, \mathbf{b}\} = \dim \text{span}\{\mathbf{a}_1, \mathbf{a}_2, \ldots, \mathbf{a}_n\}$$
$$\Leftrightarrow \text{rank}(A|\mathbf{b}) = \text{rank } A.$$

(iii) Observe that $N = \alpha^{-1}(\mathbf{0})$, so, by the rank-nullity theorem (Theorem 5.5.4), $N$ is a subspace of $\mathbb{F}^m$ with

$$\dim N = \dim \alpha^{-1}(\mathbf{0}) = \dim \mathbb{F}^n - \dim \alpha(\mathbb{F}^n) = n - \text{rank } A.$$

The rest of part (iii) can be checked directly or obtained from several of our earlier results.                                                                                      □

**Exercise 5.5.12** *Compare Theorem 5.5.11 with the results obtained in Chapter 1.*

Mathematicians of an earlier generation might complain that Theorem 5.5.11 just restates 'what every gentleman knows' in fine language. There is an element of truth in this, but it is instructive to look at how the same topic was treated in a textbook, at much the same level as this one, a century ago.

Chrystal's *Algebra* [11] is an excellent text by an excellent mathematician. Here is how he states a result corresponding to part of Theorem 5.5.11.

If the reader now reconsider the course of reasoning through which we have led him in the cases of equations of the first degree in one, two and three variables respectively, he will see that the spirit of that reasoning is general; and that by pursuing the same course step by step we should arrive at the following general conclusion:–

A system of $n - r$ equations of the first degree in $n$ variables has in general a solution involving $r$ arbitrary constants; in other words has an $r$-fold infinity of of different solutions.

(Chrystal *Algebra*, Volume 1, Chapter XVI, section 14, slightly modified [11])

From our point of view, the problem with Chrystal's formulation lies with the words *in general*. Chrystal was perfectly aware that examples like

$$x + y + z + w = 1$$
$$x + 2y + 3z + 4w = 1$$
$$2x + 3y + 4z + 5w = 1$$

or

$$x + y = 1$$
$$x + y + z + w = 1$$
$$2x + 2y + z + w = 2$$

are exceptions to his 'general rule', but would have considered them easily spotted pathologies.

For Chrystal and his fellow mathematicians, systems of linear equations were peripheral to mathematics. They formed a good introduction to algebra for undergraduates, but a professional mathematician was unlikely to meet them except in very simple cases. Since then, the invention of electronic computers has moved the solution of very large systems of linear equations (where 'pathologies' are not easy to spot) to the centre stage. At the same time, mathematicians have discovered that many problems in analysis may be treated by methods analogous to those used for systems of linear equations. The 'nit picking precision' and 'unnecessary abstraction' of results like Theorem 5.5.11 are the result of real needs and not mere fashion.

**Exercise 5.5.13** (*i*) *Write down the appropriate definition of the* row rank *of a matrix.*

(*ii*) *Show that the row rank and column rank of a matrix are unaltered by elementary row and column operations.*

[*There are many ways of doing this. You may find it helpful to observe that if* $\mathbf{a}_1, \mathbf{a}_2, \ldots, \mathbf{a}_k$ *are row vectors in* $\mathbb{F}^m$ *and B is a non-singular* $m \times m$ *matrix then* $\mathbf{a}_1 B, \mathbf{a}_2 B, \ldots, \mathbf{a}_k B$ *are linearly independent if and only if* $\mathbf{a}_1, \mathbf{a}_2, \ldots, \mathbf{a}_k$ *are.*]

(*iii*) *Use Theorem 1.3.6 (or a similar result) to deduce that the row rank of a matrix equals its column rank. For this reason we can refer to the* rank *of a matrix rather than its* column rank.

[*We give a less computational proof in Exercise 11.4.18.*]

**Exercise 5.5.14** *Suppose that a and b are real. Find the rank r of the matrix*

$$\begin{pmatrix} a & 0 & b & 0 \\ 0 & a & 0 & b \\ a & 0 & 0 & b \\ 0 & b & 0 & a \end{pmatrix}$$

*and, when* $r \neq 0$, *exhibit a non-singular* $r \times r$ *submatrix.*

Here is another application of the rank-nullity theorem.

**Lemma 5.5.15** *Let U, V and W be finite dimensional vector spaces over* $\mathbb{F}$ *and let* $\alpha : V \to W$ *and* $\beta : U \to V$ *be linear maps. Then*

$$\min\{\text{rank}\,\alpha, \text{rank}\,\beta\} \geq \text{rank}\,\alpha\beta \geq \text{rank}\,\alpha + \text{rank}\,\beta - \dim U.$$

*Proof* Let

$$Z = \beta U = \{\beta \mathbf{u} : \mathbf{u} \in U\}$$

and define $\alpha|_Z : Z \to W$, the restriction of $\alpha$ to $Z$, in the usual way, by setting $\alpha|_Z \mathbf{z} = \alpha \mathbf{z}$ for all $\mathbf{z} \in Z$.

Since $Z \subseteq V$,

$$\text{rank}\, \alpha\beta = \text{rank}\, \alpha|_Z = \dim \alpha(Z) \leq \dim \alpha(V) = \text{rank}\, \alpha.$$

By the rank-nullity theorem

$$\text{rank}\, \beta = \dim Z = \text{rank}\, \alpha|_Z + \text{nullity}\, \alpha|_Z \geq \text{rank}\, \alpha|_Z = \text{rank}\, \alpha\beta.$$

Applying the rank-nullity theorem twice,

$$\begin{aligned}
\text{rank}\, \alpha\beta = \text{rank}\, \alpha|_Z &= \dim Z - \text{nullity}\, \alpha|_Z \\
&= \text{rank}\, \beta - \text{nullity}\, \alpha|_Z = \dim U - \text{nullity}\, \beta - \text{nullity}\, \alpha|_Z.
\end{aligned}$$ ★

Since $Z \subseteq V$,

$$\{\mathbf{z} \in Z : \alpha \mathbf{z} = \mathbf{0}\} \subseteq \{\mathbf{v} \in V : \alpha \mathbf{v} = \mathbf{0}\}$$

we have nullity $\alpha \geq$ nullity $\alpha|_Z$ and, using ★,

$$\text{rank}\, \alpha\beta \geq \dim U - \text{nullity}\, \beta - \text{nullity}\, \alpha = \text{rank}\, \alpha + \text{rank}\, \beta - \dim U,$$

as required.                                                                 □

**Exercise 5.5.16** *By considering the product of appropriate $n \times n$ diagonal matrices $A$ and $B$, or otherwise, show that, given any integers $n, r, s$ and $t$ with*

$$n \geq \max\{r, s\} \geq \min\{r, s\} \geq t \geq \max\{r + s - n, 0\},$$

*we can find linear maps $\alpha, \beta : \mathbb{F}^n \to \mathbb{F}^n$ such that* rank $\alpha = r$, rank $\beta = s$ *and* rank $\alpha\beta = t$.

If the reader is interested, she can glance forward to Theorem 11.2.2 which develops similar ideas.

**Exercise 5.5.17 [Fisher's inequality]** *At the start of each year, the jovial and popular Dean of Muddling (pronounced 'Chumly') College organises n parties for the m students in the College. Each student is invited to exactly k parties, and every two students are invited to exactly one party in common. Naturally, $k \geq 2$. Let $P = (p_{ij})$ be the $m \times n$ matrix defined by*

$$p_{ij} = \begin{cases} 1 & \textit{if student } i \textit{ is invited to party } j \\ 0 & \textit{otherwise.} \end{cases}$$

*Calculate the matrix $P P^T$ and find its rank. Deduce that $n \geq m$.*

*After the Master's cat has been found dyed green, maroon and purple on successive nights, the other Fellows insist that next year k = 1. What will happen? (The answer required is mathematical rather than sociological in nature.) Why does the proof that n ≥ m now fail?*

It is natural to ask about the rank of $\alpha + \beta$ when $\alpha$, $\beta \in \mathcal{L}(U, V)$. A little experimentation with diagonal matrices suggests the answer.

**Exercise 5.5.18** *(i) Suppose that U and V are finite dimensional spaces over $\mathbb{F}$ and $\alpha$, $\beta : U \to V$ are linear maps. By using Lemma 5.4.10, or otherwise, show that*

$$\min\{\dim U, \dim V, \operatorname{rank}\alpha + \operatorname{rank}\beta\} \geq \operatorname{rank}(\alpha + \beta).$$

*(ii) By considering $\alpha + \beta$ and $-\beta$, or otherwise, show that, under the conditions of (i),*

$$\operatorname{rank}(\alpha + \beta) \geq |\operatorname{rank}\alpha - \operatorname{rank}\beta|.$$

*(iii) Suppose that n, r, s and t are positive integers with*

$$\min\{n, r + s\} \geq t \geq |r - s|.$$

*Show that, given any finite dimensional vector space U of dimension n, we can find $\alpha$, $\beta \in \mathcal{L}(U, U)$ such that* $\operatorname{rank}\alpha = r$, $\operatorname{rank}\beta = s$ *and* $\operatorname{rank}(\alpha + \beta) = t$.

**Exercise 5.5.19** **[Simple magic squares]** *Consider the set $\Gamma$ of $3 \times 3$ real matrices $A = (a_{ij})$ such that all rows and columns add up to the same number. (Thus $A \in \Gamma$ if and only if there is a K such that $\sum_{r=1}^{3} a_{rj} = K$ for all j and $\sum_{r=1}^{3} a_{ir} = K$ for all i.)*

*(i) Show that $\Gamma$ is a finite dimensional real vector space with the usual matrix addition and multiplication by scalars.*

*(ii) Find the dimension of $\Gamma$. Find a basis for $\Gamma$ and show that it is indeed a basis. Do you think there is a 'natural basis'?*

*(iii) Extend your results to $n \times n$ 'simple magic squares'.*

[*We continue with these ideas in Exercise 5.7.11.*]

## 5.6 Secret sharing

The contents of this section are not meant to be taken very seriously. I suggest that the reader 'just goes with the flow' without worrying about the details. If she returns to this section when she has more experience with algebra, she will see that it is entirely rigorous.

So far we have only dealt with vector spaces and systems of equations over $\mathbb{F}$, where $\mathbb{F}$ is $\mathbb{R}$ or $\mathbb{C}$. But $\mathbb{R}$ and $\mathbb{C}$ are not the only systems in which we can add, subtract, multiply and divide in a natural manner. In particular, we can do all these things when we consider the integers modulo $p$, where $p$ is prime.

If we imitate our work on Gaussian elimination, working with the integers modulo $p$, we arrive at the following theorem.

**Theorem 5.6.1** *Let $A = (a_{ij})$ be an $n \times n$ matrix of integers and let $b_i \in \mathbb{Z}$. Then the system of equations*

$$\sum_{j=1}^{n} a_{ij} x_j \equiv b_i \qquad [1 \le i \le n]$$

*modulo $p$ has a unique[11] solution modulo $p$ if and only if $\det A \not\equiv 0$ modulo $p$.*

This observation has been made the basis of an ingenious method of secret sharing. We have all seen films in which two keys owned by separate people are required to open a safe. In the same way, we can imagine a safe with $k$ combination locks with the number required by each separate lock each known to a separate person. But what happens if one of the $k$ secret holders is unavoidably absent? In order to avoid this problem we require $n$ secret holders, any $k$ of whom acting together can open the safe, but any $k - 1$ of whom cannot do so.

Here is the neat solution found by Shamir.[12] The locksmith chooses a very large prime (as the reader probably knows, it is very easy to find large primes) and then chooses a random integer $S$ with $0 \le S \le p - 1$. She then makes a combination lock which can only be opened using $S$. Next she chooses integers $b_1, b_2, \ldots, b_{k-1}$ at random subject to $0 \le b_j \le p - 1$, and distinct integers $c_1, c_2, \ldots, c_n$ at random subject to $1 \le c_j \le p - 1$. She now sets $b_0 = S$ and computes

$$P(r) \equiv b_0 + b_1 c_r + b_2 c_r^2 + \cdots + b_{k-1} c_r^{k-1} \quad \bmod p$$

choosing $0 \le P(r) \le p - 1$. She calls on each 'key holder' in turn, telling the $r$th 'key holder' their secret number pair $(c_r, P(r))$. She then tells all the key holders the value of $p$ and burns her calculations.

Suppose that $k$ secret holders $r(1), r(2), \ldots, r(k)$ meet together. By the properties of the Vandermonde determinant (see Exercise 4.4.9)

$$\det \begin{pmatrix} 1 & c_{r(1)} & c_{r(1)}^2 & \cdots & c_{r(1)}^{k-1} \\ 1 & c_{r(2)} & c_{r(2)}^2 & \cdots & c_{r(2)}^{k-1} \\ 1 & c_{r(3)} & c_{r(3)}^2 & \cdots & c_{r(3)}^{k-1} \\ \vdots & \vdots & \vdots & \ddots & \vdots \\ 1 & c_{r(k)} & c_{r(k)}^2 & \cdots & c_{r(k)}^{k-1} \end{pmatrix} \equiv \det \begin{pmatrix} 1 & 1 & 1 & \cdots & 1 \\ c_{r(1)} & c_{r(2)} & c_{r(3)} & \cdots & c_{r(k)} \\ c_{r(1)}^2 & c_{r(2)}^2 & c_{r(3)}^2 & \cdots & c_{r(k)}^2 \\ \vdots & \vdots & \vdots & \ddots & \vdots \\ c_{r(1)}^{k-1} & c_{r(2)}^{k-1} & c_{r(3)}^{k-1} & \cdots & c_{r(k)}^{k-1} \end{pmatrix}$$

$$\equiv \prod_{1 \le j < i \le k} (c_{r(i)} - c_{r(j)}) \not\equiv 0 \quad \bmod p.$$

---

[11]  That is to say, if **x** and **x**′ are solutions, then $x_j \equiv x_j'$ modulo p.
[12]  A similar scheme was invented independently by Blakely.

Thus the system of equations

$$x_0 + c_{r(1)}x_1 + c_{r(1)}^2 x_2 + \cdots + c_{r(1)}^{k-1} x_{k-1} \equiv P(c_{r(1)})$$

$$x_0 + c_{r(2)}x_1 + c_{r(2)}^2 x_2 + \cdots + c_{r(2)}^{k-1} x_{k-1} \equiv P(c_{r(2)})$$

$$x_0 + c_{r(3)}x_1 + c_{r(3)}^2 x_2 + \cdots + c_{r(3)}^{k-1} x_{k-1} \equiv P(c_{r(3)})$$

$$\vdots$$

$$x_0 + c_{r(k)}x_1 + c_{r(k)}^2 x_2 + \cdots + c_{r(k)}^{k-1} x_{k-1} \equiv P(c_{r(k)})$$

has a unique solution $\mathbf{x}$. But we know that $\mathbf{b} = (b_0, b_1, \ldots, b_{k-1})^T$ is a solution, so $\mathbf{x} = \mathbf{b}$ and the number of the combination lock $S = b_0 = x_0$.

On the other hand

$$\det \begin{pmatrix} c_{r(1)} & c_{r(1)}^2 & \cdots & c_{r(1)}^{k-1} \\ c_{r(2)} & c_{r(2)}^2 & \cdots & c_{r(2)}^{k-1} \\ c_{r(3)} & c_{r(3)}^2 & \cdots & c_{r(3)}^{k-1} \\ \vdots & \vdots & \ddots & \vdots \\ c_{r(k-1)} & c_{r(k-1)}^2 & \cdots & c_{r(k-1)}^{k-1} \end{pmatrix} \equiv c_{r(1)}c_{r(2)}\cdots c_{r(k-1)} \prod_{1 \le j < i \le k-1} (c_{r(i)} - c_{r(j)}) \not\equiv 0$$

modulo $p$, so the system of equations

$$x_0 + c_{r(1)}x_1 + c_{r(1)}^2 x_2 + \cdots + c_{r(1)}^{k-1} x_{k-1} \equiv P(c_{r(1)})$$

$$x_0 + c_{r(2)}x_1 + c_{r(2)}^2 x_2 + \cdots + c_{r(2)}^{k-1} x_{k-1} \equiv P(c_{r(2)})$$

$$x_0 + c_{r(3)}x_1 + c_{r(3)}^2 x_2 + \cdots + c_{r(3)}^{k-1} x_{k-1} \equiv P(c_{r(3)})$$

$$\vdots$$

$$x_0 + c_{r(k-1)}x_1 + c_{r(k-1)}^2 x_2 + \cdots + c_{r(k-1)}^{k-1} x_{k-1} \equiv P(c_{r(k-1)})$$

has a solution, *whatever value of $x_0$ we take*, and there is no way that $k - 1$ secret holders can work out the number of the combination lock!

**Exercise 5.6.2** *Suppose that, with the notation above, we take $p = 7$, $k = 2$, $b_0 = 2$, $b_1 = 2$, $c_1 = 2$, $c_2 = 4$ and $c_3 = 5$. Compute $P(1)$, $P(2)$ and $P(3)$ and perform the recovery of $b_0$ from the pair $(P(1), c(1))$ and $(P(2), c(2))$ and from the pair $(P(2), c(2))$ and $(P(3), c(3))$.*

**Exercise 5.6.3** *We required that the $c_j$ be distinct. Why is it obvious that this is a good idea? At what point in the argument did we make use of the fact that the $c_j$ are distinct?*

*We required that the $c_j$ be non-zero. Why is it obvious that this is a good idea? At what point in the argument did we make use of the fact that the $c_j$ are non-zero?*

**Exercise 5.6.4** *Suppose that, with the notation above, we take $p = 6$ (so $p$ is not a prime), $k = 2$, $b_0 = 1$, $b_1 = 1$, $c_1 = 1$, $c_2 = 4$. Show that you cannot recover $b_0$ from $(P(1), c(1))$ and $(P(2), c(2))$ What part of our discussion of the case when $p$ is prime fails?*

**Exercise 5.6.5** (*i*) *Suppose that, instead of choosing the $c_j$ at random, we simply take $c_j = j$ (but choose the $b_s$ at random). Is it still true that $k$ secret holders can work out the combination but $k - 1$ cannot?*

(*ii*) *Suppose that, instead of choosing the $b_s$ at random, we simply take $b_s = s$ for $s \neq 0$ (but choose the $c_j$ at random). Is it still true that $k$ secret holders can work out the combination but $k - 1$ cannot?*

[*Note, however, that a guiding principle of cryptography is that, if something can be kept secret, it should be kept secret.*]

**Exercise 5.6.6** *The Dark Lord Y'Trinti has acquired the services of the dwarf Trigon who can engrave pairs of very large integers on very small rings. The Dark Lord instructs Trigon to use the method of secret sharing described above to engrave n rings in such a way that anyone who acquires k of the rings and knows the Prime Perilous p can deduce the Integer N of Power but owning k − 1 rings will give no information whatever.*

*For reasons to be explained in the prequel, Trigon engraves an $(n + 1)$st ring with random integers. A band of heroes (who know the Prime Perilous and all the information given in this exercise) sets out to recover the rings. What, if anything, can they say, with very high probability, about the Integer of Power if they have k rings (possibly including the fake)? What can they say if they have $k + 1$ rings? What if they have $k + 2$ rings?*

## 5.7 Further exercises

**Exercise 5.7.1** Let $A$ and $B$ be $n \times n$ matrices. State and prove necessary and sufficient conditions involving the row ranks of $A$ and the $n \times 2n$ matrix $(A \ B)$ for the existence of an $n \times n$ matrix $X$ with $AX = B$. When is $X$ unique and why?

Find $X$ when

$$A = \begin{pmatrix} 4 & 1 & 1 \\ 1 & 2 & 1 \\ 0 & 3 & 1 \end{pmatrix} \quad \text{and} \quad B = \begin{pmatrix} 1 & 1 & 1 \\ 0 & 1 & 0 \\ 3 & 1 & 2 \end{pmatrix}.$$

**Exercise 5.7.2** Let $V$ be a vector space over $\mathbb{F}$ with basis $\mathbf{e}_1, \mathbf{e}_2, \dots, \mathbf{e}_n$ where $n \geq 2$. For which values of $n$, if any, are the following bases of $V$?

(i) $\mathbf{e}_1 - \mathbf{e}_2, \mathbf{e}_2 - \mathbf{e}_3, \dots, \mathbf{e}_{n-1} - \mathbf{e}_n, \mathbf{e}_n - \mathbf{e}_1$.

(ii) $\mathbf{e}_1 + \mathbf{e}_2, \mathbf{e}_2 + \mathbf{e}_3, \dots, \mathbf{e}_{n-1} + \mathbf{e}_n, \mathbf{e}_n + \mathbf{e}_1$.

Prove your answers.

**Exercise 5.7.3** Consider the vector space $\mathcal{P}$ of real polynomials $P : \mathbb{R} \to \mathbb{R}$ with the usual operations. Which of the following define linear maps from $\mathcal{P}$ to $\mathcal{P}$? Give reasons for your answers.

(i) $(Dp)(t) = p'(t)$.

(ii) $(Sp)(t) = p(t^2 + 1)$.

(iii) $(Tp)(t) = p(t)^2 + 1$.

(iv) $(Ep)(t) = p(e^t)$.

(v) $(Jp)(t) = \int_0^t p(s)\,ds$.

(vi) $(Kp)(t) = 1 + \int_0^t p(s)\,ds$.

(vii) $(Lp)(t) = p(0) + \int_0^t p(s)\,ds$.

(viii) $(Mp)(t) = p(t^2) - tp(t)$.

(ix) $R$ and $Q$, where $Rp$ and $Qp$ are polynomials satisfying the conditions that $p(t) = (t^2+1)(Qp)(t) + (Rp)(t)$ with $Rp$ having degree at most 1.

**Exercise 5.7.4** Let $A$, $B$ and $C$ be subspaces of a finite dimensional vector space $V$ over $\mathbb{F}$ and let $\alpha : U \to U$ be linear. Which of the following statements are always true and which may be false? Give proofs or counterexamples as appropriate.

(i) If $\dim(A \cap B) = \dim(A + B)$, then $A = B$.

(ii) $\alpha(A \cap B) = \alpha A \cap \alpha B$.

(iii) $(B + C) \cap (C + A) \cap (A + B) = (B \cap C) + (C \cap A) + (A \cap B)$.

**Exercise 5.7.5** Suppose that $W$ is a vector space over $\mathbb{F}$ with subspaces $U$ and $V$. If $U \cup V$ is a subspace, show that $U \supseteq V$ and/or $V \supseteq U$.

**Exercise 5.7.6** Show that $\mathbb{C}$ is a vector space over $\mathbb{R}$ if we use the usual definitions of addition and multiplication. Prove that it has dimension 2.

State and prove a similar result about $\mathbb{C}^n$.

**Exercise 5.7.7** (i) By expanding $(1 + 1)(\mathbf{x} + \mathbf{y})$ in two different ways, show that condition (ii) in the definition of a vector space (Definition 5.2.2) is redundant (that is to say, can be deduced from the other axioms).

(ii) Let $U = \mathbb{R}^2$ and define $\lambda \bullet \mathbf{x} = \lambda \mathbf{x} \cdot \mathbf{e}$ where we use the standard inner product from Section 2.3 and $\mathbf{e}$ is a fixed unit vector. Show that, if we replace scalar multiplication $(\lambda, \mathbf{x}) \mapsto \lambda \mathbf{x}$ by the 'new scalar multiplication' $(\lambda, \mathbf{x}) \mapsto \lambda \bullet \mathbf{x}$, the new system obeys all the axioms for a vector space except that there is no $\lambda \in \mathbb{R}$ with $\lambda \bullet \mathbf{x} = \mathbf{x}$ for all $\mathbf{x} \in U$.

**Exercise 5.7.8** Let $P$, $Q$ and $R$ be $n \times n$ matrices. Use elementary row operations to show the $2n \times 2n$ matrices

$$\begin{pmatrix} PQ & 0 \\ Q & QR \end{pmatrix} \quad \text{and} \quad \begin{pmatrix} 0 & PQR \\ Q & 0 \end{pmatrix}$$

have the same row rank. Hence show that

$$\operatorname{rank}(PQ) + \operatorname{rank}(QR) \le \operatorname{rank} Q + \operatorname{rank} PQR.$$

**Exercise 5.7.9** If $A$ and $B$ are $2 \times 2$ matrices over $\mathbb{R}$, is it necessarily true that

$$\operatorname{rank} AB = \operatorname{rank} BA?$$

Give reasons.

**Exercise 5.7.10** If $A$ is an $n \times n$ matrix and $n \geq 2$, show that

$$\text{rank Adj } A = \begin{cases} n & \text{if rank } A = n, \\ 1 & \text{if rank } A = n - 1, \\ 0 & \text{if rank } A < n - 1. \end{cases}$$

Show that, if $n \geq 3$, and $A$ is not invertible, then $\text{Adj Adj } A = 0$. Is this true if $n = 2$? Give a proof or counterexample.

**Exercise 5.7.11** We continue the discussion of Exercise 5.5.19 by looking at 'diagonal magic squares'.

(i) Let $\Gamma'$ be the set of $3 \times 3$ real matrices $A = (a_{ij})$ such that all rows, columns and diagonals add up to the same value. (Thus $A \in \Gamma$ if and only if there is a $K$ such that

$$\sum_{i=1}^{3} a_{ij} = \sum_{i=1}^{3} a_{ji} = \sum_{i=1}^{3} a_{ii} = \sum_{i=1}^{3} a_{i,3-i} = K$$

for all $j$.) Show that $\Gamma'$ is a finite dimensional real vector space with the usual matrix addition and multiplication by scalars and find its dimension.

(ii) Think about the problem of extending these results to $n \times n$ diagonal magic squares and, if you feel it would be a useful exercise, carry out such an extension.

[Outside textbooks on linear algebra, 'magic square' means a 'diagonal magic square' with integer entries. I think part (ii) requires courage rather than insight, but there is a solution in an article '*Vector spaces of magic squares*' by J. E. Ward [32].]

**Exercise 5.7.12** Consider $\mathcal{P}$ the set of polynomials in one real variable with real coefficients. Show that $\mathcal{P}$ is a subspace of the vector space of maps $f : \mathbb{R} \to \mathbb{R}$ and so a vector space over $\mathbb{R}$.

Show that the maps $T$ and $D$ defined by

$$T\left(\sum_{j=0}^{n} a_j t^j\right) = \sum_{j=0}^{n} \frac{a_j}{j+1} t^{j+1} \quad \text{and} \quad D\left(\sum_{j=0}^{n} a_j t^j\right) = \sum_{j=1}^{n} j a_j t^{j-1}$$

are linear.

(i) Show that $T$ is injective, but not surjective.

(ii) Show that $D$ is surjective, but not injective. What is the kernel of $D$?

(iii) Show that $DT$ is the identity, but $TD$ is not.

(iv) Which polynomials $p$, if any, have the property that $p(D) = 0$? (In other words, find all $p(t) = \sum_{j=0}^{n} b_j t^j$ such that the linear map $\sum_{j=0}^{n} b_j D^j = 0$.)

(v) Suppose that $V$ is a subspace of $\mathcal{P}$ such that $f \in V \Rightarrow Tf \in V$. Show that $V$ cannot be finite dimensional.

(vi) Let $W$ be a subspace of $\mathcal{P}$. Show that $W$ is finite dimensional if and only if there exists an $m$ such that $D^m f = 0$ for all $f \in W$.

**Exercise 5.7.13** Let $U$ be a finite dimensional vector space over $\mathbb{F}$ and let $\alpha$, $\beta : U \to U$ be linear. State and prove necessary and sufficient conditions involving $\alpha(U)$ and $\beta(U)$ for the existence of a linear map $\gamma : U \to U$ with $\alpha\gamma = \beta$. When is $\gamma$ unique and why?

Explain how this links with the necessary and sufficient condition of Exercise 5.7.1.

Generalise the result of this question and its parallel in Exercise 5.7.1 to the case when $U, V, W$ are finite dimensional vector spaces and $\alpha : U \to W$, $\beta : V \to W$ are linear.

**Exercise 5.7.14 [The circulant determinant]** We work over $\mathbb{C}$. Consider the *circulant matrix*

$$C = \begin{pmatrix} x_0 & x_1 & x_2 & \cdots & x_n \\ x_n & x_0 & x_1 & \cdots & x_{n-1} \\ x_{n-1} & x_n & x_0 & \cdots & x_{n-2} \\ \vdots & \vdots & \vdots & \ddots & \vdots \\ x_1 & x_2 & x_3 & \cdots & x_0 \end{pmatrix}.$$

By considering factors of polynomials in $n$ variables, or otherwise, show that

$$\det C = \prod_{j=0}^{n} f(\zeta^j),$$

where $f(t) = \sum_{j=0}^{n} x_j t^j$ and $\zeta = \exp\left(2\pi i/(n+1)\right)$.

**Exercise 5.7.15** If $A$ and $B$ are $n \times n$ matrices of complex numbers, show that

$$\det \begin{pmatrix} A & -B \\ B & A \end{pmatrix} = \det(A + iB)\det(A - iB).$$

**Exercise 5.7.16** If $J$ is a real $m \times m$ matrix satisfying $J^2 = -I$, show that $m = 2n$ for some integer $n$ and there exists an invertible matrix $P$ with

$$P^{-1}JP = \begin{pmatrix} 0 & I \\ -I & 0 \end{pmatrix},$$

where the matrix entries are themselves $n \times n$ matrices.

Find the dimension of the space of $2n \times 2n$ real matrices such that

$$A^T J + J A = 0.$$

# 6

# Linear maps from $\mathbb{F}^n$ to itself

## 6.1 Linear maps, bases and matrices

We already know that matrices can be associated with linear maps. In this section we show how to associate every linear map $\alpha : \mathbb{F}^n \to \mathbb{F}^n$ with an $n \times n$ matrix.

**Definition 6.1.1** *Let $\alpha : \mathbb{F}^n \to \mathbb{F}^n$ be linear and let $\mathbf{e}_1, \mathbf{e}_2, \ldots, \mathbf{e}_n$ be a basis. Then the matrix $A = (a_{ij})$ of $\alpha$ with respect to this basis is given by the rule*

$$\alpha(\mathbf{e}_j) = \sum_{i=1}^{n} a_{ij} \mathbf{e}_i.$$

At first sight, this definition looks a little odd. The reader may ask why '$a_{ij}\mathbf{e}_i$' and not '$a_{ji}\mathbf{e}_i$'? Observe that, if $\mathbf{x} = \sum_{j=1}^{n} x_j \mathbf{e}_j$ and $\alpha(\mathbf{x}) = \mathbf{y} = \sum_{i=1}^{n} y_i \mathbf{e}_i$, then

$$y_i = \sum_{j=1}^{n} a_{ij} x_j.$$

Thus coordinates and bases must go *opposite ways*. The definition chosen is conventional,[1] but represents a universal convention and must be learnt.

The reader may also ask why our definition introduces a general basis $\mathbf{e}_1, \mathbf{e}_2, \ldots, \mathbf{e}_n$ rather than sticking to a fixed basis. The answer is that different problems may be most easily tackled by using different bases. For example, in many problems in mechanics it is easier to take one basis vector along the vertical (since that is the direction of gravity) but in others it may be better to take one basis vector parallel to the Earth's axis of rotation. (For another example see Exercise 6.8.1.)

If we do use the so-called standard basis, then the following observation is quite useful.

**Exercise 6.1.2** *Let us work in the column vector space $\mathbb{F}^n$. If $\mathbf{e}_1, \mathbf{e}_2, \ldots, \mathbf{e}_n$ is the standard basis (that is to say, $\mathbf{e}_j$ is the column vector with 1 in the $j$th place and zero elsewhere),*

---

[1] An Englishman is asked why he has never visited France. 'I know that they drive on the right there, so I tried it one day in London. Never again!'

*then the matrix A of a linear map* $\alpha : \mathbb{F}^n \to \mathbb{F}^n$ *with respect to this basis has* $\alpha(\mathbf{e}_j)$ *as its jth column.*

Our definition of 'the matrix associated with a linear map $\alpha : \mathbb{F}^n \to \mathbb{F}^n$' meshes well with our definition of matrix multiplication.

**Exercise 6.1.3** *Let* $\alpha, \beta : \mathbb{F}^n \to \mathbb{F}^n$ *be linear and let* $\mathbf{e}_1, \mathbf{e}_2, \ldots, \mathbf{e}_n$ *be a basis. If* $\alpha$ *has matrix A and* $\beta$ *has matrix B with respect to the stated basis, then* $\alpha\beta$ *has matrix AB with respect to the stated basis.*

Exercise 6.1.3 allows us to translate results on linear maps from $\mathbb{F}^n$ to itself into results on square matrices and vice versa. Thus we can deduce the result

$$(A + B)C = AB + AC$$

from the result

$$(\alpha + \beta)\gamma = \alpha\gamma + \alpha\gamma$$

or vice versa. On the whole, I prefer to deduce results on matrices from results on linear maps in accordance with the following motto:

**linear maps for understanding, matrices for computation.**

Since we allow different bases and since different bases assign different matrices to the same linear map, we need a way of translating from one basis to another.

**Theorem 6.1.4 [Change of basis]** *Let* $\alpha : \mathbb{F}^n \to \mathbb{F}^n$ *be a linear map. If* $\alpha$ *has matrix* $A = (a_{ij})$ *with respect to a basis* $\mathbf{e}_1, \mathbf{e}_2, \ldots, \mathbf{e}_n$ *and matrix* $B = (b_{ij})$ *with respect to a basis* $\mathbf{f}_1, \mathbf{f}_2, \ldots, \mathbf{f}_n,$ *then there is an invertible* $n \times n$ *matrix P such that*

$$B = P^{-1}AP.$$

*The matrix* $P = (p_{ij})$ *is given by the rule*

$$\mathbf{f}_j = \sum_{i=1}^{n} p_{ij}\mathbf{e}_i.$$

*Proof* Since the $\mathbf{e}_i$ form a basis, we can find unique $p_{ij} \in \mathbb{F}$ such that

$$\mathbf{f}_j = \sum_{i=1}^{n} p_{ij}\mathbf{e}_i.$$

Similarly, since the $\mathbf{f}_i$ form a basis, we can find unique $q_{ij} \in \mathbb{F}$ such that

$$\mathbf{e}_j = \sum_{i=1}^{n} q_{ij}\mathbf{f}_i.$$

Thus, using the definitions of $A$ and $B$ and the linearity of $\alpha$,

$$\sum_{i=1}^{n} b_{ij}\mathbf{f}_i = \alpha(\mathbf{f}_j) = \alpha\left(\sum_{r=1}^{n} p_{rj}\mathbf{e}_r\right)$$

$$= \sum_{r=1}^{n} p_{rj}\alpha\mathbf{e}_r = \sum_{r=1}^{n} p_{rj}\left(\sum_{s=1}^{n} a_{sr}\mathbf{e}_s\right)$$

$$= \sum_{r=1}^{n} p_{rj}\left(\sum_{s=1}^{n} a_{sr}\left(\sum_{i=1}^{n} q_{is}\mathbf{f}_i\right)\right)$$

$$= \sum_{i=1}^{n}\left(\sum_{s=1}^{n}\sum_{r=1}^{n} q_{is}a_{sr}p_{rj}\right)\mathbf{f}_i.$$

Since the $\mathbf{f}_i$ form a basis,

$$b_{ij} = \sum_{s=1}^{n}\sum_{r=1}^{n} q_{is}a_{sr}p_{rj} = \sum_{r=1}^{n}\left(\sum_{s=1}^{n} q_{is}a_{sr}\right)p_{rj}$$

and so

$$B = Q(AP) = QAP.$$

Since the result is true for any linear map, it is true, in particular, for the identity map $\iota$. Here $A = B = I$, so $I = QP$ and we see that $P$ is invertible with inverse $P^{-1} = Q$. $\quad\square$

**Exercise 6.1.5** *Write out the proof of Theorem 6.1.4 using the summation convention.*

Theorem 6.1.4 is associated with a definition.

**Definition 6.1.6** *Let $A$ and $B$ be $n \times n$ matrices. We say that $A$ and $B$ are* similar *(or* conjugate[2]*) if there exists an invertible $n \times n$ matrix $P$ such that $B = P^{-1}AP$.*

**Exercise 6.1.7** (i) *Show that, if two $n \times n$ matrices $A$ and $B$ are similar, then, given a basis*

$$\mathbf{e}_1, \mathbf{e}_2, \ldots, \mathbf{e}_n,$$

*we can find a basis*

$$\mathbf{f}_1, \mathbf{f}_2, \ldots, \mathbf{f}_n$$

*such that $A$ and $B$ represent the same linear map with respect to the two bases.*

(ii) *Show that two $n \times n$ matrices are similar if and only if they represent the same linear map with respect to two bases.*

(iii) *Show that similarity is an equivalence relation by using the definition directly. (See Exercise 6.8.34 if you need to recall the definition of an equivalence relation.)*

(iv) *Show that similarity is an equivalence relation by using part (ii).*

---

[2] The word 'similar' is overused and the word 'conjugate' fits in well with the rest of algebra, but the majority of authors use 'similar'.

Exercise 6.8.25 proves the useful, and not entirely obvious, fact that, if two $n \times n$ matrices with real entries are similar when considered as matrices over $\mathbb{C}$, they are similar when considered as matrices over $\mathbb{R}$.

In practice it may be tedious to compute $P$ and still more tedious[3] to compute $P^{-1}$.

**Exercise 6.1.8** *Let us work in the column vector space $\mathbb{F}^n$. If $\mathbf{e}_1, \mathbf{e}_2, \ldots, \mathbf{e}_n$ is the standard basis (that is to say, $\mathbf{e}_j$ is the column vector with $1$ in the $j$th place and zero elsewhere) and $\mathbf{f}_1, \mathbf{f}_2, \ldots, \mathbf{f}_n$, is another basis, explain why the matrix $P$ in Theorem 6.1.4 has $\mathbf{f}_j$ as its $j$th column.*

**Exercise 6.1.9** *Although we wish to avoid explicit computation as much as possible, the reader ought, perhaps, to do at least one example. Suppose that $\alpha : \mathbb{R}^3 \to \mathbb{R}^3$ has matrix*

$$\begin{pmatrix} 1 & 1 & 2 \\ -1 & 2 & 1 \\ 0 & 1 & 3 \end{pmatrix}$$

*with respect to the standard basis $\mathbf{e}_1 = (1, 0, 0)^T$, $\mathbf{e}_2 = (0, 1, 0)^T$, $\mathbf{e}_3 = (0, 0, 1)^T$. Find the matrix associated with $\alpha$ for the basis $\mathbf{f}_1 = (1, 0, 0)^T$, $\mathbf{f}_2 = (1, 1, 0)^T$, $\mathbf{f}_3 = (1, 1, 1)^T$.*

Although Theorem 6.1.4 is rarely used computationally, it is extremely important from a theoretical view.

**Theorem 6.1.10** *Let $\alpha : \mathbb{F}^n \to \mathbb{F}^n$ be a linear map. If $\alpha$ has matrix $A = (a_{ij})$ with respect to a basis $\mathbf{e}_1, \mathbf{e}_2, \ldots, \mathbf{e}_n$ and matrix $B = (b_{ij})$ with respect to a basis $\mathbf{f}_1, \mathbf{f}_2, \ldots, \mathbf{f}_n$, then $\det A = \det B$.*

*Proof* By Theorem 6.1.4, there is an invertible $n \times n$ matrix such that $B = P^{-1}AP$. Thus

$$\det B = \det P^{-1} \det A \det P = (\det P)^{-1} \det A \det P = \det A.$$

$\square$

Theorem 6.1.10 allows us to make the following definition.

**Definition 6.1.11** *Let $\alpha : \mathbb{F}^n \to \mathbb{F}^n$ be a linear map. If $\alpha$ has matrix $A = (a_{ij})$ with respect to a basis $\mathbf{e}_1, \mathbf{e}_2, \ldots, \mathbf{e}_n$, then we define $\det \alpha = \det A$.*

**Exercise 6.1.12** *Explain why we needed Theorem 6.1.10 in order to make this definition.*

From the point of view of Chapter 4, we would expect Theorem 6.1.10 to hold. The determinant $\det \alpha$ is the scale factor for the change in volume occurring when we apply the linear map $\alpha$ and this cannot depend on the choice of basis. However, as we pointed out earlier, this kind of argument, which appears plausible for linear maps involving $\mathbb{R}^2$ and $\mathbb{R}^3$, is less convincing when applied to $\mathbb{R}^n$ and does not make a great deal of sense when applied to $\mathbb{C}^n$. Pure mathematicians have had to look rather deeper in order to find a fully

---

[3] If you are faced with an exam question which seems to require the computation of inverses, it is worth taking a little time to check that this is actually the case.

satisfactory 'basis free' treatment of determinants and we shall not consider the matter in this book.

## 6.2 Eigenvectors and eigenvalues

As we emphasised in the previous section, the same linear map $\alpha : \mathbb{F}^n \to \mathbb{F}^n$ will be represented by different matrices with respect to different bases. Sometimes we can find a basis with respect to which the representing matrix takes a very simple form. A particularly useful technique for doing this is given by the notion of an eigenvector.[4]

**Definition 6.2.1** *If $\alpha : \mathbb{F}^n \to \mathbb{F}^n$ is linear and $\alpha(\mathbf{u}) = \lambda\mathbf{u}$ for some vector $\mathbf{u} \neq \mathbf{0}$ and some $\lambda \in \mathbb{F}$, we say that $\mathbf{u}$ is an* eigenvector *of $\alpha$ with* eigenvalue *$\lambda$.*

Note that, though an *eigenvalue* may be zero, the zero vector cannot be an *eigenvector*.

When we deal with finite dimensional vector spaces, there is a strong link between eigenvalues and determinants.[5]

**Theorem 6.2.2** *If $\alpha : \mathbb{F}^n \to \mathbb{F}^n$ is linear, then $\lambda$ is an eigenvalue of $\alpha$ if and only if $\det(\lambda\iota - \alpha) = 0$.*

*Proof* Observe that

$$\lambda \text{ is an eigenvalue of } \alpha$$
$$\Leftrightarrow (\alpha - \lambda\iota)\mathbf{u} = \mathbf{0} \text{ has a non-trivial solution}$$
$$\Leftrightarrow (\alpha - \lambda\iota) \text{ is not invertible}$$
$$\Leftrightarrow \det(\alpha - \lambda\iota) = 0$$
$$\Leftrightarrow \det(\lambda\iota - \alpha) = 0$$

as stated. $\qquad\qquad\square$

We call the polynomial $\chi_\alpha(t) = \det(t\iota - \alpha)$ the *characteristic polynomial*[6] of $\alpha$.

**Exercise 6.2.3** *(i) Verify that*

$$\det\left(t\begin{pmatrix} 1 & 0 \\ 0 & 1 \end{pmatrix} - \begin{pmatrix} a & b \\ c & d \end{pmatrix}\right) = t^2 - (\operatorname{Tr} A)t + \det A,$$

*where $\operatorname{Tr} A = a + d$.*

*(ii) If $A = (a_{ij})$ is a $3 \times 3$ matrix, show that*

$$\det(tI - A) = t^3 - (\operatorname{Tr} A)t^2 + ct - \det A,$$

*where $\operatorname{Tr} A = a_{11} + a_{22} + a_{33}$ and $c$ depends on $A$, but need not be calculated explicitly.*

---

[4] The development of quantum theory involved eigenvalues, eigenfunctions, eigenstates and similar eigenobjects. Eigenwords followed the physics from German into English, but kept their link with German grammar in which the adjective is strongly bound to the noun.

[5] However, the notion of eigenvalue generalises to infinite dimensional vector spaces and the definition of determinant does not, so it is important to use Definition 6.2.1 as our definition rather than some other definition involving determinants.

[6] Older texts sometimes talk about characteristic values and characteristic vectors rather than eigenvalues and eigenvectors.

**Exercise 6.2.4** *Let $A = (a_{ij})$ be an $n \times n$ matrix. Write $b_{ii}(t) = t - a_{ii}$ and $b_{ij}(t) = -a_{ij}$ if $i \neq j$. If*

$$\sigma : \{1, 2, \ldots, n\} \to \{1, 2, \ldots, n\}$$

*is a bijection (that is to say, $\sigma$ is a permutation) show that $P_\sigma(t) = \prod_{i=1}^{n} b_{i\sigma(i)}(t)$ is a polynomial of degree at most $n - 2$ unless $\sigma$ is the identity map. Deduce that*

$$\det(tI - A) = \prod_{i=1}^{n}(t - a_{ii}) + Q(t),$$

*where $Q$ is a polynomial of degree at most $n - 2$.*
  *Conclude that*

$$\det(tI - A) = t^n + c_{n-1}t^{n-1} + c_{n-2}t^{n-2} + \cdots + c_0$$

*with $c_{n-1} = -\operatorname{Tr} A = -\sum_{i=1}^{n} a_{ii}$. By taking $t = 0$, or otherwise, show that $c_0 = (-1)^n \det A$. (Tr $A$ is called the trace of $A$.)*

In order to exploit Theorem 6.2.2 fully, we need two deep theorems from analysis.

**Theorem 6.2.5** *If $P$ is polynomial of odd degree with real coefficients, then $P$ has at least one real root.*

(Theorem 6.2.5 is a very special case of the intermediate value theorem.)

**Theorem 6.2.6 [Fundamental Theorem of Algebra]** *If $P$ is polynomial of degree at least 1 with coefficients in $\mathbb{C}$, then $P$ has at least one root in $\mathbb{C}$.*

Exercise 6.8.38 gives the proof, which the reader may well have met before, of the associated factorisation theorem.

**Theorem 6.2.7 [Factorisation of polynomials over $\mathbb{C}$]** *If $P$ is polynomial of degree $n$ with coefficients in $\mathbb{C}$, then we can find $c \in \mathbb{C}$ and $\lambda_j \in \mathbb{C}$ such that*

$$P(t) = c \prod_{j=1}^{n}(t - \lambda_j).$$

The following lemma shows that our machinery gives results which are not immediately obvious.

**Lemma 6.2.8** *Any linear map $\alpha : \mathbb{R}^3 \to \mathbb{R}^3$ has an eigenvector. It follows that there exists some line $l$ through $\mathbf{0}$ with $\alpha(l) \subseteq l$.*

*Proof* Since $\det(t\iota - \alpha)$ is a real cubic, the equation $\det(t\iota - \alpha) = 0$ has a real root, say $\lambda$. We know that $\lambda$ is an eigenvalue and so has an associated eigenvector $\mathbf{u}$, say. Let

$$l = \{s\mathbf{u} \,:\, s \in \mathbb{R}\}.$$

$\square$

**Exercise 6.2.9**  *Generalise Lemma 6.2.8 to linear maps $\alpha : \mathbb{R}^{2n+1} \to \mathbb{R}^{2n+1}$.*

If we consider real vector spaces of even dimension, it is easy to find linear maps with no eigenvectors.

**Example 6.2.10**  *Consider the linear map $\rho_\theta : \mathbb{R}^2 \to \mathbb{R}^2$ whose matrix with respect to the standard basis $(1, 0)^T, (0, 1)^T$ is*

$$R_\theta = \begin{pmatrix} \cos\theta & -\sin\theta \\ \sin\theta & \cos\theta \end{pmatrix}.$$

*The map $\rho_\theta$ has no eigenvectors unless $\theta \equiv 0 \mod \pi$. If $\theta \equiv 0 \mod 2\pi$, every non-zero vector is an eigenvector with eigenvalue 1. If $\theta \equiv \pi \mod 2\pi$, every non-zero vector is an eigenvector with eigenvalue $-1$.*

**Exercise 6.2.11**  *Prove the results of Example 6.2.10 by first showing that the equation $\det(t\iota - \rho_\theta) = 0$ has no real roots unless $\theta \equiv 0 \mod \pi$ and then looking at the cases when $\theta \equiv 0 \mod \pi$.*

The reader will probably recognise $\rho_\theta$ as a rotation through an angle $\theta$. (If not, she can wait for the discussion in Section 7.3.) She should convince herself that the result is obvious if we interpret it in terms of rotations. So far, we have emphasised the similarities between $\mathbb{R}^n$ and $\mathbb{C}^n$. The next result gives a striking example of an essential difference.

**Lemma 6.2.12**  *Any linear map $\alpha : \mathbb{C}^n \to \mathbb{C}^n$ has an eigenvector. It follows that there exists a one dimensional complex subspace*

$$l = \{w\mathbf{e} : w \in \mathbb{C}\}$$

*(where $\mathbf{e} \neq \mathbf{0}$) with $\alpha(l) \subseteq l$.*

*Proof*  The proof, which resembles the proof of Lemma 6.2.8, is left as an exercise for the reader.                                                                                          $\square$

The following observation is sometimes useful when dealing with singular $n \times n$ matrices.

**Lemma 6.2.13**  *If $A$ is an $n \times n$ matrix over $\mathbb{F}$, then there exists a $\delta > 0$ such that $A + tI$ is non-singular for all $0 \neq |t| < \delta$.*

*Proof*  Since $P_A(t) = \det(tI + A)$ is a non-trivial polynomial, it has only finitely many roots and so there exists a $\delta > 0$ such that $P_A(t) \neq 0$ and so $A + tI$ is non-singular for all $0 \neq |t| < \delta$.                                                                          $\square$

**Exercise 6.2.14**  *We use the hypotheses and notation of Lemma 6.2.13 and its proof.*
   *(i) Show that $P_A(t) = (-1)^n \chi_A(-t)$ where $\chi_A(t) = \det(tI - A)$.*
   *(ii) Prove the following very simple consequence of Lemma 6.2.13. We can find $t_n \to 0$ such that $t_n I + A$ is non-singular.*

**Exercise 6.2.15** *Suppose that $A$ and $B$ are $n \times n$ matrices and $A$ is non-singular. Use the observation that $\det A \det(t A^{-1} - B) = \det(t I - AB)$ to show that $\chi_{AB} = \chi_{BA}$ (where $\chi_C$ denotes the characteristic polynomial of $C$). Use Exercise 6.2.14 (ii) to show that the condition that $A$ is non-singular can be removed.*

*Explain why, if $\chi_A = \chi_B$, then $\operatorname{Tr} A = \operatorname{Tr} B$. Deduce that $\operatorname{Tr} AB = \operatorname{Tr} BA$. Which of the following two statements (if any) are true for all $n \times n$ matrices $A$, $B$ and $C$?*

*(i) $\chi_{ABC} = \chi_{ACB}$.*

*(ii) $\chi_{ABC} = \chi_{BCA}$.*

*Justify your answers.*

**Exercise 6.2.16** *Let us say that an $n \times n$ matrix $A$ is simple magic if the sum of the elements of each row and the sum of the elements of each column all take the same value. (In other words, $\sum_{i=1}^{n} a_{iu} = \sum_{j=1}^{n} a_{vj} = \kappa$ for all $u$ and $v$ and some $\kappa$.) Identify an eigenvector of $A$.*

*If $A$ is simple magic and $BA = AB = I$, show that $B$ is simple magic. Deduce that, if $A$ is simple magic and invertible, then $\operatorname{Adj} A$ is simple magic. Show, more generally, that, if $A$ is simple magic, so is $\operatorname{Adj} A$.*

We give other applications of Exercise 6.2.14 (ii) in Exercises 6.8.8 and 6.8.11.

## 6.3 Diagonalisation and eigenvectors

As we said in the previous section, a linear map $\alpha : \mathbb{F}^n \to \mathbb{F}^n$ may have many different matrices associated with it according to our choice of basis. We asked whether there are any bases with respect to which the associated matrix takes a particularly simple form. We now specialise our question and ask whether there are any bases with respect to which the associated matrix is diagonal. The next result is essentially a tautology, but shows that the answer is closely bound up with the notion of an eigenvector.

**Theorem 6.3.1** *Suppose that $\alpha : \mathbb{F}^n \to \mathbb{F}^n$ is linear. Then $\alpha$ has diagonal matrix $D$ with respect to a basis $\mathbf{e}_1, \mathbf{e}_2, \dots, \mathbf{e}_n$ if and only if the $\mathbf{e}_j$ are eigenvectors. The diagonal entries $d_{ii}$ of $D$ are the eigenvalues of the $\mathbf{e}_i$.*

*Proof* If $\alpha$ has matrix $A = (a_{ij})$ with respect to a basis $\mathbf{e}_1, \mathbf{e}_2, \dots, \mathbf{e}_n$ then, by definition,

$$\alpha \mathbf{e}_j = \sum_{i=1}^{n} a_{ij} \mathbf{e}_i.$$

Thus $A$ is diagonal with $a_{ii} = d_{ii}$ if and only if

$$\alpha \mathbf{e}_j = d_{jj} \mathbf{e}_j,$$

that is to say, each $\mathbf{e}_j$ is an eigenvector of $\alpha$ with associated eigenvalue $d_{jj}$. $\square$

The next exercise prepares the way for a slightly more difficult result.

**Exercise 6.3.2** (*i*) *Let* $\alpha : \mathbb{F}^2 \to \mathbb{F}^2$ *be a linear map. Suppose that* $\mathbf{e}_1$ *and* $\mathbf{e}_2$ *are eigenvectors of* $\alpha$ *with distinct eigenvalues* $\lambda_1$ *and* $\lambda_2$. *Suppose that*

$$x_1 \mathbf{e}_1 + x_2 \mathbf{e}_2 = \mathbf{0}. \tag{6.1}$$

*By applying* $\alpha$ *to both sides, deduce that*

$$\lambda_1 x_1 \mathbf{e}_1 + \lambda_2 x_2 \mathbf{e}_2 = \mathbf{0}. \tag{6.2}$$

*By subtracting (2) from* $\lambda_2$ *times (1) and using the fact that* $\lambda_1 \neq \lambda_2$, *deduce that* $x_1 = 0$. *Show that* $x_2 = 0$ *and conclude that* $\mathbf{e}_1$ *and* $\mathbf{e}_2$ *are linearly independent.*

(*ii*) *Obtain the result of (i) by applying* $\alpha - \lambda_2 \iota$ *to both sides of (1).*

(*iii*) *Let* $\alpha : \mathbb{F}^3 \to \mathbb{F}^3$ *be a linear map. Suppose that* $\mathbf{e}_1$, $\mathbf{e}_2$ *and* $\mathbf{e}_3$ *are eigenvectors of* $\alpha$ *with distinct eigenvalues* $\lambda_1$, $\lambda_2$ *and* $\lambda_3$. *Suppose that*

$$x_1 \mathbf{e}_1 + x_2 \mathbf{e}_2 + x_3 \mathbf{e}_3 = \mathbf{0}.$$

*By first applying* $\alpha - \lambda_3 \iota$ *to both sides of the equation and then applying* $\alpha - \lambda_2 \iota$ *to both sides of the result, show that* $x_3 = 0$.

*Show that* $\mathbf{e}_1$, $\mathbf{e}_2$ *and* $\mathbf{e}_3$ *are linearly independent.*

**Theorem 6.3.3** *If a linear map* $\alpha : \mathbb{F}^n \to \mathbb{F}^n$ *has n distinct eigenvalues, then the associated eigenvectors form a basis and* $\alpha$ *has a diagonal matrix with respect to this basis.*

*Proof* Let the eigenvectors be $\mathbf{e}_j$ with eigenvalues $\lambda_j$. We observe that

$$(\alpha - \lambda \iota)\mathbf{e}_j = (\lambda_j - \lambda)\mathbf{e}_j.$$

Suppose that

$$\sum_{k=1}^{n} x_k \mathbf{e}_k = \mathbf{0}.$$

Applying

$$\beta_n = (\alpha - \lambda_1 \iota)(\alpha - \lambda_2 \iota) \dots (\alpha - \lambda_{n-1} \iota)$$

to both sides of the equation, we get

$$x_n \prod_{j=1}^{n-1} (\lambda_n - \lambda_j)\mathbf{e}_n = \mathbf{0}.$$

Since an eigenvector must be non-zero, it follows that

$$x_n \prod_{j=1}^{n-1} (\lambda_n - \lambda_j) = 0$$

and, since $\lambda_n - \lambda_j \neq 0$ for $1 \leq j \leq n-1$, we have $x_n = 0$. A similar argument shows that $x_j = 0$ for each $1 \leq j \leq n$ and so $\mathbf{e}_1, \mathbf{e}_2, \dots, \mathbf{e}_n$ are linearly independent. Since every linearly independent collection of $n$ vectors in $\mathbb{F}^n$ forms a basis, the eigenvectors form a basis and the result follows. $\qquad\square$

## 6.4 Linear maps from $\mathbb{C}^2$ to itself

Theorem 6.3.3 gives a sufficient but not a necessary condition for a linear map to be diagonalisable. The identity map $\iota : \mathbb{F}^n \to \mathbb{F}^n$ has only one eigenvalue but has the diagonal matrix $I$ with respect to any basis. On the other hand, even when we work in $\mathbb{C}^n$ rather than $\mathbb{R}^n$, not every linear map is diagonalisable.

**Example 6.4.1** *Let* $\mathbf{u}_1, \mathbf{u}_2$ *be a basis for* $\mathbb{F}^2$. *The linear map* $\beta : \mathbb{F}^2 \to \mathbb{F}^2$ *given by*

$$\beta(x_1\mathbf{u}_1 + x_2\mathbf{u}_2) = x_2\mathbf{u}_1$$

*is non-diagonalisable.*

*Proof* Suppose that $\beta$ is diagonalisable with respect to some basis. Then $\beta$ would have matrix representation

$$D = \begin{pmatrix} d_1 & 0 \\ 0 & d_2 \end{pmatrix},$$

say, with respect to that basis and $\beta^2$ would have matrix representation

$$D^2 = \begin{pmatrix} d_1^2 & 0 \\ 0 & d_2^2 \end{pmatrix}$$

with respect to that basis.

However,

$$\beta^2(x_1\mathbf{u}_1 + x_2\mathbf{u}_2) = \beta(x_2\mathbf{u}_1) = 0$$

for all $x_j$, so $\beta^2 = 0$ and $\beta^2$ has matrix representation

$$\begin{pmatrix} 0 & 0 \\ 0 & 0 \end{pmatrix}$$

with respect to every basis. We deduce that $d_1^2 = d_2^2 = 0$, so $d_1 = d_2 = 0$ and $\beta = 0$ which is absurd. Thus $\beta$ is not diagonalisable. $\qquad\square$

**Exercise 6.4.2** *Here is a slightly different proof that the mapping* $\beta$ *of Example 6.4.1 is not diagonalisable.*

    *(i) Find the characteristic polynomial of* $\beta$ *and show that 0 is the only eigenvalue of* $\beta$.
    *(ii) Find all the eigenvectors of* $\beta$ *and show that they do not span* $\mathbb{F}^n$.

Fortunately, the map just given is the 'typical' non-diagonalisable linear map for $\mathbb{C}^2$.

**Theorem 6.4.3** *If* $\alpha : \mathbb{C}^2 \to \mathbb{C}^2$ *is linear, then exactly one of the following three statements must be true.*

*(i) $\alpha$ has two distinct eigenvalues $\lambda$ and $\mu$ and we can take a basis of eigenvectors $\mathbf{e}_1$, $\mathbf{e}_2$ for $\mathbb{C}^2$. With respect to this basis, $\alpha$ has matrix*

$$\begin{pmatrix} \lambda & 0 \\ 0 & \mu \end{pmatrix}.$$

*(ii) $\alpha$ has only one distinct eigenvalue $\lambda$, but is diagonalisable. Then $\alpha = \lambda \iota$ and has matrix*

$$\begin{pmatrix} \lambda & 0 \\ 0 & \lambda \end{pmatrix}$$

*with respect to any basis.*

*(iii) $\alpha$ has only one distinct eigenvalue $\lambda$ and is not diagonalisable. Then there exists a basis $\mathbf{e}_1$, $\mathbf{e}_2$ for $\mathbb{C}^2$ with respect to which $\alpha$ has matrix*

$$\begin{pmatrix} \lambda & 1 \\ 0 & \lambda \end{pmatrix}.$$

*Note that $\mathbf{e}_1$ is an eigenvector with eigenvalue $\lambda$ but $\mathbf{e}_2$ is not.*

*Proof* As the reader will remember from Lemma 6.2.12, the fundamental theorem of algebra tells us that $\alpha$ must have at least one eigenvalue.

Case (i) is covered by Theorem 6.3.3, so we need only consider the case when $\alpha$ has only one distinct eigenvalue $\lambda$.

If $\alpha$ has matrix representation $A$ with respect to some basis, then $\alpha - \lambda \iota$ has matrix representation $A - \lambda I$ and, conversely, if $\alpha - \lambda \iota$ has matrix representation $A - \lambda I$ with respect to some basis, then $\alpha$ has matrix representation $A$. Further $\alpha - \lambda \iota$ has only one distinct eigenvalue 0. Thus we need only consider the case when $\lambda = 0$.

Now suppose that $\alpha$ has only one distinct eigenvalue 0. If $\alpha = 0$, then we have case (ii). If $\alpha \neq 0$, there must be a vector $\mathbf{e}_2$ such that $\alpha \mathbf{e}_2 \neq 0$. Write $\mathbf{e}_1 = \alpha \mathbf{e}_2$. We show that $\mathbf{e}_1$ and $\mathbf{e}_2$ are linearly independent. Suppose that

$$x_1 \mathbf{e}_1 + x_2 \mathbf{e}_2 = \mathbf{0}. \qquad\qquad \bigstar$$

Applying $\alpha$ to both sides of $\bigstar$, we get

$$x_1 \alpha(\mathbf{e}_1) + x_2 \mathbf{e}_1 = \mathbf{0}.$$

If $x_1 \neq 0$, then $\mathbf{e}_1$ is an eigenvector with eigenvalue $-x_2/x_1$. Thus $x_2 = 0$ and $\bigstar$ tells us that $x_1 = 0$, contrary to our initial assumption. The only consistent possibility is that $x_1 = 0$ and so, from $\bigstar$, $x_2 = 0$. We have shown that $\mathbf{e}_2$ and $\mathbf{e}_2$ are linearly independent and so form a basis for $\mathbb{F}^2$.

Let $\mathbf{u}$ be an eigenvector. Since $\mathbf{e}_1$ and $\mathbf{e}_2$ form a basis, we can find $y_1$, $y_2 \in \mathbb{C}$ such that

$$\mathbf{u} = y_1 \mathbf{e}_1 + y_2 \mathbf{e}_2.$$

Applying $\alpha$ to both sides, we get

$$\mathbf{0} = y_1 \alpha \mathbf{e}_1 + y_2 \mathbf{e}_1.$$

If $y_1 \neq 0$ and $y_2 \neq 0$, then $\mathbf{e}_1$ is an eigenvector with eigenvalue $-y_2/y_1$ which is absurd. If $y_1 = 0$ and $y_2 = 0$, we get $\mathbf{u} = \mathbf{0}$, which is impossible since eigenvectors are non-zero. If $y_1 = 0$ and $y_2 \neq 0$, we get $\mathbf{e}_1 = \mathbf{0}$ which is also impossible.

The only remaining possibility[7] is that $y_1 \neq 0$, $y_2 = 0$ and $\alpha \mathbf{e}_1 = \mathbf{0}$, that is to say, case (iii) holds. $\qquad\square$

The general case of a linear map $\alpha : \mathbb{C}^n \to \mathbb{C}^n$ is substantially more complicated. The possible outcomes are classified using the 'Jordan normal form' which is studied in Section 12.4.

It is unfortunate that we cannot diagonalise all linear maps $\alpha : \mathbb{C}^n \to \mathbb{C}^n$, but the reader should remember that the only cases in which diagonalisation may fail occur when the characteristic polynomial does not have $n$ distinct roots.

We have the following corollary to Theorem 6.4.3.

**Example 6.4.4 [Cayley–Hamilton in two dimensions]** *If* $\alpha : \mathbb{C}^2 \to \mathbb{C}^2$ *is a linear map, let us write* $Q(t) = \det(t\iota - \alpha)$ *for the characteristic polynomial of* $\alpha$. *Then we have*

$$Q(t) = t^2 + at + b$$

*where* $a$, $b \in \mathbb{C}$. *The Cayley–Hamilton theorem states that*

$$\alpha^2 + a\alpha + b\iota = 0$$

*or, more briefly,[8] that* $Q(\alpha) = 0$.

We give the proof as an exercise.

**Exercise 6.4.5** *(i) Suppose that* $\alpha : \mathbb{C}^2 \to \mathbb{C}^2$ *has matrix*

$$A = \begin{pmatrix} \lambda & 0 \\ 0 & \mu \end{pmatrix}$$

*(where* $\lambda$ *and* $\mu$ *need not be distinct) with respect to some basis. Find the characteristic polynomial* $Q(t) = t^2 + at + b$ *of* $\alpha$ *and show that* $A^2 + aA + bI = 0$. *Deduce that*

$$\alpha^2 + a\alpha + b\iota = 0.$$

*(ii) Repeat the calculations of part (i) in the case when* $\alpha : \mathbb{C}^2 \to \mathbb{C}^2$ *has matrix*

$$A = \begin{pmatrix} \lambda & 1 \\ 0 & \lambda \end{pmatrix}$$

*with respect to some basis.*

*(iii) Use Theorem 6.4.3 to obtain the result of Example 6.4.4.*

We shall extend this result to higher dimensions in Section 12.2.

---

[7] How often have I said to you that, when you have eliminated the impossible, whatever remains, however improbable, must be the truth.

(Conan Doyle *The Sign of Four* [12])

[8] But more confusingly for the novice.

The change of basis formula enables us to translate Theorem 6.4.3 into a theorem on $2 \times 2$ complex matrices.

**Theorem 6.4.6** *We work in $\mathbb{C}$. If $A$ is a $2 \times 2$ matrix, then exactly one of the following three things must happen.*
  *(i) There exists a $2 \times 2$ invertible matrix $P$ such that*

$$P^{-1}AP = \begin{pmatrix} \lambda & 0 \\ 0 & \mu \end{pmatrix}$$

*with $\lambda \neq \mu$.*
  *(ii) $A = \lambda I$.*
  *(iii) There exists a $2 \times 2$ invertible matrix $P$ such that*

$$P^{-1}AP = \begin{pmatrix} \lambda & 1 \\ 0 & \lambda \end{pmatrix}.$$

Here is a slightly stronger version of this result.

**Exercise 6.4.7** *By considering matrices of the form $vP$ with $v \in \mathbb{C}$, show that we can choose the matrix $P$ in Theorem 6.4.6 so that $\det P = 1$.*

Theorem 6.4.6 gives us a new way of looking at simultaneous linear differential equations of the form

$$x_1'(t) = a_{11}x_1(t) + a_{12}x_2(t)$$
$$x_2'(t) = a_{21}x_1(t) + a_{22}x_2(t),$$

where $x_1$ and $x_2$ are differentiable functions and the $a_{ij}$ are constants. If we write

$$A = \begin{pmatrix} a_{11} & a_{12} \\ a_{21} & a_{22} \end{pmatrix},$$

then, by Theorem 6.4.6, we can find an invertible $2 \times 2$ matrix $P$ such that $P^{-1}AP = B$, where $B$ takes one of the following forms:

$$\begin{pmatrix} \lambda_1 & 0 \\ 0 & \lambda_2 \end{pmatrix} \text{ with } \lambda_1 \neq \lambda_2, \quad \begin{pmatrix} \lambda & 0 \\ 0 & \lambda \end{pmatrix}, \quad \text{or} \quad \begin{pmatrix} \lambda & 1 \\ 0 & \lambda \end{pmatrix}.$$

If we set

$$\begin{pmatrix} X_1(t) \\ X_2(t) \end{pmatrix} = P^{-1} \begin{pmatrix} x_1(t) \\ x_2(t) \end{pmatrix},$$

then

$$\begin{pmatrix} \dot{X}_1(t) \\ \dot{X}_2(t) \end{pmatrix} = P^{-1} \begin{pmatrix} \dot{x}_1(t) \\ \dot{x}_2(t) \end{pmatrix} = P^{-1}A \begin{pmatrix} x_1(t) \\ x_2(t) \end{pmatrix} = P^{-1}AP \begin{pmatrix} X_1(t) \\ X_2(t) \end{pmatrix} = B \begin{pmatrix} X_1(t) \\ X_2(t) \end{pmatrix}.$$

Thus, if

$$B = \begin{pmatrix} \lambda_1 & 0 \\ 0 & \lambda_2 \end{pmatrix} \quad \text{with } \lambda_1 \neq \lambda_2,$$

then

$$\dot{X}_1(t) = \lambda_1 X_1(t)$$
$$\dot{X}_2(t) = \lambda_2 X_2(t),$$

and so, by elementary analysis (see Exercise 6.4.8),

$$X_1(t) = C_1 e^{\lambda_1 t}, \quad X_2(t) = C_2 e^{\lambda_2 t}$$

for some arbitrary constants $C_1$ and $C_2$. It follows that

$$\begin{pmatrix} x_1(t) \\ x_2(t) \end{pmatrix} = P \begin{pmatrix} X_1(t) \\ X_2(t) \end{pmatrix} = P \begin{pmatrix} C_1 e^{\lambda_1 t} \\ C_2 e^{\lambda_2 t} \end{pmatrix}$$

for arbitrary constants $C_1$ and $C_2$.

If

$$B = \begin{pmatrix} \lambda & 0 \\ 0 & \lambda \end{pmatrix} = \lambda I,$$

then $A = \lambda I$, $\dot{x}_j(t) = \lambda x_j(t)$ and $x_j(t) = C_j e^{\lambda t}$ $[j = 1, 2]$ for arbitrary constants $C_1$ and $C_2$.

If

$$B = \begin{pmatrix} \lambda & 1 \\ 0 & \lambda \end{pmatrix},$$

then

$$\dot{X}_1(t) = \lambda_1 X_1(t) + X_2(t)$$
$$\dot{X}_2(t) = \lambda_2 X_2(t),$$

and so, by elementary analysis,

$$X_2(t) = C_2 e^{\lambda t},$$

for some arbitrary constant $C_2$ and

$$\dot{X}_1(t) = \lambda X_1(t) + C_2 e^{\lambda t}.$$

It follows (see Exercise 6.4.8) that

$$X_1(t) = (C_1 + C_2 t) e^{\lambda t}$$

for some arbitrary constant $C_1$ and

$$\begin{pmatrix} x_1(t) \\ x_2(t) \end{pmatrix} = P \begin{pmatrix} X_1(t) \\ X_2(t) \end{pmatrix} = P \begin{pmatrix} (C_1 + C_2 t) e^{\lambda t} \\ C_2 e^{\lambda t} \end{pmatrix}.$$

**Exercise 6.4.8** (*Most readers will already know the contents of this exercise.*)
 (i) *If $\dot{x}(t) = \lambda x(t)$, show that*

$$\frac{d}{dt}(e^{-\lambda t}x(t)) = 0.$$

*Deduce that $e^{-\lambda t}x(t) = C$ for some constant $C$ and so $x(t) = Ce^{\lambda t}$.*
 (ii) *If $\dot{x}(t) = \lambda x(t) + Ke^{\lambda t}$, show that*

$$\frac{d}{dt}\left(e^{-\lambda t}x(t) - Kt\right) = 0$$

*and deduce that $x(t) = (C + Kt)e^{\lambda t}$ for some constant $C$.*

**Exercise 6.4.9**   *Consider the differential equation*

$$\ddot{x}(t) + a\dot{x}(t) + bx(t) = 0.$$

*Show that, if we write $x_1(t) = x(t)$ and $x_2(t) = \dot{x}(t)$, we obtain the equivalent system*

$$\dot{x}_1(t) = x_2(t)$$
$$\dot{x}_2(t) = -bx_1(t) - ax_2(t).$$

*Show, by direct calculation, that the eigenvalues of the matrix*

$$A = \begin{pmatrix} 0 & 1 \\ -b & -a \end{pmatrix}$$

*are the roots of the, so-called,* auxiliary polynomial

$$\lambda^2 + b\lambda + a.$$

At the end of this discussion, the reader may ask whether we can solve any system of differential equations that we could not solve before. The answer is, of course, no, but we have learnt a new way of looking at linear differential equations and two ways of looking at something may be better than just one.

## 6.5 Diagonalising square matrices

If we use the change of basis formula to translate our results on linear maps $\alpha : \mathbb{F}^n \to \mathbb{F}^n$ into theorems about $n \times n$ matrices, Theorem 6.3.3 takes the following form.

**Theorem 6.5.1**   *We work in $\mathbb{F}$. If $A$ is an $n \times n$ matrix and the polynomial $Q(t) = \det(tI - A)$ has $n$ distinct roots in $\mathbb{F}$, then there exists an $n \times n$ invertible matrix $P$ such that $P^{-1}AP$ is diagonal.*

**Exercise 6.5.2** *Let $A = (a_{ij})$ be an $n \times n$ matrix with entries in $\mathbb{C}$. Consider the system of simultaneous linear differential equations*

$$\dot{x}_j(t) = \sum_{j=1}^{n} a_{ij} x_i(t).$$

*If $A$ has $n$ distinct eigenvalues $\lambda_1, \lambda_2, \ldots, \lambda_n$, show that*

$$x_1(t) = \sum_{j=1}^{n} \mu_j e^{\lambda_j t}$$

*for some constants $\mu_j$.*

*By considering the special case when $A$ is diagonal show that, in some cases, it may not be possible to choose the $\mu_j$ freely.*

Bearing in mind our motto 'linear maps for understanding, matrices for computation' we may ask how to convert Theorem 6.5.1 from a statement of theory to a concrete computation.

It will be helpful to make the following definitions, transferring notions already familiar for linear maps to the context of matrices.

**Definition 6.5.3** *We work over $\mathbb{F}$. If $A$ is an $n \times n$ matrix, we say that the* characteristic polynomial $\chi_A$ *of $A$ is given by*

$$\chi_A(t) = \det(tI - A).$$

*If $\mathbf{u}$ is a non-zero column vector, we say that $\mathbf{u}$ is an* eigenvector *of $A$ with associated eigenvalue $\lambda$ if*

$$A\mathbf{u} = \lambda\mathbf{u}.$$

*We say that $A$ is* diagonalisable *if we can find an invertible $n \times n$ matrix $P$ and an $n \times n$ diagonal matrix $D$ with $P^{-1}AP = D$.*

Suppose that we wish to 'diagonalise' an $n \times n$ matrix $A$. The first step is to look at the roots of the characteristic polynomial

$$\chi_A(t) = \det(tI - A).$$

If we work over $\mathbb{R}$ and some of the roots of $\chi_A$ are not real, we know at once that $A$ is not diagonalisable (over $\mathbb{R}$). If we work over $\mathbb{C}$ or if we work over $\mathbb{R}$ and all the roots are real, we can move on to the next stage. Either the characteristic polynomial has $n$ distinct roots or it does not. We shall discuss the case when it does not in Section 12.4. If it does, we know that $A$ is diagonalisable. If we find the $n$ distinct roots (easier said than done outside the artificial conditions of the examination room) $\lambda_1, \lambda_2, \ldots, \lambda_n$, we know, without further computation, that there exists a non-singular $P$ such that $P^{-1}AP = D$ where $D$ is a diagonal matrix with diagonal entries $\lambda_j$. Often knowledge of $D$ is sufficient for our purposes,[9] but, if not, we proceed to find $P$ as follows. For each $\lambda_j$, we know that the system of $n$ linear equations in $n$ unknowns given by

$$(A - \lambda_j I)\mathbf{x} = \mathbf{0}$$

(where $\mathbf{x}$ is a column vector) has non-zero solutions. Let $\mathbf{e}_j$ be one of them so that

$$A\mathbf{e}_j = \lambda_j \mathbf{e}_j.$$

If $P$ is the $n \times n$ matrix with $j$th column $\mathbf{e}_j$ and $\mathbf{u}_j$ is the column vector with 1 in the $j$th place and 0 elsewhere, then $P\mathbf{u}_j = \mathbf{e}_j$ and so $P^{-1}\mathbf{e}_j = \mathbf{u}_j$. It follows that

$$P^{-1}AP\mathbf{u}_j = P^{-1}A\mathbf{e}_j = \lambda_j P^{-1}\mathbf{e}_j = \lambda_j \mathbf{u}_j = D\mathbf{u}_j$$

for all $1 \le j \le n$ and so

$$P^{-1}AP = D.$$

If we need to know $P^{-1}$, we calculate it by inverting $P$ in some standard way.

**Example 6.5.4** *Diagonalise the matrix*

$$R_\theta = \begin{pmatrix} \cos\theta & -\sin\theta \\ \sin\theta & \cos\theta \end{pmatrix}$$

*(with $\theta$ real) over* $\mathbb{C}$.

*Calculation* We have

$$\det(tI - R_\theta) = \det \begin{pmatrix} t - \cos\theta & \sin\theta \\ -\sin\theta & t - \cos\theta \end{pmatrix}$$
$$= (t - \cos\theta)^2 + \sin^2\theta$$
$$= \big((t - \cos\theta) + i\sin\theta\big)\big((t - \cos\theta) - i\sin\theta\big)$$
$$= (t - e^{i\theta})(t - e^{-i\theta}).$$

If $\theta \equiv 0 \mod \pi$, then the characteristic polynomial has a repeated root. In the case $\theta \equiv 0 \mod 2\pi$, $R_\theta = I$. In the case $\theta \equiv \pi \mod 2\pi$, $R_\theta = -I$. In both cases, $R_\theta$ is already in diagonal form.

From now on we take $\theta \not\equiv 0 \mod \pi$, so that the characteristic polynomial has two distinct roots $e^{i\theta}$ and $e^{-i\theta}$. Without further calculation, we know that $R_\theta$ can be diagonalised to obtain the diagonal matrix

$$D = \begin{pmatrix} e^{i\theta} & 0 \\ 0 & e^{-i\theta} \end{pmatrix}.$$

---

[9] It is always worthwhile to pause before indulging in extensive calculation and ask why we need the result.

We now look for an eigenvector corresponding to the eigenvalue $e^{i\theta}$. We need to find a non-zero solution to the equation $R_\theta \mathbf{z} = e^{i\theta}\mathbf{z}$, that is to say, to the system of equations

$$(\cos\theta)z_1 - (\sin\theta)z_2 = (\cos\theta + i\sin\theta)z_1$$
$$(\sin\theta)z_1 + (\cos\theta)z_2 = (\cos\theta + i\sin\theta)z_2.$$

Since $\theta \neq 0 \mod \pi$, we have $\sin\theta \neq 0$ and our system of equations collapses to the single equation

$$z_2 = -iz_1.$$

We can choose any non-zero multiple of $(1, -i)^T$ as an appropriate eigenvector, but we shall simply take our eigenvector as

$$\mathbf{u} = \begin{pmatrix} 1 \\ -i \end{pmatrix}.$$

To find an eigenvector corresponding to the eigenvalue $e^{-i\theta}$, we look at the equation $R_\theta \mathbf{z} = e^{-i\theta}\mathbf{z}$, that is to say,

$$(\cos\theta)z_1 - (\sin\theta)z_2 = (\cos\theta - i\sin\theta)z_1$$
$$(\sin\theta)z_1 + (\cos\theta)z_2 = (\cos\theta - i\sin\theta)z_2,$$

which reduces to

$$z_1 = -iz_2.$$

We take our eigenvector to be

$$\mathbf{v} = \begin{pmatrix} 1 \\ -i \end{pmatrix}.$$

With these choices

$$P = (\mathbf{u}|\mathbf{v}) = \begin{pmatrix} 1 & -i \\ -i & 1 \end{pmatrix}.$$

Using one of the methods for inverting matrices (all are easy in the $2 \times 2$ case), we get

$$P^{-1} = \frac{1}{2}\begin{pmatrix} 1 & i \\ i & 1 \end{pmatrix}$$

and

$$P^{-1}R_\theta P = D.$$

A feeling for symmetry suggests that, instead of using $P$, we should use

$$Q = \frac{1}{\sqrt{2}}P,$$

so that

$$Q = \frac{1}{\sqrt{2}} \begin{pmatrix} 1 & -i \\ -i & 1 \end{pmatrix} \quad \text{and} \quad Q^{-1} = \frac{1}{\sqrt{2}} \begin{pmatrix} 1 & i \\ i & 1 \end{pmatrix}.$$

This just corresponds to a different choice of eigenvectors.                                □

**Exercise 6.5.5** *Check, by doing the matrix multiplication, that*

$$Q^{-1} R_\theta Q = D.$$

The reader will probably find herself computing eigenvalues and eigenvectors in several different courses. This is a useful exercise for familiarising oneself with matrices, determinants and eigenvalues, but, like most exercises, slightly artificial.[10] Obviously, the matrix will have been chosen to make calculation easier. If you have to find the roots of a cubic, it will often turn out that the numbers have been chosen so that one root is a small integer. More seriously and less obviously, polynomials of high order need not behave well and the kind of computation which works for $3 \times 3$ matrices may be unsuitable for $n \times n$ matrices when $n$ is large. To see one of the problems that may arise, look at Exercise 6.8.31.

### 6.6 Iteration's artful aid

During the last 150 years, mathematicians have become increasingly interested in iteration. If we have a map $\alpha : X \to X$, what can we say about $\alpha^q$, the map obtained by applying $\alpha$ $q$ times, when $q$ is large? If $X$ is a vector space and $\alpha$ a linear map, then eigenvectors provide a powerful tool for investigation.

**Lemma 6.6.1** *We consider $n \times n$ matrices over $\mathbb{F}$. If $D$ is diagonal, $P$ invertible and $PAP^{-1} = D$, then*

$$A^q = PD^q P^{-1}.$$

*Proof* This is immediate. Observe that $A = PDP^{-1}$ so

$$A^q = (PDP^{-1})(PAP^{-1})\ldots(PAP^{-1})$$
$$= PD(PP^{-1})D(PP^{-1})\ldots(PP^{-1})DP = PD^q P^{-1}.$$

                                                                                □

**Exercise 6.6.2** *(i) Why is it easy to compute $D^q$?*

*(ii) Explain the result of Lemma 6.6.1 by considering the matrix of the linear maps $\alpha$ and $\alpha^q$ with respect to two different bases.*

Usually, it is more instructive to look at the eigenvectors themselves.

---

[10] Though not as artificial as pressing an 'eigenvalue button' on a calculator and thinking that one gains understanding thereby.

**Lemma 6.6.3** *Suppose that* $\alpha : \mathbb{F}^n \to \mathbb{F}^n$ *is a linear map with a basis of eigenvectors* $\mathbf{e}_1$, $\mathbf{e}_2, \ldots, \mathbf{e}_n$ *with eigenvalues* $\lambda_1, \lambda_2, \ldots, \lambda_n$. *Then*

$$\alpha^q \sum_{j=1}^{n} x_j \mathbf{e}_j = \sum_{j=1}^{n} \lambda_j^q x_j \mathbf{e}_j.$$

*Proof* Immediate. $\qquad\qquad\qquad\qquad\qquad\qquad\qquad\qquad\qquad\qquad\qquad$ $\square$

It is natural to introduce the following definition.

**Definition 6.6.4** *If* $\mathbf{u}(1), \mathbf{u}(2), \ldots$ *is a sequence of vectors in* $\mathbb{F}^n$ *and* $\mathbf{u} \in \mathbb{F}^n$, *we say that* $\mathbf{u}(r) \to \mathbf{u}$ coordinatewise *if* $u_i(r) \to u_i$ *as* $r \to \infty$ *for each* $1 \leq i \leq n$.

**Lemma 6.6.5** *Suppose that* $\alpha : \mathbb{F}^n \to \mathbb{F}^n$ *is a linear map with a basis of eigenvectors* $\mathbf{e}_1$, $\mathbf{e}_2, \ldots, \mathbf{e}_n$ *with eigenvalues* $\lambda_1, \lambda_2, \ldots, \lambda_n$. *If* $|\lambda_1| > |\lambda_j|$ *for* $2 \leq j \leq n$, *then*

$$\lambda_1^{-q} \alpha^q \sum_{j=1}^{n} x_j \mathbf{e}_j \to x_1 \mathbf{e}_1$$

*coordinatewise as* $q \to \infty$.

Speaking very informally, repeated iteration brings out the eigenvector corresponding to the largest eigenvalue in absolute value.

The hypotheses of Lemma 6.6.5 demand that $\alpha$ be diagonalisable and that the largest eigenvalue in absolute value should be unique. In the special case $\alpha : \mathbb{C}^2 \to \mathbb{C}^2$ we can use Theorem 6.4.3 to say rather more.

**Example 6.6.6** *Suppose that* $\alpha : \mathbb{C}^2 \to \mathbb{C}^2$ *is linear. Exactly one of the following things must happen.*

*(i) There is a basis* $\mathbf{e}_1$, $\mathbf{e}_2$ *with respect to which A has matrix*

$$\begin{pmatrix} \lambda & 0 \\ 0 & \mu \end{pmatrix}$$

*with* $|\lambda| > |\mu|$. *Then*

$$\lambda^{-q} \alpha^q (x_1 \mathbf{e}_1 + x_2 \mathbf{e}_2) \to x_1 \mathbf{e}_1$$

*coordinatewise.*

*(i)' There is a basis* $\mathbf{e}_1$, $\mathbf{e}_2$ *with respect to which* $\alpha$ *has matrix*

$$\begin{pmatrix} \lambda & 0 \\ 0 & \mu \end{pmatrix}$$

*with* $|\lambda| = |\mu|$ *but* $\lambda \neq \mu$. *Then* $\lambda^{-q} \alpha^q (x_1 \mathbf{e}_1 + x_2 \mathbf{e}_2)$ *fails to converge coordinatewise except in the special cases when* $x_2 = 0$.

*(ii)* $\alpha = \lambda \iota$ *with* $\lambda \neq 0$, *so*

$$\lambda^{-q} \alpha^q \mathbf{u} = \mathbf{u} \to \mathbf{u}$$

*coordinatewise for all* $\mathbf{u}$.

*(ii)′* $\alpha = 0$ *and*

$$\alpha^q \mathbf{u} = \mathbf{0} \to \mathbf{0}$$

*coordinatewise for all* $\mathbf{u}$.

*(iii) There is a basis* $\mathbf{e}_1$, $\mathbf{e}_2$ *with respect to which* $\alpha$ *has matrix*

$$\begin{pmatrix} \lambda & 1 \\ 0 & \lambda \end{pmatrix}$$

*with* $\lambda \neq 0$. *Then*

$$\alpha^q (x_1 \mathbf{e}_1 + x_2 \mathbf{e}_2) = (\lambda^q x_1 + q\lambda^{q-1} x_2)\mathbf{e}_1 + \lambda^q x_2 \mathbf{e}_2$$

*for all* $q \geq 0$ *and so*

$$q^{-1}\lambda^{-q}\alpha^q (x_1 \mathbf{e}_1 + x_2 \mathbf{e}_2) \to \lambda^{-1} x_2 \mathbf{e}_1$$

*coordinatewise.*

*(iii)′ There is a basis* $\mathbf{e}_1$, $\mathbf{e}_2$ *with respect to which* $\alpha$ *has matrix*

$$\begin{pmatrix} 0 & 1 \\ 0 & 0 \end{pmatrix}.$$

*Then*

$$\alpha^q \mathbf{u} = \mathbf{0}$$

*for all* $q \geq 2$ *and so*

$$\alpha^q \mathbf{u} \to \mathbf{0}$$

*coordinatewise for all* $\mathbf{u}$.

**Exercise 6.6.7** *Check the statements in Example 6.6.6. Pay particular attention to part (iii).*

As an application, we consider sequences generated by linear difference equations. A typical example is given by the Fibonacci sequence[11]

$$1, \ 1, \ 2, \ 3, \ 5, \ 8, \ 13, \ 21, \ 34, \ 55, \ldots$$

where the $n$th term $F_n$ is defined by the equation

$$F_n = F_{n-1} + F_{n-2}$$

and we impose the condition $F_1 = F_2 = 1$.

**Exercise 6.6.8** *Find* $F_{10}$. *Show that* $F_0 = 0$ *and compute* $F_{-1}$ *and* $F_{-2}$. *Show that* $F_n = (-1)^{n+1} F_{-n}$.

More generally, we look at sequences $u_n$ of complex numbers satisfying

$$u_n + au_{n-1} + bu_{n-2} = 0$$

with $b \neq 0$.

---

[11] 'Have you ever formed any theory, why in spire of leaves … the angles go 1/2, 1/3, 2/5, 3/8, etc. … . It seems to me most marvellous.' (Darwin [13]).

**Exercise 6.6.9** *What can we say if $b = 0$ but $a \neq 0$? What can we say if $a = b = 0$?*

Analogy with linear differential equations suggests that we look at vectors

$$\mathbf{u}(n) = \begin{pmatrix} u_n \\ u_{n+1} \end{pmatrix}.$$

We then have

$$\mathbf{u}(n+1) = A\mathbf{u}(n), \quad \text{where } A = \begin{pmatrix} 0 & 1 \\ -a & -b \end{pmatrix}$$

and so

$$\begin{pmatrix} u_n \\ u_{n+1} \end{pmatrix} = A^n \begin{pmatrix} u_0 \\ u_1 \end{pmatrix}.$$

The eigenvalues of $A$ are the roots of

$$Q(t) = \det(tI - A) = \det \begin{pmatrix} t & -1 \\ a & t+b \end{pmatrix} = t^2 + bt + a.$$

Since $a \neq 0$, the eigenvalues are non-zero.

If $Q$ has two distinct roots $\lambda$ and $\mu$ with corresponding eigenvectors $\mathbf{z}$ and $\mathbf{w}$, then we can find real numbers $p$ and $q$ such that

$$\begin{pmatrix} u_0 \\ u_1 \end{pmatrix} = p\mathbf{z} + q\mathbf{w},$$

so

$$\begin{pmatrix} u_n \\ u_{n+1} \end{pmatrix} = A^n(p\mathbf{z} + q\mathbf{w}) = \lambda^n p\mathbf{z} + \mu^n q\mathbf{w}$$

and, looking at the first entry of the column vector,

$$u_n = z_1 p\lambda^n + w_1 q\mu^n.$$

Thus $u_n = c\lambda^n + c'\mu^n$ for some constants $c$ and $c'$ depending on $u_0$ and $u_1$.

**Exercise 6.6.10** *If $Q$ has only one distinct root $\lambda$ show by a similar argument that*

$$u_n = (c + c'n)\lambda^n$$

*for some constants $c$ and $c'$ depending on $u_0$ and $u_1$.*

**Exercise 6.6.11** *Suppose that the polynomial $P(t) = t^m + \sum_{j=0}^{m-1} a_j t^j$ has $m$ distinct non-zero roots $\lambda_1, \lambda_2, \ldots, \lambda_m$. Show, by an argument modelled on that just given in the case $m = 2$, that any sequence $u_n$ satisfying the difference equation*

$$u_r + \sum_{j=0}^{m-1} a_j u_{j-m+r} = 0 \qquad \bigstar$$

*must have the form*

$$u_r = \sum_{k=1}^{m} c_k \lambda_k^r \qquad\qquad \bigstar\bigstar$$

*for some constants* $c_k$. *Show, conversely, by direct substitution in* $\bigstar$, *that any expression of the form* $\bigstar\bigstar$ *satisfies* $\bigstar$.

*If* $|\lambda_1| > |\lambda_k|$ *for* $2 \leq k \leq m$ *and* $c_1 \neq 0$, *show that* $u_r \neq 0$ *for* $r$ *large and*

$$\frac{u_r}{u_{r-1}} \to \lambda_1$$

*as* $r \to \infty$. *Formulate a similar theorem for the case* $r \to -\infty$.
*[If you prefer not to use too much notation, just do the case* $m = 3$.]*

**Exercise 6.6.12** *Our results take a particularly pretty form when applied to the Fibonacci sequence.*

(i) *If we write* $\tau = (1 + \sqrt{5})/2$, *show that* $\tau^{-1} = (-1 + \sqrt{5})/2$. *(The number* $\tau$ *is sometimes called the golden ratio.)*

(ii) *Find the eigenvalues and associated eigenvectors for*

$$A = \begin{pmatrix} 0 & 1 \\ 1 & 1 \end{pmatrix}.$$

*Write* $(0, 1)^T$ *as the sum of eigenvectors.*

(iii) *Use the method outlined in our discussion of difference equations to obtain* $F_n = c_1 \tau^n + c_2 \tau^{-n}$, *where* $c_1$ *and* $c_2$ *are to be found explicitly.*

(iv) *Show that* $F(n)$ *is the closest integer to* $\tau^n / \sqrt{5}$. *Use more explicit calculations when* $n$ *is small to show that* $F(n)$ *is the closest integer to* $\tau^n / \sqrt{5}$ *for all* $n \geq 0$.

(v) *Show that*

$$A^n = \begin{pmatrix} F_{n-1} & F_n \\ F_n & F_{n+1} \end{pmatrix}$$

*for all* $n \geq 1$. *Deduce, by taking determinants, that*

$$F_{n-1} F_{n+1} - F_n^2 = (-1)^n$$

*for all* $n \geq 1$. *(This is Casini's identity.)*

(vi) *Are the results of* (v) *true for all* $n$? *Why?*

(vii) *Use the observation that* $A^n A^n = A^{2n}$ *to show that*

$$F_n^2 + F_{n-1}^2 = F_{2n-1}.$$

**Exercise 6.6.13** *Here is another example of iteration. Suppose that we have* $m$ *airports called, rather unimaginatively,* $1, 2, \ldots, m$. *Some are linked by direct flights and some are not. We are initially interested in whether it is possible to get from* $i$ *to* $j$ *in* $r$ *flights.*

(i) *Let* $D = (d_{ij})$ *be the* $m \times m$ *matrix such that* $d_{ij} = 1$ *if* $i \neq j$ *and there is a direct flight from* $i$ *to* $j$ *and* $d_{ij} = 0$ *otherwise. If we write* $(d_{ij}^{(n)}) = D^n$, *show that* $d_{ij}^{(n)}$ *is the* *number of journeys from i to j which involve exactly n flights. In particular, there is a journey from i to j which involves exactly n flights if and only if $d_{ij}^{(n)} > 0$.*

*(ii) Let $\tilde{D} = (\tilde{d}_{ij})$ be the $m \times m$ matrix such that $\tilde{d}_{ij} = 1$ if there is a direct flight from i to j or if $i = j$ and $\tilde{d}_{ij} = 0$ otherwise. If we write $(\tilde{d}_{ij}^{(n)}) = \tilde{D}^n$, interpret the meaning of $\tilde{d}_{ij}^{(n)}$. Produce useful information along the lines of the last sentence of (i).*

*(iii) A peasant must row a wolf, a goat and a cabbage across a river in a boat that will only carry one passenger at a time. If he leaves the wolf with the goat, then the wolf will eat the goat. If he leaves the goat with the cabbage, then the goat will eat the cabbage. The cabbage represents no threat to the wolf, nor the wolf to the cabbage. Explain (there is no need to carry out the calculation) how to use the ideas above to find the smallest number of trips the peasant must make. If the problem is insoluble will your method reveal the fact and, if so, how?*

## 6.7 LU factorisation

Although diagonal matrices are very easy to handle, there are other convenient forms of matrices. People who actually have to do computations are particularly fond of triangular matrices (see Definition 4.5.2).

We have met such matrices several times before, but we start by recalling some elementary properties.

**Exercise 6.7.1** *(i) If L is lower triangular, show that $\det L = \prod_{j=1}^{n} l_{jj}$.*

*(ii) Show that a lower triangular matrix is invertible if and only if all its diagonal entries are non-zero.*

*(iii) If L is lower triangular, show that the roots of the characteristic polynomial are the diagonal entries $l_{jj}$ (multiple roots occurring with the correct multiplicity). Can you identify one eigenvector of L explicitly?*

If $L$ is an invertible lower triangular $n \times n$ matrix, then, as we have noted before, it is very easy to solve the system of linear equations

$$L\mathbf{x} = \mathbf{y},$$

since they take the form

$$l_{11}x_1 = y_1$$
$$l_{21}x_1 + l_{22}x_2 = y_2$$
$$l_{31}x_1 + l_{32}x_2 + l_{33}x_3 = y_3$$
$$\vdots$$
$$l_{n1}x_1 + l_{n2}x_2 + l_{n3}x_3 + \cdots + l_{nn}x_n = y_n.$$

We first compute $x_1 = l_{11}^{-1} y_1$. Knowing $x_1$, we can now compute

$$x_2 = l_{22}^{-1}(y_2 - l_{21}x_1).$$

We now know $x_1$ and $x_2$ and can compute

$$x_3 = l_{33}^{-1}(y_3 - l_{31}x_1 - l_{32}x_3),$$

and so on.

**Exercise 6.7.2**  *Suppose that $U$ is an invertible upper triangular $n \times n$ matrix. Show how to solve $U\mathbf{x} = \mathbf{y}$ in an efficient manner.*

**Exercise 6.7.3**  *(i) If $L$ is an invertible lower triangular matrix, show that $L^{-1}$ is a lower triangular matrix.*
*(ii) Show that the product of two lower triangular matrices is lower triangular.*

Our main result is given in the next theorem. The method of proof is as important as the proof itself.

**Theorem 6.7.4**  *If $A$ is an $n \times n$ invertible matrix, then, possibly after interchange of columns, we can find a lower triangular $n \times n$ matrix $L$ with all diagonal entries 1 and a non-singular upper triangular invertible $n \times n$ matrix $U$ such that*

$$A = LU.$$

*Proof*  We use induction on $n$. If we deal with $1 \times 1$ matrices, then we have the trivial equality $(a)(1) = (a)$, so the result holds for $n = 1$.

We now suppose that the result is true for $(n-1) \times (n-1)$ matrices and that $A$ is an invertible $n \times n$ matrix.

Since $A$ is invertible, at least one element of its first row must be non-zero. By interchanging columns,[12] we may assume that $a_{11} \neq 0$. If we now take

$$\mathbf{l} = \begin{pmatrix} l_{11} \\ l_{21} \\ \vdots \\ l_{n1} \end{pmatrix} = \begin{pmatrix} 1 \\ a_{11}^{-1}a_{21} \\ \vdots \\ a_{11}^{-1}a_{n1} \end{pmatrix} \quad \text{and} \quad \mathbf{u} = \begin{pmatrix} u_{11} \\ u_{12} \\ \vdots \\ u_{1n} \end{pmatrix} = \begin{pmatrix} a_{11} \\ a_{12} \\ \vdots \\ a_{1n} \end{pmatrix},$$

then $\mathbf{l}\mathbf{u}^T$ is an $n \times n$ matrix whose first row and column coincide with the first row and column of $A$.

---

[12] Remembering the method of Gaussian elimination, the reader may suspect, correctly, that in numerical computation it may be wise to ensure that $|a_{11}| \geq |a_{1j}|$ for $1 \leq j \leq n$. Note that, if we do interchange columns, we must keep a record of what we have done.

We thus have

$$A - \mathbf{l}\mathbf{u}^T = \begin{pmatrix} 0 & 0 & 0 & \cdots & 0 \\ 0 & b_{22} & b_{23} & \cdots & b_{2n} \\ 0 & b_{32} & b_{33} & \cdots & b_{3n} \\ \vdots & \vdots & \vdots & & \vdots \\ 0 & b_{n2} & b_{n3} & \cdots & b_{nn} \end{pmatrix}.$$

We take $B$ to be the $(n-1) \times (n-1)$ matrix $(b_{ij})_{2 \leq i, j \leq n}$ with

$$b_{ij} = a_{ij} - \frac{a_{i1} a_{1j}}{a_{11}}.$$

Using various results about calculating determinants, in particular the rule that adding multiples of the first row to other rows of a square matrix leaves the determinant unchanged, we see that

$$a_{11} \det B = \det \begin{pmatrix} a_{11} & 0 & 0 & \cdots & 0 \\ 0 & b_{22} & b_{23} & \cdots & b_{2n} \\ 0 & b_{32} & b_{33} & \cdots & b_{3n} \\ \vdots & \vdots & \vdots & & \vdots \\ 0 & b_{n2} & b_{n3} & \cdots & b_{nn} \end{pmatrix}$$

$$= \det \begin{pmatrix} a_{11} & a_{12} & a_{13} & \cdots & a_{1n} \\ 0 & b_{22} & b_{23} & \cdots & b_{2n} \\ 0 & b_{32} & b_{33} & \cdots & b_{3n} \\ \vdots & \vdots & \vdots & & \vdots \\ 0 & b_{n2} & b_{n3} & \cdots & b_{nn} \end{pmatrix} = \det A.$$

Since $\det A \neq 0$, it follows that $\det B \neq 0$ and, by the inductive hypothesis, we can find an $n-1 \times n-1$ lower triangular matrix $\tilde{L} = (l_{ij})_{2 \leq i, j \leq n}$ with $l_{ii} = 1$ $[2 \leq i \leq n]$ and a non-singular $n-1 \times n-1$ upper triangular matrix $\tilde{U} = (u_{ij})_{2 \leq i, j \leq n}$ such that $B = \tilde{L}\tilde{U}$.

If we now define $l_{1j} = 0$ for $2 \leq j \leq n$ and $u_{i1} = 0$ for $2 \leq i \leq n$, then $L = (l_{ij})_{1 \leq i, j \leq n}$ is an $n \times n$ lower triangular matrix with $l_{ii} = 1$ $[1 \leq i \leq n]$, and $U = (u_{ij})_{1 \leq i, j \leq n}$ is an $n \times n$ non-singular upper triangular matrix (recall that a triangular matrix is non-singular if and only if its diagonal entries are non-zero) with

$$LU = A.$$

The induction is complete. □

This theorem is sometimes attributed to Turing. Certainly Turing was one of the first people to realise the importance of $LU$ factorisation for the new era of numerical analysis made possible by the electronic computer.

Observe that the method of proof of our theorem gives a method for actually finding $L$ and $U$. I suggest that the reader studies Example 6.7.5 and then does some calculations of her own (for example, she could do Exercise 6.8.26) before returning to the proof. She may well then find that matters are a lot simpler than at first appears.

**Example 6.7.5** *Find the LU factorisation of*

$$A = \begin{pmatrix} 2 & 1 & 1 \\ 4 & 1 & 0 \\ -2 & 2 & 1 \end{pmatrix}.$$

*Solution* Observe that

$$\begin{pmatrix} 1 \\ 2 \\ -1 \end{pmatrix} \begin{pmatrix} 2 & 1 & 1 \end{pmatrix} = \begin{pmatrix} 2 & 1 & 1 \\ 4 & 2 & 2 \\ -2 & -1 & -1 \end{pmatrix}$$

and

$$\begin{pmatrix} 2 & 1 & 1 \\ 4 & 1 & 0 \\ -2 & 2 & 1 \end{pmatrix} - \begin{pmatrix} 2 & 1 & 1 \\ 4 & 2 & 2 \\ -2 & -1 & -1 \end{pmatrix} = \begin{pmatrix} 0 & 0 & 0 \\ 0 & -1 & -2 \\ 0 & 3 & 2 \end{pmatrix}.$$

Next observe that

$$\begin{pmatrix} 1 \\ -3 \end{pmatrix} \begin{pmatrix} -1 & -2 \end{pmatrix} = \begin{pmatrix} -1 & -2 \\ 3 & 6 \end{pmatrix}$$

and

$$\begin{pmatrix} -1 & -2 \\ 3 & 2 \end{pmatrix} - \begin{pmatrix} -1 & -2 \\ 3 & 6 \end{pmatrix} = \begin{pmatrix} 0 & 0 \\ 0 & -4 \end{pmatrix}.$$

Since $(1)(-4) = (-4)$, we see that $A = LU$ with

$$L = \begin{pmatrix} 1 & 0 & 0 \\ 2 & 1 & 0 \\ -1 & -3 & 1 \end{pmatrix} \quad \text{and} \quad U = \begin{pmatrix} 2 & 1 & 1 \\ 0 & -1 & -2 \\ 0 & 0 & -4 \end{pmatrix}.$$

$\square$

**Exercise 6.7.6** *Check that, if we take L and U as in Example 6.7.5, it is, indeed, true that* $LU = A$. *(It is usually wise to perform this check.)*

**Exercise 6.7.7** *We have assumed that* $\det A \neq 0$. *If we try our method of LU factorisation (with column interchange) in the case* $\det A = 0$, *what will happen?*

Once we have an $LU$ factorisation, it is very easy to solve systems of simultaneous equations. Suppose that $A = LU$, where $L$ and $U$ are non-singular lower and upper triangular $n \times n$ matrices. Then

$$A\mathbf{x} = \mathbf{y} \iff LU\mathbf{x} = \mathbf{y} \iff U\mathbf{x} = \mathbf{u} \quad \text{where } L\mathbf{u} = \mathbf{y}.$$

Thus we need only solve the triangular system $L\mathbf{u} = \mathbf{y}$ and then solve the triangular system $U\mathbf{x} = \mathbf{u}$.

**Example 6.7.8** *Solve the system of equations*

$$2x + y + z = 1$$
$$4x + y \phantom{{}+ z} = -3$$
$$-2x + 2y + z = 6$$

*using LU factorisation.*

*Solution* Using the result of Example 6.7.5, we see that we need to solve the systems of equations

$$u \phantom{{}+ 2v + w} = 1 \qquad\qquad 2x + y + z = u$$
$$2u + v \phantom{{}+ w} = -3 \qquad \text{and} \qquad -y - z = v$$
$$-u - 3v + w = 6 \qquad\qquad -4z = w.$$

Solving the first system step by step, we get $u = 1$, $v = -5$ and $w = -8$. Thus we need to solve

$$2x + y + z = 1$$
$$-y - z = -5$$
$$-4z = -8$$

and a step by step solution gives $z = 2$, $y = 1$ and $x = -1$.                    □

The work required to obtain an *LU* factorisation is essentially the same as that required to solve the system of equations by Gaussian elimination. However, if we have to solve the same system of equations

$$A\mathbf{x} = \mathbf{y}$$

*repeatedly* with the same $A$ but different $\mathbf{y}$, we need only perform the factorisation *once* and this represents a major economy of effort.

**Exercise 6.7.9** *Show that the equation*

$$LU = \begin{pmatrix} 0 & 1 \\ 1 & 0 \end{pmatrix}$$

*has no solution with L a lower triangular matrix and U an upper triangular matrix. Why does this not contradict Theorem 6.7.4?*

**Exercise 6.7.10** *Suppose that $L_1$ and $L_2$ are lower triangular $n \times n$ matrices with all diagonal entries 1 and $U_1$ and $U_2$ are non-singular upper triangular $n \times n$ matrices. Show, by induction on n, or otherwise, that if*

$$L_1 U_1 = L_2 U_2$$

then $L_1 = L_2$ and $U_1 = U_2$. (*This does not show that the LU factorisation given above is unique, since we allow the interchange of columns.*[13])

**Exercise 6.7.11** *Suppose that L is a lower triangular $n \times n$ matrix with all diagonal entries 1 and U is an $n \times n$ upper triangular matrix. If $A = LU$, give an efficient method of finding* det *A. Give a reasonably efficient method for finding $A^{-1}$.*

**Exercise 6.7.12** (*i*) *If A is a non-singular $n \times n$ matrix, show that (rearranging columns if necessary) we can find a lower triangular matrix L with all diagonal entries 1 and an upper triangular matrix U with all diagonal entries 1 together with a non-singular diagonal matrix D such that*

$$A = LDU.$$

*State and prove an appropriate result along the lines of Exercise 6.7.10.*

(*ii*) *In this section we proved results on LU factorisation. Do there exist corresponding results on UL factorisation? Give reasons.*

(*iii*) *Is it true that (after reordering columns if necessary) every invertible $n \times n$ matrix A can be written in the form $A = BC$ where $B = (b_{ij})$ and $C = (c_{ij})$ are $n \times n$ matrices with $b_{ij} = 0$ if $i > j$ and $c_{ij} = 0$ if $j > i$? Give reasons.*

## 6.8 Further exercises

**Exercise 6.8.1 [The Lorentz transformation]** Let $\mathbf{r}, \mathbf{v} \in \mathbb{R}^3$ with $\|\mathbf{v}\| < c$ and set $\gamma = c(c^2 - \|\mathbf{v}\|^2)^{-1/2}$. If

$$\mathbf{r}' = \mathbf{r} + \left( \frac{(\gamma - 1)\mathbf{v} \cdot \mathbf{r}}{\|\mathbf{v}\|^2} - \gamma t \right) \mathbf{v}, \quad t' = \gamma \left( t - \frac{\mathbf{v} \cdot \mathbf{r}}{c^2} \right),$$

prove the reciprocal relations

$$\mathbf{r} = \mathbf{r}' + \left( \frac{(\gamma - 1)(-\mathbf{v}) \cdot \mathbf{r}'}{\| - \mathbf{v}\|^2} - \gamma t' \right)(-\mathbf{v}), \quad t = \gamma \left( t' - \frac{(-\mathbf{v}) \cdot \mathbf{r}'}{c^2} \right),$$

that is to say, the relations

$$\mathbf{r} = \mathbf{r}' + \left( \frac{(\gamma - 1)\mathbf{v} \cdot \mathbf{r}'}{\|\mathbf{v}\|^2} + \gamma t' \right) \mathbf{v}, \quad t = \gamma \left( t' + \frac{\mathbf{v} \cdot \mathbf{r}'}{c^2} \right).$$

[Courageous students will tackle the calculations head on. Less gung ho students may choose an appropriate coordinate system. Both sets of students should then try the alternative method.]

---

[13] Some writers do not allow the interchange of columns. *LU* factorisation is then unique, but may not exist, even if the matrix to be factorised is invertible.

**Exercise 6.8.2** In this exercise we work in $\mathbb{R}$. Find the eigenvalues and eigenvectors of

$$A = \begin{pmatrix} 3 & -1 & 2 \\ 0 & 4-s & 2s-2 \\ 0 & -2s+2 & 4s-1 \end{pmatrix}$$

for all values of $s$.

For which values of $s$ is $A$ diagonalisable? Give reasons for your answer.

**Exercise 6.8.3** The $3 \times 3$ matrix $A$ satisfies $A^3 = A$. What are the possible values of $\det A$? Write down examples to show that each possibility occurs.

Write down an example of a $3 \times 3$ matrix $B$ with $B^3 = 0$ but $B \neq 0$.

If $A$ satisfies the conditions of the first paragraph and $B$ those of the second, what are the possible values of $\det AB$? Give reasons.

**Exercise 6.8.4** We work in $\mathbb{R}^3$ (using column vectors) with the standard basis

$$\mathbf{e}_1 = (1, 0, 0)^T, \quad \mathbf{e}_1 = (0, 1, 0)^T, \quad \mathbf{e}_3 = (0, 0, 1)^T.$$

Consider a non-singular linear map $\alpha : \mathbb{R}^3 \to \mathbb{R}^3$ with matrix $A$ with respect to the standard basis. If $\Gamma$ is a plane through the origin with equation $\mathbf{a} \cdot \mathbf{x} = 0$ for some $\mathbf{a} \neq \mathbf{0}$, show that $\alpha\Gamma$ is the plane through the origin with equation $((A^T)^{-1}\mathbf{a}) \cdot \mathbf{x} = 0$. Deduce the existence of a plane through the origin such that $\alpha(\Gamma) = \Gamma$.

Show, by means of an example, that there may not be any line $l$ through the origin with $l \subseteq \Gamma$ and $\alpha l = l$.

**Exercise 6.8.5** We work with $n \times n$ matrices over $\mathbb{F}$.

(i) Let $P$ be a permutation matrix and $E$ a non-singular diagonal matrix. If $D$ is diagonal, show that $(PE)^{-1}D(PE)$ is.

(ii) If $D$ is a diagonal matrix with all diagonal entries distinct and $B$ is a non-singular matrix such that $B^{-1}DB$ is diagonal, show, by considering eigenvectors, or otherwise, that $B = PE$ where $P$ is a permutation matrix and $E$ a non-singular diagonal matrix.

(iii) Let $A$ have $n$ distinct eigenvalues. If $Q$ is non-singular and $Q^{-1}AQ$ is diagonal, show that the non-singular matrix $R$ is such that $R^{-1}AR$ is diagonal if and only if $R = PEQ$ where $P$ is a permutation matrix and $E$ a non-singular diagonal matrix.

(iv) Does (ii) remain true if we drop the condition that all the diagonal entries of $D$ are distinct? Give a proof or a counterexample.

**Exercise 6.8.6** (i) Show that a $2 \times 2$ complex matrix $A$ satisfies the condition $A^2 = 0$ if and only if it takes one of the forms

$$\begin{pmatrix} a & a\lambda \\ -a\lambda^{-1} & -a \end{pmatrix} \quad \text{or} \quad \begin{pmatrix} 0 & a \\ 0 & 0 \end{pmatrix} \quad \text{or} \quad \begin{pmatrix} 0 & 0 \\ a & 0 \end{pmatrix}$$

with $a$, $\lambda \in \mathbb{C}$ and $\lambda \neq 0$.

(ii) We work with $2 \times 2$ complex matrices. Is it always true that, if $A$ and $B$ satisfy $A^2 + B^2 = 0$, then $(A + B)^2 = 0$? Is it always true that, if $A$ and $B$ are not diagonalisable, then $A + B$ is not diagonalisable? Is it always true that, if $A$ and $B$ are diagonalisable, then $A + B$ is diagonalisable? Give proofs or counterexamples.

(iii) Let

$$A = \begin{pmatrix} a & b \\ b & c \end{pmatrix}, \quad B = \begin{pmatrix} a & b \\ b^* & c \end{pmatrix}$$

with $a, c \in \mathbb{R}, b \in \mathbb{C}$. Is it always true that $A$ is diagonalisable over $\mathbb{C}$? Is it always true that $B$ is diagonalisable over $\mathbb{C}$? Give proofs or counterexamples.

**Exercise 6.8.7** We work with $2 \times 2$ complex matrices. Are the following statements true or false? Give a proof or counterexample.

(i) $AB = 0 \Rightarrow BA = 0$.

(ii) If $AB = 0$ and $B \neq 0$, then there exists a $C \neq 0$ such that $AC = CA = 0$.

**Exercise 6.8.8** (i) Suppose that $A, B, C$ and $D$ are $n \times n$ matrices over $\mathbb{F}$ such that $AB = BA$ and $A$ is invertible. By considering

$$\begin{pmatrix} I & 0 \\ X & I \end{pmatrix} \begin{pmatrix} A & B \\ C & D \end{pmatrix},$$

for suitable $X$, or otherwise, show that

$$\det \begin{pmatrix} A & B \\ C & D \end{pmatrix} = \det(AD - BC).$$

(iii) Use Exercise 6.2.14 (ii) to show that the condition $A$ invertible can be removed in part (i).

(iv) Can the condition $AB = BA$ be removed? Give a proof or counterexample.

**Exercise 6.8.9** Explain briefly why the set $M_n$ of all $n \times n$ matrices over $\mathbb{F}$ is a vector space under the usual operations. What is the dimension of $M_n$? Give reasons.

If $A \in M_n$ we define $L_A, R_A : M_n \to M_n$ by $L_A X = AX$ and $R_A X = XA$ for all $X \in M_n$. Show that $L_A$ and $R_A$ are linear,

$$\det L_A = (\det A)^n = \det R_A \quad \text{and} \quad \det(L_A - R_A) = 0.$$

[Hint: Find an appropriate basis.]

**Exercise 6.8.10** By first considering the case when $A$ and $B$ are non-singular, or otherwise, show that, if $A$ and $B$ are $n \times n$ matrices over $\mathbb{F}$, then

$$\text{Adj } AB = \text{Adj } B \text{ Adj } A.$$

**Exercise 6.8.11 [Sylvester's determinant identity]**

(i) Suppose that $A$ and $B$ are $n \times n$ matrices over $\mathbb{F}$ and $A$ is invertible. Show that

$$\det(I + AB) = \det(I + BA).$$

(ii) Use Exercise 6.2.14 to show that, if $A$ and $B$ are $n \times n$ matrices over $\mathbb{F}$, then

$$\det(I + AB) = \det(I + BA).$$

(iii) Suppose that $n \geq m$, $A$ is an $n \times m$ matrix and $B$ an $m \times n$ matrix. By considering the $n \times n$ matrices

$$\begin{pmatrix} A & 0 \end{pmatrix} \quad \text{and} \quad \begin{pmatrix} B \\ 0 \end{pmatrix}$$

obtained by adding columns of zeros to $A$ and rows of zeros to $B$, show that

$$\det(I_n + AB) = \det(I_m + BA)$$

where, as usual, $I_r$ denotes the $r \times r$ identity matrix.
[This result is called Sylvester's determinant identity.]

(iv) If $\mathbf{u}$ and $\mathbf{v}$ are column vectors with $n$ entries show that

$$\det(I + \mathbf{u}\mathbf{v}^T) = 1 + \mathbf{u} \cdot \mathbf{v}.$$

If, in addition, $A$ is an $n \times n$ invertible matrix show that

$$\det(A + \mathbf{u}\mathbf{v}^T) = (1 + \mathbf{v}^T A^{-1} \mathbf{u}) \det A.$$

**Exercise 6.8.12 [Alternative proof of Sylvester's identity]** Suppose that $n \geq m$, $A$ is an $n \times m$ matrix and $B$ an $m \times n$ matrix. Show that

$$\begin{pmatrix} I_n & 0 \\ B & I_m \end{pmatrix} \begin{pmatrix} I_n & 0 \\ 0 & I_m - BA \end{pmatrix} \begin{pmatrix} I_n & A \\ 0 & I_m \end{pmatrix} = \begin{pmatrix} I_n & A \\ B & I_m \end{pmatrix}$$

$$= \begin{pmatrix} I_n & A \\ 0 & I_m \end{pmatrix} \begin{pmatrix} I_n - AB & 0 \\ 0 & I_m \end{pmatrix} \begin{pmatrix} I_n & 0 \\ B & I_m \end{pmatrix}$$

and deduce Sylvester's identity (Exercise 6.8.11).

**Exercise 6.8.13** Let $A$ be the $n \times n$ matrix all of whose entries are 1. Show that $A$ is diagonalisable and find an associated diagonal matrix.

**Exercise 6.8.14** Let $C$ be an $n \times n$ matrix such that $C^m = I$ for some integer $m \geq 1$. Show that

$$I + C + C^2 + \cdots + C^{m-1} = 0 \Leftrightarrow \det(I - C) \neq 0.$$

**Exercise 6.8.15** We work over $\mathbb{F}$. Let $A$ be an $n \times n$ matrix over $\mathbb{F}$. Show directly from the definition of an eigenvalue (in particular, without using determinants) that the following results hold.

(i) If $\lambda$ is an eigenvalue of $A$, then $\lambda^r$ is an eigenvalue of $A^r$ for all integers $r \geq 1$.

(ii) If $A$ is invertible and $\lambda$ is an eigenvalue of $A$, then $\lambda \neq 0$ and $\lambda^r$ is an eigenvalue of $A^r$ for all integers $r$. (We use the convention that $A^{-r} = (A^{-1})^r$ for $r \geq 1$ and that $A^0 = I$.)

We now consider the characteristic polynomial $\chi_A(t) = \det(tI - A)$. Use standard properties of determinants to prove the following results.

(iii) $\chi_{A^2}(t^2) = \chi_A(t)\chi_A(-t)$.

(iv) If $A$ is invertible, then

$$\chi_{A^{-1}}(t) = (\det A)^{-1}(-1)^n t^n \chi_A(t^{-1})$$

for all $t \neq 0$.

Suppose that $\mathbb{F} = \mathbb{C}$. If $A^2$ has an eigenvalue $\mu$, does it follow that $A$ has an eigenvalue $\lambda$ with $\mu = \lambda^2$?

Suppose that $\mathbb{F} = \mathbb{C}$ and $n$ is a strictly positive integer. If $A^n$ has an eigenvalue $\mu$, does it follow that $A$ has an eigenvalue $\lambda$ with $\mu = \lambda^n$?

Suppose that $\mathbb{F} = \mathbb{R}$. If $A^2$ has an eigenvalue $\mu$, does it follow that $A$ has an eigenvalue $\lambda$ with $\mu = \lambda^2$?

Give proofs or counterexamples as appropriate.

**Exercise 6.8.16** Let $A$ be an $n \times m$ matrix of row rank $r$. Show that (possibly after reordering the columns of $A$) we can find $B$ an $n \times r$ matrix and $C$ an $r \times m$ matrix such that $A = BC$. Give an example to show why it may be necessary to reorder the columns of $A$.

Explain why $r$ is the least integer $s$ such that $A = B'C'$ where $B'$ is an $m \times s$ matrix and $C'$ is an $s \times n$ matrix. Use the relation $(BC)^T = C^T B^T$ to give another proof that the row rank of $A$ equals its column rank.

**Exercise 6.8.17** (Requires elementary knowledge of linear differential equations.)
(i) Consider the differential equation

$$\begin{pmatrix} \dot{x} \\ \dot{y} \\ \dot{z} \end{pmatrix} = \begin{pmatrix} -1 & 2 & -1 \\ 1 & 0 & -1 \\ 1 & -2 & 1 \end{pmatrix} \begin{pmatrix} x \\ y \\ z \end{pmatrix}.$$

Obtain the general solution in the form $\mathbf{x}(t) = \gamma_1(t)\mathbf{u}_1 + \gamma_2(t)\mathbf{u}_2 + \gamma_3(t)\mathbf{u}_3$ where the $\mathbf{u}_j$ (to be found explicitly) form a basis of eigenvectors of the matrix.

(ii) For each real $\lambda$, find the general solution of

$$\begin{pmatrix} \dot{x} \\ \dot{y} \\ \dot{z} \end{pmatrix} = \begin{pmatrix} -1 & 2 & -1 \\ 1 & 0 & -1 \\ 1 & -2 & 1 \end{pmatrix} \begin{pmatrix} x \\ y \\ z \end{pmatrix} + 2\begin{pmatrix} -\lambda \\ 1 \\ \lambda \end{pmatrix} e^{2t}.$$

What particular phenomenon occurs when $\lambda = 1$?

(iii) Let $\lambda = -1$. Find a solution of the equation in (ii) which has $\mathbf{x} = (0, 1, 0)^T$ when $t = 0$.

**Exercise 6.8.18** Explain why (if we are allowed to renumber columns) we can perform an $LU$ factorisation of an $n \times n$ matrix $A$ so as to obtain $A = LU$ where $L$ is lower triangular with diagonal elements 1 and all entries of modulus at most 1 and $U$ is upper triangular.

If all the elements of $A$ have modulus at most 1 show that all the entries of $U$ have modulus at most $2^{n-1}$.

Perform the *LU* factorisation (without row pivoting) on the $3 \times 3$ matrix

$$A = \begin{pmatrix} 1 & 0 & 0 & 1 \\ -1 & 1 & 0 & 1 \\ -1 & -1 & 1 & 1 \\ -1 & -1 & -1 & 1 \end{pmatrix}$$

so as to obtain $A = LU$ where $L$ is lower triangular with diagonal elements 1 and all entries of modulus at most 1 and $U$ is upper triangular. By generalising this idea, show that the result of the previous paragraph is best possible for every $n$.

**Exercise 6.8.19** (This requires a little group theory and, in particular, knowledge of the Möbius group $\mathcal{M}$.)

Let $SL(\mathbb{C}^2)$ be the set of $2 \times 2$ complex matrices $A$ with $\det A = 1$.

(i) Show that $SL(\mathbb{C}^2)$ is a group under matrix multiplication.

(ii) Let $\mathcal{M}$ be the group of maps $T : \mathbb{C} \cup \{\infty\} \to \mathbb{C} \cup \{\infty\}$ given by

$$Tz = \frac{az+b}{cz+d}$$

where $ad - bc \neq 0$. Show that the map $\theta : SL(\mathbb{C}^2) \to \mathcal{M}$ given by

$$\theta \begin{pmatrix} a & b \\ c & d \end{pmatrix} (z) = \frac{az+b}{cz+d}$$

is a group homomorphism. Show further that $\theta$ is surjective and has kernel $\{I, -I\}$.

(iii) Use (ii) and Exercise 6.4.7 to show that, given $T \in \mathcal{M}$, one of the following statements must be true.

(1) $Tz = z$ for all $z \in \mathbb{C} \cup \{\infty\}$.

(2) There exists an $S \in \mathcal{M}$ such that $S^{-1}TSz = \lambda z$ for all $z \in \mathbb{C}$ and some $\lambda \neq 0$.

(3) There exists an $S \in \mathcal{M}$ such that either $S^{-1}TSz = z + 1$ for all $z \in \mathbb{C}$ or $S^{-1}TSz = z - 1$ for all $z \in \mathbb{C}$.

**Exercise 6.8.20 [The trace]** If $A = (a_{ij})$ is an $n \times n$ matrix over $\mathbb{F}$, then we define the trace $\operatorname{Tr} A$ by $\operatorname{Tr} A = a_{ii}$ (using the summation convention). There are many ways of showing that

$$\operatorname{Tr} B^{-1}AB = \operatorname{Tr} A \qquad\qquad \bigstar$$

whenever $B$ is an invertible $n \times n$ matrix. You are asked to consider four of them in this question.

(i) Prove $\bigstar$ directly using the summation convention. (To avoid confusion, I suggest you write $C = B^{-1}$.)

(ii) If $E$ is an $n \times m$ matrix and $F$ is an $m \times n$ matrix explain why $\operatorname{Tr} EF$ and $\operatorname{Tr} FE$ are defined. Show that

$$\operatorname{Tr} EF = \operatorname{Tr} FE,$$

and use this result to obtain $\bigstar$.

(iii) Show that $-\operatorname{Tr} A$ is the coefficient of $t^{n-1}$ in the characteristic polynomial $\det(tI - A)$. Write $\det\left(B^{-1}(tI - A)B\right)$ in two different ways to obtain ★.

(iv) Show that if $A$ is the matrix of the linear map $\alpha$ with respect to some basis, then $-\operatorname{Tr} A$ is the coefficient of $t^{n-1}$ in the characteristic polynomial $\det(t\iota - \alpha)$ and explain why this observation gives ★.

Explain why ★ means that, if $U$ is a finite dimensional space and $\alpha : U \to U$ a linear map, we may define

$$\operatorname{Tr} \alpha = \operatorname{Tr} A$$

where $A$ is the matrix of $\alpha$ with respect to some given basis.

**Exercise 6.8.21** If $A$ and $B$ are $n \times n$ matrices over $\mathbb{F}$, we write $[A, B] = AB - BA$. (We call $[A, B]$ the *commutator* of $A$ and $B$.) Show, using the summation convention, that $\operatorname{Tr}[A, B] = 0$.

If $E(r, s)$ is the $n \times n$ matrix with 1 in the $(r, s)$th place and zeros everywhere else, compute $[E(r, s), E(u, v)]$. Show that, given an $n \times n$ diagonal matrix $D$, with $\operatorname{Tr} D = 0$ we can find $A$ and $B$ such that $[A, B] = D$.

Deduce that if $C$ is a diagonalisable $n \times n$ matrix with $\operatorname{Tr} C = 0$, then we can find $F$ and $G$ such that $[F, G] = C$.

Suppose that we work over $\mathbb{C}$. By using Theorem 6.4.3, or otherwise, show that if $C$ is any $2 \times 2$ matrix with $\operatorname{Tr} C = 0$, then we can find $F$ and $G$ such that $[F, G] = C$. [In Exercise 12.6.24 we will use sharper tools to show that the result of the last paragraph holds for $n \times n$ matrices.]

**Exercise 6.8.22** We work with $n \times n$ matrices over $\mathbb{F}$. Show that, if $\operatorname{Tr} AX = 0$ for every $X$, then $A = 0$.

Is it true that, if $\det AX = 0$ for every $X$, then $A = 0$? Give a proof or counterexample.

**Exercise 6.8.23** If $C$ is a $2 \times 2$ matrix over $\mathbb{F}$ with $\operatorname{Tr} C = 0$, show that $C^2$ is a multiple of $I$ (that is to say, $C^2 = \lambda I$ for some $\lambda \in \mathbb{F}$). Conclude that, if $A$ and $B$ are $2 \times 2$ matrices, then $[A, B]^2$ is a multiple of $I$.
[Hint: Suppose first that $\mathbb{F} = \mathbb{C}$ and use Theorem 6.4.3.]

If $A$ and $B$ are $2 \times 2$ matrices, does it follow that $[A, B]$ is a multiple of $I$? If $A$ and $B$ are $4 \times 4$ matrices does it follow that $[A, B]^2$ is a multiple of $I$? Give proofs or counterexamples.

**Exercise 6.8.24** (i) Suppose that $V$ is a vector space over $\mathbb{F}$. If $\alpha$, $\beta$, $\gamma : V \to V$ are linear maps such that

$$\alpha\beta = \beta\gamma = \iota$$

(where $\iota$ is the identity map), show by looking at $\alpha(\beta\gamma)$, or otherwise, that $\alpha = \gamma$.

(ii) Show that $\mathcal{P}$, the space of polynomials with real coefficients, is a vector space over $\mathbb{R}$. If $\beta(P)(t) = tP(t)$ show that $\beta : \mathcal{P} \to \mathcal{P}$ is a linear map. Show that there exists a linear map $\alpha : \mathcal{P} \to \mathcal{P}$ such that $\alpha\beta = \iota$, but that there does not exist a linear map $\gamma : \mathcal{P} \to \mathcal{P}$ such that $\beta\gamma = \iota$.

(iii) If $\mathcal{P}$ is as in (ii), find a linear map $\beta : \mathcal{P} \to \mathcal{P}$ such that there exists a linear map $\gamma : \mathcal{P} \to \mathcal{P}$ with $\beta\gamma = \iota$, but that there does not exist a linear map $\alpha : \mathcal{P} \to \mathcal{P}$ such that $\alpha\beta = \iota$.

(iv) Let $c = \mathbb{F}^{\mathbb{N}}$ be the vector space introduced in Exercise 5.2.9 (iv). Find a linear map $\beta : c \to c$ such that there exists a linear map $\gamma : c \to c$ with $\gamma\beta = \iota$, but that there does not exist a linear map $\alpha : c \to c$ such that $\beta\alpha = \iota$. Find a linear map $\theta : c \to c$ such that there exists a linear map $\phi : c \to c$ with $\theta\phi = \iota$, but that there does not exist a linear map $\alpha : c \to c$ such that $\alpha\theta = \iota$.

(v) Do there exist a finite dimensional vector space $V$ and linear maps $\alpha, \beta : V \to V$ such that $\alpha\beta = \iota$, but not a linear map $\gamma : V \to V$ such that $\beta\gamma = \iota$? Give reasons for your answer.

**Exercise 6.8.25** Let $C$ be an $n \times n$ matrix over $\mathbb{C}$. We write $C = A + iB$ with $A$ and $B$ real $n \times n$ matrices. By considering the polynomial $P(z) = \det(A + zB)$, show that, if $C$ is invertible, there must exist a real $t$ such that $A + tB$ is invertible. Hence, show that, if $R$ and $S$ are $n \times n$ matrices with *real* entries which are similar when considered over $\mathbb{C}$ (i.e. there exists an invertible matrix $C$ with entries in $\mathbb{C}$ such that $R = C^{-1}SC$), then they are similar when considered over $\mathbb{R}$ (i.e. there exists an invertible matrix $P$ with entries in $\mathbb{R}$ such that $R = P^{-1}SP$).

**Exercise 6.8.26** Find an $LU$ factorisation of the matrix

$$A = \begin{pmatrix} 2 & -1 & 3 & 2 \\ -4 & 3 & -4 & -2 \\ 4 & -2 & 3 & 6 \\ -6 & 5 & -8 & 1 \end{pmatrix}$$

and use it to solve $A\mathbf{x} = \mathbf{b}$ where

$$\mathbf{b} = \begin{pmatrix} -2 \\ 2 \\ 4 \\ 11 \end{pmatrix}.$$

**Exercise 6.8.27** Let

$$A = \begin{pmatrix} 1 & a & a^2 & a^3 \\ a^3 & 1 & a & a^2 \\ a^2 & a^3 & 1 & a \\ a & a^2 & a^3 & 1 \end{pmatrix}.$$

Find the $LU$ factorisation of $A$ and compute $\det A$. Generalise your results to $n \times n$ matrices.

**Exercise 6.8.28** Suppose that $A$ is an $n \times n$ matrix and there exists a polynomial $f$ such that $B = f(A)$. Show that $BA = AB$.

Suppose that $A$ is an $n \times n$ matrix with $n$ distinct eigenvalues. Show that if $B$ commutes with $A$ then, if $A$ is diagonal with respect to some basis, so is $B$. By considering an appropriate Vandermonde determinant (see Exercise 4.4.9) show that there exists a polynomial $f$ of degree at most $n - 1$ such that $B = f(A)$.

Suppose that we only know that $A$ is an $n \times n$ diagonalisable matrix. Is it always true that if $B$ commutes with $A$ then $B$ is diagonalisable? Is it always true that, if $B$ commutes with $A$, then there exists a polynomial $f$ such that $B = f(A)$? Give reasons.

**Exercise 6.8.29** Matrices of the form

$$A = \begin{pmatrix} a_0 & a_1 & a_2 & \cdots & a_n \\ a_n & a_0 & a_1 & \cdots & a_{n-1} \\ a_{n-1} & a_n & a_0 & \cdots & a_{n-2} \\ \vdots & & & \ddots & \vdots \\ a_1 & a_2 & a_3 & \cdots & a_0 \end{pmatrix}$$

are called circulant matrices. Show that, for certain values of $\eta$, to be found, $e(\eta) = (1, \eta, \eta^2, \ldots, \eta^n)^T$ is an eigenvector of $A$.

Explain why the eigenvectors you have found form a basis for $\mathbb{C}^{n+1}$. Use this result to evaluate $\det A$. (This gives another way of obtaining the result of Exercise 5.7.14.) Check the result of Exercise 6.8.27 using the formula just obtained.

By using the basis discussed above, or otherwise, show that if $A$ and $B$ are circulants of the same size, then $AB = BA$.

**Exercise 6.8.30** We work in $\mathbb{R}$. Let $A$ be a diagonalisable and invertible $n \times n$ matrix and $B$ an $n \times n$ matrix such that $AB = tBA$ for some real number $t > 1$.

(i) Show that $B$ is nilpotent, that is to say, $B^k = 0$ for some positive integer $k$.

(ii) Suppose that we can find a vector $\mathbf{v} \in \mathbb{R}^n$ such that $A\mathbf{v} = \mathbf{v}$ and $B^{n-1}\mathbf{v} \neq \mathbf{0}$. Find the eigenvalues of $A$.

What is the largest subset of $X$ of $\mathbb{R}$ such that, if $A$ is a diagonalisable and invertible $n \times n$ matrix and $B$ an $n \times n$ matrix such that $AB = sBA$ for some $s \in X$, then $B$ must be nilpotent? Prove your answer.

**Exercise 6.8.31** The object of this exercise is to show why finding eigenvalues of a large matrix is not just a matter of finding a large fast computer.

Consider the $n \times n$ complex matrix $A = (a_{ij})$ given by

$$a_{j\,j+1} = 1 \qquad\qquad \text{for } 1 \leq j \leq n - 1$$
$$a_{n1} = \kappa^n$$
$$a_{ij} = 0 \qquad\qquad \text{otherwise,}$$

where $\kappa \in \mathbb{C}$ is non-zero. Thus, when $n = 2$ and $n = 3$, we get the matrices

$$\begin{pmatrix} 0 & 1 \\ \kappa^2 & 0 \end{pmatrix} \quad \text{and} \quad \begin{pmatrix} 0 & 1 & 0 \\ 0 & 0 & 1 \\ \kappa^3 & 0 & 0 \end{pmatrix}.$$

(i) Find the eigenvalues and associated eigenvectors of $A$ for $n = 2$ and $n = 3$. (Note that we are working over $\mathbb{C}$, so we must consider complex roots.)

(ii) By guessing and then verifying your answers, or otherwise, find the eigenvalues and associated eigenvectors of $A$ for for all $n \geq 2$.

(iii) Suppose that your computer works to 15 decimal places and that $n = 100$. You decide to find the eigenvalues of $A$ in the cases $\kappa = 2^{-1}$ and $\kappa = 3^{-1}$. Explain why at least one (and more probably both) attempts will deliver answers which bear no relation to the true answers.

**Exercise 6.8.32 [Bézout's theorem]** In this question we work in $\mathbb{Z}$. Let $r$ and $s$ be non-zero integers. Show that the set

$$\Gamma = \{ur + vs : u, v \in \mathbb{Z} \text{ and } ur + vs > 0\}$$

is a non-empty subset of the strictly positive integers. Conclude that $\Gamma$ has a least element $c$.

Show that we can find an integer $a$ such that

$$0 \leq r + ac < c.$$

Explain why either $r + ac \in \Gamma$ or $r + ac = 0$ and deduce that $r = -ac$. Thus $c$ divides $r$ and similarly $c$ divides $s$. Deduce that $c$ divides the highest common divisor of $r$ and $s$. Use the definition of $\Gamma$ to show that the highest common divisor of $r$ and $s$ divides $c$ and so $c$ must be the highest common divisor of $r$ and $s$.

Conclude that, if $d$ is the highest common divisor of $r$ and $s$, then there exist $u$ and $v$ such that

$$d = ur + vs.$$

If $p$ is a prime and $r$ is an integer not divisible by $p$, show, by setting $p = s$, that there exists an integer $u$ such that

$$1 \equiv ur \pmod{p}.$$

**Exercise 6.8.33 [Fermat's little theorem]** If $p$ is a prime and $k$ is an integer with $1 \leq k \leq p - 1$, show that $\binom{p}{k}$ is divisible by $p$. Deduce that

$$(r + 1)^p \equiv r^p + 1 \pmod{p}$$

for all integers $u$. Hence, or otherwise, deduce Fermat's little theorem

$$r^p \equiv r \pmod{p}$$

for all integers $r$.

By multiplying both sides of the equation by the integer $u$ defined in the last sentence of the previous question, show that

$$r \not\equiv 0 \quad (\text{mod } p) \Leftrightarrow r^{p-1} \equiv 1 \quad (\text{mod } p)$$

and deduce that that (if $r \not\equiv 0$) we have $u \equiv r^{p-2}$ (mod $p$)

**Exercise 6.8.34** This question is intended as a revision of the notion of equivalence relations and classes.

Recall that, if $A$ is a non-empty set, a *relation* is a set $R \subseteq A \times A$. We write $aRb$ if $(a, b) \in R$.

(I) We say that $R$ is *reflexive* if $aRa$ for all $a \in A$.

(II) We say that $R$ is *symmetric* if $aRb \Rightarrow bRa$.

(III) We say that $R$ is *transitive* if $aRb \Rightarrow bRa$.

By considering possible $R$ when $A = \{1, 2, 3\}$, show that each of the eight possible combinations of the type 'not reflexive, symmetric, not transitive' can occur.

A relation is called an *equivalence relation* if it is reflexive, symmetric and transitive. A collection $\mathcal{F}$ of subsets of $A$ is called a *partition* of $A$ if the following conditions hold.

(i) $\bigcup_{F \in \mathcal{F}} F = A$.

(ii) If $F, G \in \mathcal{F}$, then $F \cap G \neq \emptyset \Rightarrow F = G$.

Show that, if $\mathcal{F}$ is a partition of $A$, the relation $R_{\mathcal{F}}$ defined by

$$aR_{\mathcal{F}}b \Leftrightarrow a, b \in F \quad \text{for some } F \in \mathcal{F}$$

is an equivalence relation.

Show that, if $R$ is an equivalence relation on $A$, the collection $A/R$ of sets of the form

$$[a] = \{x \in A : aRx\}$$

is a partition of $A$. We say that $A/R$ is the *quotient of $A$ by $R$*. We call the elements $[a] \in A/R$ *equivalence classes*.

**Exercise 6.8.35** In this question we show how to construct $\mathbb{Z}_p$ using equivalence classes.

(i) Let $n$ be an integer with $n \geq 2$. Show that the relation $R_n$ on $\mathbb{Z}$ defined by

$$uR_nv \Leftrightarrow u - v \text{ divisible by } n$$

is an equivalence relation.

(ii) If we consider equivalence classes in $\mathbb{Z}/R_n$, show that

$$[u] = [u'], \quad [v] = [v'] \Rightarrow [u + v] = [u' + v'], \quad [uv] = [u'v'].$$

Explain briefly why this allows us to make the definitions

$$[u] + [v] = [u + v], \quad [u] \times [v] = [uv].$$

We write $\mathbb{Z}_n = \mathbb{Z}/R_n$ and equip $\mathbb{Z}_n$ with the addition and multiplication just defined.

(iii) If $n = 6$, show that $[2], [3] \neq [0]$ but $[2] \times [3] = [0]$.

(iv) If $p$ is a prime, verify that $\mathbb{Z}_p$ satisfies the axioms for a field set out in Definition 13.2.1. (You you should find the verifications trivial apart from (viii) which uses Exercise 6.8.32.) In practice, we often write $r = [r]$.

(v) Show that $\mathbb{Z}_n$ is a field if and only if $n$ is a prime.

**Exercise 6.8.36** (This question requires the notion of an equivalence class. See Exercise 6.8.34.)

Let $\mathbf{u}$ be a non-zero vector in $\mathbb{R}^n$. Write $\mathbf{x} \sim \mathbf{y}$ if $\mathbf{x} = \mathbf{y} + \lambda \mathbf{u}$ for some $\lambda \in \mathbb{R}$. Show that $\sim$ is an equivalence relation on $\mathbb{R}^n$ and identify the equivalence classes

$$\mathbf{l_y} = \{\mathbf{x} : \mathbf{x} \sim \mathbf{y}\}$$

geometrically. Identify $\mathbf{l_0}$ specifically.

If $\mathcal{L}$ is the collection of equivalence classes show that the definitions

$$\mathbf{l_a} + \mathbf{l_b} = \mathbf{l_{a+b}}, \quad \lambda \mathbf{l_a} = \mathbf{l_{\lambda a}}$$

give well defined operations. Verify that $\mathcal{L}$ is a vector space. (If you only wish to do some of the verifications, prove the associative law for addition

$$\mathbf{l_a} + (\mathbf{l_b} + \mathbf{l_c}) = (\mathbf{l_a} + \mathbf{l_b}) + \mathbf{l_c}$$

and the existence of a zero vector to be identified explicitly.)

If $\mathbf{u}, \mathbf{b}_1, \mathbf{b}_2, \ldots, \mathbf{b}_{n-1}$ form a basis for $\mathbb{R}^n$, show that $\mathbf{l_{b_1}}, \mathbf{l_{b_2}}, \ldots, \mathbf{l_{b_{n-1}}}$ form a basis for $\mathcal{L}$.

Suppose that $n = 3$, $\mathbf{u} = (1, 3, -1)^T$, and $\alpha : \mathcal{L} \to \mathcal{L}$ is the linear map with

$$\alpha \mathbf{l}_{(1,-1,0)^T} = \mathbf{l}_{(2,4,-1)^T}, \quad \alpha \mathbf{l}_{(1,1,0)^T} = \mathbf{l}_{(-4,-2,1)^T}.$$

Find the matrix of $\alpha$ with respect to the basis $\mathbf{l}_{(1,0,0)^T}, \mathbf{l}_{(0,1,0)^T}$.

**Exercise 6.8.37** (It looks quite hard to set a difficult question on equivalence relations, but, in my opinion, the Cambridge examiners have managed it at least once. This exercise is included for interest and will not be used elsewhere.)

If $R$ and $S$ are two equivalence relations on the same set $A$, we define

$$R \circ S = \{(x, z) \in A \times A :$$

$$\text{there exists a } y \in A \text{ such that } (x, y) \in R \text{ and } (y, z) \in S\}.$$

Show that the following conditions are equivalent.

(i) $R \circ S$ is a symmetric relation on $A$.

(ii) $R \circ S$ is a transitive relation on $A$.

(iii) $R \circ S$ is the smallest equivalence relation on $A$ containing both $R$ and $S$.

Show also that these conditions hold if $A = \mathbb{Z}$ and $R$ and $S$ are the relations of congruence modulo $m$ and modulo $n$ for some integers $m, n \geq 2$.

**Exercise 6.8.38 [Roots of equations]** We work over $\mathbb{C}$.

(i) By induction on the degree of $P$, or otherwise, show that, if $P$ is a polynomial of degree at least 1 and $a \in \mathbb{C}$, we can find a polynomial $Q$ of degree 1 less than the degree of $P$ and an $r \in \mathbb{C}$ such that

$$P(t) = (t - a)Q(t) + r.$$

(ii) By considering the effect of setting $t = a$ in (i), show that, if $P$ is a polynomial with root $a$, we can find a polynomial $Q$ of degree 1 less than the degree of $P$ such that

$$P(t) = (t - a)Q(t).$$

(iii) Use induction and the Fundamental Theorem of Algebra (Theorem 6.2.6) to show that, if $P$ is polynomial of degree at least 1, then we can find $K \in \mathbb{C}$ and $\lambda_j \in \mathbb{C}$ such that

$$P(t) = K \prod_{j=1}^{n} (t - \lambda_j).$$

(iv) Show that a polynomial of degree $n$ can have at most $n$ distinct roots. For each $n \geq 1$, give a polynomial of degree $n$ with only one distinct root.

**Exercise 6.8.39** Linear differential equations are very important, but there are many other kinds of differential equations and analogy with the linear case may then lead us astray. Consider the first order differential equation

$$f'(x) = 3f(x)^{2/3}. \qquad\qquad \bigstar$$

Show that, if $u \leq v$, the function

$$f(x) = \begin{cases} (x - u)^3 & \text{if } x \leq u \\ 0 & \text{if } u < x < v \\ (x - v)^3 & \text{if } v \leq x \end{cases}$$

is a once differentiable function satisfying $\bigstar$. (Notice that there are *two* constants involved in specifying $f$.) Can you spot any other solutions?

**Exercise 6.8.40** Let us fix a basis for $\mathbb{R}^n$. Which of the following are subgroups of $GL(\mathbb{R}^n)$ for $n \geq 2$? Give proofs or counterexamples.

(i) The set $H_1$ of $\alpha \in GL(\mathbb{R}^n)$ with matrix $A = (a_{ij})$ where $a_{11} = 1$.

(ii) The set $H_2$ of $\alpha \in GL(\mathbb{R}^n)$ with $\det \alpha > 0$.

(iii) The set $H_3$ of $\alpha \in GL(\mathbb{R}^n)$ with $\det \alpha$ a non-zero integer.

(iv) The set $H_4$ of $\alpha \in GL(\mathbb{R}^n)$ with matrix $A = (a_{ij})$, where $a_{ij} \in \mathbb{Z}$ and $\det A = 1$.

(v) The set $H_5$ of $\alpha \in GL(\mathbb{R}^n)$ with matrix $A = (a_{ij})$ such that exactly one element in each row and column is non-zero.

(vi) The set $H_6$ of $\alpha \in GL(\mathbb{R}^n)$ with lower triangular matrix.

**Exercise 6.8.41** Let us fix a basis for $\mathbb{C}^n$. Which of the following are subgroups of $GL(\mathbb{C}^n)$ for $n \geq 1$? Give proofs or counterexamples.

(i) The set of $\alpha \in GL(\mathbb{C}^n)$ with matrix $A = (a_{ij})$, where the real and imaginary parts of the $a_{ij}$ are integers.

(ii) The set of $\alpha \in GL(\mathbb{C}^n)$ with matrix $A = (a_{ij})$, where the real and imaginary parts of the $a_{ij}$ are integers and $(\det A)^4 = 1$.

(iii) The set of $\alpha \in GL(\mathbb{C}^n)$ with matrix $A = (a_{ij})$, where the real and imaginary parts of the $a_{ij}$ are integers and $|\det A| = 1$.

The set $T$ consists of all $2 \times 2$ complex matrices of the form

$$A = \begin{pmatrix} z & w \\ -w^* & z^* \end{pmatrix}$$

with $z$ and $w$ having integer real and imaginary parts. If $Q$ consists of all elements of $T$ with inverses in $T$, show that the $\alpha$ with matrices in $Q$ form a subgroup of $GL(\mathbb{C}^2)$ with eight elements. Show that $Q$ contains elements $\alpha$ and $\beta$ with $\alpha\beta \neq \beta\alpha$. Show that $Q$ contains six elements $\gamma$ with $\gamma^4 = \iota$ but $\gamma^2 \neq \iota$. ($Q$ is called the quaternion group.)

# 7

# Distance preserving linear maps

## 7.1 Orthonormal bases

We start with a trivial example.

**Example 7.1.1** *A restaurant serves n different dishes. The 'meal vector' of a customer is the column vector $\mathbf{x} = (x_1, x_2, \ldots, x_n)^T$, where $x_j$ is the quantity of the jth dish ordered. At the end of the meal, the waiter uses the linear map $P : \mathbb{R}^n \to \mathbb{R}$ to obtain $P(\mathbf{x})$ the amount (in pounds) the customer must pay.*

Although the 'meal vectors' live in $\mathbb{R}^n$, it is not very useful to talk about the distance between two meals. There are many other examples where it is counter-productive to saddle $\mathbb{R}^n$ with things like distance and angle.

Equally, there are other occasions (particularly in the study of the real world) when it makes sense to consider $\mathbb{R}^n$ equipped with the inner product

$$\langle \mathbf{x}, \mathbf{y} \rangle = \mathbf{x} \cdot \mathbf{y} = \sum_{r=1}^{n} x_r y_r,$$

which we studied in Section 2.3 and the associated Euclidean norm

$$\|\mathbf{x}\| = \langle \mathbf{x}, \mathbf{x} \rangle^{1/2}.$$

We change from the 'dot product notation' $\mathbf{x} \cdot \mathbf{y}$ to the 'bracket notation' $\langle \mathbf{x}, \mathbf{y} \rangle$ partly to expose the reader to both notations and partly because the new notation seems clearer in certain expressions. The reader may wish to reread Section 2.3 before continuing.

Recall that we said that two vectors $\mathbf{x}$ and $\mathbf{y}$ are *orthogonal* (or *perpendicular*) if $\langle \mathbf{x}, \mathbf{y} \rangle = 0$.

**Definition 7.1.2** *We say that $\mathbf{x}$, $\mathbf{y} \in \mathbb{R}^n$ are* orthonormal *if $\mathbf{x}$ and $\mathbf{y}$ are orthogonal and both have norm 1. We say that a set of vectors is* orthonormal *if any two distinct elements are orthonormal.*

Informally, $\mathbf{x}$ and $\mathbf{y}$ are orthonormal if they 'have length 1 and are at right angles'.
The following observations are simple but important.

**Lemma 7.1.3** *We work in* $\mathbb{R}^n$.

(i) *If* $\mathbf{e}_1, \mathbf{e}_2, \ldots, \mathbf{e}_k$ *are orthonormal and*

$$\mathbf{x} = \sum_{j=1}^{k} \lambda_j \mathbf{e}_j,$$

*for some* $\lambda_j \in \mathbb{R}$, *then* $\lambda_j = \langle \mathbf{x}, \mathbf{e}_j \rangle$ *for* $1 \le j \le k$.

(ii) *If* $\mathbf{e}_1, \mathbf{e}_2, \ldots, \mathbf{e}_k$ *are orthonormal, then they are linearly independent.*

(iii) *Any collection of* $n$ *orthonormal vectors forms a basis.*

(iv) *If* $\mathbf{e}_1, \mathbf{e}_2, \ldots, \mathbf{e}_n$ *are orthonormal and* $\mathbf{x} \in \mathbb{R}^n$, *then*

$$\mathbf{x} = \sum_{j=1}^{n} \langle \mathbf{x}, \mathbf{e}_j \rangle \mathbf{e}_j.$$

*Proof* (i) Observe that

$$\langle \mathbf{x}, \mathbf{e}_r \rangle = \left\langle \sum_{j=1}^{k} \lambda_j \mathbf{e}_j, \mathbf{e}_r \right\rangle = \sum_{j=1}^{k} \lambda_j \langle \mathbf{e}_j, \mathbf{e}_r \rangle = \lambda_j.$$

(ii) If $\sum_{j=1}^{k} \lambda_j \mathbf{e}_j = \mathbf{0}$, then part (i) tells us that $\lambda_j = \langle \mathbf{0}, \mathbf{e}_j \rangle = 0$ for $1 \le j \le k$.

(iii) Recall that any $n$ linearly independent vectors form a basis for $\mathbb{R}^n$.

(iv) Use parts (iii) and (i). $\qquad\qquad\square$

**Exercise 7.1.4** (i) *If* $U$ *is a subspace of* $\mathbb{R}^n$ *of dimension* $k$, *show that any collection of* $k$ *orthonormal vectors in* $U$ *forms a basis for* $U$.

(ii) *If* $\mathbf{e}_1, \mathbf{e}_2, \ldots, \mathbf{e}_k$ *form an orthonormal basis for the subspace* $U$ *in* (i) *and* $\mathbf{x} \in U$, *show that*

$$\mathbf{x} = \sum_{j=1}^{k} \langle \mathbf{x}, \mathbf{e}_j \rangle \mathbf{e}_j.$$

If a basis for some subspace $U$ of $\mathbb{R}^n$ consists of orthonormal vectors we say that it is an *orthonormal basis* for $U$.

The next set of results are used in many areas of mathematics.

**Theorem 7.1.5 [The Gram–Schmidt method]** *We work in* $\mathbb{R}^n$.

(i) *If* $\mathbf{e}_1, \mathbf{e}_2, \ldots, \mathbf{e}_k$ *are orthonormal and* $\mathbf{x} \in \mathbb{R}^n$, *then*

$$\mathbf{v} = \mathbf{x} - \sum_{j=1}^{k} \langle \mathbf{x}, \mathbf{e}_j \rangle \mathbf{e}_j$$

*is orthogonal to each of* $\mathbf{e}_1, \mathbf{e}_2, \ldots, \mathbf{e}_k$.

(ii) *If* $\mathbf{e}_1, \mathbf{e}_2, \ldots, \mathbf{e}_k$ *are orthonormal and* $\mathbf{x} \in \mathbb{R}^n$, *then either*

$$\mathbf{x} \in \mathrm{span}\{\mathbf{e}_1, \mathbf{e}_2, \ldots, \mathbf{e}_k\}$$

*or the vector* $\mathbf{v}$ *defined in* (i) *is non-zero and, writing* $\mathbf{e}_{k+1} = \|\mathbf{v}\|^{-1}\mathbf{v}$, *we know that* $\mathbf{e}_1$, $\mathbf{e}_2, \ldots, \mathbf{e}_{k+1}$ *are orthonormal and*

$$\mathbf{x} \in \text{span}\{\mathbf{e}_1, \mathbf{e}_2, \ldots, \mathbf{e}_{k+1}\}.$$

(iii) *Suppose that* $1 \leq k \leq q \leq n$. *If* $U$ *is a subspace of* $\mathbb{R}^n$ *of dimension* $q$ *and* $\mathbf{e}_1$, $\mathbf{e}_2$, $\ldots, \mathbf{e}_k$ *are orthonormal vectors in* $U$, *we can find an orthonormal basis* $\mathbf{e}_1$, $\mathbf{e}_2, \ldots, \mathbf{e}_q$ *for* $U$.

(iv) *Every subspace of* $\mathbb{R}^n$ *has an orthonormal basis.*

*Proof* (i) Observe that

$$\langle \mathbf{v}, \mathbf{e}_r \rangle = \langle \mathbf{x}, \mathbf{e}_r \rangle - \sum_{j=1}^{k} \langle \mathbf{x}, \mathbf{e}_r \rangle \langle \mathbf{e}_j, \mathbf{e}_r \rangle = \langle \mathbf{x}, \mathbf{e}_r \rangle - \langle \mathbf{x}, \mathbf{e}_r \rangle = 0$$

for all $1 \leq r \leq k$.

(ii) If $\mathbf{v} = \mathbf{0}$, then

$$\mathbf{x} = \sum_{j=1}^{k} \langle \mathbf{x}, \mathbf{e}_j \rangle \mathbf{e}_j \in \text{span}\{\mathbf{e}_1, \mathbf{e}_2, \ldots, \mathbf{e}_k\}.$$

If $\mathbf{v} \neq \mathbf{0}$, then $\|\mathbf{v}\| \neq 0$ and

$$\mathbf{x} = \mathbf{v} + \sum_{j=1}^{k} \langle \mathbf{x}, \mathbf{e}_j \rangle \mathbf{e}_j$$

$$= \|\mathbf{v}\| \mathbf{e}_{k+1} + \sum_{j=1}^{k} \langle \mathbf{x}, \mathbf{e}_j \rangle \mathbf{e}_j \in \text{span}\{\mathbf{e}_1, \mathbf{e}_2, \ldots, \mathbf{e}_{k+1}\}.$$

(iii) If $\mathbf{e}_1, \mathbf{e}_2, \ldots, \mathbf{e}_k$ do not form a basis for $U$, then we can find

$$\mathbf{x} \in U \setminus \text{span}\{\mathbf{e}_1, \mathbf{e}_2, \ldots, \mathbf{e}_k\}.$$

Defining $\mathbf{v}$ as in part (i), we see that $\mathbf{v} \in U$ and so the vector $\mathbf{e}_{k+1}$ defined in (ii) lies in $U$. Thus we have found orthonormal vectors $\mathbf{e}_1, \mathbf{e}_2, \ldots, \mathbf{e}_{k+1}$ in $U$. If they form a basis for $U$ we stop. If not, we repeat the process. Since no set of $q + 1$ vectors in $U$ can be orthonormal (because no set of $q + 1$ vectors in $U$ can be linearly independent), the process must terminate with an orthonormal basis for $U$ of the required form.

(iv) This follows from (iii).                                                          $\square$

Note that the method of proof for Theorem 7.1.5 not only proves the existence of the appropriate orthonormal bases, but gives a method for finding them.

**Exercise 7.1.6** *Work with row vectors. Find an orthonormal basis* $\mathbf{e}_1$, $\mathbf{e}_2$, $\mathbf{e}_3$ *for* $\mathbb{R}^3$ *with* $\mathbf{e}_1 = 3^{-1/2}(1, 1, 1)$. *Show that it is not unique by writing down another such basis.*

Here is an important consequence of the results just proved.

**Theorem 7.1.7** (*i*) *If $U$ is a subspace of $\mathbb{R}^n$ and $\mathbf{a} \in \mathbb{R}^n$, then there exists a unique point $\mathbf{b} \in U$ such that*

$$\|\mathbf{b} - \mathbf{a}\| \leq \|\mathbf{u} - \mathbf{a}\|$$

*for all $\mathbf{u} \in U$.*

*Moreover, $\mathbf{b}$ is the unique point in $U$ such that $\langle \mathbf{b} - \mathbf{a}, \mathbf{u} \rangle = 0$ for all $\mathbf{u} \in U$.*

In two dimensions, this corresponds to the classical theorem that, if a point $B$ does not lie on a line $l$, then there exists a unique line $l'$ perpendicular to $l$ passing through $B$. The point of intersection of $l$ with $l'$ is the closest point in $l$ to $B$. More briefly, the foot of the perpendicular dropped from $B$ to $l$ is the closest point in $l$ to $B$.

**Exercise 7.1.8** *State the corresponding results for three dimensions when $U$ has dimension 2 and when $U$ has dimension 1.*

*Proof of Theorem 7.1.7* By Theorem 7.1.5, we know that we can find an orthonormal basis $\mathbf{e}_1, \mathbf{e}_2, \ldots, \mathbf{e}_q$ for $U$. If $\mathbf{u} \in U$, then $\mathbf{u} = \sum_{j=1}^{q} \lambda_j \mathbf{e}_j$ and

$$\|\mathbf{u} - \mathbf{a}\|^2 = \left\langle \sum_{j=1}^{q} \lambda_j \mathbf{e}_j - \mathbf{a}, \sum_{j=1}^{q} \lambda_j \mathbf{e}_j - \mathbf{a} \right\rangle$$

$$= \sum_{j=1}^{q} \lambda_j^2 - 2 \sum_{j=1}^{q} \lambda_j \langle \mathbf{a}, \mathbf{e}_j \rangle + \|\mathbf{a}\|^2$$

$$= \sum_{j=1}^{q} \left( \lambda_j - \langle \mathbf{a}, \mathbf{e}_j \rangle \right)^2 + \|\mathbf{a}\|^2 - \sum_{j=1}^{q} \langle \mathbf{a}, \mathbf{e}_j \rangle^2.$$

Thus $\|\mathbf{u} - \mathbf{a}\|$ attains its minimum if and only if $\lambda_j = \langle \mathbf{a}, \mathbf{e}_j \rangle$. The first paragraph follows on setting

$$\mathbf{b} = \sum_{j=1}^{q} \langle \mathbf{a}, \mathbf{e}_j \rangle \mathbf{e}_j.$$

To check the second paragraph, observe that, if $\langle \mathbf{c} - \mathbf{a}, \mathbf{u} \rangle = 0$ for all $\mathbf{u} \in U$, then, in particular, $\langle \mathbf{c} - \mathbf{a}, \mathbf{e}_j \rangle = 0$ for all $1 \leq j \leq q$. Thus

$$\langle \mathbf{c}, \mathbf{e}_j \rangle = \langle \mathbf{a}, \mathbf{e}_j \rangle$$

for all $1 \leq j \leq q$ and so $\mathbf{c} = \mathbf{b}$. Conversely,

$$\left\langle \mathbf{b} - \mathbf{a}, \sum_{j=1}^{q} \lambda_j \mathbf{e}_j \right\rangle = \sum_{j=1}^{q} \lambda_j \left( \langle \mathbf{b}, \mathbf{e}_j \rangle - \langle \mathbf{a}, \mathbf{e}_j \rangle \right) = \sum_{j=1}^{q} \lambda_j 0 = 0$$

for all $\lambda_j \in \mathbb{R}$, so $\langle \mathbf{b} - \mathbf{a}, \mathbf{u} \rangle = 0$ for all $\mathbf{u} \in U$.                                        $\square$

The next exercise is a simple rerun of the proof above, but reappears as the important *Bessel's inequality* in the study of differential equations and elsewhere.

**Exercise 7.1.9**  *We work in $\mathbb{R}^n$ and take $\mathbf{e}_1, \mathbf{e}_2, \ldots, \mathbf{e}_k$ to be orthonormal vectors.*
  *(i) We have*

$$\left\| \mathbf{x} - \sum_{j=1}^{k} \lambda_j \mathbf{e}_j \right\|^2 \geq \|\mathbf{x}\|^2 - \sum_{j=1}^{k} \langle \mathbf{x}, \mathbf{e}_j \rangle^2$$

*with equality if and only if $\lambda_j = \langle \mathbf{x}, \mathbf{e}_j \rangle$ for each $j$.*
  *(ii) (A simple form of Bessel's inequality.) We have*

$$\|\mathbf{x}\|^2 \geq \sum_{j=1}^{k} \langle \mathbf{x}, \mathbf{e}_j \rangle^2$$

*with equality if and only if $\mathbf{x} \in \text{span}\{\mathbf{e}_1, \mathbf{e}_2, \ldots, \mathbf{e}_k\}$.*
*[We shall discuss the full form of Bessel's inequality in Theorem 14.1.15.]*

Exercise 7.6.7 gives another elegant geometric fact which can be obtained from Theorem 7.1.5.

The results of the following exercise will be used later.

**Exercise 7.1.10**  *If $U$ is a subspace of $\mathbb{R}^n$, show that*

$$U^{\perp} = \{\mathbf{v} \in \mathbb{R}^n : \langle \mathbf{v}, \mathbf{u} \rangle = 0 \quad \text{for all } \mathbf{u} \in U\}$$

*is a subspace of $\mathbb{R}^n$. Show, by using Theorem 7.1.7, that every $\mathbf{a} \in \mathbf{R}^n$ can be written in one and only one way as $\mathbf{x} = \mathbf{u} + \mathbf{v}$ with $\mathbf{u} \in U$, $\mathbf{v} \in U^{\perp}$. Deduce that*

$$\dim U + \dim U^{\perp} = n.$$

## 7.2 Orthogonal maps and matrices

Recall from Definition 4.3.8 that, if $A$ is the $n \times n$ matrix $(a_{ij})$, then $A^T$ (the transpose of $A$) is the $n \times n$ matrix $(b_{ij})$ with $b_{ij} = a_{ji}$ [$1 \leq i, j \leq n$].

**Lemma 7.2.1**  *If the linear map $\alpha : \mathbb{R}^n \to \mathbb{R}^n$ has matrix $A$ with respect to some orthonormal basis and $\alpha^* : \mathbb{R}^n \to \mathbb{R}^n$ is the linear map with matrix $A^T$ with respect to the same basis, then*

$$\langle \alpha \mathbf{x}, \mathbf{y} \rangle = \langle \mathbf{x}, \alpha^* \mathbf{y} \rangle$$

*for all $\mathbf{x}, \mathbf{y} \in \mathbb{R}^n$.*
  *Further, if the linear map $\beta : \mathbb{R}^n \to \mathbb{R}^n$ satisfies*

$$\langle \alpha \mathbf{x}, \mathbf{y} \rangle = \langle \mathbf{x}, \beta \mathbf{y} \rangle$$

*for all $\mathbf{x}, \mathbf{y} \in \mathbb{R}^n$, then $\beta = \alpha^*$.*

*Proof*  Suppose that $\alpha$ has matrix $A = (a_{ij})$ with respect to some orthonormal basis $\mathbf{e}_1, \mathbf{e}_2,$ $\ldots, \mathbf{e}_n$ and $\alpha^* : \mathbb{R}^n \to \mathbb{R}^n$ is the linear map with matrix $A^T$ with respect to the same basis.

If $\mathbf{x} = \sum_{i=1}^{n} x_i \mathbf{e}_i$ and $\mathbf{y} = \sum_{i=1}^{n} y_i \mathbf{e}_i$ for some $x_i$, $y_i \in \mathbb{R}$, then, writing $c_{ij} = a_{ji}$ and using the summation convention,

$$\langle \alpha \mathbf{x}, \mathbf{y} \rangle = a_{ij} x_j y_i = x_j c_{ji} y_i = \langle \mathbf{x}, \alpha^* \mathbf{y} \rangle.$$

To obtain the conclusion of the second paragraph, observe that

$$\langle \mathbf{x}, \beta \mathbf{y} \rangle = \langle \alpha \mathbf{x}, \mathbf{y} \rangle = \langle \mathbf{x}, \alpha^* \mathbf{y} \rangle$$

for all $\mathbf{x}$ and all $\mathbf{y}$ so (by Lemma 2.3.8)

$$\beta \mathbf{y} = \alpha^* \mathbf{y}$$

for all $\mathbf{y}$ so, by the definition of the equality of functions, $\beta = \alpha^*$. □

**Exercise 7.2.2** *Prove the first paragraph of Lemma 7.2.1 without using the summation convention.*

Lemma 7.2.1 enables us to make the following definition.

**Definition 7.2.3** *If $\alpha : \mathbb{R}^n \to \mathbb{R}^n$ is a linear map, we define the adjoint $\alpha^*$ to be the unique linear map such that*

$$\langle \alpha \mathbf{x}, \mathbf{y} \rangle = \langle \mathbf{x}, \alpha^* \mathbf{y} \rangle$$

*for all $\mathbf{x}$, $\mathbf{y} \in \mathbb{R}^n$.*

Lemma 7.2.1 now yields the following result.

**Lemma 7.2.4** *If $\alpha : \mathbb{R}^n \to \mathbb{R}^n$ is a linear map with adjoint $\alpha^*$, then, if $\alpha$ has matrix $A$ with respect to some orthonormal basis, it follows that $\alpha^*$ has matrix $A^T$ with respect to the same basis.*

Lemma 7.2.4 suggests the notation $\alpha^T = \alpha^*$ which is, indeed, sometimes used, but does not mesh well with the ideas developed in the second part of this text.

**Lemma 7.2.5** *Let $\alpha$, $\beta : \mathbb{R}^n \to \mathbb{R}^n$ be linear and let $\lambda$, $\mu \in \mathbb{R}$. Then the following results hold.*
*(i) $(\alpha \beta)^* = \beta^* \alpha^*$.*
*(ii) $\alpha^{**} = \alpha$, where we write $\alpha^{**} = (\alpha^*)^*$.*
*(iii) $(\lambda \alpha + \mu \beta)^* = \lambda \alpha^* + \mu \beta^*$.*
*(iv) $\iota^* = \iota$.*

*Proof* (i) Observe that

$$\langle (\alpha \beta)^* \mathbf{x}, \mathbf{y} \rangle = \langle \mathbf{x}, (\alpha \beta) \mathbf{y} \rangle = \langle \mathbf{x}, \alpha (\beta \mathbf{y}) \rangle = \langle \alpha (\beta \mathbf{y}), \mathbf{x} \rangle$$
$$= \langle \beta \mathbf{y}, \alpha^* \mathbf{x} \rangle = \langle \mathbf{y}, \beta^* (\alpha^* \mathbf{x}) \rangle = \langle \mathbf{y}, (\beta^* \alpha^*) \mathbf{x} \rangle = \langle (\beta^* \alpha^*) \mathbf{x}, \mathbf{y} \rangle$$

for all $\mathbf{x}$ and all $\mathbf{y}$, so (by Lemma 2.3.8)

$$(\alpha \beta)^* \mathbf{x} = (\beta^* \alpha^*) \mathbf{x}$$

for all $\mathbf{x}$ and, by the definition of the equality of functions, $(\alpha\beta)^* = \beta^*\alpha^*$.

(ii) Observe that

$$\langle \alpha^{**}\mathbf{x}, \mathbf{y} \rangle = \langle \mathbf{x}, \alpha^*\mathbf{y} \rangle = \langle \alpha^*\mathbf{y}, \mathbf{x} \rangle = \langle \mathbf{y}, \alpha\mathbf{x} \rangle = \langle \alpha\mathbf{x}, \mathbf{y} \rangle$$

for all $\mathbf{x}$ and all $\mathbf{y}$, so

$$\alpha^{**}\mathbf{x} = \alpha\mathbf{x}$$

for all $\mathbf{x}$ and $\alpha^{**} = \alpha$.

(iii) and (iv) Left as an exercise for the reader.                                □

**Exercise 7.2.6** *Let $A$ and $B$ be $n \times n$ real matrices and let $\lambda$, $\mu \in \mathbb{R}$. Prove the following results, first by using Lemmas 7.2.5 and 7.2.4 and then by direct computations.*

*(i) $(AB)^T = B^T A^T$.*

*(ii) $A^{TT} = A$.*

*(iii) $(\lambda A + \mu B)^T = \lambda A^T + \mu B^T$.*

*(iv) $I^T = I$.*

The reader may, quite reasonably, ask why we did not prove the matrix results first and then use them to obtain the results on linear maps. The answer that this procedure would tell us that the results were true, but not why they were true, may strike the reader as mere verbiage. She may be happier to be told that the coordinate free proofs we have given turn out to generalise in a way that the coordinate dependent proofs do not.

We can now characterise those linear maps which preserve length.

**Theorem 7.2.7** *Let $\alpha : \mathbb{R}^n \to \mathbb{R}^n$ be linear. The following statements are equivalent.*

*(i) $\|\alpha\mathbf{x}\| = \|\mathbf{x}\|$ for all $\mathbf{x} \in \mathbb{R}^n$.*

*(ii) $\langle \alpha\mathbf{x}, \alpha\mathbf{y} \rangle = \langle \mathbf{x}, \mathbf{y} \rangle$ for all $\mathbf{x}, \mathbf{y} \in \mathbb{R}^n$.*

*(iii) $\alpha^*\alpha = \iota$.*

*(iv) $\alpha$ is invertible with inverse $\alpha^*$.*

*(v) If $\alpha$ has matrix $A$ with respect to some orthonormal basis, then $A^T A = I$.*

*Proof* (i)$\Rightarrow$(ii). If (i) holds, then the useful *polarisation* identity

$$4\langle \mathbf{u}, \mathbf{v} \rangle = \|\mathbf{u} + \mathbf{v}\|^2 - \|\mathbf{u} - \mathbf{v}\|^2$$

gives

$$4\langle \alpha\mathbf{x}, \alpha\mathbf{y} \rangle = \|\alpha\mathbf{x} + \alpha\mathbf{y}\|^2 - \|\alpha\mathbf{x} - \alpha\mathbf{y}\|^2 = \|\alpha(\mathbf{x} + \mathbf{y})\|^2 - \|\alpha(\mathbf{x} - \mathbf{y})\|^2$$
$$= \|\mathbf{x} + \mathbf{y}\|^2 - \|\mathbf{x} - \mathbf{y}\|^2 = 4\langle \mathbf{x}, \mathbf{y} \rangle$$

and we are done.

(ii)$\Rightarrow$(iii). If (ii) holds, then

$$\langle (\alpha^*\alpha)\mathbf{x}, \mathbf{y} \rangle = \langle \alpha^*(\alpha\mathbf{x}), \mathbf{y} \rangle = \langle \alpha\mathbf{x}, \alpha\mathbf{y} \rangle = \langle \mathbf{x}, \mathbf{y} \rangle$$

for all $\mathbf{x}$ and all $\mathbf{y}$, so

$$(\alpha^*\alpha)\mathbf{x} = \mathbf{x}$$

for all $\mathbf{x}$ and $\alpha^*\alpha = \iota$.

(iii)$\Rightarrow$(i). If (iii) holds, then

$$\|\alpha\mathbf{x}\|^2 = \langle \alpha\mathbf{x}, \alpha\mathbf{x} \rangle = \langle \alpha^*(\alpha\mathbf{x}), \mathbf{x} \rangle = \langle \mathbf{x}, \mathbf{x} \rangle = \|\mathbf{x}\|^2$$

as required.

Conditions (iv) and (v) are automatically equivalent to (iii). $\qquad\square$

If, as I shall tend to do, we think of the linear maps as central, we refer to the collection of distance preserving (or *isometric*) linear[1] maps by the name $O(\mathbb{R}^n)$. If we think of the matrices as central, we refer to the collection of real $n \times n$ matrices $A$ with $AA^T = I$ by the name $O(\mathbb{R}^n)$. In practice, most people use whichever convention is most convenient at the time and no confusion results. A real $n \times n$ matrix $A$ with $AA^T = I$ is called an *orthogonal* matrix.

**Lemma 7.2.8** $O(\mathbb{R}^n)$ *is a subgroup of* $GL(\mathbb{R}^n)$.

*Proof* We check the conditions of Definition 5.3.15.
   (i) $\iota^* = \iota$, so $\iota^*\iota = \iota^2 = \iota$ and $\iota \in O(\mathbb{R}^n)$.
   (ii) If $\alpha \in O(\mathbb{R}^n)$, then $\alpha^{-1} = \alpha^*$, so

$$(\alpha^{-1})^*\alpha^{-1} = (\alpha^*)^*\alpha^* = \alpha\alpha^* = \alpha\alpha^{-1} = \iota$$

and $\alpha^{-1} \in O(\mathbb{R}^n)$.
   (iii) If $\alpha, \beta \in O(\mathbb{R}^n)$, then

$$(\alpha\beta)^*(\alpha\beta) = (\beta^*\alpha^*)(\alpha\beta) = \beta^*(\alpha^*\alpha)\beta = \beta^*\iota\beta = \iota$$

and so $\alpha\beta \in O(\mathbb{R}^n)$. $\qquad\square$

We call $O(\mathbb{R}^n)$ the *orthogonal group*.

The following remark is often useful as a check in computation.

**Lemma 7.2.9** *The following three conditions on a real* $n \times n$ *matrix* $A$ *are equivalent.*
   (i) $A \in O(\mathbb{R}^n)$.
   (ii) *The columns of* $A$ *are orthonormal.*
   (iii) *The rows of* $A$ *are orthonormal.*

*Proof* We leave the proof as a simple exercise for the reader. $\qquad\square$

**Exercise 7.2.10** *Prove Lemma 7.2.9.*

*Are the following statements about a real* $n \times n$ *matrix* $A$ *true? Give proofs or counterexamples as appropriate.*

---

[1] We look at general distance preserving maps in Exercise 7.6.8.

(i) *If all the rows of A are orthogonal, then the columns of A are orthogonal.*

(ii) *If all the rows of A are row vectors of norm (i.e. Euclidean length) 1, then the columns of A are column vectors of norm 1.*

(iii) *If $A \in O(\mathbb{R}^n)$, then any $n - 1$ rows determine the remaining row uniquely.*

(iv) *If $A \in O(\mathbb{R}^n)$ and $\det A = 1$, then any $n - 1$ rows determine the remaining row uniquely.*

We recall that the determinant of a square matrix can be evaluated by row or by column expansion and so

$$\det A^T = \det A.$$

The next lemma is an immediate consequence.

**Lemma 7.2.11** *If $\alpha : \mathbb{R}^n \to \mathbb{R}^n$ is linear, then $\det \alpha^* = \det \alpha$.*

*Proof* We leave the proof as a simple exercise for the reader.                  □

**Lemma 7.2.12** *If $\alpha \in O(\mathbb{R}^n)$, then $\det \alpha = 1$ or $\det \alpha = -1$.*

*Proof* Observe that

$$1 = \det \iota = \det(\alpha^* \alpha) = \det \alpha^* \det \alpha = (\det \alpha)^2.$$

                                                                                  □

**Exercise 7.2.13** *Write down a $2 \times 2$ real matrix A with $\det A = 1$ which is not orthogonal. Write down a $2 \times 2$ real matrix B with $\det B = -1$ which is not orthogonal. Prove your assertions.*

If we think in terms of linear maps, we define

$$SO(\mathbb{R}^n) = \{\alpha \in O(\mathbb{R}^n) : \det \alpha = 1\}.$$

If we think in terms of matrices, we define

$$SO(\mathbb{R}^n) = \{A \in O(\mathbb{R}^n) : \det A = 1\}.$$

**Lemma 7.2.14** *$SO(\mathbb{R}^n)$ is a subgroup of $O(\mathbb{R}^n)$.*

The proof is left to the reader. We call $SO(\mathbb{R}^n)$ the *special orthogonal group*.

We restate these ideas for the three dimensional case, using the summation convention.

**Lemma 7.2.15** *The matrix $L \in O(\mathbb{R}^3)$ if and only if, using the summation convention,*

$$l_{ik}l_{jk} = \delta_{ij}.$$

*If $L \in O(\mathbb{R}^3)$*

$$\epsilon_{ijk}l_{ir}l_{js}l_{kt} = \pm\epsilon_{rst}$$

*with the positive sign if $L \in SO(\mathbb{R}^3)$ and the negative sign otherwise.*

The proof is left to the reader.

## 7.3 Rotations and reflections in $\mathbb{R}^2$ and $\mathbb{R}^3$

In this section we shall look at matrix representations of $O(\mathbb{R}^n)$ and $SO(\mathbb{R}^n)$ when $n = 2$ and $n = 3$. We start by looking at the two dimensional case.

**Theorem 7.3.1** (*i*) *If the linear map* $\alpha : \mathbb{R}^2 \to \mathbb{R}^2$ *has matrix*

$$A = \begin{pmatrix} \cos\theta & -\sin\theta \\ \sin\theta & \cos\theta \end{pmatrix}$$

*relative to some orthonormal basis, then* $\alpha \in SO(\mathbb{R}^2)$.

  (*ii*) *If* $\alpha \in SO(\mathbb{R}^2)$, *then its matrix* $A$ *relative to a* given *orthonormal basis takes the form*

$$A = \begin{pmatrix} \cos\theta & -\sin\theta \\ \sin\theta & \cos\theta \end{pmatrix}$$

*for some unique* $\theta$ (*depending on the basis*) *with* $-\pi \leq \theta < \pi$.

  (*iii*) *If* $\alpha \in SO(\mathbb{R}^2)$, *then there exists a unique* $\theta$ *with* $0 \leq \theta \leq \pi$ *such that, with respect to* any *orthonormal basis, the matrix of* $\alpha$ *takes one of the two following forms:*

$$\begin{pmatrix} \cos\theta & -\sin\theta \\ \sin\theta & \cos\theta \end{pmatrix} \quad or \quad \begin{pmatrix} \cos(-\theta) & -\sin(-\theta) \\ \sin(-\theta) & \cos(-\theta) \end{pmatrix}.$$

Observe that in part (ii) we have a *given* orthonormal basis, but that the more precise part (iii) refers to *any* orthonormal basis. When the reader does Exercise 7.3.2, she will see why we had to allow two possible forms.

*Proof* (i) Direct computation which is left to the reader.

  (ii) Let

$$A = \begin{pmatrix} a & b \\ c & d \end{pmatrix}.$$

We have

$$\begin{pmatrix} 1 & 0 \\ 0 & 1 \end{pmatrix} = I = AA^T = \begin{pmatrix} a & b \\ c & d \end{pmatrix} \begin{pmatrix} a & c \\ b & d \end{pmatrix} = \begin{pmatrix} a^2 + b^2 & ac + bd \\ ac + bd & c^2 + d^2 \end{pmatrix}.$$

Thus $a^2 + b^2 = 1$ and, since $a$ and $b$ are real, we can take $a = \cos\theta$, $b = -\sin\theta$ for some real $\theta$. Similarly, we can can take $a = \sin\phi$, $b = \cos\phi$ for some real $\phi$. Since $ac + bd = 0$, we know that

$$\sin(\theta - \phi) = \cos\theta \sin\phi - \sin\theta \cos\phi = 0$$

and so $\theta - \phi \equiv 0$ modulo $\pi$.

  We also know that

$$1 = \det A = ad - bc = \cos\theta \cos\phi + \sin\theta \sin\phi = \cos(\theta - \phi),$$

so $\theta - \phi \equiv 0$ modulo $2\pi$ and

$$A = \begin{pmatrix} \cos\theta & -\sin\theta \\ \sin\theta & \cos\theta \end{pmatrix}.$$

We know that, if $a^2 + b^2 = 1$, the equation $\cos\theta = a$, $\sin\theta = b$ has exactly one solution with $-\pi \le \theta < \pi$, so the uniqueness follows.

(iii) Suppose that $\alpha$ has matrix representations

$$A = \begin{pmatrix} \cos\theta & -\sin\theta \\ \sin\theta & \cos\theta \end{pmatrix} \quad \text{and} \quad B = \begin{pmatrix} \cos\phi & -\sin\phi \\ \sin\phi & \cos\phi \end{pmatrix}$$

with respect to two orthonormal bases. Since the characteristic polynomial does not depend on the choice of basis,

$$\det(tI - A) = \det(tI - B).$$

We observe that

$$\det(tI - A) = \det\begin{pmatrix} t - \cos\theta & \sin\theta \\ -\sin\theta & t - \cos\theta \end{pmatrix} = (t - \cos\theta)^2 + \sin\theta^2$$

$$= t^2 - 2t\cos\theta + \cos^2\theta + \sin\theta^2 = t^2 - 2t\cos\theta + 1$$

and so

$$t^2 - 2t\cos\theta + 1 = t^2 - 2t\cos\phi + 1,$$

for all $t$. Thus $\cos\theta = \cos\phi$ and so $\sin\theta = \pm\sin\phi$. The result follows. $\qquad\square$

**Exercise 7.3.2** *Suppose that $e_1$ and $e_2$ form an orthonormal basis for $\mathbb{R}^2$ and the linear map $\alpha : \mathbb{R}^2 \to \mathbb{R}^2$ has matrix*

$$A = \begin{pmatrix} \cos\theta & -\sin\theta \\ \sin\theta & \cos\theta \end{pmatrix}$$

*with respect to this basis.*

*Show that $e_1$ and $-e_2$ form an orthonormal basis for $\mathbb{R}^2$ and find the matrix of $\alpha$ with respect to this new basis.*

**Theorem 7.3.3** *(i) If $\alpha \in O(\mathbb{R}^2) \setminus SO(\mathbb{R}^2)$, then its matrix $A$ relative to an orthonormal basis takes the form*

$$A = \begin{pmatrix} \cos\phi & \sin\phi \\ \sin\phi & -\cos\phi \end{pmatrix}$$

*for some unique $\phi$ with $-\pi \le \phi < \pi$.*

*(ii) If $\alpha \in O(\mathbb{R}^2) \setminus SO(\mathbb{R}^2)$, then there exists an orthonormal basis with respect to which $\alpha$ has matrix*

$$B = \begin{pmatrix} -1 & 0 \\ 0 & 1 \end{pmatrix}.$$

*(iii) If the linear map* $\alpha : \mathbb{R}^2 \to \mathbb{R}^2$ *has matrix*

$$A = \begin{pmatrix} \cos\theta & \sin\theta \\ \sin\theta & -\cos\theta \end{pmatrix}$$

*relative to some orthonormal basis, then* $\alpha \in O(\mathbb{R}^2) \setminus SO(\mathbb{R}^2)$.

*Proof* (i) Exactly as in the proof of Theorem 7.3.1 (ii), the condition $AA^T = I$ tells us that

$$A = \begin{pmatrix} \cos\theta & -\sin\theta \\ \sin\phi & \cos\phi \end{pmatrix}.$$

Since $\det A = -1$, we have

$$-1 = \cos\theta\cos\phi + \sin\theta\sin\phi = \cos(\theta - \phi),$$

so $\theta - \phi \equiv -\pi$ modulo $2\pi$ and

$$A = \begin{pmatrix} \cos\phi & \sin\phi \\ \sin\phi & -\cos\phi \end{pmatrix}.$$

(ii) By part (i),

$$\det(t\iota - \alpha) = \begin{pmatrix} t - \cos\theta & \sin\theta \\ \sin\theta & t + \cos\theta \end{pmatrix} = t^2 - \cos^2\theta - \sin^2\theta = t^2 - 1 = (t+1)(t-1).$$

Thus $\alpha$ has eigenvalues $-1$ and $1$. Let $e_1$ and $e_2$ be associated eigenvectors of norm 1 with

$$\alpha e_1 = -e_1 \quad \text{and} \quad \alpha e_2 = -e_2.$$

Since $\alpha$ preserves the inner product,

$$\langle e_1, e_2 \rangle = \langle \alpha e_1, \alpha e_2 \rangle = \langle -e_1, e_2 \rangle = -\langle e_1, e_2 \rangle$$

so $e_1$ and $e_2$ form an orthonormal basis with respect to which $\alpha$ has matrix

$$B = \begin{pmatrix} -1 & 0 \\ 0 & 1 \end{pmatrix}.$$

(iii) Direct calculation which is left to the reader.                          $\square$

**Exercise 7.3.4** (*i*) *Let* $e_1$ *and* $e_2$ *form an orthonormal basis for* $\mathbb{R}^2$. *Convince yourself that the linear map with matrix representation*

$$\begin{pmatrix} \cos\theta & -\sin\theta \\ \sin\theta & \cos\theta \end{pmatrix}$$

*represents a rotation though* $\theta$ *about* $\mathbf{0}$ *and the linear map with matrix representation*

$$\begin{pmatrix} -1 & 0 \\ 0 & 1 \end{pmatrix}$$

*represents what the mathematician in the street would call a reflection in the line* $\{te_1 : t \in e_1\}$.

[*Of course, unless you have a formal definition of reflection and rotation, you cannot* prove *this result.*]

(ii) *Let* $\mathbf{u}_1$ *and* $\mathbf{u}_2$ *form another orthonormal basis for* $\mathbb{R}^2$. *Suppose that the reflection* $\alpha$ *has matrix*

$$\begin{pmatrix} -1 & 0 \\ 0 & 1 \end{pmatrix}$$

*with respect to the basis* $\mathbf{e}_1$, $\mathbf{e}_2$ *and matrix*

$$\begin{pmatrix} \cos\theta & \sin\theta \\ \sin\theta & -\cos\theta \end{pmatrix}$$

*with respect to the basis* $\mathbf{u}_1$, $\mathbf{u}_2$. *Prove a formula relating* $\theta$ *to* $\phi$ *where* $\phi$ *is the (appropriately chosen) angle between* $\mathbf{e}_1$ *and* $\mathbf{u}_1$.

The reader may feel that we have gone about things the wrong way and have merely found matrix representations for rotation and reflection. However, we have done more, since we have shown that there are no other distance preserving linear maps.

We can push matters a little further and deal with $O(\mathbb{R}^3)$ and $SO(\mathbb{R}^3)$ in a similar manner.

**Theorem 7.3.5** *If* $\alpha \in SO(\mathbb{R}^3)$, *then we can find an orthonormal basis such that* $\alpha$ *has matrix representation*

$$A = \begin{pmatrix} 1 & 0 & 0 \\ 0 & \cos\theta & -\sin\theta \\ 0 & \sin\theta & \cos\theta \end{pmatrix}.$$

*If* $\alpha \in O(\mathbb{R}^3) \setminus SO(\mathbb{R}^3)$, *then we can find an orthonormal basis such that* $\alpha$ *has matrix representation*

$$A = \begin{pmatrix} -1 & 0 & 0 \\ 0 & \cos\theta & -\sin\theta \\ 0 & \sin\theta & \cos\theta \end{pmatrix}.$$

*Proof* Suppose that $\alpha \in O(\mathbb{R}^3)$. Since every real cubic has a real root, the characteristic polynomial $\det(t\iota - \alpha)$ has a real root and so $\alpha$ has an eigenvalue $\lambda$ with a corresponding eigenvector $\mathbf{e}_1$ of norm 1. Since $\alpha$ preserves length,

$$|\lambda| = \|\lambda\mathbf{e}_1\| = \|\alpha\mathbf{e}_1\| = \|\alpha\mathbf{e}_1\| = 1$$

and so $\lambda = \pm 1$.

Now consider the subspace

$$\mathbf{e}_1^\perp = \{\mathbf{x} : \langle\mathbf{e}_1, \mathbf{x}\rangle = 0\}.$$

This has dimension 2 (see Lemma 7.1.10) and, since $\alpha$ preserves the inner product,

$$\mathbf{x} \in \mathbf{e}_1^\perp \Rightarrow \langle\mathbf{e}_1, \alpha\mathbf{x}\rangle = \lambda^{-1}\langle\alpha\mathbf{e}_1, \alpha\mathbf{x}\rangle = \lambda^{-1}\langle\mathbf{e}_1, \mathbf{x}\rangle = 0 \Rightarrow \alpha\mathbf{x} \in \mathbf{e}_1^\perp$$

so $\alpha$ maps elements of $\mathbf{e}_1^\perp$ to elements of $\mathbf{e}_1^\perp$.

Thus the restriction of $\alpha$ to $\mathbf{e}_1^\perp$ is a norm preserving linear map on the two dimensional inner product space $\mathbf{e}_1^\perp$. It follows that we can find orthonormal vectors $\mathbf{e}_2$ and $\mathbf{e}_3$ such that *either*

$$\alpha \mathbf{e}_2 = \cos\theta \mathbf{e}_2 + \sin\theta \mathbf{e}_3 \quad \text{and} \quad \alpha \mathbf{e}_3 = -\sin\theta \mathbf{e}_2 + \cos\theta \mathbf{e}_3$$

for some $\theta$ *or*

$$\alpha \mathbf{e}_2 = -\mathbf{e}_2 \quad \text{and} \quad \alpha \mathbf{e}_3 = \mathbf{e}_3.$$

We observe that $\mathbf{e}_1, \mathbf{e}_2, \mathbf{e}_3$ form an orthonormal basis for $\mathbb{R}^3$ with respect to which $\alpha$ has matrix taking one of the following forms

$$A_1 = \begin{pmatrix} 1 & 0 & 0 \\ 0 & \cos\theta & -\sin\theta \\ 0 & \sin\theta & \cos\theta \end{pmatrix}, \quad A_2 = \begin{pmatrix} -1 & 0 & 0 \\ 0 & \cos\theta & -\sin\theta \\ 0 & \sin\theta & \cos\theta \end{pmatrix},$$

$$A_3 = \begin{pmatrix} 1 & 0 & 0 \\ 0 & -1 & 0 \\ 0 & 0 & 1 \end{pmatrix}, \quad A_4 = \begin{pmatrix} -1 & 0 & 0 \\ 0 & -1 & 0 \\ 0 & 0 & 1 \end{pmatrix}.$$

In the case when we obtain $A_3$, if we take our basis vectors in a different order, we can produce an orthonormal basis with respect to which $\alpha$ has matrix

$$\begin{pmatrix} -1 & 0 & 0 \\ 0 & 1 & 0 \\ 0 & 0 & 1 \end{pmatrix} = \begin{pmatrix} 1 & 0 & 0 \\ 0 & \cos 0 & -\sin 0 \\ 0 & \sin 0 & \cos 0 \end{pmatrix}.$$

In the case when we obtain $A_4$, if we take our basis vectors in a different order, we can produce an orthonormal basis with respect to which $\alpha$ has matrix

$$\begin{pmatrix} 1 & 0 & 0 \\ 0 & -1 & 0 \\ 0 & 0 & -1 \end{pmatrix} = \begin{pmatrix} 1 & 0 & 0 \\ 0 & \cos\pi & -\sin\pi \\ 0 & \sin\pi & \cos\pi \end{pmatrix}.$$

Thus we know that there is always an orthogonal basis with respect to which $\alpha$ has one of the matrices $A_1$ or $A_2$. By direct calculation, $\det A_1 = 1$ and $\det A_2 = -1$, so we are done. $\quad\square$

We have shown that, if $\alpha \in SO(\mathbb{R}^3)$, then there is an orthonormal basis $\mathbf{e}_1, \mathbf{e}_2, \mathbf{e}_3$ with respect to which $\alpha$ has matrix

$$\begin{pmatrix} 1 & 0 & 0 \\ 0 & \cos\theta & -\sin\theta \\ 0 & \sin\theta & \cos\theta \end{pmatrix}.$$

This is naturally interpreted as saying that $\alpha$ is a rotation through angle $\theta$ about an axis along $\mathbf{e}_1$. This result is sometimes stated as saying that 'every rotation has an axis'.

However, if $\alpha \in O(\mathbb{R}^3) \setminus SO(\mathbb{R}^3)$, so that there is an orthonormal basis $\mathbf{e}_1, \mathbf{e}_2, \mathbf{e}_3$ with respect to which $\alpha$ has matrix

$$\begin{pmatrix} -1 & 0 & 0 \\ 0 & \cos\theta & -\sin\theta \\ 0 & \sin\theta & \cos\theta \end{pmatrix}, \qquad \qquad \bigstar$$

then $\alpha$ is clearly not a rotation.

It is natural to call $SO(3)$ the set of rotations of $\mathbb{R}^3$.

**Exercise 7.3.6** *By considering eigenvectors, or otherwise, show that the $\alpha$ just considered, with matrix given in $\bigstar$, is a reflection in a plane only if $\theta \equiv 0$ modulo $2\pi$ and a reflection in the origin[2] only if $\theta \equiv \pi$ modulo $2\pi$.*

A still more interesting example occurs if we consider a linear map $\alpha : \mathbb{R}^4 \to \mathbb{R}^4$ whose matrix with respect to some orthonormal basis is given by

$$\begin{pmatrix} \cos\theta & -\sin\theta & 0 & 0 \\ \sin\theta & \cos\theta & 0 & 0 \\ 0 & 0 & \cos\phi & -\sin\phi \\ 0 & 0 & \sin\phi & \cos\phi \end{pmatrix}.$$

Direct calculation gives $\alpha \in SO(\mathbb{R}^4)$ but, unless $\theta$ and $\phi$ take special values, there is no 'axis of rotation' and no 'angle of rotation'. (Exercise 7.6.18 goes deeper into the matter.)

**Exercise 7.3.7** *Show that the $\alpha$ just considered has no eigenvalues (over $\mathbb{R}$) unless $\theta$ or $\phi$ take special values to be determined.*

In classical physics we only work in three dimensions, so the results of this section are sufficient. However, if we wish to look at higher dimensions, we need a different approach.

## 7.4 Reflections in $\mathbb{R}^n$

The following approach goes back to Euler. We start with a natural generalisation of the notion of reflection to all dimensions.

**Definition 7.4.1** *If $\mathbf{n}$ is a vector of norm 1, the map $\rho : \mathbb{R}^n \to \mathbb{R}^n$ given by*

$$\rho\mathbf{x} = \mathbf{x} - 2\langle\mathbf{x}, \mathbf{n}\rangle\mathbf{n}$$

*is said to be a* reflection *in*

$$\pi = \{\mathbf{x} : \langle\mathbf{x}, \mathbf{n}\rangle = 0\}.$$

**Lemma 7.4.2** *The following two statements about a map $\rho : \mathbb{R}^n \to \mathbb{R}^n$ are equivalent.*

---

[2] Ignore this if you do not know the terminology.

(i) $\rho$ *is a reflection in*

$$\pi = \{\mathbf{x} : \langle \mathbf{x}, \mathbf{n} \rangle = 0\}$$

*where* $\mathbf{n}$ *has norm 1.*

(ii) $\rho$ *is a linear map and there is an orthonormal basis* $\mathbf{e}_1, \mathbf{e}_2, \ldots, \mathbf{e}_n$ *with respect to which* $\rho$ *has a diagonal matrix* $D$ *with* $d_{11} = -1$, $d_{ii} = 1$ *for all* $2 \le i \le n$.

*Proof* (i)$\Rightarrow$(ii). Suppose that (i) is true. Set $\mathbf{e}_1 = \mathbf{n}$ and choose an orthonormal basis $\mathbf{e}_1, \mathbf{e}_2, \ldots, \mathbf{e}_n$. Simple calculations show that

$$\rho \mathbf{e}_j = \begin{cases} \mathbf{e}_1 - 2\mathbf{e}_1 = -\mathbf{e}_1 & \text{if } j = 1, \\ \mathbf{e}_j - 0\mathbf{e}_j = \mathbf{e}_j & \text{otherwise.} \end{cases}$$

Thus $\rho$ has matrix $D$ with respect to the given basis.

(ii)$\Rightarrow$(i). Suppose that (ii) is true. Set $\mathbf{n} = \mathbf{e}_1$. If $\mathbf{x} \in \mathbb{R}^n$, we can write $\mathbf{x} = \sum_{j=1}^n x_j \mathbf{e}_j$ for some $x_j \in \mathbb{R}^n$. We then have

$$\rho \mathbf{x} = \rho \left( \sum_{j=1}^n x_j \mathbf{e}_j \right) = \sum_{j=1}^n x_j \rho \mathbf{e}_j = -x_1 \mathbf{e}_1 + \sum_{j=2}^n x_j \mathbf{e}_j,$$

and

$$\mathbf{x} - 2\langle \mathbf{x}, \mathbf{n} \rangle \mathbf{n} = \sum_{j=1}^n x_j \mathbf{e}_j - 2 \left\langle \sum_{j=1}^n x_j \mathbf{e}_j, \mathbf{e}_1 \right\rangle \mathbf{e}_1$$

$$= \sum_{j=1}^n x_j \mathbf{e}_j - 2x_1 \mathbf{e}_1 = -x_1 \mathbf{e}_1 + \sum_{j=2}^n x_j \mathbf{e}_j,$$

so that

$$\rho(\mathbf{x}) = \mathbf{x} - 2\langle \mathbf{x}, \mathbf{n} \rangle \mathbf{n}$$

as stated. $\qquad\square$

**Exercise 7.4.3** *We work in* $\mathbb{R}^n$. *Suppose that* $\mathbf{n}$ *is a vector of norm 1. If* $\mathbf{x} \in \mathbb{R}^n$, *show that*

$$\mathbf{x} = \mathbf{u} + \mathbf{v}$$

*where* $\mathbf{u} = \langle \mathbf{x}, \mathbf{n} \rangle \mathbf{n}$ *and* $\mathbf{v} \perp \mathbf{n}$. *Show that, if* $\rho$ *is the reflection given in Definition 7.4.1,*

$$\rho \mathbf{x} = -\mathbf{u} + \mathbf{v}.$$

**Lemma 7.4.4** *If* $\rho$ *is a reflection, then* $\rho$ *has the following properties.*
  (i) $\rho^2 = \iota$.
  (ii) $\rho \in O(\mathbb{R}^n)$.
  (iii) $\det \rho = -1$.

*Proof* The results follow immediately from condition (ii) of Lemma 7.4.2 on observing that $D^2 = I$, $D^T = D$ and $\det D = -1$. $\qquad\square$

The main result of this section depends on the following observation.

**Lemma 7.4.5** *If $\|\mathbf{a}\| = \|\mathbf{b}\|$ and $\mathbf{a} \neq \mathbf{b}$, then we can find a unit vector $\mathbf{n}$ such that the associated reflection $\rho$ has the property that $\rho\mathbf{a} = \mathbf{b}$. Moreover, we can choose $\mathbf{n}$ in such a way that, whenever $\mathbf{u}$ is perpendicular to both $\mathbf{a}$ and $\mathbf{b}$, we have $\rho\mathbf{u} = \mathbf{u}$.*

**Exercise 7.4.6** *Prove the first part of Lemma 7.4.5 geometrically in the case when $n = 2$.*

Once we have done Exercise 7.4.6, it is more or less clear how to attack the general case.

*Proof of Lemma 7.4.5* Let $\mathbf{n} = \|\mathbf{a} - \mathbf{b}\|^{-1}(\mathbf{a} - \mathbf{b})$ and set

$$\rho(\mathbf{x}) = \mathbf{x} - 2\langle \mathbf{x}, \mathbf{n}\rangle\mathbf{n}.$$

Then, since $\|\mathbf{a}\| = \|\mathbf{b}\|$,

$$\langle \mathbf{a}, \mathbf{a} - \mathbf{b}\rangle = \|\mathbf{a}\|^2 - \langle \mathbf{a}, \mathbf{b}\rangle = \frac{1}{2}(\|\mathbf{a}\|^2 - 2\langle \mathbf{a}, \mathbf{b}\rangle + \|\mathbf{b}\|^2) = \frac{1}{2}\|\mathbf{a} - \mathbf{b}\|^2,$$

and so

$$\rho(\mathbf{a}) = \mathbf{a} - 2\|\mathbf{a} - \mathbf{b}\|^{-2}\langle \mathbf{a}, \mathbf{a} - \mathbf{b}\rangle(\mathbf{a} - \mathbf{b}) = \mathbf{a} + (\mathbf{b} - \mathbf{a}) = \mathbf{b}.$$

If $\mathbf{u}$ is perpendicular to both $\mathbf{a}$ and $\mathbf{b}$, then $\mathbf{u}$ is perpendicular to $\mathbf{n}$ and

$$\rho(\mathbf{u}) = \mathbf{u} - (2 \times 0)\mathbf{u} = \mathbf{u}$$

as required. $\square$

We can use this result to 'fix vectors' as follows.

**Lemma 7.4.7** *Suppose that $\beta \in O(\mathbb{R}^n)$ and $\beta$ fixes the orthonormal vectors $\mathbf{e}_1$, $\mathbf{e}_2$, ..., $\mathbf{e}_k$ (that is to say, $\beta(\mathbf{e}_r) = \mathbf{e}_r$ for $1 \leq r \leq k$). Then either $\beta = \iota$ or we can find a $\mathbf{e}_{k+1}$ and a reflection $\rho$ such that $\mathbf{e}_1$, $\mathbf{e}_2$, ..., $\mathbf{e}_{k+1}$ are orthonormal and $\rho\beta(\mathbf{e}_r) = \mathbf{e}_r$ for $1 \leq r \leq k + 1$.*

*Proof* If $\beta \neq \iota$, then there must exist an $\mathbf{x} \in \mathbb{R}^n$ such that $\beta\mathbf{x} \neq \mathbf{x}$ and so, setting

$$\mathbf{v} = \mathbf{x} - \sum_{j=1}^{k}\langle \mathbf{x}, \mathbf{e}_j\rangle\mathbf{e}_j \quad \text{and} \quad \mathbf{e}_{r+1} = \|\mathbf{v}\|^{-1}\mathbf{v},$$

we see that there exists a vector $\mathbf{e}_{r+1}$ of norm 1 perpendicular to $\mathbf{e}_1$, $\mathbf{e}_2$, ..., $\mathbf{e}_k$ such that $\beta\mathbf{e}_{r+1} \neq \mathbf{e}_{r+1}$.

Since $\beta$ preserves norm and inner product (recall Theorem 7.2.7, if necessary), we know that $\beta\mathbf{e}_{r+1}$ has norm 1 and is perpendicular to $\mathbf{e}_1$, $\mathbf{e}_2$, ..., $\mathbf{e}_k$. By Lemma 7.4.5 we can find a reflection $\rho$ such that

$$\rho(\beta\mathbf{e}_{r+1}) = \mathbf{e}_{r+1} \quad \text{and} \quad \rho\mathbf{e}_j = \mathbf{e}_j \quad \text{for all } 1 \leq j \leq k.$$

Automatically $(\rho\beta)\mathbf{e}_j = \mathbf{e}_j$ for for all $1 \leq j \leq k + 1$. $\square$

**Theorem 7.4.8** *If $\alpha \in O(\mathbb{R}^n)$, then we can find reflections $\rho_1$, $\rho_2$, ..., $\rho_k$ with $0 \le k \le n$ such that*

$$\alpha = \rho_1 \rho_2 \ldots \rho_k.$$

In other words, every norm preserving linear map $\alpha : \mathbb{R}^n \to \mathbb{R}^n$ is the product of at most $n$ reflections. (We adopt the convention that the product of no reflections is the identity map.)

*Proof* We know that $\mathbb{R}^n$ can contain no more than $n$ orthonormal vectors. Thus by applying Lemma 7.4.7 at most $n$ times, we can find reflections $\rho_1$, $\rho_2$, ..., $\rho_k$ with $0 \le k \le n$ such that

$$\rho_k \rho_{k-1} \ldots \rho_1 \alpha = \iota$$

and so

$$\alpha = (\rho_1 \rho_2 \ldots \rho_k)(\rho_k \rho_{k-1} \ldots \rho_1)\alpha = \rho_1 \rho_2 \ldots \rho_k.$$

$\square$

**Exercise 7.4.9** *If $\alpha$ is a rotation through angle $\theta$ in $\mathbb{R}^2$, find, with proof, all the pairs of reflections $\rho_1$, $\rho_2$ with $\alpha = \rho_1 \rho_2$. (It may be helpful to think geometrically.)*

We have a simple corollary.

**Lemma 7.4.10** *Consider a linear map $\alpha : \mathbb{R}^n \to \mathbb{R}^n$. We have $\alpha \in SO(\mathbb{R}^n)$ if and only if $\alpha$ is the product of an even number of reflections. We have $\alpha \in O(\mathbb{R}^n) \setminus SO(\mathbb{R}^n)$ if and only if $\alpha$ is the product of an odd number of reflections.*

*Proof* Take determinants. $\square$

Exercise 7.6.18 shows how to use the ideas of this section to obtain a nice matrix representation (with respect to some orthonormal basis) of any orthogonal linear map.

## 7.5 *QR* factorisation

(Note that, in this section, we deal with $n \times m$ matrices to conform with standard statistical notation in which we have $n$ observations and $m$ explanatory variables.)

If we measure the height of a mountain once, we get a single number which we call the height of the mountain. If we measure the height of a mountain several times, we get several different numbers. None the less, although we have replaced a consistent system of one apparent height for an inconsistent system of several apparent heights, we believe that taking more measurements has given us better information.[3]

---

[3] These are deep philosophical and psychological waters, but it is unlikely that those who believe that we are *worse off* with more information will read this book.

In the same way, if we are told that two distinct points $(u_1, v_1)$ and $(u_2, v_2)$ are close to the line

$$\{(u, v) \in \mathbb{R}^2 : u + hv = k\}$$

but not told the values of $h$ and $k$, we can solve the equations

$$u_1 + h'v_1 = k'$$
$$u_2 + h'v_2 = k'$$

to obtain $h'$ and $k'$ which we hope are close to $h$ and $k$. However, if we are given two new points $(u_3, v_3)$ and $(u_4, v_4)$ which are close to the line, we get a system of equations

$$u_1 + h'v_1 = k'$$
$$u_2 + h'v_2 = k'$$
$$u_3 + h'v_3 = k'$$
$$u_4 + h'v_4 = k'$$

which will, in general, be inconsistent. In spite of this we believe that we must be better off with more information.

The situation may be generalised as follows. (Note that we change our notation quite substantially.) Suppose that $A$ is an $n \times m$ matrix[4] of rank $m$ and $\mathbf{b}$ is a column vector of length $n$. Suppose that we have good reason to believe that $A$ differs from a matrix $A'$ and $\mathbf{b}$ from a vector $\mathbf{b}'$ only because of errors in measurement and that there exists a column vector $\mathbf{x}'$ such that

$$A'\mathbf{x}' = \mathbf{b}'.$$

How should we estimate $\mathbf{x}'$ from $A$ and $\mathbf{b}$? In the absence of further information, it seems reasonable to choose a value of $\mathbf{x}$ which minimises

$$\|A\mathbf{x} - \mathbf{b}\|.$$

**Exercise 7.5.1** (*i*) *If $m = 1 \leq n$, and $a_{i1} = 1$, show that we will choose $\mathbf{x} = (x)$ where*

$$x = n^{-1} \sum_{i=1}^{n} b_i.$$

*How does this relate to our example of the height of a mountain? Is our choice reasonable?*

(*ii*) *Suppose that $m = 2 \leq n$, $a_{i1} = 1$, $a_{i2} = v_i$ and the $v_i$ are distinct. Suppose, in addition, that $\sum_{i=1}^{n} v_i = 0$ (this is simply a change of origin to simplify the algebra). By*

---

[4] In practice, it is usually desirable, not only that $n$ should be large, but also that $m$ should be very small. If the reader remembers nothing else from this book except the paragraph that follows, her time will not have been wasted. In it, Freeman Dyson recalls a meeting with the great physicist Fermi who told him that certain of his calculations lacked physical meaning.

'In desperation I asked Fermi whether he was not impressed by the agreement between our calculated numbers and his measured numbers. He replied, "How many arbitrary parameters did you use for your calculations?" I thought for a moment about our cut-off procedures and said, "Four." He said, "I remember my friend Johnny von Neumann used to say, with four parameters I can fit an elephant, and with five I can make him wiggle his trunk". With that, the conversation was over.' [15]

*using calculus, or otherwise, show that we will choose* $\mathbf{x} = (\mu, \kappa)$ *where*

$$\mu = n^{-1} \sum_{i=1}^{n} b_i, \quad \kappa = \frac{\sum_{i=1}^{n} v_i b_i}{\sum_{i=1}^{n} v_i^2}.$$

*Taking* $x_1 = k$, $x_2 = h$, *explain how this relates to our example of a straight line? Is our choice reasonable? (Do not puzzle too long over this if you cannot come to a conclusion.)*

There are of course many other more or less reasonable choices we could make. For example, we could decide to minimise

$$\sum_{i=1}^{n} \left| \sum_{j=1}^{m} a_{ij} x_j - b_i \right| \quad \text{or} \quad \max_{1 \le i \le n} \left| \sum_{j=1}^{m} a_{ij} x_j - b_i \right|.$$

**Exercise 7.5.2** *Examine the problem of minimising the various penalty functions*

$$\sum_{i=1}^{n} \left( \sum_{j=1}^{m} a_{ij} x_j - b_i \right)^2, \quad \sum_{i=1}^{n} \left| \sum_{j=1}^{m} a_{ij} x_j - b_i \right| \quad \text{and} \quad \max_{1 \le i \le n} \left| \sum_{j=1}^{m} a_{ij} x_j - b_i \right|.$$

*You should look at the case when m and n are small but bear in mind that the procedures you suggest should work when n is large.*

If the reader puts some work in to the previous exercise, she will see that *computational ease* should play a major role in our choice of penalty function and that, judged by this criterion,

$$\sum_{i=1}^{n} \left( \sum_{j=1}^{m} a_{ij} x_j - b_i \right)^2$$

is particularly adapted for calculation.[5]

It is one thing to state the objective of minimising $\| A\mathbf{x} - \mathbf{b} \|$, it is another to achieve it. The key lies in the Gram–Schmidt method discussed earlier. If we write the *columns* of $A$ as *column vectors* $\mathbf{a}_1, \mathbf{a}_2, \ldots, \mathbf{a}_m$, then the Gram–Schmidt method gives us orthonormal column vectors $\mathbf{e}_1, \mathbf{e}_2, \ldots, \mathbf{e}_m$ such that

$$\mathbf{a}_1 = r_{11} \mathbf{e}_1$$
$$\mathbf{a}_2 = r_{12} \mathbf{e}_1 + r_{22} \mathbf{e}_2$$
$$\mathbf{a}_3 = r_{13} \mathbf{e}_1 + r_{23} \mathbf{e}_2 + r_{33} \mathbf{e}_3$$
$$\vdots$$
$$\mathbf{a}_m = r_{1m} \mathbf{e}_1 + r_{2m} \mathbf{e}_2 + r_{3m} \mathbf{e}_3 + \cdots + r_{mm} \mathbf{e}_m$$

---

[5] The reader may feel that the matter is rather trivial. She must then explain why the great mathematicians Gauss and Legendre engaged in a priority dispute over the invention of the 'method of least squares'.

for some $r_{ij}$ $[1 \le j \le i \le m]$ with $r_{ii} \ne 0$ for $1 \le i \le m$. If we now set $r_{ij} = 0$ when $i < j$ we have

$$\mathbf{a}_j = \sum_{i=1}^{m} r_{ij} \mathbf{e}_i. \qquad \bigstar$$

Using the Gram–Schmidt method again, we can now find $\mathbf{e}_{m+1}, \mathbf{e}_{m+2}, \ldots, \mathbf{e}_n$ so that the vectors $\mathbf{e}_j$ with $1 \le j \le n$ form an orthonormal basis for the space $\mathbb{R}^n$ of column vectors. If we take $Q$ to be the $n \times n$ matrix with $j$th column $\mathbf{e}_j$, then $Q$ is orthogonal. If we take $r_{ij} = 0$ for $m < i \le n$, $1 \le j \le m$ and let $R$ be the $n \times m$ matrix $R$ as $(r_{ij})$ then $R$ is 'thin upper triangular'[6] and condition $\bigstar$ gives

$$A = QR.$$

**Exercise 7.5.3** (*i*) *In order to simplify matters, we assume throughout this section, with the exception outlined in the next sentence, that* rank $A = m$. *In this exercise and Exercise 7.5.4, we look at what happens if we drop this assumption. Show that, in general, if $A$ is an $n \times m$ matrix (where $m \le n$) then, possibly after rearranging the order of columns in $A$, we can still find an $n \times n$ orthogonal matrix $Q$ and an $n \times m$ thin upper triangular matrix $R = (r_{ij})$ such that*

$$a_{ij} = \sum_{k=1}^{n} q_{ik} r_{kj}$$

*or, more briefly*

$$A = QR.$$

(*ii*) *Suppose that $A = QR$ with $Q$ an $n \times n$ orthogonal matrix and $R$ an $n \times m$ right triangular matrix $[m \le n]$. State and prove a necessary and sufficient condition for $R$ to satisfy $r_{ii} \ne 0$ for $1 \le i \le m$.*

How does the factorisation $A = QR$ help us with our problem? Observe that, since orthogonal transformations preserve length,

$$\|A\mathbf{x} - \mathbf{b}\| = \|QR\mathbf{x} - \mathbf{b}\| = \|Q^T(QR\mathbf{x} - \mathbf{b})\| = \|R\mathbf{x} - \mathbf{c}\|,$$

where $\mathbf{c} = Q^T \mathbf{b}$. Our problem thus reduces to minimising $\|R\mathbf{x} - \mathbf{c}\|$.

Since

$$\|R\mathbf{x} - \mathbf{c}\|^2 = \sum_{i=1}^{n} \left( \sum_{j=1}^{m} r_{ij} x_j - c_i \right)^2$$

$$= \sum_{i=1}^{m} \left( \sum_{j=1}^{m} r_{ij} x_j - c_i \right)^2 + \sum_{i=m+1}^{n} c_i^2,$$

---

[6] The descriptions 'right triangular' and 'upper triangular' are firmly embedded in the literature as describing *square* $n \times n$ matrices and there seems to be no agreement on what to call $n \times m$ matrices $R$ of the type described here.

we see that the unique vector $\mathbf{x}$ which minimises $\| R\mathbf{x} - \mathbf{c} \|$ is the solution of

$$\sum_{i=1}^{m} r_{ij} x_j = c_i$$

for $1 \le j \le m$. We have completely solved the problem we set ourselves.

**Exercise 7.5.4** (*i*) *Suppose that $A$ is an $n \times m$ matrix with $m \le n$, but the rank of $A$ is strictly less than $m$. Let $\mathbf{b}$ be a column vector with $n$ entries. Explain, on general grounds, why there will always be an $\mathbf{x}$ which minimises $\| A\mathbf{x} - \mathbf{b} \|$. Explain why it will not be unique.*

(*ii*) *In this section we looked at $QR$ factorisation. Do there exist corresponding results on $QL$, $RQ$ and $LQ$ factorisation? If they exist, do they lend themselves as easily to the discussion (beginning 'How does the factorisation') which preceded this exercise?*

**Exercise 7.5.5 [The Householder transformation]** *Suppose that $\mathbf{a}$ and $\mathbf{b}$ are non-zero column vectors in $\mathbb{R}^n$. Explain why, if $\| \mathbf{a} \| \ne \| \mathbf{b} \|$, there cannot exist a reflection $\rho$ with $\rho(\mathbf{a}) = \mathbf{b}$. If $\| \mathbf{a} \| = \| \mathbf{b} \|$ and $\mathbf{b} \ne \pm \mathbf{a}$ set $\mathbf{c} = (\mathbf{a} - \mathbf{b})/2$. Find $\lambda$ and $\mu$ such that*

$$\rho\mathbf{x} = \lambda\mathbf{x} + \mu\langle \mathbf{c}, \mathbf{x}\rangle\mathbf{c}$$

*describes a reflection with $\rho\mathbf{a} = \mathbf{b}$ and $\rho\mathbf{x} = \mathbf{x}$ whenever $\langle \mathbf{x}, \mathbf{a}\rangle = \langle \mathbf{x}, \mathbf{b}\rangle = 0$. (Having found, or guessed, $\lambda$ and $\mu$ you should check that $\rho$ does indeed have the stated properties.) Write down the matrix $T = (t_{ij})$ associated with $\rho$ (for the standard basis). (You may use the summation convention if you wish.)*

*If $A$ is a matrix (not necessarily a square matrix) with first column $\mathbf{a}$ show that $TA$ is a matrix with first column $\mathbf{b}$. By taking $\mathbf{b} = (\| \mathbf{a} \|, 0, 0, \dots, 0)^T$ show that we can find a matrix $T_1$ representing a reflection (or the identity) such that $T_1 A$ has all the entries in the first column $0$ except possibly the first. Now show that we can find a matrix $T_2$ with first row and column consisting of zeros apart from the $(1, 1)$th place which has value $1$ such that $T_2$ represents a reflection (or the identity) and $T_2 T_1 A$ has all entries zero in the first two columns except possibly the $(1, 1)$, $(1, 2)$ and $(2, 2)$th.*

*Continuing in this way, show that we can find an $m$ and reflection (or identity) matrices $T_j$ such that*

$$T_{m-1} T_{m-2} \dots T_1 A = R$$

*is thin upper triangular. Explain why $Q = T_1 T_2 \dots T_{m-1}$ is orthonormal and $A = QR$.*

*This is a perfectly practical method of performing $QR$ factorisation. The $T_j$ are called Householder transformations or Householder reflections.*[7]

**Exercise 7.5.6** *Reduce the matrix*

$$A = \begin{pmatrix} 1 & 2 & 0 \\ 2 & 2 & 1 \\ 2 & 3 & 1 \end{pmatrix}$$

---

[7] Some people are so lost to any sense of decency that they refer to Householder transformations as 'rotations'. The reader should *never* do this. The Householder transformations are *reflections*.

to upper triangular form using a Householder reflection (and, possibly, interchange of rows). (*The numbers have been chosen so that one Householder reflection suffices.*)

**Exercise 7.5.7** *Use an appropriate QR factorisation via the Householder transformation to find the 'best fit solution' (in the sum of squares sense) to*

$$\begin{pmatrix} 1 & 3 \\ 0 & 2 \\ 0 & 2 \\ 0 & -1 \end{pmatrix} \mathbf{x} = \begin{pmatrix} 4 \\ 1 \\ 4 \\ 1 \end{pmatrix}.$$

*Verify your answer by using calculus (or completing the square) to find the* $\mathbf{x}$ *which minimises*

$$(x_1 + 3x_2 - 4)^2 + (2x_2 - 1)^2 + (2x_2 - 4)^2 + (x_2 + 1)^2.$$

Exercise 16.5.35 gives another treatment of $QR$ factorisation based on the Cholesky decomposition which we meet in Theorem 16.3.10, but I think the treatment given in this section is more transparent.

## 7.6 Further exercises

**Exercise 7.6.1** We work in $\mathbb{R}^3$ with the standard coordinate system. Write down the matrices $R_\alpha$ and $R_\beta$ representing rotation through angles $\alpha$ and $\beta$ about the $x_3$ axis. By considering $R_\alpha R_\beta$, show that

$$\cos(\alpha + \beta) = \cos \alpha \cos \beta - \sin \alpha \sin \beta,$$
$$\sin(\alpha + \beta) = \sin \alpha \cos \beta + \cos \alpha \sin \beta.$$

Write down the matrix $R_\gamma$ representing rotation through an angle $\gamma$ about the $x_1$ axis. Compute $R_\gamma R_\alpha$ and $R_\gamma R_\alpha$ checking explicitly that your answers lie in $O(\mathbb{R}^3)$. Find necessary and sufficient conditions for $R_\gamma R_\alpha$ and $R_\gamma R_\alpha$ to be equal.

**Exercise 7.6.2** Consider the matrices

$$M = \begin{pmatrix} 0 & 1 & 0 \\ 0 & 0 & 1 \\ 1 & 0 & 0 \end{pmatrix}, \quad N = \begin{pmatrix} 1 & -2 & -2 \\ 0 & 1 & -2 \\ 0 & 0 & 1 \end{pmatrix}, \quad P = \frac{1}{3} \begin{pmatrix} 1 & -2 & -2 \\ -2 & 1 & -2 \\ -2 & -2 & 1 \end{pmatrix}.$$

For each matrix, find as many linearly independent eigenvectors as possible with eigenvalues 1.

Show that one of the matrices represents a rotation and find the axis and angle of rotation. Show that another represents a reflection and find the plane of reflection. Show that the third is neither a rotation nor a reflection.

State, with reasons, which of the matrices are diagonalisable over $\mathbb{R}$ and which are diagonalisable over $\mathbb{C}$.

**Exercise 7.6.3** Let

$$A = \begin{pmatrix} \frac{1}{2} & \frac{1}{2} & \sqrt{\frac{1}{2}} \\ \frac{1}{2} & \frac{1}{2} & -\sqrt{\frac{1}{2}} \\ -\sqrt{\frac{1}{2}} & \sqrt{\frac{1}{2}} & 0 \end{pmatrix} \quad \text{and} \quad B = \begin{pmatrix} \frac{1}{2} & \frac{1}{2} & \sqrt{\frac{1}{2}} \\ \frac{1}{2} & \frac{1}{2} & -\sqrt{\frac{1}{2}} \\ \sqrt{\frac{1}{2}} & -\sqrt{\frac{1}{2}} & 0 \end{pmatrix}.$$

Show that $A$ represents a rotation and find the axis and angle of rotation. Show that $B$ is orthonormal but neither a rotation nor a reflection.

**Exercise 7.6.4** In this exercise we consider $n \times n$ real matrices. We say that $S$ is *skew-symmetric* if $S^T = -S$.

If $S$ is skew-symmetric and $I + S$ is non-singular, show that the matrix

$$A = (I + S)^{-1}(I - S)$$

is orthogonal and $\det A = 1$, that is to say, $A \in SO(\mathbb{R}^n)$.

Show that, if $A$ is orthogonal and $I + A$ is non-singular, then we can find a skew-symmetric matrix $S$ such that $I + S$ is non-singular and $A = (I + S)^{-1}(I - S)$.

The first paragraph tells us that, if $A$ is expressible in a certain way, then $A \in SO(\mathbb{R}^n)$. The second paragraph tells us that any $A \in O(\mathbb{R}^n)$ with $I + A$ non-singular can be expressed in this way. Why are the two paragraphs compatible with the observation that $O(\mathbb{R}^n) \neq SO(\mathbb{R}^n)$?

Write out the matrix $A$ when $A = (I + S)^{-1}(I - S)$ and

$$S = \begin{pmatrix} 0 & r \\ -r & 0 \end{pmatrix}.$$

**Exercise 7.6.5** Let $\alpha : \mathbb{R}^m \to \mathbb{R}^m$ be a linear map. Prove that

$$E = \{\mathbf{x} \in \mathbb{R}^m : \|\alpha^n \mathbf{x}\| \to 0 \quad \text{as } n \to \infty\}$$

and

$$F = \{\mathbf{x} \in \mathbb{R}^m : \sup_{n \geq 1} \|\alpha^n \mathbf{x}\| < \infty\}\}$$

are subspaces of $\mathbb{R}^n$.

Give an example of an $\alpha$ with $E \neq \{\mathbf{0}\}$, $F \neq E$ and $\mathbb{R}^m \neq F$.

**Exercise 7.6.6** (i) We can look at $O(\mathbb{R}^2)$ in a slightly different way by *defining* it to be the set of

$$A = \begin{pmatrix} a & b \\ c & d \end{pmatrix}$$

which have the property that, if $\mathbf{y} = A\mathbf{x}$, then $y_1^2 + y^2 = x_1^2 + x_2^2$.

By considering $\mathbf{x} = (1, 0)^T$, show that, if $A \in O(\mathbb{R}^2)$, then there is a real $\theta$ such that $a = \cos\theta$, $b = \sin\theta$. By using other test vectors, show that

$$A = \begin{pmatrix} \cos\theta & -\sin\theta \\ \sin\theta & \cos\theta \end{pmatrix} \quad \text{or} \quad A = \begin{pmatrix} \cos\theta & \sin\theta \\ \sin\theta & -\cos\theta \end{pmatrix}.$$

Show, conversely, that any $A$ of these forms is in the set $O(\mathbb{R}^2)$ as just defined.

(ii) Now let $\mathcal{L}(\mathbb{R}^2)$ be the set of $A \in GL(\mathbb{R}^2)$ with the property that, if $\mathbf{y} = A\mathbf{x}$, then $y_1^2 - y_2^2 = x_1^2 - x_2^2$. Characterise $\mathcal{L}(\mathbb{R}^2)$ in the manner of (i).
[In Special Relativity 'ordinary distance' $x^2 + y^2 + z^2$ is replaced by 'space-time distance' $x^2 + y^2 + z^2 - ct^2$. Groups like $\mathcal{L}$ are called Lorentz groups after the great Dutch physicist who first formulated the transformation rules (see Exercise 6.8.1) which underlie the Special Theory of Relativity.]

(iii) The rest of this question requires elementary group theory. Let $SO(\mathbb{R}^2)$ be the collection of $A \in O(\mathbb{R}^2)$ with

$$A = \begin{pmatrix} \cos\theta & -\sin\theta \\ \sin\theta & \cos\theta \end{pmatrix}$$

for some $\theta$. Show that $SO(\mathbb{R}^2)$ is a normal subgroup of $O(\mathbb{R}^2)$ and $SO(\mathbb{R}^2)$ is the union the two disjoint cosets $SO(\mathbb{R}^2)$ and $R(SO(\mathbb{R}^2))$ with

$$R = \begin{pmatrix} 1 & 0 \\ 0 & -1 \end{pmatrix}.$$

(iv) Let $\mathcal{L}_0$ be the collection of matrices $A$ with

$$A = \begin{pmatrix} \cosh t & \sinh t \\ \sinh t & \cosh t \end{pmatrix},$$

for some real $t$. Show that $\mathcal{L}_0$ is a normal subgroup of $\mathcal{L}$ and $\mathcal{L}$ is the union the four disjoint cosets $E_j\mathcal{L}$, where

$$E_1 = I, \quad E_2 = -I, \quad E_3 = \begin{pmatrix} 1 & 0 \\ 0 & -1 \end{pmatrix}, \quad E_4 = \begin{pmatrix} -1 & 0 \\ 0 & 1 \end{pmatrix}.$$

**Exercise 7.6.7** (i) If $U$ is a subspace of $\mathbb{R}^n$ of dimension $n - 1$, $\mathbf{a}$, $\mathbf{b} \in \mathbb{R}^n$ and $\mathbf{a}$ and $\mathbf{b}$ do not lie in $U$, show that there exists a unique point $\mathbf{c} \in U$ such that

$$\|\mathbf{c} - \mathbf{a}\| + \|\mathbf{c} - \mathbf{b}\| \le \|\mathbf{u} - \mathbf{a}\| + \|\mathbf{u} - \mathbf{b}\|$$

for all $\mathbf{u} \in U$.

Show that $\mathbf{c}$ is the unique point in $U$ such that

$$\|\mathbf{c} - \mathbf{b}\|\langle\mathbf{c} - \mathbf{a}, \mathbf{u}\rangle + \|\mathbf{c} - \mathbf{a}\|\langle\mathbf{c} - \mathbf{b}, \mathbf{u}\rangle = 0$$

for all $\mathbf{u} \in U$.

(ii) When a ray of light is reflected in a mirror the 'angle of incidence equals the angle of reflection and so light chooses the shortest path'. What is the relevance of part (i) to this statement?

(iii) How much of part (i) remains true if $\mathbf{a} \in U$ and $\mathbf{b} \notin U$? How much of part (i) remains true if $\mathbf{a}, \mathbf{b} \in U$?

**Exercise 7.6.8** In this question we find the most general distance preserving map (or *isometry*) $\alpha : \mathbb{R}^2 \to \mathbb{R}^2$.

(i) Show that, if $\alpha$ is distance preserving, we can write $\alpha = \rho\beta$, where $\rho\mathbf{x} = \mathbf{a} + \mathbf{x}$ for some fixed $\mathbf{a} \in \mathbb{R}^2$ and $\beta$ is a distance preserving map with $\beta(\mathbf{0}) = \mathbf{0}$.

(ii) Suppose that $\beta$ is a distance preserving map with $\beta(\mathbf{0}) = \mathbf{0}$. By thinking about the equality case in the triangle inequality, show that $\beta(\lambda\mathbf{x}) = \lambda\beta(\mathbf{x})$ for all $\lambda$ with $0 \leq \lambda \leq 1$ and all $\mathbf{x}$. Deduce, first, that $\beta(\lambda\mathbf{x}) = \lambda\beta(\mathbf{x})$ for all $\lambda$ with $0 \leq \lambda$ and all $\mathbf{x}$ and, then, that $\beta(\lambda\mathbf{x}) = \lambda\beta(\mathbf{x})$ for all $\lambda \in \mathbb{R}$ and all $\mathbf{x}$.

(iii) Let $\beta$ be as in (ii). Show that $\beta\left(\frac{1}{2}(\mathbf{x} + \mathbf{y})\right) = \frac{1}{2}(\beta\mathbf{x} + \beta\mathbf{y})$ and thus

$$\beta(\mathbf{x} + \mathbf{y}) = \beta\mathbf{x} + \beta\mathbf{y}.$$

Now use the equality $4\langle \mathbf{c}, \mathbf{d} \rangle = \|\mathbf{c} + \mathbf{d}\|^2 + \|\mathbf{c} - \mathbf{d}\|^2$ to show that

$$\langle \beta(\mathbf{x}), \beta(\mathbf{y}) \rangle = \langle \mathbf{x}, \mathbf{y} \rangle$$

for all $\mathbf{x}, \mathbf{y} \in \mathbb{R}^2$. Deduce that we can write $\beta = \tau\gamma$, where $\tau \in O(\mathbb{R}^2)$ and $\gamma$ is a distance preserving map which fixes the points $(0, 0)$, $(1, 0)$ and $(1, 0)$.

(iv) Show that, if $\gamma$ has the properties stated in (iii), then $\gamma$ is the identity map. Conclude that the most general distance preserving map has the form

$$\alpha\mathbf{x} = \mathbf{a} + \tau\mathbf{x}$$

with $\tau \in O(\mathbb{R}^2)$.

(v) State the corresponding result for $\mathbb{R}^3$ and provide a brief sketch of a proof.

**Exercise 7.6.9** (i) Consider the maps $T : \mathbb{R}^2 \to \mathbb{R}^2$ given by

$$T\mathbf{x} = A\mathbf{x} + \mathbf{b},$$

where $A$ is orthogonal. (We saw, in the previous question, that these are the isometries of $\mathbb{R}^2$.) We say that $\mathbf{a}$ is a fixed point of $T$ if $T\mathbf{a} = \mathbf{a}$.

If $A \notin SO(\mathbb{R}^2)$ (so $T$ is an orientation reversing isometry), identify the fixed points of $A$. If $A \in SO(\mathbb{R}^2)$ (so $T$ is an orientation preserving isometry), show that $T$ has a fixed point unless $A$ is a particular matrix. If $A$ is that matrix, show that $T$ has no fixed point unless $\mathbf{b}$ takes a particular value.

(ii) Consider the maps $T : \mathbb{R}^3 \to \mathbb{R}^3$ given by

$$T\mathbf{x} = A\mathbf{x} + \mathbf{b}$$

where $A$ is orthogonal. (We saw in the previous question that these are the isometries of $\mathbb{R}^3$.)

Make precise and prove the following statement. 'An orientation preserving isometry of $\mathbb{R}^3$ usually has no fixed point, but an orientation reversing isometry usually does.' Identify the exceptional cases.

**Exercise 7.6.10** (i) We work in $\mathbb{R}^3$ with *row* vectors. We have defined reflection in a plane passing through $\mathbf{0}$, but not reflection in a general plane. Explain why we should expect reflection in a plane $\pi$ passing through $\mathbf{a}$ to be a map $S$ given by

$$S\mathbf{x} = \mathbf{a} + R(\mathbf{x} - \mathbf{a}),$$

where $R$ is a reflection in a plane passing through $\mathbf{0}$. Show algebraically that $S$ does not depend on the choice of $\mathbf{a} \in \pi$. We call $S$ a *general reflection*. Write down a similar definition for a general rotation.

(ii) If $S$ is a general reflection, show that

$$\det \begin{pmatrix} S\mathbf{e} - S\mathbf{h} \\ S\mathbf{f} - S\mathbf{h} \\ S\mathbf{g} - S\mathbf{h} \end{pmatrix} = -\det \begin{pmatrix} \mathbf{e} - \mathbf{h} \\ \mathbf{f} - \mathbf{h} \\ \mathbf{g} - \mathbf{h} \end{pmatrix}.$$

(iii) Suppose that $S_1$ and $S_2$ are general reflections. Show that $S_1 S_2$ is either a translation $\mathbf{x} \mapsto \mathbf{c} + \mathbf{x}$ or a general rotation. (It may help to think geometrically, but the final proof should be algebraic.) Show that every translation and every general rotation is the composition of two general reflections. Show, by considering when the product of two general reflections has a fixed point, or otherwise, that only the identity is both a translation and a general rotation.

(iv) Show that every isometry is the product of at most four general reflections.

(v) Consider the map $M : \mathbb{R}^3 \to \mathbb{R}^3$ given by $(x, y, z) = (-x, -y, z + 1)$. Show that $M$ is an isometry. Show that $M$ is not the composition of two or fewer general reflections. By using (ii), or otherwise, show that $M$ is not the composition of three or fewer general reflections.

**Exercise 7.6.11 [Cauchy–Riemann equations]** (This exercise requires some knowledge of partial derivatives.) Suppose that $u$, $v : \mathbb{R}^2 \to \mathbb{R}$ are well behaved functions. Explain in general terms (this is not a book on analysis) why

$$\begin{pmatrix} u(x + \delta x, y + \delta y) \\ v(x + \delta x, y + \delta y) \end{pmatrix} - \begin{pmatrix} u(x, y) \\ v(x, y) \end{pmatrix} = \begin{pmatrix} \frac{\partial u}{\partial x} & \frac{\partial u}{\partial y} \\ \frac{\partial v}{\partial x} & \frac{\partial v}{\partial y} \end{pmatrix} \begin{pmatrix} \delta x \\ \delta y \end{pmatrix} + \text{error term}$$

with the error term decreasing faster than $\max(|\delta x|, |\delta y|)$.

A well behaved function $f : \mathbb{C} \to \mathbb{C}$ is called analytic if

$$f(z + \delta z) - f(z) = f'(z)\delta z + \text{error term}$$

with the error term decreasing faster than $|\delta z|$. Let us write

$$u(x, y) = \Re f(x + iy), \quad v(x, y) = \Im f(x + iy).$$

Show that, if $f'(z) = re^{i\theta}$, with $r \geq 0$ and $\theta$ real, we must have

$$\begin{pmatrix} \frac{\partial u}{\partial x} & \frac{\partial u}{\partial y} \\ \frac{\partial v}{\partial x} & \frac{\partial v}{\partial y} \end{pmatrix} = r \begin{pmatrix} \cos\theta & -\sin\theta \\ \sin\theta & \cos\theta \end{pmatrix}. \qquad \bigstar$$

Deduce the famous *Cauchy–Riemann equations*

$$\frac{\partial u}{\partial x} = \frac{\partial v}{\partial y}, \quad \frac{\partial u}{\partial y} = -\frac{\partial v}{\partial x}.$$

Interpret $\bigstar$ geometrically.

**Exercise 7.6.12** Use a sequence of Householder transformations to find the matrix $R$ in a $QR$ factorisation of the matrix

$$A = \begin{pmatrix} 1 & 1 & 1 \\ 2 & 12 & 1 \\ 2 & 13 & 3 \end{pmatrix}$$

and so solve the equation

$$A\mathbf{x} = \begin{pmatrix} 1 \\ -9 \\ -8 \end{pmatrix}.$$

[You are not asked to find $Q$ explicitly.]

**Exercise 7.6.13** **[Hadamard's inequality]** We know (or suspect) that the area of a parallelogram with given non-zero side lengths is greatest when the parallelogram is a rectangle and the volume of a parallelepiped with given non-zero side lengths is greatest when the edges meet at right angles. Check that the second statement is equivalent to the statement that if $A$ is a $3 \times 3$ real matrix with columns $\mathbf{a}_1$, $\mathbf{a}_2$, $\mathbf{a}_3$, then

$$|\det A| \leq \|\mathbf{a}_1\| \|\mathbf{a}_2\| \|\mathbf{a}_3\|$$

and formulate a similar inequality for $2 \times 2$ matrices.

In higher dimensions our hold on the idea of volume is less strong, but, if the reader keeps the three dimensional case in mind, she will see that the following argument is very natural. Let $A$ be an $n \times n$ real matrix with columns $\mathbf{a}_1$, $\mathbf{a}_2$, ..., $\mathbf{a}_n$. It is reasonable to use Gram–Schmidt orthogonalisation to find an orthonormal basis $\mathbf{q}_j$ with

$$\mathbf{a}_r \in \operatorname{span}\{\mathbf{q}_1, \mathbf{q}_2, \ldots, \mathbf{q}_r\}.$$

In terms of matrices, we consider the factorisation

$$A = QR$$

where $Q$ is an orthogonal matrix and $R$ is an upper triangular matrix given by $r_{ij} = \langle \mathbf{a}_i, \mathbf{q}_j \rangle$. Use the fact that $(\det A)^2 = \det A^T \det A$ to show that

$$(\det A)^2 = (\det R)^2$$

and deduce that

$$(\det A)^2 \leq \prod_{j=1}^{n} \|\mathbf{a}_j\|^2$$

with equality if and only if $r_{ij} = 0$ for all $i \neq j$.

Deduce the following version of Hadamard's inequality.

$$|\det A| \leq \prod_{j=1}^{n} \|\mathbf{a}_j\|$$

with equality if and only if one of the columns is the zero vector or all the columns of $A$ are orthonormal.

**Exercise 7.6.14** Use the result of the previous question to prove the following version of Hadamard's inequality. If $A$ is an $n \times n$ real matrix with all entries $|a_{ij}| \leq K$, then

$$|\det A| \leq K^n n^{n/2}.$$

For the rest of the question we take $K = 1$. Show that

$$|\det A| = n^{n/2}$$

if and only if every entry $a_{ij} = \pm 1$ and $A$ is a scalar multiple of an orthogonal matrix. An $n \times n$ matrix with these properties is called a *Hadamard* matrix.

Show that there are no $k \times k$ Hadamard matrices with $k$ odd and $k \geq 3$.

By looking at matrices $H_0 = (1)$ and

$$H_n = \begin{pmatrix} H_{n-1} & H_{n-1} \\ -H_{n-1} & H_{n-1} \end{pmatrix}$$

show that there are $2^k \times 2^k$ Hadamard matrices for all $k$.

[It is known that, if $H$ is a $k \times k$ Hadamard matrix, then $k = 1$, $k = 2$ or $k$ is a multiple of 4. It is, I believe, still unknown whether there exist Hadamard matrices for all $k$ a multiple of 4.]

**Exercise 7.6.15** This question links Exercise 4.5.16 with the previous two exercises.

Let $\mathcal{M}_n$ be the collection of $n \times n$ real matrices $A = (a_{ij})$ with $|a_{ij}| \leq 1$. If you know enough analysis, explain why

$$\rho(n) = \max_{A \in \mathcal{M}_n} |\operatorname{perm} A| \quad \text{and} \quad \tau(n) = \max_{A \in \mathcal{M}_n} |\det A|$$

exist. (Otherwise, take this as obvious.)

By using Stirling's formula, or otherwise, show that, given any $\epsilon > 0$, we have

$$\tau(n) \leq \rho(n)^{1/2+\epsilon}$$

for all sufficiently large $n$. Show also that

$$\tau(2^m) \geq \rho(2^m)^{1/2-\epsilon}$$

for all sufficiently large $m$.

Find an $A \in \mathcal{M}_2$ such that perm $A = 1$ and det $A = 0$. Find a $B \in \mathcal{M}_2$ such that perm $B = 0$ and det $B = 1$.

**Exercise 7.6.16** Write down the matrix $S$ corresponding to a rotation through $\pi/2$ about the $x$ axis (with the standard coordinate system) and the matrix $T$ corresponding to a rotation through $\pi/2$ about the $y$ axis. Show, by calculating $TS$ and $ST$ explicitly, that $TS \neq ST$. Confirm this by experimenting with an orange or something similar.

In the rest of the question, $\theta$ and $\phi$ will be real numbers with $|\theta|$, $|\phi| < \epsilon$. We write

$$A = O(\epsilon^r)$$

if the $3 \times 3$ matrix $A = A(\theta, \psi) = \big(a_{ij}(\theta, \psi)\big)$ satisfies the condition

$$\epsilon^{-r} \max_{|\theta|,|\psi| \le \epsilon} |a_{ij}(\theta, \psi)| \quad \text{remains bounded as } \epsilon \to 0 \text{ through positive values}$$

for all $(i, j)$. Show that, if $S_\theta$ is a rotation through an angle $\theta$ about the $x$-axis, then $I - S_\theta = O(\epsilon)$. Deduce that, if $R_\theta$ is a rotation through an angle $\theta$ about any fixed axis, $I - R_\theta = O(\epsilon)$.

If $A$, $B = O(\epsilon)$ show that

$$(I + A)(I + B) - (I + B)(I + A) = O(\epsilon^2).$$

Hence, show that if $R_\theta$ is a rotation through an angle $\theta$ about some fixed axis and $S_\phi$ is a rotation through $\phi$ about some fixed axis, then

$$R_\theta S_\phi - S_\phi R_\theta = O(\epsilon^2).$$

In the jargon of the trade, 'infinitesimal rotations commute'.

**Exercise 7.6.17** We use the standard coordinate system for $\mathbb{R}^3$. A rotation through $\pi/4$ about the $x$ axis is followed by a rotation through $\pi/4$ about the $z$ axis. Show that this is equivalent to a single rotation about an axis inclined at equal angles

$$\cos^{-1} \frac{1}{\sqrt{(5 - 2\sqrt{2})}}$$

to the $x$ and $z$ axes.

**Exercise 7.6.18** (i) Let $n \geq 2$. If $\alpha : \mathbb{R}^n \to \mathbb{R}^n$ is an orthogonal map, show that one of the following statements must be true.
  (a) $\alpha = \iota$.
  (b) $\alpha$ is a reflection.
  (c) We can find two orthonormal vectors $\mathbf{e}_1$ and $\mathbf{e}_2$ together with a real $\theta$ such that

$$\alpha \mathbf{e}_1 = \cos \theta \mathbf{e}_1 + \sin \theta \mathbf{e}_2 \quad \text{and} \quad \alpha \mathbf{e}_2 = -\sin \theta \mathbf{e}_1 + \cos \theta \mathbf{e}_2.$$

(ii) Let $n \geq 3$. If $\alpha : \mathbb{R}^n \to \mathbb{R}^n$ is an orthogonal map, show that there is an orthonormal basis of $\mathbb{R}^n$ with respect to which $\alpha$ has matrix

$$\begin{pmatrix} C & O_{2,n-2} \\ O_{n-2,2} & B \end{pmatrix}$$

where $O_{r,s}$ is an $r \times s$ matrix of zeros, $B$ is an $(n-2) \times (n-2)$ orthogonal matrix and

$$C = \begin{pmatrix} \cos \theta & -\sin \theta \\ \sin \theta & \cos \theta \end{pmatrix}$$

for some real $\theta$.

(iii) Show that, if $n = 4$, then, if $\alpha$ is special orthogonal, we can find an orthonormal basis of $\mathbb{R}^4$ with respect to which $\alpha$ has matrix

$$\begin{pmatrix} \cos \theta_1 & -\sin \theta_1 & 0 & 0 \\ \sin \theta_1 & \cos \theta_1 & 0 & 0 \\ 0 & 0 & \cos \theta_2 & -\sin \theta_2 \\ 0 & 0 & \sin \theta_2 & \cos \theta_2 \end{pmatrix},$$

for some real $\theta_1$ and $\theta_2$, whilst, if $\beta$ is orthogonal but not special orthogonal, we can find an orthonormal basis of $\mathbb{R}^4$ with respect to which $\beta$ has matrix

$$\begin{pmatrix} -1 & 0 & 0 & 0 \\ 0 & 1 & 0 & 0 \\ 0 & 0 & \cos \theta & -\sin \theta \\ 0 & 0 & \sin \theta & \cos \theta \end{pmatrix},$$

for some real $\theta$.

(iv) What happens if we take $n = 5$? What happens for general $n$?

**Exercise 7.6.19** (i) If $A \in O(\mathbb{R}^2)$ is such that $AB = BA$ for all $B \in O(\mathbb{R}^2)$, show that $A = \pm I$.

(ii) If $A \in O(\mathbb{R}^n)$ is such that $AB = BA$ for all $B \in O(\mathbb{R}^n)$, show that $A = \pm I$.

(iii) Show that, if $A, B \in SO(\mathbb{R}^2)$, then $AB = BA$.

(iv) If $A \in SO(\mathbb{R}^3)$ is such that $AB = BA$ for all $B \in SO(\mathbb{R}^n)$, show that $A = I$.

(v) If $n \geq 3$ and $A \in SO(\mathbb{R}^n)$ is such that $AB = BA$ for all $B \in SO(\mathbb{R}^n)$, show that $A = I$ if $n$ is odd and $A = \pm I$ if $n$ is even.

**Exercise 7.6.20** We work over $\mathbb{R}$.

Let $SL_n$ be the collection of $n \times n$ matrices $A$ with $\det A = 1$. Show that $SL_n$ is a group under matrix multiplication.

Let $\mathrm{Sp}_{2n}$ be the collection of $2n \times 2n$ matrices which satisfy $M^T J M = J$ where

$$J = \begin{pmatrix} O_n & I_n \\ -I_n & O_n \end{pmatrix}$$

with $I_n$ the $n \times n$ identity matrix and $0_n$ the $n \times n$ zero matrix. Prove the following results.

(i) $M \in \mathrm{Sp}_{2n} \Rightarrow \det M = \pm 1$.

(ii) $\mathrm{Sp}_{2n}$ is a group under matrix multiplication.

(iii) $M \in \mathrm{Sp}_{2n} \Rightarrow M^T \in \mathrm{Sp}_{2n}$.

(iv) Show that $\mathrm{Sp}_2 = SL_2$, but $\mathrm{Sp}_4 \neq SL_4$.

(v) Is the map $\theta : \mathrm{Sp}_{2n} \rightarrow \mathrm{Sp}_{2n}$ given by $\theta M = M^T$ a group isomorphism? Give reasons.

# 8

# Diagonalisation for orthonormal bases

## 8.1 Symmetric maps

In an earlier chapter we dealt with diagonalisation with respect to *some basis*. Once we introduce the notion of inner product, we are more interested in diagonalisation with respect to *some orthonormal basis*.

**Definition 8.1.1** *A linear map* $\alpha : \mathbb{R}^n \to \mathbb{R}^n$ *is said to be* diagonalisable *with respect to an orthonormal basis* $\mathbf{e}_1, \mathbf{e}_2, \ldots, \mathbf{e}_n$ *if we can find* $\lambda_j \in \mathbb{R}$ *such that* $\alpha \mathbf{e}_j = \lambda_j \mathbf{e}_j$ *for* $1 \leq j \leq n$.

The following observation is trivial but useful.

**Lemma 8.1.2** *A linear map* $\alpha : \mathbb{R}^n \to \mathbb{R}^n$ *is diagonalisable with respect to an orthonormal basis if and only if we can find an orthonormal basis of eigenvectors.*

*Proof* Left to the reader. (Compare Theorem 6.3.1.) □

We need the following definitions.

**Definition 8.1.3** (*i*) *A linear map* $\alpha : \mathbb{R}^n \to \mathbb{R}^n$ *is said to be* symmetric *if* $\langle \alpha \mathbf{x}, \mathbf{y} \rangle = \langle \mathbf{x}, \alpha \mathbf{y} \rangle$ *for all* $\mathbf{x}, \mathbf{y} \in \mathbb{R}^n$.
(*ii*) *An* $n \times n$ *real matrix* $A$ *is said to be* symmetric *if* $A^T = A$.

**Lemma 8.1.4** (*i*) *If the linear map* $\alpha : \mathbb{R}^n \to \mathbb{R}^n$ *is symmetric, then it has a symmetric matrix with respect to any orthonormal basis.*
(*ii*) *If a linear map* $\alpha : \mathbb{R}^n \to \mathbb{R}^n$ *has a symmetric matrix with respect to some orthonormal basis, then it is symmetric.*

*Proof* (i) Simple verification. Suppose that the vectors $\mathbf{e}_j$ form an orthonormal basis. We observe that

$$a_{ij} = \left\langle \sum_{r=1}^n a_{rj}\mathbf{e}_r, \mathbf{e}_i \right\rangle = \langle \alpha \mathbf{e}_j, \mathbf{e}_i \rangle = \langle \mathbf{e}_j, \alpha \mathbf{e}_i \rangle = \left\langle \mathbf{e}_j, \sum_{r=1}^n a_{ri}\mathbf{e}_r, \right\rangle = a_{ji}.$$

(ii) Simple verification which is left to the reader. □

We note the following simple consequence.

**Lemma 8.1.5** *If the linear map* $\alpha : \mathbb{R}^n \to \mathbb{R}^n$ *is diagonalisable with respect to some orthonormal basis, then* $\alpha$ *is symmetric.*

*Proof* If $\alpha$ is diagonalisable with respect to some orthonormal basis, then, since a diagonal matrix is symmetric, Lemma 8.1.4 (ii) tells us that $\alpha$ is symmetric. ☐

We shall see that the converse is true (that is to say, every symmetric map is diagonalisable with respect to some orthonormal basis), but the proof will require some work.

The reader may wonder whether symmetric linear maps and matrices are not too special to be worth studying. However, mathematics is full of symmetric matrices like the Hessian

$$H = \left( \frac{\partial^2 f}{\partial x_i \partial x_j} \right),$$

which occurs in the study of maxima and minima of functions $f : \mathbb{R}^n \to \mathbb{R}$, and the covariance matrix

$$E = (\mathbb{E}X_i X_j)$$

in statistics.[1] In addition, the infinite dimensional analogues of symmetric linear maps play an important role in quantum mechanics.

If we only allow orthonormal bases, Theorem 6.1.4 takes a very elegant form.

**Theorem 8.1.6** **[Change of orthonormal basis]** *Let* $\alpha : \mathbb{R}^n \to \mathbb{R}^n$ *be a linear map. If* $\alpha$ *has matrix* $A = (a_{ij})$ *with respect to an orthonormal basis* $\mathbf{e}_1, \mathbf{e}_2, \ldots, \mathbf{e}_n$ *and matrix* $B = (b_{ij})$ *with respect to an orthonormal basis* $\mathbf{f}_1, \mathbf{f}_2, \ldots, \mathbf{f}_n$, *then there is an orthogonal* $n \times n$ *matrix* $P$ *such that*

$$B = P^T A P.$$

*The matrix* $P = (p_{ij})$ *is given by the rule*

$$p_{ij} = \langle \mathbf{e}_i, \mathbf{f}_j \rangle.$$

*Proof* Observe that, if we write

$$\mathbf{f}_j = \sum_{k=1}^n p_{kj} \mathbf{e}_k \quad \text{and} \quad \mathbf{e}_j = \sum_{k=1}^n q_{kj} \mathbf{f}_k,$$

then Theorem 6.1.4 tells us that $P$ is invertible with $P^{-1} = Q$ and $B = P^{-1} A P$. We now note that

$$\langle \mathbf{e}_i, \mathbf{f}_j \rangle = p_{ij} \quad \text{and} \quad \langle \mathbf{f}_i, \mathbf{e}_j \rangle = q_{ij},$$

whence

$$q_{ij} = \langle \mathbf{e}_j, \mathbf{f}_i \rangle = p_{ji}$$

and $P^{-1} = Q = P^T$ as required. ☐

---

[1] These are just introduced as examples. The reader is not required to know anything about them.

At some stage, the reader will see that it is *obvious* that a change of orthonormal basis will leave inner products (and so lengths) unaltered and $P$ must therefore *obviously* be orthogonal. However, it is useful to wear both braces and a belt.

Part (ii) of the next exercise provides an improvement of Theorem 8.1.6 which is sometimes useful.

**Exercise 8.1.7** (*i*) *Show, by using results on matrices, that, if $P$ is an orthogonal matrix, then $P^T A P$ is a symmetric matrix if and only if $A$ is.*

(*ii*) *Let $D$ be the $n \times n$ diagonal matrix $(d_{ij})$ with $d_{11} = -1$, $d_{ii} = 1$ for $2 \leq i \leq n$ and $d_{ij} = 0$ otherwise. By considering $Q = PD$, or otherwise, show that, if there exists a $P \in O(\mathbb{R}^n)$ such that $P^T AT$ is a diagonal matrix, then there exists a $Q \in SO(\mathbb{R}^n)$ such that $Q^T AQ$ is a diagonal matrix.*

When we talk about Cartesian tensors we shall need the following remarks.

**Lemma 8.1.8** (*i*) *If $\mathbf{e}_1$, $\mathbf{e}_2$, ..., $\mathbf{e}_n$ and $\mathbf{f}_1$, $\mathbf{f}_2$, ..., $\mathbf{f}_n$ are orthonormal bases, then there is an orthogonal $n \times n$ matrix $L$ such that, if*

$$\mathbf{x} = \sum_{r=1}^{n} x_r \mathbf{e}_r = \sum_{r=1}^{n} x'_r \mathbf{f}_r,$$

*then $x'_i = \sum_{j=1}^{n} l_{ij} x_j$. The matrix $L = (l_{ij})$ is given by the rule*

$$l_{ij} = \langle \mathbf{e}_i, \mathbf{f}_j \rangle.$$

(*ii*) *Suppose that $\mathbf{e}_1$, $\mathbf{e}_2$, ..., $\mathbf{e}_n$ is an orthonormal basis and there is an orthogonal $n \times n$ matrix $L$ and vectors $\mathbf{f}_1$, $\mathbf{f}_2$, ..., $\mathbf{f}_n$ such that, if $x'_i = \sum_{j=1}^{n} l_{ij} x_j$, then*

$$\sum_{r=1}^{n} x_r \mathbf{e}_r = \sum_{r=1}^{n} x'_r \mathbf{f}_r.$$

*Then $\mathbf{f}_1$, $\mathbf{f}_2$, ..., $\mathbf{f}_n$ form an orthonormal basis.*

*Proof* (i) The proof is very close to that in Theorem 8.1.6. Observe, that if

$$\mathbf{x} = \sum_{r=1}^{n} x_r \mathbf{e}_r = \sum_{r=1}^{n} x'_r \mathbf{f}_r,$$

then

$$x'_i = \langle \mathbf{x}, \mathbf{f}_i \rangle = \sum_{r=1}^{n} x_r \langle \mathbf{e}_r, \mathbf{f}_i \rangle$$

and so

$$x'_i = \sum_{r=1}^{n} l_{ir} x_r.$$

with $l_{ir} = \langle \mathbf{e}_i, \mathbf{f}_r \rangle$. Theorem 8.1.6 tells us that $LL^T = I$.

(ii) If we set $x_j = l_{sj}$, then

$$x_i' = \sum_{j=1}^{n} l_{ij} x_j = \sum_{j=1}^{n} l_{ij} l_{sj} = \delta_{is}$$

and so

$$\sum_{r=1}^{n} l_{sr} \mathbf{e}_r = \sum_{r=1}^{n} \delta_{ir} \mathbf{f}_r = \mathbf{f}_i.$$

Thus

$$\langle \mathbf{f}_i, \mathbf{f}_j \rangle = \left\langle \sum_{r=1}^{n} l_{ir} \mathbf{e}_r, \sum_{s=1}^{n} l_{sj} \mathbf{e}_s \right\rangle$$

$$= \sum_{r=1}^{n} \sum_{s=1}^{n} l_{ir} l_{js} \langle \mathbf{e}_r, \mathbf{e}_s \rangle$$

$$= \sum_{r=1}^{n} l_{ir} l_{jr} = \delta_{ij}$$

as required. $\qquad\square$

Once again, I suspect that, with experience, the reader will come to see Lemma 8.1.8 as 'geometrically obvious'.

In situations like Lemma 8.1.8 we speak of an 'orthonormal change of coordinates'.

We shall be particularly interested in the case when $n = 3$. If the reader recalls the discussion of Section 7.3, she will consider it reasonable to refer to the case when $L \in SO(\mathbb{R}^3)$ as a 'rotation of the coordinate system'.

## 8.2 Eigenvectors for symmetric linear maps

We start with an important observation.

**Lemma 8.2.1** *Let* $\alpha : \mathbb{R}^n \to \mathbb{R}^n$ *be a symmetric linear map. If* $\mathbf{u}$ *and* $\mathbf{v}$ *are eigenvectors with distinct eigenvalues, then they are perpendicular.*

*Proof* We give the *same* proof using three *different* notations.

(1) If $\alpha \mathbf{u} = \lambda \mathbf{u}$ and $\alpha \mathbf{v} = \mu \mathbf{v}$ with $\lambda \neq \mu$, then

$$\lambda \langle \mathbf{u}, \mathbf{v} \rangle = \langle \lambda \mathbf{u}, \mathbf{v} \rangle = \langle \alpha \mathbf{u}, \mathbf{v} \rangle = \langle \mathbf{u}, \alpha \mathbf{v} \rangle = \langle \mathbf{u}, \mu \mathbf{v} \rangle = \mu \langle \mathbf{u}, \mathbf{v} \rangle.$$

Since $\lambda \neq \mu$, we have $\langle \mathbf{u}, \mathbf{v} \rangle = 0$.

(2) Suppose that $A$ is a symmetric $n \times n$ matrix and $\mathbf{u}$ and $\mathbf{v}$ are column vectors such that $A\mathbf{u} = \lambda \mathbf{u}$ and $A\mathbf{v} = \mu \mathbf{v}$ with $\lambda \neq \mu$. Then

$$\lambda \mathbf{u}^T \mathbf{v} = (A\mathbf{u})^T \mathbf{v} = \mathbf{u}^T A^T \mathbf{v} = \mathbf{u}^T A \mathbf{v} = \mathbf{u}^T (\mu \mathbf{v}) = \mu \mathbf{u}^T \mathbf{v}$$

so $\langle \mathbf{u}, \mathbf{v} \rangle = \mathbf{u}^T \mathbf{v} = 0$.

(3) Let us use the summation convention with range $1, 2, \ldots, n$. If $a_{kj} = a_{jk}, a_{kj}u_j = \lambda u_k$ and $a_{kj}v_j = \mu v_k$, but $\lambda \neq \mu$, then

$$\lambda u_k v_k = a_{kj}u_j v_k = a_{jk}u_j v_k = u_j a_{jk}v_k = \mu u_j v_j = \mu u_k v_k,$$

so $u_k v_k = 0$. $\qquad \square$

**Exercise 8.2.2** *Write out proof (3) in full without using the summation convention.*

It is not immediately obvious that a symmetric linear map must have any real eigenvalues.[2]

**Lemma 8.2.3** *If $\alpha : \mathbb{R}^n \to \mathbb{R}^n$ is a symmetric linear map, then all the roots of the characteristic polynomial $\det(t\iota - \alpha)$ are real.*

This result is usually stated as 'all the eigenvalues of a symmetric linear map are real'.

*Proof* Consider the matrix $A = (a_{ij})$ of $\alpha$ with respect to an orthonormal basis. The entries of $A$ are real, but we choose to work in $\mathbb{C}$ rather than $\mathbb{R}$. Suppose that $\lambda$ is a root of the characteristic polynomial $\det(tI - A)$. We know there is a non-zero column vector $\mathbf{z} \in \mathbb{C}^n$ such that $A\mathbf{z} = \lambda \mathbf{z}$. If we write $\mathbf{z} = (z_1, z_2, \ldots, z_n)^T$ and then set $\mathbf{z}^* = (z_1^*, z_2^*, \ldots, z_n^*)^T$ (where $z_j^*$ is the complex conjugate of $z_j$), we have $(A\mathbf{z})^* = \lambda^* \mathbf{z}^*$ so, taking our cue from the third method of proof of Lemma 8.2.1, we note that (using the summation convention)

$$\lambda z_k z_k^* = a_{kj}z_j z_k^* = a_{jk}^* z_j z_k^* = z_j(a_{jk}z_k)^* = \lambda^* z_j z_j^*,$$

so $\lambda = \lambda^*$. Thus $\lambda$ is real and the result follows. $\qquad \square$

This proof may look a little more natural after the reader has studied Section 8.4. A proof which does not use complex numbers (but requires substantial command of analysis) is given in Exercise 8.5.8.

We have the following immediate consequence. (The later Theorem 8.2.5 is stronger, but harder to prove.)

**Lemma 8.2.4** *(i) If $\alpha : \mathbb{R}^n \to \mathbb{R}^n$ is a symmetric linear map and all the roots of the characteristic polynomial $\det(t\iota - \alpha)$ are distinct, then $\alpha$ is diagonalisable with respect to some orthonormal basis.*

*(ii) If $A$ is an $n \times n$ real symmetric matrix and all the roots of the characteristic polynomial $\det(tI - A)$ are distinct, then we can find an orthogonal matrix $P$ and a diagonal matrix $D$ such that*

$$P^T A P = D.$$

*Proof* (i) By Lemma 8.2.3, $\alpha$ has $n$ distinct eigenvalues $\lambda_j \in \mathbb{R}$. If we choose $\mathbf{e}_j$ to be an eigenvector of norm 1 corresponding to $\lambda_j$, then, by Lemma 8.2.1, we obtain $n$ orthonormal

---

[2] When these ideas first arose in connection with differential equations, the analogue of Lemma 8.2.3 was not proved until twenty years after the analogue of Lemma 8.2.1.

vectors which must form an orthonormal basis of $\mathbb{R}^n$. With respect to this basis, $\alpha$ is represented by a diagonal matrix with $j$th diagonal entry $\lambda_j$.

(ii) This is just the translation of (i) into matrix language. $\qquad\square$

Because problems in applied mathematics often involve symmetries, we cannot dismiss the possibility that the characteristic polynomial has repeated roots as irrelevant. Fortunately, as we said after looking at Lemma 8.1.5, symmetric maps are always diagonalisable. The first proof of this fact was due to Hermite.

**Theorem 8.2.5** *(i) If $\alpha : \mathbb{R}^n \to \mathbb{R}^n$ is a symmetric linear map, then $\alpha$ is diagonalisable with respect to some orthonormal basis.*

*(ii) If $A$ is an $n \times n$ real symmetric matrix, then we can find an orthogonal matrix $P$ and a diagonal matrix $D$ such that*

$$P^T A P = D.$$

*Proof* (i) We prove the result by induction on $n$.

If $n = 1$, then, since every $1 \times 1$ matrix is diagonal, the result is trivial.

Suppose now that the result is true for $n = m$ and that $\alpha : \mathbb{R}^{m+1} \to \mathbb{R}^{m+1}$ is a symmetric linear map. We know that the characteristic polynomial must have a root and that all its roots are real. Thus we can can find an eigenvalue $\lambda_1 \in \mathbb{R}$ and a corresponding eigenvector $\mathbf{e}_1$ of norm 1. Consider the subspace

$$\mathbf{e}_1^\perp = \{\mathbf{u} : \langle \mathbf{e}_1, \mathbf{u} \rangle = 0\}.$$

We observe (and this is the key to the proof) that

$$\mathbf{u} \in \mathbf{e}_1^\perp \Rightarrow \langle \mathbf{e}_1, \alpha\mathbf{u} \rangle = \langle \alpha\mathbf{e}_1, \mathbf{u} \rangle = \lambda_1\langle \mathbf{e}_1, \mathbf{u} \rangle = 0 \Rightarrow \alpha\mathbf{u} \in \mathbf{e}_1^\perp.$$

Thus we can define $\alpha|_{\mathbf{e}_1^\perp} : \mathbf{e}_1^\perp \to \mathbf{e}_1^\perp$ to be the restriction of $\alpha$ to $\mathbf{e}_1^\perp$. We observe that $\alpha|_{\mathbf{e}_1^\perp}$ is symmetric and $\mathbf{e}_1^\perp$ has dimension $m$ so, by the inductive hypothesis, we can find $m$ orthonormal eigenvectors of $\alpha|_{\mathbf{e}_1^\perp}$ in $\mathbf{e}_1^\perp$. Let us call them $\mathbf{e}_2, \mathbf{e}_3, \ldots, \mathbf{e}_{m+1}$. We observe that $\mathbf{e}_1, \mathbf{e}_2, \ldots, \mathbf{e}_{m+1}$ are orthonormal eigenvectors of $\alpha$ and so $\alpha$ is diagonalisable. The induction is complete.

(ii) This is just the translation of (i) into matrix language. $\qquad\square$

**Exercise 8.2.6** *Let*

$$A = \begin{pmatrix} 2 & 0 \\ 0 & 1 \end{pmatrix} \quad and \quad P = \begin{pmatrix} 1 & 0 \\ 1 & 1 \end{pmatrix}.$$

*Compute $PAP^{-1}$ and observe that it is not a symmetric matrix, although $A$ is. Why does this not contradict the results of this chapter?*

**Exercise 8.2.7** *We have shown that every real symmetric matrix is diagonalisable. Give an example of a non-zero symmetric $2 \times 2$ matrix $A$ with complex entries whose only eigenvalues are zero. Explain why such a matrix cannot be diagonalised over $\mathbb{C}$.*

*We shall see that, if we look at complex matrices, the correct analogue of a real symmetric matrix is a* Hermitian *matrix. (See Exercises 8.4.15 and 8.4.18.)*

Moving from theory to practice, we see that the diagonalisation (using an orthogonal matrix) follows the same pattern as ordinary diagonalisation (using an invertible matrix). The first step is to look at the roots of the characteristic polynomial

$$\chi_A(t) = \det(tI - A).$$

By Lemma 8.2.3, we know that all the roots are real. If we can find the $n$ roots (in examinations, $n$ will usually be 2 or 3 and the resulting quadratics and cubics will have nice roots) $\lambda_1, \lambda_2, \ldots, \lambda_n$ (repeating repeated roots the required number of times), then we know, without further calculation, that there exists an orthogonal matrix $P$ with

$$P^T A P = D,$$

where $D$ is the diagonal matrix with diagonal entries $d_{ii} = \lambda_i$.

If we need to go further, we proceed as follows. If $\lambda_j$ is not a repeated root, we know that the system of $n$ linear equations in $n$ unknowns given by

$$(A - \lambda_j I)\mathbf{x} = \mathbf{0}$$

(with $\mathbf{x}$ a column vector) defines a one dimensional subspace of $\mathbb{R}^n$. We choose a non-zero vector $\mathbf{u}_j$ from that subspace and normalise by setting

$$\mathbf{e}_j = \|\mathbf{u}_j\|^{-1}\mathbf{u}_j.$$

If $\lambda_j$ is a repeated root,[3] we may suppose that it is a $k$ times repeated root and $\lambda_j = \lambda_{j+1} = \ldots = \lambda_{j+k-1}$. We know that the system of $n$ linear equations in $n$ unknowns given by

$$(A - \lambda_j I)\mathbf{x} = \mathbf{0}$$

(with $\mathbf{x} \in \mathbb{R}^n$ a column vector) defines a $k$-dimensional subspace of $\mathbb{R}^n$. Pick $k$ orthonormal vectors $\mathbf{e}_j, \mathbf{e}_{j+1}, \ldots, \mathbf{e}_{j+k-1}$ in the subspace.[4]

Unless we are unusually confident of our arithmetic, we conclude our calculations by checking that, as Lemma 8.2.1 predicts,

$$\langle \mathbf{e}_i, \mathbf{e}_j \rangle = \delta_{ij}.$$

If $P$ is the $n \times n$ matrix with $j$th column $\mathbf{e}_j$, then, from the formula just given, $P$ is orthogonal (i.e., $PP^T = I$ and so $P^{-1} = P^T$). We note that, if we write $\mathbf{v}_j$ for the unit vector with 1 in the $j$th place, 0 elsewhere, then

$$P^T A P \mathbf{v}_j = P^{-1} A \mathbf{e}_j = \lambda_j P^{-1} \mathbf{e}_j = \lambda_j \mathbf{v}_j = D\mathbf{v}_j$$

---

[3] Sometimes, people refer to 'repeated eigenvalues'. However, it is not the *eigenvalues* which are repeated, but the *roots* of the characteristic polynomial. (We return to the matter much later in Exercise 12.4.14.)

[4] This looks rather daunting, but turns out to be quite easy. You should remember the Franco–British conference where a French delegate objected that 'The British proposal might work in practice, but would not work in theory'.

for all $1 \leq j \leq n$ and so

$$P^T A P = D.$$

Our construction gives $P \in O(\mathbb{R}^n)$, but does not guarantee that $P \in SO(\mathbb{R}^n)$. If det $P = 1$, then $P \in SO(\mathbb{R}^n)$. If det $P = -1$, then replacing $e_1$ by $-e_1$ gives a new $P$ in $SO(\mathbb{R}^n)$.

Here are a couple of simple worked examples.

**Example 8.2.8** (*i*) *Diagonalise*

$$A = \begin{pmatrix} 1 & 1 & 0 \\ 1 & 0 & 1 \\ 0 & 1 & 1 \end{pmatrix}$$

*using an appropriate orthogonal matrix.*

(*ii*) *Diagonalise*

$$B = \begin{pmatrix} 1 & 0 & 0 \\ 0 & 0 & 1 \\ 0 & 1 & 0 \end{pmatrix}$$

*using an appropriate orthogonal matrix.*

*Solution.* (i) We have

$$\det(tI - A) = \det \begin{pmatrix} t-1 & -1 & 0 \\ -1 & t & -1 \\ 0 & -1 & t-1 \end{pmatrix}$$

$$= (t-1) \det \begin{pmatrix} t & -1 \\ -1 & t-1 \end{pmatrix} + \det \begin{pmatrix} -1 & -1 \\ 0 & t-1 \end{pmatrix}$$

$$= (t-1)(t^2 - t - 1) - (t-1) = (t-1)(t^2 - t - 2)$$

$$= (t-1)(t+1)(t-2).$$

Thus the eigenvalues are $1, -1$ and $2$.

We have

$$A\mathbf{x} = \mathbf{x} \Leftrightarrow \begin{cases} x+y & = x \\ x & +z = y \\ & y+z = z \end{cases} \Leftrightarrow \begin{cases} y = 0 \\ x+z = 0. \end{cases}$$

Thus $e_1 = 2^{-1/2}(1, 0, -1)^T$ is an eigenvector of norm 1 with eigenvalue 1.

We have

$$A\mathbf{x} = -\mathbf{x} \Leftrightarrow \begin{cases} x+y & = -x \\ x & +z = -y \\ & y+z = -z \end{cases} \Leftrightarrow \begin{cases} y = -2x \\ y = -2z. \end{cases}$$

Thus $e_2 = 6^{-1/2}(-1, 2, -1)^T$ is an eigenvector of norm 1 with eigenvalue 1.

We have

$$Ax = 2x \Leftrightarrow \begin{cases} x + y & = 2x \\ x & + z = 2y \\ & y + z = 2z \end{cases} \Leftrightarrow \begin{cases} y = x \\ y = z. \end{cases}$$

Thus $e_3 = 3^{-1/2}(1, 1, 1)^T$ is an eigenvector of norm 1 with eigenvalue 2.

If we set

$$P = (e_1 | e_2 | e_3) = \begin{pmatrix} 2^{-1/2} & -6^{-1/2} & 3^{-1/2} \\ 0 & 2 \times 6^{-1/2} & 3^{-1/2} \\ 2^{-1/2} & -6^{-1/2} & 3^{-1/2} \end{pmatrix},$$

then $P$ is orthogonal and

$$P^T A P = \begin{pmatrix} 1 & 0 & 0 \\ 0 & -1 & 0 \\ 0 & 0 & 2 \end{pmatrix}.$$

(ii) We have

$$\det(tI - B) = \det \begin{pmatrix} t - 1 & 0 & 0 \\ 0 & t & -1 \\ 0 & -1 & t \end{pmatrix} = (t - 1) \det \begin{pmatrix} t & -1 \\ -1 & t \end{pmatrix}$$

$$= (t - 1)(t^2 - 1) = (t - 1)^2(t + 1).$$

Thus the eigenvalues are 1 and $-1$.

We have

$$Bx = x \Leftrightarrow \begin{cases} x = x \\ z = y \\ y = z \end{cases} \Leftrightarrow z = y.$$

By inspection, we find two orthonormal eigenvectors $e_1 = (1, 0, 0)^T$ and $e_2 = 2^{-1/2}(0, 1, 1)^T$ corresponding to the eigenvalue 1 which span the space of solutions of $Bx = x$.

We have

$$Bx = -x \Leftrightarrow \begin{cases} x = -x \\ z = -y \\ y = -z \end{cases} \Leftrightarrow \begin{cases} x = 0 \\ y = -z. \end{cases}$$

Thus $e_3 = 2^{-1/2}(0, -1, 1)^T$ is an eigenvector of norm 1 with eigenvalue $-1$.

If we set

$$Q = (\mathbf{e}_1|\mathbf{e}_2|\mathbf{e}_3) = \begin{pmatrix} 1 & 0 & 0 \\ 0 & 2^{-1/2} & 2^{1/2} \\ 0 & -2^{-1/2} & 2^{-1/2} \end{pmatrix},$$

then $Q$ is orthogonal and

$$Q^T B Q = \begin{pmatrix} 1 & 0 & 0 \\ 0 & 1 & 0 \\ 0 & 0 & -1 \end{pmatrix}.$$

$\square$

**Exercise 8.2.9** *Why is part (ii) more or less obvious geometrically?*

**Exercise 8.2.10** *If A is an $n \times n$ real symmetric matrix show that either (i) there exist exactly $2^n n!$ distinct orthogonal matrices P with $P^T A P$ diagonal or (ii) there exist infinitely many distinct orthogonal matrices P with $P^T A P$ diagonal. When does case (ii) occur?*

## 8.3 Stationary points

If $f : \mathbb{R}^2 \to \mathbb{R}$ is a well behaved function, then Taylor's theorem tells us that $f$ behaves locally like a quadratic function. Thus, near $\mathbf{0} = (0, 0)$,

$$f(x, y) \approx c + (ax + by) + \frac{1}{2}(ux^2 + 2vxy + wy^2).$$

The formal theorem, which we shall not prove, runs as follows.

**Theorem 8.3.1** *If $f : \mathbb{R}^2 \to \mathbb{R}$ is three times continuously differentiable in the neighbourhood of $(0, 0)$, then*

$$f(h, k) = f(0, 0) + \left( \frac{\partial f}{\partial x}(0, 0)h + \frac{\partial f}{\partial y}(0, 0)k \right)$$
$$+ \frac{1}{2} \left( \frac{\partial^2 f}{\partial x^2}(0, 0)h^2 + 2\frac{\partial^2 f}{\partial x \partial y}(0, 0)hk + \frac{\partial^2 f}{\partial y^2}(0, 0)k^2 \right) + \epsilon(h, k)(h^2 + k^2),$$

*where $\epsilon(h, k) \to 0$ as $(h^2 + k^2)^{1/2} \to 0$.*

If $a = b = 0$ we say that we have a *stationary point*.

Let us investigate the behaviour near $(0, 0)$ of the polynomial in two variables given by

$$p(x, y) = c + (ax + by) + \frac{1}{2}(ux^2 + 2vxy + wy^2).$$

If $a \neq 0$ or $b \neq 0$, the term $ax + by$ dominates the term $\frac{1}{2}(ux^2 + 2vxy + wy^2)$ and $p$ cannot have a maximum or minimum at $(0, 0)$.

If $a = b = 0$, then

$$2(p(x, y) - p(0, 0)) = ux^2 + 2vxy + wy^2 = (x, y)\begin{pmatrix} u & v \\ v & w \end{pmatrix}\begin{pmatrix} x \\ y \end{pmatrix}$$

or, more briefly using *column vectors*,[5]

$$2(p(\mathbf{x}) - p(\mathbf{0})) = \mathbf{x}^T A\mathbf{x},$$

where $A$ is the symmetric matrix given by

$$A = \begin{pmatrix} u & v \\ v & w \end{pmatrix}.$$

We know that there exists a matrix $R \in SO(\mathbb{R}^2)$ such that

$$RAR^T = \begin{pmatrix} u & v \\ v & w \end{pmatrix} = D = \begin{pmatrix} \lambda_1 & 0 \\ 0 & \lambda_2 \end{pmatrix}$$

and $A = R^T DR$. If we put

$$\begin{pmatrix} X \\ Y \end{pmatrix} = R\begin{pmatrix} x \\ y \end{pmatrix},$$

then

$$\mathbf{x}^T A\mathbf{x} = \mathbf{x}^T R^T DR\mathbf{x} = \mathbf{X}^T D\mathbf{X} = \lambda_1 X^2 + \lambda_2 Y^2.$$

(We could say that 'by rotating axes we reduce our system to diagonal form'.)

If $\lambda_1, \lambda_2 > 0$, we see that $p$ has a minimum at $\mathbf{0}$ and, if $\lambda_1, \lambda_2 < 0$, we see that $p$ has a maximum at $\mathbf{0}$. If $\lambda_1 > 0 > \lambda_2$, then $\lambda_1 X^2$ has a minimum at $X = 0$ and $\lambda_2 Y^2$ has a maximum at $Y = 0$. The surface

$$\{\mathbf{X} : \lambda_1 X^2 + \lambda_2 Y^2\}$$

looks like a *saddle* or *pass* near $\mathbf{0}$. Inhabitants of the lowland town at $\mathbf{X} = (-1, 0)^T$ ascend the path $X = t, Y = 0$ as $t$ runs from $-1$ to $0$ and then descend the path $X = t, Y = 0$ as $t$ runs from $0$ to $1$ to reach another lowland town at $\mathbf{X} = (1, 0)^T$. Inhabitants of the mountain village at $\mathbf{X} = (0, -1)^T$ descend the path $X = 0, Y = t$ as $t$ runs from $-1$ to $0$ and then ascend the path $X = 0, Y = t$ as $t$ runs from $0$ to $1$ to reach another mountain village at $\mathbf{X} = (0, 1)^T$. We refer to the origin as a minimum, maximum or saddle point. The cases when one or more of the eigenvalues vanish must be dealt with by further investigation when they arise.

**Exercise 8.3.2** *Let*

$$A = \begin{pmatrix} u & v \\ v & w \end{pmatrix}.$$

---

[5] This change reflects a culture clash between analysts and algebraists. Analysts tend to prefer row vectors and algebraists column vectors.

(i) Show that the eigenvalues of A are non-zero and of opposite sign if and only if det $A < 0$.

(ii) Show that A has a zero eigenvalue if and only if det $A = 0$.

(iii) If det $A > 0$, show that the eigenvalues of A are strictly positive if and only if Tr $A > 0$. (By definition, Tr $A = u + w$, see Exercise 6.2.3.)

(iv) If det $A > 0$, show that $u \neq 0$. Show further that, if $u > 0$, the eigenvalues of A are strictly positive and that, if $u < 0$, the eigenvalues of A are both strictly negative.
[Note that the results of this exercise do not carry over as they stand to $n \times n$ matrices. We discuss the more general problem in Section 16.3.]

**Exercise 8.3.3** *Extend the ideas of this section to functions of n variables.*
[This will be done in various ways in the second part of the book, but it is a useful way of fixing ideas for the reader to run through this exercise now, even if she only does it informally without writing things down.]

**Exercise 8.3.4** *Suppose that a, b, c $\in \mathbb{R}$. Show that the set*

$$\{(x, y) \in \mathbb{R}^2 : ax^2 + 2bxy + cy^2 = d\}$$

*is an ellipse, a point, the empty set, a hyperbola, a pair of lines meeting at $(0, 0)$, a pair of parallel lines, a single line or the whole plane.*
[This is an exercise in careful enumeration of possibilities.]

**Exercise 8.3.5** *Show that the equation*

$$8x^2 - 2\sqrt{6}xy + 7y^2 = 10$$

*represents an ellipse and find its axes of symmetry.*

## 8.4 Complex inner product

If we try to find an appropriate 'inner product' for $\mathbb{C}^n$, we cannot use our 'geometric intuition', but we can use our 'algebraic intuition' to try to discover a 'complex inner product' that will mimic the real inner product 'as closely as possible'. It is quite possible that our first few guesses will not work very well, but experience will show that the following definition has many desirable properties.

**Definition 8.4.1** *If z, w $\in \mathbb{C}^n$, we set*

$$\langle \mathbf{z}, \mathbf{w} \rangle = \sum_{r=1}^{n} z_r w_r^*.$$

We develop the properties of this inner product in a series of exercises which should provide a useful test of the reader's understanding of the proofs in the last two chapters.

**Exercise 8.4.2** *If* $\mathbf{z}, \mathbf{w}, \mathbf{u} \in \mathbb{C}^n$ *and* $\lambda \in \mathbb{C}$, *show that the following results hold.*

(*i*) $\langle \mathbf{z}, \mathbf{z} \rangle$ *is always real and positive.*

(*ii*) $\langle \mathbf{z}, \mathbf{z} \rangle = 0$ *if and only if* $\mathbf{z} = \mathbf{0}$.

(*iii*) $\langle \lambda \mathbf{z}, \mathbf{w} \rangle = \lambda \langle \mathbf{z}, \mathbf{w} \rangle$.

(*iv*) $\langle \mathbf{z} + \mathbf{u}, \mathbf{w} \rangle = \langle \mathbf{z}, \mathbf{w} \rangle + \langle \mathbf{u}, \mathbf{w} \rangle$.

(*v*) $\langle \mathbf{w}, \mathbf{z} \rangle = \langle \mathbf{z}, \mathbf{w} \rangle^*$.

Rule (v) is a warning that we must tread carefully with our new complex inner product and not expect it to behave quite as simply as the old real inner product. However, it turns out that

$$\|\mathbf{z}\| = \langle \mathbf{z}, \mathbf{z} \rangle^{1/2}$$

behaves just as we wish it to behave. (This is not really surprising, if we write $z_r = x_r + iy_r$ with $x_r$ and $y_r$ real, we get

$$\|\mathbf{z}\|^2 = \sum_{r=1}^n x_r^2 + \sum_{r=1}^n y_r^2$$

which is clearly well behaved.)

**Exercise 8.4.3 [Cauchy–Schwarz]** *If* $\mathbf{z}, \mathbf{w} \in \mathbb{C}^n$, *show that*

$$|\langle \mathbf{z}, \mathbf{w} \rangle| \le \|\mathbf{z}\| \|\mathbf{w}\|.$$

*Show that* $|\langle \mathbf{z}, \mathbf{w} \rangle| = \|\mathbf{z}\| \|\mathbf{w}\|$ *if and only if we can find* $\lambda$, $\mu \in \mathbb{C}$ *not both zero such that* $\lambda \mathbf{z} = \mu \mathbf{w}$.

*[One way of proceeding is first to prove the result when* $\langle \mathbf{z}, \mathbf{w} \rangle$ *is real and positive and then to consider* $\langle e^{i\theta} \mathbf{z}, \mathbf{w} \rangle$.]

**Exercise 8.4.4** *If* $\mathbf{z}, \mathbf{w} \in \mathbb{C}^n$ *and* $\lambda$, $\mu \in \mathbb{C}$, *show that the following results hold.*

(*i*) $\|\mathbf{z}\| \ge 0$.

(*ii*) $\|\mathbf{z}\| = 0$ *if and only if* $\mathbf{z} = \mathbf{0}$.

(*iii*) $\|\lambda \mathbf{z}\| = |\lambda| \|\mathbf{z}\|$.

(*iv*) $\|\mathbf{z} + \mathbf{w}\| \le \|\mathbf{z}\| + \|\mathbf{w}\|$.

**Definition 8.4.5** (*i*) *We say that* $\mathbf{z}, \mathbf{w} \in \mathbb{C}^n$ *are* orthogonal *if* $\langle \mathbf{z}, \mathbf{w} \rangle = 0$.

(*ii*) *We say that* $\mathbf{z}, \mathbf{w} \in \mathbb{C}^n$ *are* orthonormal *if* $\mathbf{z}$ *and* $\mathbf{w}$ *are orthogonal and* $\|\mathbf{z}\| = \|\mathbf{w}\| = 1$.

(*iii*) *We say that a set* $E$ *of vectors is* orthonormal *if any two distinct members of* $E$ *are orthonormal.*

**Exercise 8.4.6** (*i*) *Show that any collection of* $n$ *orthonormal vectors in* $\mathbb{C}^n$ *form a basis.*

(*ii*) *If* $\mathbf{e}_1, \mathbf{e}_2, \ldots, \mathbf{e}_n$ *are orthonormal vectors in* $\mathbb{C}^n$ *and* $\mathbf{z} \in \mathbb{C}^n$, *show that*

$$\mathbf{z} = \sum_{j=1}^n \langle \mathbf{z}, \mathbf{e}_j \rangle \mathbf{e}_j.$$

(*iii*) *Suppose that* $\mathbf{e}_1, \mathbf{e}_2, \ldots, \mathbf{e}_n$ *are orthonormal vectors in* $\mathbb{C}^m$ *and* $\mathbf{z} \in \mathbb{C}^m$. *Does the relation* $\mathbf{z} = \sum_{j=1}^{n} \langle \mathbf{e}_j, \mathbf{z} \rangle \mathbf{e}_j$ *always hold? Give a proof or a counterexample.*

**Exercise 8.4.7** *Suppose that* $1 \le k \le q \le n$. *If* $U$ *is a subspace of* $\mathbb{C}^n$ *of dimension* $q$ *and* $\mathbf{e}_1, \mathbf{e}_2, \ldots, \mathbf{e}_k$ *are orthonormal vectors in* $U$, *show that we can find an orthonormal basis* $\mathbf{e}_1, \mathbf{e}_2, \ldots, \mathbf{e}_q$ *for* $U$.

**Exercise 8.4.8** *We work in* $\mathbb{C}^n$ *and take* $\mathbf{e}_1, \mathbf{e}_2, \ldots, \mathbf{e}_k$ *to be orthonormal vectors. Show that the following results hold.*

(*i*) *We have*

$$\left\| \mathbf{z} - \sum_{j=1}^{k} \lambda_j \mathbf{e}_j \right\|^2 \ge \|\mathbf{z}\|^2 - \sum_{j=1}^{k} |\langle \mathbf{z}, \mathbf{e}_j \rangle|^2,$$

*with equality if and only if* $\lambda_j = \langle \mathbf{z}, \mathbf{e}_j \rangle$.

(*ii*) (*A simple form of Bessel's inequality.*) *We have*

$$\|\mathbf{z}\|^2 \ge \sum_{j=1}^{k} \langle \mathbf{z}, \mathbf{e}_j \rangle^2,$$

*with equality if and only if* $\mathbf{z} \in \mathrm{span}\{\mathbf{e}_1, \mathbf{e}_2, \ldots, \mathbf{e}_k\}$.

**Exercise 8.4.9** *If* $\mathbf{z}, \mathbf{w} \in \mathbb{C}^n$, *prove the* polarisation identity

$$\|\mathbf{z} + \mathbf{w}\|^2 - \|\mathbf{z} - \mathbf{w}\|^2 + i\|\mathbf{z} + i\mathbf{w}\|^2 - i\|\mathbf{z} - i\mathbf{w}\|^2 = 4\langle \mathbf{z}, \mathbf{w} \rangle.$$

**Exercise 8.4.10** *If* $\alpha : \mathbb{C}^n \to \mathbb{C}^n$ *is a linear map, show that there is a unique linear map* $\alpha^* : \mathbb{C}^n \to \mathbb{C}^n$ *such that*

$$\langle \alpha \mathbf{z}, \mathbf{w} \rangle = \langle \mathbf{z}, \alpha^* \mathbf{w} \rangle$$

*for all* $\mathbf{z}, \mathbf{w} \in \mathbb{C}^n$.

*Show, if you have not already done so, that, if* $\alpha$ *has matrix* $A = (a_{ij})$ *with respect to some orthonormal basis, then* $\alpha^*$ *has matrix* $A^* = (b_{ij})$ *with* $b_{ij} = a_{ji}^*$ *(the complex conjugate of* $a_{ij}$) *with respect to the same basis.*

*Show that* $\det \alpha^* = (\det \alpha)^*$.

We call $\alpha^*$ *the* adjoint *of* $\alpha$ *and* $A^*$ *the* adjoint *of* $A$.

**Exercise 8.4.11** *Let* $\alpha : \mathbb{C}^n \to \mathbb{C}^n$ *be linear. Show that the following statements are equivalent.*

(*i*) $\|\alpha \mathbf{z}\| = \|\mathbf{z}\|$ *for all* $\mathbf{z} \in \mathbb{C}^n$.

(*ii*) $\langle \alpha \mathbf{z}, \alpha \mathbf{w} \rangle = \langle \mathbf{z}, \mathbf{w} \rangle$ *for all* $\mathbf{z}, \mathbf{w} \in \mathbb{C}^n$.

(*iii*) $\alpha^* \alpha = \iota$.

(*iv*) $\alpha$ *is invertible with inverse* $\alpha^*$.

(*v*) *If* $\alpha$ *has matrix* $A$ *with respect to some orthonormal basis, then* $A^* A = I$.

*(vi) If α has matrix A with respect to some orthonormal basis, then the columns of A are orthonormal.*

If $\alpha\alpha^* = \iota$, we say that $\alpha$ is *unitary*. We write $U(\mathbb{C}^n)$ for the set of unitary linear maps $\alpha : \mathbb{C}^n \to \mathbb{C}^n$. We use the same nomenclature for the corresponding matrix ideas.

**Exercise 8.4.12**  *(i) Show that $U(\mathbb{C}^n)$ is a subgroup of $GL(\mathbb{C}^n)$.*

*(ii) If $\alpha \in U(\mathbb{C}^n)$, show that $|\det\alpha| = 1$. Is the converse true? Give a proof or counterexample.*

**Exercise 8.4.13**  *Find all diagonal orthogonal $n \times n$ real matrices. Find all diagonal unitary $n \times n$ matrices.*

*Show that, given $\theta \in \mathbb{R}$, we can find $\alpha \in U(\mathbb{C}^n)$ such that $\det\alpha = e^{i\theta}$.*

Exercise 8.4.13 marks the beginning rather than the end of the study of $U(\mathbb{C}^n)$, but we shall not proceed further in this direction. We write $SU(\mathbb{C}^n)$ for the set of $\alpha \in U(\mathbb{C}^n)$ with $\det\alpha = 1$.

**Exercise 8.4.14**  *Show that $SU(\mathbb{C}^n)$ is a subgroup of $GL(\mathbb{C}^n)$.*

The generalisation of the symmetric matrix has the expected form. (If you have any problems with the exercises look at the corresponding proofs for symmetric matrices.)

**Exercise 8.4.15**  *Let $\alpha : \mathbb{C}^n \to \mathbb{C}^n$ be linear. Show that the following statements are equivalent.*

*(i) $\langle \alpha\mathbf{z}, \mathbf{w} \rangle = \langle \mathbf{z}, \alpha\mathbf{w} \rangle$ for all $\mathbf{w}, \mathbf{z} \in \mathbb{C}^n$.*

*(ii) If $\alpha$ has matrix A with respect to some orthonormal basis, then $A = A^*$.*

We call $\alpha$ and $A$, having the properties just described, *Hermitian* or *self-adjoint*.

**Exercise 8.4.16**  *Show that, if A is Hermitian, then $\det A$ is real. Is the converse true? Give a proof or a counterexample.*

**Exercise 8.4.17**  *If $\alpha : \mathbb{C}^n \to \mathbb{C}^n$ is Hermitian, prove the following results:*

*(i) All the eigenvalues of $\alpha$ are real.*

*(ii) The eigenvectors corresponding to distinct eigenvalues of $\alpha$ are orthogonal.*

**Exercise 8.4.18**  *(i) Show that the map $\alpha : \mathbb{C}^n \to \mathbb{C}^n$ is Hermitian if and only if there exists an orthonormal basis of eigenvectors of $\mathbb{C}^n$ with respect to which $\alpha$ has a diagonal matrix with real entries.*

*(ii) Show that the $n \times n$ complex matrix A is Hermitian if and only if there exists a matrix $P \in SU(\mathbb{C}^n)$ such that $P^*AP$ is diagonal with real entries.*

**Exercise 8.4.19**  *If*

$$A = \begin{pmatrix} 5 & 2i \\ -2i & 2 \end{pmatrix}$$

*find a unitary matrix U such that $U^*AU$ is a diagonal matrix.*

**Exercise 8.4.20** *Suppose that* $\gamma : \mathbb{C}^n \to \mathbb{C}^n$ *is unitary. Show that there exist unique Hermitian linear maps* $\alpha$, $\beta : \mathbb{C}^n \to \mathbb{C}^n$ *such that*

$$\gamma = \alpha + i\beta.$$

*Show that* $\alpha\beta = \beta\alpha$ *and* $\alpha^2 + \beta^2 = \iota$.

*What familiar ideas reappear if you take* $n = 1$? *(If you cannot do the first part, this will act as a hint.)*

## 8.5 Further exercises

**Exercise 8.5.1** The following idea goes back to the time of Fourier and has very important generalisations.

Let $A$ be an $n \times n$ matrix over $\mathbb{R}$ such that there exists a basis of (column) eigenvectors $\mathbf{e}_i$ for $\mathbb{R}^n$ with associated eigenvalues $\lambda_i$.

If $\mathbf{y} = \sum_{j=1}^{n} Y_j \mathbf{e}_j$ and $\mu$ is not an eigenvalue, show that

$$A\mathbf{x} - \mu\mathbf{x} = \mathbf{y}$$

has a unique solution and find it in the form $\mathbf{x} = \sum_{j=1}^{n} X_j \mathbf{e}_j$. What happens if $\mu$ is an eigenvalue?

Now suppose that $A$ is symmetric and the $\mathbf{e}_i$ are orthonormal. Find $X_j$ in terms of $\langle \mathbf{y}, \mathbf{e}_j \rangle$. If $n = 3$

$$A = \begin{pmatrix} 1 & 0 & 0 \\ 0 & 1 & 1 \\ 0 & 1 & 1 \end{pmatrix}, \quad \mathbf{y} = \begin{pmatrix} 1 \\ 2 \\ 1 \end{pmatrix} \quad \text{and } \mu = 3$$

find appropriate $\mathbf{e}_j$ and $X_j$. Hence find $\mathbf{x}$ as a column vector $(x_1, x_2, x_3)^T$.

**Exercise 8.5.2** Consider the symmetric $2 \times 2$ real matrix

$$A = \begin{pmatrix} a & b \\ b & c \end{pmatrix}.$$

Are the following statements always true? Give proofs or counterexamples.

(i) If $A$ has all its eigenvalues strictly positive, then $a$, $b > 0$.
(ii) If $A$ has all its eigenvalues strictly positive, then $c > 0$.
(iii) If $a$, $b$, $c > 0$, then $A$ has at least one strictly positive eigenvalue.
(iv) If $a$, $b$, $c > 0$, then all the eigenvalues of $A$ are strictly positive.

[Hint: You may find it useful to look at $\mathbf{x}^T A \mathbf{x}$.]

**Exercise 8.5.3** Which of the following are subgroups of the group $GL(\mathbb{R}^n)$ of $n \times n$ real invertible matrices for $n \geq 2$? Give proofs or counterexamples.

(i) The lower triangular matrices with non-zero entries on the diagonal.
(ii) The symmetric matrices with all eigenvalues non-zero.
(iii) The diagonalisable matrices with all eigenvalues non-zero.

[Hint: It may be helpful to ask which $2 \times 2$ lower triangular matrices are diagonalisable.]

**Exercise 8.5.4** Suppose that a particle of mass $m$ is constrained to move on the curve $z = \frac{1}{2}\kappa x^2$, where $z$ is the vertical axis and $x$ is a horizontal axis (we take $\kappa > 0$). We wish to find the equation of motion for small oscillations about equilibrium. The kinetic energy $E$ is given exactly by $E = \frac{1}{2}m(\dot{x}^2 + \dot{z}^2)$, but, since we deal only with small oscillations, we may take $E = \frac{1}{2}m\dot{x}^2$. The potential energy $U = mgz = \frac{1}{2}mg\kappa x^2$. Conservation of energy tells us that $U + E$ is constant. By differentiating $U + E$, obtain an equation relating $\ddot{x}$ and $x$ and solve it.

A particle is placed in a bowl of the form

$$z = \frac{1}{2}k(x^2 + 2\lambda xy + y^2)$$

with $k > 0$ and $|\lambda| < 1$. Here $x$, $y$, $z$ are rectangular coordinates and $z$ is vertical. By using an appropriate coordinate system, find the general equation for $(x(t), y(t))$ for small oscillations. (If you can produce a solution with four arbitrary constants, you may assume that you have the most general solution.)

Find $(x(t), y(t))$ if the particle starts from rest at $x = a$, $y = 0$, $t = 0$ and both $k$ and $a$ are very small compared with 1. If $|\lambda|$ is very small compared with 1 but non-zero, show that the motion first approximates to motion along the $x$ axis and then, after a long time $\tau$, to be found, to circular motion and then, after a further time $\tau$ has elapsed, to motion along the $y$ axis and so on.

**Exercise 8.5.5** We work over $\mathbb{R}$. Consider the $n \times n$ matrix $A = I + \mathbf{u}\mathbf{u}^T$ where $\mathbf{u}$ is a column vector in $\mathbb{R}^n$. By identifying an appropriate basis, or otherwise, find simple expressions for $\det A$ and $A^{-1}$.

Verify your answers by direct calculation when $n = 2$ and $\mathbf{u} = (u, v)^T$.

**Exercise 8.5.6** Are the following statements true for a *symmetric* $3 \times 3$ real matrix $A = (a_{ij})$ with $a_{ij} = a_{ji}$ Give proofs or counterexamples.

(i) If $A$ has all its eigenvalues strictly positive, then $a_{rr} > 0$ for all $r$.

(ii) If $a_{rr}$ is strictly positive for all $r$, then at least one eigenvalue is strictly positive.

(iii) If $a_{rr}$ is strictly positive for all $r$, then all the eigenvalues of $A$ are strictly positive.

(iv) If $\det A > 0$, then at least one of the eigenvalues of $A$ is strictly positive.

(v) If $\operatorname{Tr} A > 0$, then at least one of the eigenvalues of $A$ is strictly positive.

(vii) If $\det A$, $\operatorname{Tr} A > 0$, then all the eigenvalues of $A$ are strictly positive.

If a $4 \times 4$ symmetric matrix $B$ has $\det B > 0$, does it follow that $B$ has a positive eigenvalue? Give a proof or a counterexample.

**Exercise 8.5.7** Find analogues for the results of Exercise 7.6.4 for $n \times n$ complex matrices $S$ with the property that $S^* = -S$ (such matrices are called skew-Hermitian).

**Exercise 8.5.8** Let $\alpha : \mathbb{R}^n \to \mathbb{R}^n$ be a symmetric linear map. If you know enough analysis, prove that there exists a $\mathbf{u} \in \mathbb{R}^n$ with $\|\mathbf{u}\| \leq 1$ such that

$$|\langle \alpha \mathbf{v}, \mathbf{v} \rangle| \leq |\langle \alpha \mathbf{u}, \mathbf{u} \rangle|$$

whenever $\|\mathbf{v}\| \le 1$. Otherwise, accept the result as obvious. By replacing $\alpha$ by $-\alpha$, if necessary, we may suppose that

$$\langle \alpha \mathbf{v}, \mathbf{v} \rangle \le \langle \alpha \mathbf{u}, \mathbf{u} \rangle$$

whenever $\|\mathbf{v}\| \le 1$.

(i) If $\mathbf{h} \perp \mathbf{u}$, $\|\mathbf{h}\| = 1$ and $\delta \in \mathbb{R}$, show that

$$\|\mathbf{u} + \delta \mathbf{h}\| = 1 + \delta^2$$

and deduce that

$$\langle \alpha(\mathbf{u} + \delta \mathbf{h}), \mathbf{u} + \delta \mathbf{h} \rangle \le (1 + \delta^2)\langle \alpha \mathbf{u}, \mathbf{u} \rangle.$$

(ii) Use (i) to show that there is a constant $A$, depending only on $\mathbf{u}$ and $\mathbf{h}$, such that

$$2\langle \alpha \mathbf{u}, \mathbf{h} \rangle \delta \le A \delta^2.$$

By considering what happens when $\delta$ is small and positive or small and negative, show that

$$\langle \alpha \mathbf{u}, \mathbf{h} \rangle = 0.$$

(iii) We have shown that

$$\mathbf{h} \perp \mathbf{u} \Rightarrow \mathbf{h} \perp \alpha \mathbf{u}.$$

Deduce that $\alpha \mathbf{u} = \lambda \mathbf{u}$ for some $\lambda \in \mathbb{R}$.

[Exercise 15.5.7 runs through a similar argument.]

**Exercise 8.5.9** Let $V$ be a finite dimensional vector space over $\mathbb{C}$ and $\alpha : V \to V$ a linear map such that $\alpha^r = \iota$ for some integer $r \ge 1$. We write $\zeta = \exp(2\pi i / r)$.

If $\mathbf{x}$ is any element of $V$, show that

$$(\zeta^k \alpha^{r-1} + \zeta^{2k} \alpha^{r-2} + \cdots + \zeta^{(r-1)k} \alpha + \iota)\mathbf{x}$$

is either the zero vector or an eigenvector of $\alpha$. Hence show that $\mathbf{x}$ is the sum of eigenvectors. Deduce that $V$ has a basis of eigenvectors of $\alpha$ and that any $n \times n$ complex matrix $A$ with $A^r = I$ is diagonalisable.

For each $r \ge 1$, give an example of a $2 \times 2$ complex matrix such that $A^s \ne I$ for $1 \le s \le r - 1$ but $A^r = I$.

**Exercise 8.5.10** Let $A$ be a $3 \times 3$ antisymmetric matrix (that is to say, $A^T = -A$) with real entries. Show that $iA$ is Hermitian and deduce that, if we work over $\mathbb{C}$, there is a non-zero vector $\mathbf{w}$ such that $A\mathbf{z} = -i\theta \mathbf{z}$ with $\theta$ real.

We now work over $\mathbb{R}$. Show that there exist non-zero real vectors $\mathbf{x}$ and $\mathbf{y}$ and a real number $\theta$ such that

$$A\mathbf{x} = \theta \mathbf{y} \quad \text{and} \quad A\mathbf{y} = -\theta \mathbf{x}.$$

Show further that $A$ has a real eigenvector $\mathbf{u}$ with eigenvalue $0$ perpendicular to $\mathbf{x}$ and $\mathbf{y}$.

**Exercise 8.5.11** Let $A$ be an $n \times n$ antisymmetric matrix (that is to say, $A^T = -A$) with real entries. Show that $\mathbf{v}^T A \mathbf{v} = 0$ for all real column vectors $\mathbf{v} \in \mathbb{R}^n$.

Now suppose that $\lambda = \mu + i\nu$ is a complex eigenvalue with associated complex eigenvector $\mathbf{z} = \mathbf{x} + i\mathbf{y}$ where $\mu$ and $\nu$ are real and $\mathbf{x}$ and $\mathbf{y}$ are real column vectors.

(i) Find expressions for $A\mathbf{x}$ and $A\mathbf{y}$ in terms of $\mathbf{x}$, $\mathbf{y}$, $\mu$ and $\nu$.

(ii) By considering $\mathbf{x}^T A \mathbf{x} + \mathbf{y}^T A \mathbf{y}$, or otherwise, show that $\mu = 0$.

Now suppose that $\nu \neq 0$ (i.e. $\lambda \neq 0$).

(iii) Show that $\mathbf{x}^T \mathbf{y} = 0$ (i.e. $\mathbf{x}$ and $\mathbf{y}$ are orthogonal) and $\|\mathbf{x}\| = \|\mathbf{y}\|$.

# 9

# Cartesian tensors

## 9.1 Physical vectors

When we discussed the use of vectors in geometry, I did not set up the axioms of geometry, but appealed to the reader's knowledge and intuition concerning planes and lines. It is useful and instructive to see how Euclidean geometry can be developed from axioms, but it would have taken us far away from our main topic.

In the next two chapters we shall develop the idea of a Cartesian tensor. Cartesian tensors are mainly used in physics, so we shall encounter 'point masses', 'smoothly varying functions of position' and similar slightly louche characters. In addition, matters which would be made explicit by a pure mathematician will be allowed to remain implicit. Repeated trials have shown that it is rarely useful or instructive to try to develop physics from axioms and it seems foolish to expound a theory in a different language to that spoken by its users.

If the reader is unwilling to adopt a less rigorous approach than that used elsewhere in this book, she may simply omit these chapters which will not be used later. She should, however, recall that

... a well-schooled man is one who searches for that degree of precision in each kind of study which the nature of the subject at hand admits.

(Aristotle *Nicomachean Ethics* [2])

Unless otherwise explicitly stated, we will work in the three dimensional space $\mathbb{R}^3$ with the standard inner product and use the summation convention. When we talk of a coordinate system we will mean a Cartesian coordinate system with perpendicular axes.

One way of making progress in mathematics is to show that objects which have been considered to be of the same type are, in fact, of different types. Another is to show that objects which have been considered to be of different types can be considered to be of the same type. I hope that by the time she has finished this book the reader will see that there is no *universal* idea of a vector but there is instead a family of related ideas.

In Chapter 2 we considered *position vectors*

$$\mathbf{x} = (x_1, x_2, x_3)$$

which gave the position of points in space. In the next two chapters we shall consider *physical vectors*

$$\mathsf{u} = (u_1, u_2, u_3)$$

which give the measurements of physical objects like velocity or the strength of a magnetic field.[1]

In the *Principia*, Newton writes that the laws governing the descent of a stone must be the same in Europe and America. We might interpret this as saying that the laws of physics are translation invariant. Of course, in science, experiment must have the last word and we can never totally exclude the possibility that we are wrong. However, we can say that we would be most reluctant to accept a physical theory which was not translation invariant.

In the same way, we would be most reluctant to accept a physical theory which was not rotation invariant. We expect the laws of physics to look the same whether we stand on our head or our heels.

There are no landmarks in space; one portion of space is exactly like every other portion, so that we cannot tell where we are. We are, as it were, on an unruffled sea, without stars, compass, soundings, wind or tide, and we cannot tell in which direction we are going. We have no log which we can cast out to take dead reckoning by; we may compute our rate of motion with respect to the neighbouring bodies, but we do not know how these bodies may be moving in space.

(Maxwell *Matter and Motion* [22])

If we believe that our theories must be rotation invariant then it is natural to seek a notation which reflects this invariance. The system of *Cartesian tensors* enables us to write down our laws in a way that is automatically rotation invariant.

The first thing to decide is what we mean by a rotation. The Oxford English Dictionary tells us that it is 'The action of moving round a centre, or of turning round (and round) on an axis; also, the action of producing a motion of this kind'. In Chapter 7 we discussed distance preserving linear maps in $\mathbb{R}^3$ and showed that it was natural to call $SO(\mathbb{R}^3)$ the set of rotations of $\mathbb{R}^3$. Since our view of the matter is a great deal clearer than that of the Oxford English Dictionary, we shall use the word 'rotation' as a synonym for 'member of $SO(\mathbb{R}^3)$'. It makes very little difference whether we rotate our laboratory (imagined far out in space) or our coordinate system. It is more usual to rotate our coordinate system in the manner of Lemma 8.1.8.

We know that, if *position vector* **x** is transformed to **x**′ by a rotation of the coordinate system[2] with associated matrix $L \in SO(\mathbb{R}^3)$, then (using the summation convention)

$$x_i' = l_{ij} x_j.$$

We shall say that an observed ordered triple $\mathsf{u} = (u_1, u_2, u_3)$ (think of three dials showing certain values) is a *physical vector* or *Cartesian tensor of order* 1 if, when we rotate the

---

[1] In order to emphasise that these are a new sort of object we initially use a different font, but, within a few pages, we shall drop this convention.

[2] This notation means that ′ is no longer available to denote differentiation. We shall use $\dot{a}$ for the derivative of $a$ with respect to $t$.

coordinate system $S$ in the manner just indicated, to get a new coordinate system $S'$, we get a new ordered triple u' with

$$u'_i = l_{ij} u_j. \qquad \qquad \bigstar$$

In other words, a Cartesian tensor of order 1 behaves like a position vector under rotation of the coordinate system.

**Lemma 9.1.1** *With the notation just introduced*

$$u_i = l_{ji} u'_j.$$

*Proof* Observe that, by definition, $LL^T = I$ so, using the summation convention,

$$l_{ji} u'_j = l_{ji} l_{jk} u_k = \delta_{ki} u_k = u_i.$$

$\qquad\qquad\qquad\qquad\qquad\qquad\qquad\qquad\qquad\qquad\qquad\qquad\qquad\qquad\square$

In the next exercise the reader is asked to show that the triple $(x_1^4, x_2^4, x_3^4)$ (where **x** is a position vector) fails the test $\bigstar$ and is therefore not a tensor.

**Exercise 9.1.2** *Show that*

$$L = \begin{pmatrix} 1 & 0 & 0 \\ 0 & 2^{-1/2} & -2^{-1/2} \\ 0 & 2^{-1/2} & 2^{-1/2} \end{pmatrix} \in SO(\mathbb{R}^3),$$

*but there exists a point whose position vector satisfies* $x_i'^4 \neq l_{ij} x_j^4$.

In the same way, the ordered triple u $= (u_1, u_2, u_3)$, where $u_1$ is the temperature, $u_2$ the pressure and $u_3$ the electric potential at a point, does not obey $\bigstar$ and so cannot be a physical vector.

**Example 9.1.3** *(i) The position vector* **x** $=$ **x** *of a particle is a Cartesian tensor of order* 1. *(ii) If a particle is moving in a smooth manner along a path* **x**$(t)$, *then the velocity*

$$u(t) = \dot{x}(t) = (\dot{x}_1(t), \dot{x}_2(t), \dot{x}_3(t))$$

*is a Cartesian tensor of order* 1. *(iii) If* $\phi : \mathbb{R}^3 \to \mathbb{R}$ *is smooth, then*

$$v = \left( \frac{\partial \phi}{\partial x_1}, \frac{\partial \phi}{\partial x_2}, \frac{\partial \phi}{\partial x_3} \right)$$

*is a Cartesian tensor of order* 1.

*Proof* (i) Direct from the definition. (ii) Observe that (using the summation convention)

$$x'_i(t) = l_{ij} x_j(t).$$

so, differentiating both sides and observing that $l_{ij}$ is constant,

$$u'_i(t) = \dot{x}'_i(t) = l_{ij}\dot{x}_j(t) = l_{ij}u_j(t).$$

(iii) (This is a deeper result.) By Lemma 9.1.1

$$x_i = l_{ji}x'_j.$$

It follows, since $l_{ji}$ is constant, that

$$\frac{\partial x_i}{\partial x'_j} = l_{ji}.$$

Now we are looking at the same point in our physical system, so

$$v' = \left( \frac{\partial\phi}{\partial x'_1}, \frac{\partial\phi}{\partial x'_2}, \frac{\partial\phi}{\partial x'_3} \right),$$

and the chain rule yields

$$v'_i = \frac{\partial\phi}{\partial x'_i} = \frac{\partial\phi}{\partial x_j}\frac{\partial x_j}{\partial x'_i} = \frac{\partial\phi}{\partial x_j}\frac{\partial x_j}{\partial x'_i} = l_{ij}\frac{\partial\phi}{\partial x_j}$$

as required.                                                                           □

**Exercise 9.1.4** (*i*) *If* u *is a Cartesian tensor of order* 1, *show that* (*provided it changes smoothly in time*) *so is* u̇.

(*ii*) *Show that the object* F *occurring in the following version of Newton's third law*

$$F = m\ddot{x}$$

*is a Cartesian tensor of order* 1.

## 9.2 General Cartesian tensors

So far, we have done nothing very interesting, but, as Maxwell observed,

There are physical quantities of another kind which are related to directions in space, but which are not vectors. Stresses and strains in solid bodies are examples of these, and so are some of the properties of bodies considered in the theory of elasticity and of double refraction. Quantities of this class require for their definition *nine* numerical specifications.

(Maxwell *Treatise on Electricity and Magnetism* [21])

To deal with Maxwell's observation, we introduce a new type of object. We shall say that an observed ordered $3 \times 3$ array

$$a = \begin{pmatrix} a_{11} & a_{12} & a_{13} \\ a_{21} & a_{22} & a_{23} \\ a_{21} & a_{22} & a_{23} \end{pmatrix}$$

(think of nine dials showing certain values) is a *Cartesian tensor of order* 2 (or a Cartesian tensor of *rank* 2) if, when we rotate the coordinate system $S$ in our standard manner to get the coordinate system $S'$, we get a new ordered $3 \times 3$ array with

$$a'_{ij} = l_{ir}l_{js}a_{rs}$$

(where, as throughout this chapter, we use the summation convention).

It is not difficult to find interesting examples of Cartesian tensors of order 2.

**Exercise 9.2.1** (*i*) *If* u *and* v *are Cartesian tensors of order* 1 *and we define* a $=$ u $\otimes$ v *to be the* $3 \times 3$ *array given by*

$$a_{ij} = u_i v_j$$

*in each rotated coordinate system, then* u $\otimes$ v *is a Cartesian tensor of order* 2. (*In older texts* u $\otimes$ v *is called a* dyad.)

(*ii*) *If* u *is a smoothly varying Cartesian tensor of order* 1 *and* a *is the* $3 \times 3$ *array given by*

$$a_{ij} = \frac{\partial u_j}{\partial x_i}$$

*in each rotated coordinate system, then* a *is a Cartesian tensor of order* 2.

(*iii*) *If* $\phi : \mathbb{R}^3 \to \mathbb{R}$ *is smooth and* a *is the* $3 \times 3$ *array given by*

$$a_{ij} = \frac{\partial^2 \phi}{\partial x_i \partial x_j}$$

*in each rotated coordinate system, then* a *is a Cartesian tensor of order* 2.

**Lemma 9.2.2** *If* a *is the* $3 \times 3$ *array given by*

$$a_{ij} = \delta_{ij}$$

(*with* $\delta_{ij}$ *the standard Kronecker delta*) *in each rotated coordinate system, then* a *is a Cartesian tensor of order* 2.

*Proof* Observe that

$$l_{ir}l_{js}\delta_{rs} = l_{ir}l_{jr} = \delta_{ij}$$

as required. ∎

Whereas an ordinary tensor of order 2 may be thought of as a $3 \times 3$ array of dials which move in a complicated interdependent way when we rotate our coordinate system, the 'Kronecker tensor' of Lemma 9.2.2 consists of nine dials painted on a block of wood which remain unchanged under rotation.

## 9.3 More examples

Physicists are busy people and like to condense definitions by making implicit what pure mathematicians like to make explicit. In accordance with this policy, they would say that '$a_{ij}$ is a second order Cartesian tensor if it transforms according to the rule $a'_{ij} = l_{ir}l_{js}a_{rs}$'. Since Cartesian tensors are used more by physicists than by pure mathematicians, we shall adopt the shorter usage from now on.

We can clearly generalise further and say that $a_{ij...m}$ (with $N$ suffices) is a Cartesian tensor of *order* or *rank*[3] $N$ if it transforms according to the rule

$$a'_{ij...m} = l_{ip}l_{jq}\ldots l_{mt}a_{pq...t}.$$

Observe that a Cartesian tensor $a$ of order 0 consists of a single real number which remains unchanged under rotation of our coordinate system. As an example, the mass $m$ of a particle is a Cartesian tensor of order 0.

**Exercise 9.3.1** *Produce Cartesian tensors of order 3 along the lines of each of the three parts of Exercise 9.2.1.*

The following remarks are more or less obvious.

(1) The coordinatewise sum of two Cartesian tensors of order $N$ is a Cartesian tensor of order $N$.
(2) If $a_{ij...p}$ is a Cartesian tensor of order $N$ and $b_{qr...t}$ is a Cartesian tensor of order $M$, then the product $a_{ij...p}b_{qr...t}$ is a Cartesian tensor of order $N + M$.
(3) If $a_{ij...p}$ is a Cartesian tensor of order $N$ and we set two suffices equal (so the summation convention operates) the result is a Cartesian tensor of order $N - 2$. (This operation is called *contraction*.)
(4) If $a_{ij...p}$ is a Cartesian tensor of order $N$ whose value varies smoothly as a function of time, then $\dot{a}_{ij...p}$ is a Cartesian tensor of order $N$.
(5) If $a_{jk...p}$ is a Cartesian tensor of order $N$ whose value varies smoothly as a function of position, then

$$\frac{\partial a_{j...p}}{\partial x_i}$$

is a Cartesian tensor of order $N + 1$. (A pure mathematician would replace this by a much longer and more exact statement.)

The general proofs involve lots of notation, so I shall simply prove some typical cases.[4]

**Example 9.3.2** (i) *If $a_{ijkp}$ and $b_{ijkp}$ are Cartesian tensors of order 4, then so is $a_{ijkp} + b_{ijkp}$.*

(ii) *If $a_{ijk}$ is a Cartesian tensor of order 3 and $b_{mn}$ is a Cartesian tensor of order 2, then $a_{ijk}b_{mn}$ is a Cartesian tensor of order 5.*

---

[3] Many physicists use the word 'rank', but this clashes unpleasantly with the definition of rank used in this book.
[4] If the reader objects, then she can do the general proofs as exercises.

*(iii)* If $a_{ijkp}$ is a Cartesian tensor of order 4, then $a_{ijip}$ is a Cartesian tensor of order 2.

*(iv)* If $a_{ijk}$ is a Cartesian tensor of order 3 whose value varies smoothly as a function of time, then $\dot{a}_{ijk}$ is a Cartesian tensor of order 3.

*(v)* If $a_{jkmn}$ is a Cartesian tensor of order 4 whose value varies smoothly as a function of position, then

$$\frac{\partial a_{jkmn}}{\partial x_i}$$

is a Cartesian tensor of order 5.

*Proof* (i) Observe that

$$(a_{ijkp} + b_{ijkp})' = a'_{ijkp} + b'_{ijkp} = l_{ir}l_{js}l_{kt}l_{pu}a_{rstu} + l_{ir}l_{js}l_{kt}l_{pu}b_{rstu}$$
$$= l_{ir}l_{js}l_{kt}l_{pu}(a_{rstu} + b_{rstu}).$$

(ii) Observe that

$$(a_{ijk}b_{mn})' = a'_{ijk}b'_{mn} = l_{ir}l_{js}l_{kt}a_{rst}l_{mp}l_{nq}b_{pq} = l_{ir}l_{js}l_{kt}l_{mp}l_{nq}(a_{rst}b_{pq}).$$

(iii) Observe that

$$a'_{ijip} = l_{ir}l_{js}l_{it}l_{pu}a_{rstu} = \delta_{rt}l_{js}l_{pu}a_{rstu} = l_{js}l_{pu}a_{rsru}.$$

(iv) Left to the reader.

(v) We use the same argument as in Example 9.1.3 (iii).

By Lemma 9.1.1,

$$x_i = l_{ji}x'_j.$$

It follows, since $l_{ji}$ is constant, that

$$\frac{\partial x_i}{\partial x'_j} = l_{ji}.$$

Now we are looking at the same point in our physical system, so the chain rule yields

$$\left(\frac{\partial a_{jkmn}}{\partial x_i}\right)' = \frac{\partial a'_{jkmn}}{\partial x'_i} = \frac{\partial}{\partial x'_i}l_{jr}l_{ks}l_{mt}l_{nu}a_{rstu}$$
$$= l_{jr}l_{ks}l_{mt}l_{nu}\frac{\partial a_{rstu}}{\partial x'_i} = l_{jr}l_{ks}l_{mt}l_{nu}\frac{\partial a_{rstu}}{\partial x_v}\frac{\partial x_v}{\partial x'_i}$$
$$= l_{jr}l_{ks}l_{mt}l_{nu}l_{iv}\frac{\partial a_{rstu}}{\partial x_v}$$

as required. □

There is another way of obtaining Cartesian tensors called the *quotient rule* which is very useful. We need a trivial, but important, preliminary observation.

**Lemma 9.3.3** *Let* $\alpha_{ij...m}$ *be a* $\underbrace{3 \times 3 \times \cdots \times 3}_{n}$ *array. Then, given any specified coordinate system, we can find a tensor* $\mathbf{a}$ *with* $a_{ij...m} = \alpha_{ij...m}$ *in that system.*

*Proof* Define

$$a'_{ij...m} = l_{ir}l_{js} \ldots l_{mu}\alpha_{rs...u},$$

so the required condition for a tensor is obeyed automatically.  $\square$

**Theorem 9.3.4** **[The quotient rule]** *If* $a_{ij}b_j$ *is a Cartesian tensor of order 1 whenever* $b_j$ *is a tensor of order 1, then* $a_{ij}$ *is a Cartesian tensor of order 2.*

*Proof* Observe that, in our standard notation

$$l_{jk}a'_{ij}b_k = a'_{ij}b'_j = (a_{ij}b_j)' = l_{ir}(a_{rj}b_j) = l_{ir}a_{rk}b_k$$

and so

$$(l_{jk}a'_{ij} - l_{ir}a_{rk})b_k = 0.$$

Since we can assign $b_k$ any values we please in a particular coordinate system, we must have

$$l_{jk}a'_{ij} - l_{ir}a_{rk} = 0,$$

so

$$l_{ir}l_{mk}a_{rk} = l_{mk}(l_{ir}a_{rk}) = l_{mk}(l_{jk}a'_{ij}) = l_{mk}l_{jk}a'_{ij} = \delta_{mj}a'_{ij} = a'_{im}$$

and we have shown that $a_{ij}$ is a Cartesian tensor.  $\square$

It is clear that the quotient rule can be extended but, once again, we refrain from the notational complexity of a general proof.

**Exercise 9.3.5** (i) *If* $a_ib_i$ *is a Cartesian tensor of order 0 whenever* $b_i$ *is a Cartesian tensor of order 1, show that* $a_i$ *is a Cartesian tensor of order 1.*

(ii) *If* $a_{ij}b_ic_j$ *is a Cartesian tensor of order 0 whenever* $b_i$ *and* $c_j$ *are Cartesian tensors of order 1, show that* $a_{ij}$ *is a Cartesian tensor of order 2.*

(iii) *Show that, if* $a_{ij}u_{ij}$ *is a Cartesian tensor of order 0 whenever* $u_{ij}$ *is a Cartesian tensor of order 2, then* $a_{ij}$ *is a Cartesian tensor of order 2.*

(iv) *If* $a_{ijkm}b_{km}$ *is a Cartesian tensor of order 2 whenever* $b_{km}$ *is a Cartesian tensor of order 2, show that* $a_{ijkm}$ *is a Cartesian tensor of order 4.*

The theory of elasticity deals with two types of Cartesian tensors of order 2, the stress tensor $e_{ij}$ and the strain tensor $p_{ij}$. On general grounds we expect them to be connected by a linear relation

$$p_{ij} = c_{ijkm}e_{km}.$$

If the stress tensor $e_{km}$ could be chosen freely, the quotient rule given in Exercise 9.3.5 (iv) would tell us that $c_{ijkm}$ is a Cartesian tensor of order 4.

**Exercise 9.3.6** *In fact, matters are a little more complicated since the definition of the stress tensor $e_{km}$ yields $e_{km} = e_{mk}$. (There are no other constraints.)*
*We expect a linear relation*

$$p_{ij} = \tilde{c}_{ijkm}e_{km},$$

*but we cannot now show that $\tilde{c}_{ijkm}$ is a tensor.*

*(i) Show that, if we set $c_{ijkm} = \frac{1}{2}(\tilde{c}_{ijkm} + \tilde{c}_{ijmk})$, then we have*

$$p_{ij} = c_{ijkm}e_{km} \quad and \quad c_{ijkm} = c_{ijmk}.$$

*(ii) If $b_{km}$ is a general Cartesian tensor of order 2, show that there are tensors $e_{mk}$ and $f_{mk}$ with*

$$b_{mk} = e_{mk} + f_{mk}, \quad e_{km} = e_{mk}, \quad f_{km} = -f_{mk}.$$

*(iii) Show that, with the notation of (ii),*

$$c_{ijkm}e_{km} = c_{ijkm}b_{km}$$

*and deduce that $c_{ijkm}$ is a tensor of order 4.*

The next result is trivial but notationally complicated and will not be used in our main discussion.

**Lemma 9.3.7** *In this lemma we will not apply the summation convention to A, B, C and D. If $\mathsf{t}$, $\mathsf{u}$, $\mathsf{v}$ and $\mathsf{w}$ are Cartesian tensors of order 1, we define $\mathsf{c} = \mathsf{t} \otimes \mathsf{u} \otimes \mathsf{v} \otimes \mathsf{w}$ to be the $3 \times 3 \times 3 \times 3$ array given by*

$$c_{ijkm} = t_i u_j v_k w_m$$

*in each rotated coordinate system.*

*Let the order 1 Cartesian tensors $\mathsf{e}(A)$ [A = 1, 2, 3] correspond to the arrays*

$$(e_1(1), e_2(1), e_3(1)) = (1, 0, 0)$$
$$(e_1(2), e_2(2), e_3(2)) = (0, 1, 0)$$
$$(e_1(3), e_2(3), e_3(3)) = (0, 0, 1),$$

*in a given coordinate system S. Then any Cartesian tensor $\mathsf{a}$ of order 4 can be written uniquely as*

$$\mathsf{a} = \sum_{A=1}^{3}\sum_{B=1}^{3}\sum_{C=1}^{3}\sum_{D=1}^{3} \lambda_{ABCD}\mathsf{e}(A) \otimes \mathsf{e}(B) \otimes \mathsf{e}(C) \otimes \mathsf{e}(D) \qquad \bigstar$$

*with $\lambda_{ABCD} \in \mathbb{R}$.*

*Proof* If we work in $S$, then $\bigstar$ yields $\lambda_{ABCD} = a_{ABCD}$, where $a_{ijkm}$ is the array corresponding to $\mathsf{a}$ in $S$.

Conversely, if $\lambda_{ABCD} = a_{ABCD}$, then the $3 \times 3 \times 3 \times 3$ arrays corresponding to the tensors on either side of the equation $\bigstar$ are equal. Since two tensors whose arrays agree in any one coordinate system must be equal, the tensorial equation $\bigstar$ must hold. $\qquad\square$

**Exercise 9.3.8** *Show, in as much detail as you consider desirable, that the Cartesian tensors of order 4 form a vector space of dimension $3^4$.*

*State the general result for Cartesian tensors of order n.*

It is clear that a Cartesian tensor of order 3 is a new kind of object, but a Cartesian tensor of order 0 behaves like a scalar and a Cartesian tensor of order 1 behaves like a vector. On the principle that if it looks like a duck, swims like a duck and quacks like a duck, then it is a duck, physicists call a Cartesian tensor of order 0 a scalar and a Cartesian tensor of order 1 a vector. From now on we shall do the same, but, as we discuss in Section 10.3, not all ducks behave in the same way.

Henceforward we shall use **u** rather than u to denote Cartesian tensors of order 1.

## 9.4 The vector product

Recall that we defined the Levi-Civita symbol $\epsilon_{ijk}$ by taking $\epsilon_{123} = 1$ and using the rule that interchanging two indices changed the sign. This gave

$$\epsilon_{\alpha\beta\gamma} = \begin{cases} 1 & \text{if } (\alpha, \beta, \gamma) \in \{(1, 2, 3), (2, 3, 1), (3, 1, 2)\}, \\ -1 & \text{if } (\alpha, \beta, \gamma) \in \{(3, 2, 1), (1, 3, 2), (2, 1, 3)\}, \\ 0 & \text{otherwise.} \end{cases}$$

**Lemma 9.4.1** $\epsilon_{ijk}$ *is a tensor.*

*Proof* Observe that

$$l_{ir}l_{js}l_{kt}\epsilon_{rst} = \det \begin{pmatrix} l_{i1} & l_{i2} & l_{i3} \\ l_{j1} & l_{j2} & l_{j3} \\ l_{k1} & l_{k2} & l_{k3} \end{pmatrix} = \epsilon_{ijk} \det \begin{pmatrix} l_{11} & l_{12} & l_{13} \\ l_{21} & l_{22} & l_{23} \\ l_{31} & l_{32} & l_{33} \end{pmatrix} = \epsilon_{ijk} = \epsilon'_{ijk},$$

as required.                                                                              □

There are very few formulae which are worth memorising, but part (iv) of the next theorem may be one of them.

**Theorem 9.4.2** *(i) We have*

$$\epsilon_{ijk} = \det \begin{pmatrix} \delta_{i1} & \delta_{i2} & \delta_{i3} \\ \delta_{j1} & \delta_{j2} & \delta_{j3} \\ \delta_{k1} & \delta_{k2} & \delta_{k3} \end{pmatrix}.$$

*(ii) We have*

$$\epsilon_{ijk}\epsilon_{rst} = \det \begin{pmatrix} \delta_{ir} & \delta_{is} & \delta_{it} \\ \delta_{jr} & \delta_{js} & \delta_{jt} \\ \delta_{kr} & \delta_{ks} & \delta_{kt} \end{pmatrix}.$$

*(iii)* We have

$$\epsilon_{ijk}\epsilon_{rst} = \delta_{ir}\delta_{js}\delta_{kt} + \delta_{it}\delta_{jr}\delta_{ks} + \delta_{is}\delta_{jt}\delta_{kr} - \delta_{ir}\delta_{ks}\delta_{jt} - \delta_{it}\delta_{kr}\delta_{js} - \delta_{is}\delta_{kt}\delta_{jr}.$$

*(iv)* **[The Levi-Civita identity]** *We have*

$$\epsilon_{ijk}\epsilon_{ist} = \delta_{js}\delta_{kt} - \delta_{ks}\delta_{jt}.$$

*(v)* We have $\epsilon_{ijk}\epsilon_{ijt} = 2\delta_{kt}$ and $\epsilon_{ijk}\epsilon_{ijk} = 6$.

Note that the Levi-Civita identity of Theorem 9.4.2 (iv) asserts the equality of two Cartesian tensors of order 4, that is to say, it asserts the equality of $3 \times 3 \times 3 \times 3 = 81$ entries in two $3 \times 3 \times 3 \times 3$ arrays. Although we have chosen a slightly indirect proof of the identity, it will also yield easily to direct (but systematic) attack.

*Proof of Theorem 9.4.2.* (i) Recall that interchanging two rows of a determinant multiplies its value by $-1$ and that the determinant of the identity matrix is 1.

(ii) Recall that interchanging two columns of a determinant multiplies its value by $-1$.

(iii) Compute the determinant of (ii) in the standard manner.

(iv) By (iii),

$$\epsilon_{ijk}\epsilon_{ist} = \delta_{ii}\delta_{js}\delta_{kt} + \delta_{it}\delta_{ji}\delta_{ks} + \delta_{is}\delta_{jt}\delta_{ki} - \delta_{ii}\delta_{ks}\delta_{jt} - \delta_{it}\delta_{ki}\delta_{js} - \delta_{is}\delta_{kt}\delta_{ji}$$
$$= 3\delta_{js}\delta_{kt} + \delta_{jt}\delta_{ks} + \delta_{jt}\delta_{ks} - 3\delta_{ks}\delta_{jt} - \delta_{kt}\delta_{js} - \delta_{kt}\delta_{js} = \delta_{js}\delta_{kt} - \delta_{ks}\delta_{jt}.$$

(v) Left as an exercise for the reader using the summation convention. $\qquad\square$

**Exercise 9.4.3** *Show that $\epsilon_{ijk}\epsilon_{klm}\epsilon_{mni} = \epsilon_{nlj}$.*

**Exercise 9.4.4** *Let $\epsilon_{ijk...q}$ be the Levi-Civita symbol of order $n$ (with the obvious definition). Show that*

$$\epsilon_{ijk...q}\epsilon_{ijk...q} = n!$$

We now define the *vector product* (or *cross product*) $\mathbf{c} = \mathbf{a} \times \mathbf{b}$ of two vectors $\mathbf{a}$ and $\mathbf{b}$ by

$$c_i = \epsilon_{ijk}a_jb_k$$

or, more briefly, by saying

$$(\mathbf{a} \times \mathbf{b})_i = \epsilon_{ijk}a_jb_k.$$

Notice that many people write the vector product as $\mathbf{a} \wedge \mathbf{b}$ and sometimes talk about the 'wedge product'.

If we need to calculate a specific vector product, we can unpack our definition to obtain

$$(a_1, a_2, a_3) \times (b_1, b_2, b_3) = (a_2b_3 - a_3b_2, a_3b_1 - a_1b_3, a_1b_2 - a_2b_1). \qquad\bigstar$$

The algebraic definition using the summation convention is simple and computationally convenient, but conveys no geometric picture. To assign a geometric meaning, observe that, since tensors retain their relations under rotation, we may suppose, by rotating axes, that

$$\mathbf{a} = (a, 0, 0) \quad \text{and} \quad \mathbf{b} = (b\cos\theta, b\sin\theta, 0)$$

with $a, b > 0$ and $0 \le \theta \le \pi$. We then have

$$\mathbf{a} \times \mathbf{b} = (0, 0, ab\sin\theta)$$

so $\mathbf{a} \times \mathbf{b}$ is a vector of length $\|\mathbf{a}\|\|\mathbf{b}\||\sin\theta|$ (where, as usual, $\|\mathbf{x}\|$ denotes the length of $\mathbf{x}$) perpendicular to $\mathbf{a}$ and $\mathbf{b}$. (There are two such vectors, but we leave the discussion as to which one is chosen by our formula until Section 10.3.)

Our algebraic definition makes it easy to derive the following properties of the cross product.

**Exercise 9.4.5** *Suppose that* $\mathbf{a}$, $\mathbf{b}$ *and* $\mathbf{c}$ *are vectors and* $\lambda$ *is a scalar. Show that the following relations hold.*
  (*i*) $\lambda(\mathbf{a} \times \mathbf{b}) = (\lambda\mathbf{a}) \times \mathbf{b} = \mathbf{a} \times (\lambda\mathbf{b})$.
  (*ii*) $\mathbf{a} \times (\mathbf{b} + \mathbf{c}) = \mathbf{a} \times \mathbf{b} + \mathbf{a} \times \mathbf{c}$.
  (*iii*) $\mathbf{a} \times \mathbf{b} = -\mathbf{b} \times \mathbf{a}$.
  (*iv*) $\mathbf{a} \times \mathbf{a} = 0$.

**Exercise 9.4.6** *Is* $\mathbb{R}^3$ *a group (see Definition 5.3.13) under the vector product? Give reasons.*

We also have the *dot product*[5] $\mathbf{a} \cdot \mathbf{b}$ of two vectors $\mathbf{a}$ and $\mathbf{b}$ given by

$$\mathbf{a} \cdot \mathbf{b} = a_i b_i.$$

The reader will recognise this as our usual *inner product* under a different name.

**Exercise 9.4.7** *Suppose that* $\mathbf{a}$, $\mathbf{b}$ *and* $\mathbf{c}$ *are vectors and* $\lambda$ *is a scalar. Use the definition just given and the summation convention to show that the following relations hold.*
  (*i*) $\lambda(\mathbf{a} \cdot \mathbf{b}) = (\lambda\mathbf{a}) \cdot \mathbf{b} = \mathbf{a} \cdot (\lambda\mathbf{b})$.
  (*ii*) $\mathbf{a} \cdot (\mathbf{b} + \mathbf{c}) = \mathbf{a} \cdot \mathbf{b} + \mathbf{a} \cdot \mathbf{c}$.
  (*iii*) $\mathbf{a} \cdot \mathbf{b} = \mathbf{b} \cdot \mathbf{a}$.

**Theorem 9.4.8 [The triple vector product]** *If* $\mathbf{a}$, $\mathbf{b}$ *and* $\mathbf{c}$ *are vectors, then*

$$\mathbf{a} \times (\mathbf{b} \times \mathbf{c}) = (\mathbf{a} \cdot \mathbf{c})\mathbf{b} - (\mathbf{a} \cdot \mathbf{b})\mathbf{c}.$$

*Proof* Observe that

$$\begin{aligned}
\left(\mathbf{a} \times (\mathbf{b} \times \mathbf{c})\right)_i &= \epsilon_{ijk} a_j (\mathbf{b} \times \mathbf{c})_k = \epsilon_{ijk} a_j (\epsilon_{krs} b_r c_s) \\
&= \epsilon_{kij} \epsilon_{krs} a_j b_r c_s = (\delta_{ir}\delta_{js} - \delta_{is}\delta_{jr}) a_j b_r c_s \\
&= a_s b_i c_s - a_r b_r c_i = \left((\mathbf{a} \cdot \mathbf{c})\mathbf{b} - (\mathbf{a} \cdot \mathbf{b})\mathbf{c}\right)_i,
\end{aligned}$$

as required. $\qquad\qquad\qquad\qquad\qquad\qquad\qquad\qquad\qquad\qquad\qquad\qquad\qquad$ $\square$

---

[5] Some British mathematicians, including the present author, use a lowered dot **a.b**, but this is definitely old fashioned.

**Exercise 9.4.9** *(i) Prove Theorem 9.4.8 from the formula*

$$(a_1, a_2, a_3) \times (b_1, b_2, b_3) = (a_2b_3 - a_3b_2, a_3b_1 - a_1b_3, a_1b_2 - a_2b_1)$$

*without using the summation convention. (This is neither difficult nor time consuming, provided that you start with a sufficiently large piece of paper.)*

*(ii) Think about how you would go about proving Theorem 9.4.8 from our geometric description of the vector product. (Only write down a proof if you can find a nice one.)*

**Exercise 9.4.10** *Show that*

$$(\mathbf{a} \times \mathbf{b}) \times \mathbf{c} = (\mathbf{a} \cdot \mathbf{c})\mathbf{b} - (\mathbf{b} \cdot \mathbf{c})\mathbf{a}.$$

*Write down explicit vectors* $\mathbf{x}$, $\mathbf{y}$ *and* $\mathbf{z}$ *such that*

$$(\mathbf{x} \times \mathbf{y}) \times \mathbf{z} \neq \mathbf{x} \times (\mathbf{y} \times \mathbf{z}).$$

*[An algebraist would say that the vector product does not obey the associative rule.]*

**Exercise 9.4.11** *Show that*

$$\mathbf{a} \times (\mathbf{b} \times \mathbf{c}) + \mathbf{b} \times (\mathbf{c} \times \mathbf{a}) + \mathbf{c} \times (\mathbf{a} \times \mathbf{b}) = \mathbf{0}.$$

**Exercise 9.4.12** *Use the summation convention (recall, as usual, that suffices must not appear more than twice) to show that*

$$(\mathbf{a} \times \mathbf{b}) \cdot (\mathbf{a} \times \mathbf{b}) + (\mathbf{a} \cdot \mathbf{b})^2 = \|\mathbf{a}\|^2 \|\mathbf{b}\|^2.$$

*To what well known trigonometric formula does this correspond?*

There is another natural way to 'multiply three vectors'. We write

$$[\mathbf{a}, \mathbf{b}, \mathbf{c}] = \mathbf{a} \cdot (\mathbf{b} \times \mathbf{c})$$

and call the result the *scalar triple product* (or just *triple product*). Observe that

$$[\mathbf{a}, \mathbf{b}, \mathbf{c}] = \epsilon_{ijk} a_i b_j c_k = \det \begin{pmatrix} a_1 & a_2 & a_3 \\ b_1 & b_2 & b_3 \\ c_1 & c_2 & c_3 \end{pmatrix}.$$

In particular, interchanging two entries in the scalar triple product multiplies the result by $-1$. We may think of the scalar triple product $[\mathbf{a}, \mathbf{b}, \mathbf{c}]$ as the (signed) volume of a parallelepiped with one vertex at $\mathbf{0}$ and adjacent vertices at $\mathbf{a}$, $\mathbf{b}$ and $\mathbf{c}$.

**Exercise 9.4.13** *Give one line proofs of the relation*

$$(\mathbf{a} \times \mathbf{b}) \cdot \mathbf{a} = 0$$

*using the following ideas.*

*(i) Summation convention.*

*(ii) Orthogonality of vectors.*

*(iii) Volume of a parallelepiped.*

**Exercise 9.4.14**  *By applying the triple vector product formula to*

$$(\mathbf{a} \times \mathbf{b}) \times (\mathbf{c} \times \mathbf{d}),$$

*or otherwise, show that*

$$[\mathbf{a}, \mathbf{b}, \mathbf{c}]\mathbf{d} = [\mathbf{b}, \mathbf{c}, \mathbf{d}]\mathbf{a} + [\mathbf{c}, \mathbf{a}, \mathbf{d}]\mathbf{b} + [\mathbf{a}, \mathbf{b}, \mathbf{d}]\mathbf{c}.$$

*Show also that*

$$(\mathbf{a} \times \mathbf{b}) \times (\mathbf{a} \times \mathbf{c}) = [\mathbf{a}, \mathbf{b}, \mathbf{c}]\mathbf{a}.$$

**Exercise 9.4.15**  *If* $\mathbf{a}$, $\mathbf{b}$ *and* $\mathbf{c}$ *are linearly independent, we know that any* $\mathbf{x}$ *can be written uniquely as*

$$\mathbf{x} = \lambda\mathbf{a} + \mu\mathbf{b} + \nu\mathbf{c}$$

*with* $\lambda$, $\mu$, $\nu \in \mathbb{R}$. *Find* $\lambda$ *by considering the dot product (that is to say, inner product) of* $\mathbf{x}$ *with a vector perpendicular to* $\mathbf{b}$ *and* $\mathbf{c}$. *Write down* $\mu$ *and* $\nu$ *similarly.*

**Exercise 9.4.16**  *(i) By considering a matrix of the form* $AA^T$, *or otherwise, show that*

$$[\mathbf{a}, \mathbf{b}, \mathbf{c}]^2 = \det \begin{pmatrix} \mathbf{a} \cdot \mathbf{a} & \mathbf{a} \cdot \mathbf{b} & \mathbf{a} \cdot \mathbf{c} \\ \mathbf{b} \cdot \mathbf{a} & \mathbf{b} \cdot \mathbf{b} & \mathbf{b} \cdot \mathbf{c} \\ \mathbf{c} \cdot \mathbf{a} & \mathbf{c} \cdot \mathbf{b} & \mathbf{c} \cdot \mathbf{c} \end{pmatrix}.$$

*(ii) (Rather less interesting.) If* $\|\mathbf{a}\| = \|\mathbf{b}\| = \|\mathbf{c}\| = \|\mathbf{a} + \mathbf{b} + \mathbf{c}\| = r$, *show that*

$$[\mathbf{a}, \mathbf{b}, \mathbf{c}]^2 = 2(r^2 + \mathbf{a} \cdot \mathbf{b})(r^2 + \mathbf{b} \cdot \mathbf{c})(r^2 + \mathbf{c} \cdot \mathbf{a}).$$

**Exercise 9.4.17**  *Show that*

$$(\mathbf{a} \times \mathbf{b}) \cdot (\mathbf{c} \times \mathbf{d}) + (\mathbf{a} \times \mathbf{c}) \cdot (\mathbf{d} \times \mathbf{b}) + (\mathbf{a} \times \mathbf{d}) \cdot (\mathbf{b} \times \mathbf{c}) = 0.$$

The results in the next lemma and the following exercise are unsurprising and easy to prove.

**Lemma 9.4.18**  *If* $\mathbf{a}$, $\mathbf{b}$ *are smoothly varying vector functions of time, then*

$$\frac{d}{dt}\mathbf{a} \times \mathbf{b} = \dot{\mathbf{a}} \times \mathbf{b} + \mathbf{a} \times \dot{\mathbf{b}}.$$

*Proof* Using the summation convention,

$$\frac{d}{dt}\epsilon_{ijk}a_jb_k = \epsilon_{ijk}\frac{d}{dt}a_jb_k = \epsilon_{ijk}(\dot{a}_jb_k + a_j\dot{b}_k) = \epsilon_{ijk}\dot{a}_jb_k + \epsilon_{ijk}a_j\dot{b}_k,$$

as required.                                                                                    $\square$

**Exercise 9.4.19**  *Suppose that* $\mathbf{a}$, $\mathbf{b}$ *are smoothly varying vector functions of time and* $\phi$ *is a smoothly varying scalar function of time. Prove the following results.*

(i) $\dfrac{d}{dt}\phi\mathbf{a} = \dot{\phi}\mathbf{a} + \phi\dot{\mathbf{a}}.$

(ii) $\dfrac{d}{dt}\mathbf{a}\cdot\mathbf{b} = \dot{\mathbf{a}}\cdot\mathbf{b} + \mathbf{a}\cdot\dot{\mathbf{b}}$.

(iii) $\dfrac{d}{dt}\mathbf{a}\times\dot{\mathbf{a}} = \mathbf{a}\times\ddot{\mathbf{a}}$.

Hamilton, Tait, Maxwell and others introduced a collection of Cartesian tensors based on the 'differential operator'[6] $\nabla$. (Maxwell and his contemporaries called this symbol 'nabla' after an oriental harp 'said by Hieronymus and other authorities to have had the shape $\nabla$'. The more prosaic modern world prefers 'del'.) If $\phi$ is a smoothly varying scalar function of position, then the vector $\nabla\phi$ is given by

$$(\nabla\phi)_i = \frac{\partial\phi}{\partial x_i}.$$

If $\mathbf{u}$ is a smoothly varying vector function of position, then the scalar $\nabla\cdot\mathbf{u}$ and the vector $\nabla\times\mathbf{u}$ are given by

$$\nabla\cdot\mathbf{u} = \frac{\partial u_i}{\partial x_i} \quad\text{and}\quad (\nabla\times\mathbf{u})_i = \epsilon_{ijk}\frac{\partial u_j}{\partial x_k}.$$

The following alternative names are in common use

$$\operatorname{grad}\phi = \nabla\phi, \quad \operatorname{div}\mathbf{u} = \nabla\cdot\mathbf{u}, \quad \operatorname{curl}\mathbf{u} = \nabla\times\mathbf{u}.$$

We speak of the 'gradient of $\phi$', the 'divergence of $\mathbf{u}$' and the 'curl of $\mathbf{u}$'.

We write

$$\nabla^2\phi = \nabla\cdot(\nabla\phi) = \frac{\partial^2\phi}{\partial x_i\partial x_i}$$

and call $\nabla^2\phi$ the Laplacian of $\phi$. We also write

$$\left(\nabla^2\mathbf{u}\right)_i = \frac{\partial^2 u_i}{\partial x_j\partial x_j}.$$

The following, less important, objects occur from time to time. Let $\mathbf{a}$ be a vector function of position, $\phi$ a smooth function of position and $\mathbf{u}$ a smooth vector function of position. We define

$$(\mathbf{a}\cdot\nabla)\phi = a_i\frac{\partial\phi}{\partial x_i} \quad\text{and}\quad \left((\mathbf{a}\cdot\nabla)\mathbf{u}\right)_j = a_i\frac{\partial u_j}{\partial x_i}.$$

**Lemma 9.4.20** (i) *If $\phi$ is a smooth scalar valued function of position and $\mathbf{u}$ is a smooth vector valued function of position, then*

$$\nabla\cdot(\phi\mathbf{u}) = (\nabla\phi)\cdot\mathbf{u} + \phi\nabla\cdot\mathbf{u}.$$

(ii) *If $\phi$ is a smooth scalar valued function of position, then*

$$\nabla\times(\nabla\cdot\phi) = 0.$$

---

[6] So far as we are concerned, the phrase 'differential operator' is merely decorative. We shall not define the term or use the idea.

(ii) *If* **u** *is a smooth vector valued function of position, then*

$$\nabla \times (\nabla \mathbf{u}) = \nabla(\nabla \cdot \mathbf{u}) - \nabla^2 \mathbf{u}.$$

*Proof* (i) Observe that

$$\frac{\partial}{\partial x_i} \phi u_i = \frac{\partial \phi}{\partial x_i} u_i + \phi \frac{\partial u_i}{\partial x_i}.$$

(ii) We know that partial derivatives of a smooth function commute, so

$$\epsilon_{ijk} \frac{\partial}{\partial x_j} \left( \frac{\partial \phi}{\partial x_k} \right) = \epsilon_{ijk} \frac{\partial^2 \phi}{\partial x_j \partial x_k} = \epsilon_{ijk} \frac{\partial^2 \phi}{\partial x_k \partial x_j}$$

$$= \epsilon_{ijk} \frac{\partial}{\partial x_k} \left( \frac{\partial \phi}{\partial x_j} \right) = -\epsilon_{ijk} \frac{\partial}{\partial x_j} \left( \frac{\partial \phi}{\partial x_k} \right).$$

Thus

$$\epsilon_{ijk} \frac{\partial}{\partial x_j} \left( \frac{\partial \phi}{\partial x_k} \right) = 0,$$

which is the stated result.

(iii) Observe that, using Levi-Civita's formula,

$$\left( \nabla \times (\nabla \mathbf{u}) \right)_i = \epsilon_{ijk} \frac{\partial}{\partial x_j} \left( \epsilon_{krs} \frac{\partial u_s}{\partial x_r} \right) = \epsilon_{kij} \epsilon_{krs} \frac{\partial^2 u_s}{\partial x_j \partial x_r}$$

$$= (\delta_{ir} \delta_{js} - \delta_{is} \delta_{jr}) \frac{\partial^2 u_s}{\partial x_j \partial x_r} = \frac{\partial^2 u_j}{\partial x_j \partial x_i} - \frac{\partial^2 u_i}{\partial x_j \partial x_j}$$

$$= \frac{\partial}{\partial x_i} \frac{\partial u_j}{\partial x_j} - \frac{\partial^2 u_i}{\partial x_j \partial x_j}$$

$$= \left( \nabla(\nabla \cdot \mathbf{u}) - \nabla^2 \mathbf{u} \right)_i,$$

as required.                                                                                           □

**Exercise 9.4.21** (i) *If $\phi$ is a smooth scalar valued function of position and* **u** *is a smooth vector valued function of position, show that*

$$\nabla \times (\phi \mathbf{u}) = (\nabla \phi) \times \mathbf{u} + \phi \nabla \times \mathbf{u}.$$

(ii) *If* **u** *is a smooth vector valued function of position, show that*

$$\nabla \cdot (\nabla \times \mathbf{u}) = 0.$$

(iii) *If* **u** *and* **v** *are smooth vector valued functions of position, show that*

$$\nabla \times (\mathbf{u} \times \mathbf{v}) = (\nabla \cdot \mathbf{v}) \mathbf{u} + \mathbf{v} \cdot \nabla \mathbf{u} - (\nabla \cdot \mathbf{u}) \mathbf{v} - \mathbf{u} \cdot \nabla \mathbf{v}.$$

Although the $\nabla$ notation is very suggestive, the formulae so suggested need to be verified, since analogies with simple vectorial formulae can fail.

The power of multidimensional calculus using $\nabla$ is only really unleashed once we have the associated integral theorems (the integral divergence theorem, Stokes' theorem and so on). However, we shall give an application of the vector product to mechanics in Section 10.2.

**Exercise 9.4.22** *The vector product is so useful that we would like to generalise it from* $\mathbb{R}^3$ *to* $\mathbb{R}^n$. *If we look at the case* $n = 4$, *we see one possible generalisation. Using the summation convention with range* 1, 2, 3, 4, *we define*

$$\mathbf{x} \wedge \mathbf{y} \wedge \mathbf{z} = \mathbf{a}$$

*by*

$$a_i = \epsilon_{ijkl} x_j y_k z_l.$$

(i) *Show that, if we make the natural extension of the definition of a Cartesian tensor to four dimensions,*[7] *then, if* $\mathbf{x}$, $\mathbf{y}$, $\mathbf{z}$ *are vectors (i.e. Cartesian tensors of order* 1 *in* $\mathbb{R}^4$*), it follows that* $\mathbf{x} \wedge \mathbf{y} \wedge \mathbf{z}$ *is a vector.*

(ii) *Show that, if* $\mathbf{x}$, $\mathbf{y}$, $\mathbf{z}$, $\mathbf{u}$ *are vectors and* $\lambda$, $\mu$ *scalars, then*

$$(\lambda \mathbf{x} + \mu \mathbf{u}) \wedge \mathbf{y} \wedge \mathbf{z} = \lambda \mathbf{x} \wedge \mathbf{y} \wedge \mathbf{z} + \mu \mathbf{u} \wedge \mathbf{y} \wedge \mathbf{z}$$

*and*

$$\mathbf{x} \wedge \mathbf{y} \wedge \mathbf{z} = -\mathbf{y} \wedge \mathbf{x} \wedge \mathbf{z} = -\mathbf{z} \wedge \mathbf{y} \wedge \mathbf{x}.$$

(iii) *Write down the appropriate definition for* $n = 5$.

*In some ways this is satisfactory, but the reader will observe that, if we work in* $\mathbb{R}^n$, *the 'generalised vector product' must involve* $n - 1$ *vectors. In more advanced work, mathematicians introduce 'generalised vector products' involving* $r$ *vectors from* $\mathbb{R}^n$, *but these 'wedge products' no longer live in* $\mathbb{R}^n$.

## 9.5 Further exercises

**Exercise 9.5.1** Show directly from the definition of a Cartesian tensor that, if $T_{ij}$ is a second order Cartesian tensor, then $\partial T_{ij}/\partial x_j$ is a vector.

**Exercise 9.5.2** Show that, if $\mathbf{v}$ is a smooth function of position,

$$(\mathbf{v} \cdot \nabla)\mathbf{v} = \nabla(\tfrac{1}{2}\|\mathbf{v}\|^2) - \mathbf{v} \times (\nabla \times \mathbf{v}).$$

**Exercise 9.5.3** If we write

$$D_i = \epsilon_{ijk} x_j \frac{\partial}{\partial x_k},$$

show that

$$(D_1 D_2 - D_2 D_1)\phi = -D_3 \phi$$

---

[7] If the reader wants everything spelled out in detail, she should ignore this exercise.

for any smooth function $\phi$ of position. (If we were thinking about commutators, we would write $[D_1, D_2] = -D_3$.)

**Exercise 9.5.4** We work in $\mathbb{R}^3$. Give a geometrical interpretation of the equation

$$\mathbf{n} \cdot \mathbf{r} = b \qquad \qquad \bigstar$$

and of the equation

$$\mathbf{u} \times \mathbf{r} = \mathbf{c}, \qquad \qquad \bigstar\bigstar$$

where $b$ is a constant, $\mathbf{n}$, $\mathbf{u}$, $\mathbf{c}$ are constant vectors, $\|\mathbf{n}\| = \|\mathbf{u}\| = 1$, $\mathbf{u} \cdot \mathbf{c} = 0$ and $\mathbf{r}$ is a position vector. Your answer should contain geometric interpretations of $\mathbf{n}$ and $\mathbf{u}$.

Determine which values of $\mathbf{r}$ satisfy equations $\bigstar$ and $\bigstar\bigstar$ simultaneously, (a) assuming that $\mathbf{u} \cdot \mathbf{n} \neq 0$ and (b) assuming that $\mathbf{u} \cdot \mathbf{n} = 0$. Interpret the difference between the results for (a) and (b) geometrically.

**Exercise 9.5.5** Let $\mathbf{a}$ and $\mathbf{b}$ be linearly independent vectors in $\mathbb{R}^3$. By considering the vector product with an appropriate vector, show that, if the equation

$$\mathbf{x} + (\mathbf{a} \cdot \mathbf{x})\mathbf{a} + \mathbf{a} \times \mathbf{x} = \mathbf{b} \qquad (9.1)$$

has a solution, it must be

$$\mathbf{x} = \frac{\mathbf{b} - \mathbf{a} \times \mathbf{b}}{1 + \|\mathbf{a}\|^2}. \qquad (9.2)$$

Verify that this is indeed a solution.

Rewrite equations (1) and (2) in tensor form as

$$M_{ij}x_j = b_i \quad \text{and} \quad x_j = N_{jk}b_k.$$

Compute $M_{ij}N_{jk}$ and explain why you should expect the answer that you obtain.

**Exercise 9.5.6** Let $\mathbf{a}$, $\mathbf{b}$ and $\mathbf{c}$ be linearly independent. Show that the equation

$$(\mathbf{a} \times \mathbf{b} + \mathbf{b} \times \mathbf{c} + \mathbf{c} \times \mathbf{a}) \cdot \mathbf{x} = [\mathbf{a}, \mathbf{b}, \mathbf{c}] \qquad \qquad \bigstar$$

defines the plane through $\mathbf{a}$, $\mathbf{b}$ and $\mathbf{c}$.

What object is defined by $\bigstar$ if $\mathbf{a}$ and $\mathbf{b}$ are linearly independent, but $\mathbf{c} \in \text{span}\{\mathbf{a}, \mathbf{b}\}$? What object is defined by $\bigstar$ if $\mathbf{a} \neq \mathbf{0}$, but $\mathbf{b}$, $\mathbf{c} \in \text{span}\{\mathbf{a}\}$? What object is defined by $\bigstar$ if $\mathbf{a} = \mathbf{b} = \mathbf{c} = \mathbf{0}$? Give reasons for your answers.

**Exercise 9.5.7** Unit circles with centres at $\mathbf{r}_1$ and $\mathbf{r}_2$ are drawn in two non-parallel planes with equations $\mathbf{r} \cdot \mathbf{k}_1 = p_1$ and $\mathbf{r} \cdot \mathbf{k}_2 = p_2$ respectively (where the $\mathbf{k}_j$ are unit vectors and the $p_j \geq 0$). Show that there is a sphere passing through both circles if and only if

$$(\mathbf{r}_1 - \mathbf{r}_2) \cdot (\mathbf{k}_1 \times \mathbf{k}_2) = 0 \quad \text{and} \quad (p_1 - \mathbf{k}_1 \cdot \mathbf{r}_2)^2 = (p_2 - \mathbf{k}_2 \cdot \mathbf{r}_1)^2.$$

**Exercise 9.5.8** (i) Show that

$$(\mathbf{a} \times \mathbf{b}) \cdot (\mathbf{c} \times \mathbf{d}) = (\mathbf{a} \cdot \mathbf{c})(\mathbf{b} \cdot \mathbf{d}) - (\mathbf{a} \cdot \mathbf{d})(\mathbf{b} \cdot \mathbf{c}).$$

(ii) Deduce that

$$(\mathbf{a} \times \mathbf{b}) \cdot (\mathbf{a} \times \mathbf{c}) = \|\mathbf{a}\|^2(\mathbf{b} \cdot \mathbf{c}) - (\mathbf{a} \cdot \mathbf{b})(\mathbf{a} \cdot \mathbf{c}).$$

(iii) Show, from (ii), that

$$(\mathbf{a} \times \mathbf{b}) \cdot (\mathbf{a} \times \mathbf{b}) = \|\mathbf{a}\|^2\|\mathbf{b}\|^2 - (\mathbf{a} \cdot \mathbf{b})^2.$$

Deduce the Cauchy–Schwarz inequality in three dimensions.

**Exercise 9.5.9 [Some formulae from spherical trigonometry]**

Let him that is melancholy calculate spherical triangles.

(Robert Burton *The Anatomy of Melancholy* [7])

Consider three points $\mathcal{A}$, $\mathcal{B}$ and $\mathcal{C}$ on the unit sphere with centre $O$ at $\mathbf{0}$ in $\mathbb{R}^3$.

We write $\mathbf{a}$ for the position vector of $\mathcal{A}$. We write $a$ for the angle between the lines $O\mathcal{B}$ and $O\mathcal{C}$ satisfying $0 \leq a \leq \pi$ and $A$ for the angle between the plane $\pi_{\mathcal{A},\mathcal{B}}$ containing $O$, $\mathcal{A}$, $\mathcal{B}$ and the plane $\pi_{\mathcal{A},\mathcal{C}}$ containing $O$, $\mathcal{A}$, $\mathcal{C}$. (There are two possible choices for $A$ even if we impose the condition $0 \leq A \leq \pi$. We resolve the ambiguity later.)

(i) Show, by thinking about the angle between $\mathbf{a} \times \mathbf{b}$ and $\mathbf{a} \times \mathbf{c}$, or otherwise, that

$$(\mathbf{a} \times \mathbf{b}) \cdot (\mathbf{a} \times \mathbf{c}) = \sin c \sin b \cos A,$$

provided that we choose $A$ appropriately. By applying the formula in Exercise 9.5.8, show that

$$\cos a = \cos b \cos c \pm \sin b \sin c \cos A.$$

(ii) Explain why (if $0 < a, b, c < \pi$)

$$\sin A = \frac{|(\mathbf{a} \times \mathbf{b}) \times (\mathbf{a} \times \mathbf{c})|}{|\mathbf{a} \times \mathbf{b}||\mathbf{a} \times \mathbf{c}|}$$

and deduce that

$$\frac{\sin A}{\sin a} = \frac{\sin B}{\sin b} = \frac{\sin C}{\sin c} = \frac{6\,\mathrm{Vol}(O\,\mathcal{ABC})}{\sin a \sin b \sin c}.$$

(iii) Given the latitude and longitude of London and Los Angeles and the radius of the Earth, explain how you would calculate the distance covered in a direct flight. How would you calculate the compass direction relative to true North that the aircraft captain should choose for such a flight?

[The formulae given in parts (i) and (ii) appear in *De Triangulis*, written in 1462–1463 by Regiomontanus, but spherical trigonometry was invented by the ancient Greeks for use in astronomy. For reasons which will be clear if you did part (iii), they were interested in the case when the angle $A$ was a right angle.]

**Exercise 9.5.10 [Frenet–Serret formulae]** In this exercise we are interested in understanding what is going on, rather than worrying about rigour and special cases. You should assume that everything is sufficiently well behaved to avoid problems.

(i) Let $\mathbf{r} : \mathbb{R} \to \mathbb{R}^3$ be a smooth curve in space. If $\phi : \mathbb{R} \to \mathbb{R}$ is smooth, show that

$$\frac{d}{dt}\mathbf{r}\big(\phi(t)\big) = \phi'(t)\mathbf{r}'\big(\phi(t)\big).$$

Explain why (under reasonable conditions) we can restrict ourselves to studying $\mathbf{r}$ such that $\|\mathbf{r}'(s)\| = 1$ for all $s$. Such a function $\mathbf{r}$ is said to be parameterised by arc length. For the rest of the exercise we shall use this parameterisation.

(ii) Let $\mathbf{t}(s) = \mathbf{r}'(s)$. (We call $\mathbf{t}$ the tangent vector.) Show that $\mathbf{t}(s) \cdot \mathbf{t}'(s) = 0$ and deduce that, unless $\|\mathbf{t}'(s)\| = 0$, we can find a unit vector $\mathbf{n}(s)$ such that $\|\mathbf{t}'(s)\|\mathbf{n}(s) = \mathbf{t}'(s)$. For the rest of the question we shall assume that we can find a smooth function $\kappa : \mathbb{R} \to \mathbb{R}$ with $\kappa(s) > 0$ for all $s$ and a smoothly varying unit vector $\mathbf{n} : \mathbb{R} \to \mathbb{R}^3$ such that $\mathbf{t}'(s) = \kappa(s)\mathbf{n}(s)$. (We call $\mathbf{n}$ the unit normal.)

(iii) If $\mathbf{r}(s) = (R^{-1}\cos Rs, R^{-1}\sin Rs, 0)$, verify that we have an arc length parameterisation. Show that $\kappa(s) = R^{-1}$. For this reason $\kappa(s)^{-1}$ is called the 'radius of the circle of curvature' or just 'the radius of curvature'.

(iv) Let $\mathbf{b}(s) = \mathbf{t}(s) \times \mathbf{n}(s)$. Show that there is a unique $\tau : \mathbb{R} \to \mathbb{R}$ such that

$$\mathbf{n}'(s) = -\kappa(s)\mathbf{t}(s) + \tau(s)\mathbf{b}(s).$$

Show further that

$$\mathbf{b}'(s) = -\tau(s)\mathbf{n}(s).$$

(v) A particle travels along the curve at variable speed. If its position at time $t$ is given by $\mathbf{x}(t) = \mathbf{r}\big(\psi(t)\big)$, express the acceleration $\mathbf{x}''(t)$ in terms of (some of) $\mathbf{t}$, $\mathbf{n}$, $\mathbf{b}$ and $\kappa$, $\tau$, $\psi$ and their derivatives.[8] Why does the vector $\mathbf{b}$ rarely occur in elementary mechanics?

**Exercise 9.5.11** We work in $\mathbb{R}^3$ and do not use the summation convention. Explain geometrically why $\mathbf{e}_1$, $\mathbf{e}_2$, $\mathbf{e}_3$ are linearly independent (and so form a basis for $\mathbb{R}^3$) if and only if

$$(\mathbf{e}_1 \times \mathbf{e}_2) \cdot \mathbf{e}_3 \neq 0. \qquad\qquad \bigstar$$

For the rest of this question we assume that $\bigstar$ holds.

(i) Let $E$ be the matrix with $j$th row $\mathbf{e}_j$. By considering the matrix $E\hat{E}^T$, or otherwise, show that there is a unique matrix $\hat{E}$ with $i$th row $\hat{\mathbf{e}}_j$ such that

$$\mathbf{e}_i \cdot \hat{\mathbf{e}}_j = \delta_{ij}.$$

---

[8] The reader is warned that not all mathematicians use exactly the same definition of quantities like $\kappa(s)$. In particular, signs may change from $+$ to $-$ and vice versa.

(ii) If $\mathbf{x} \in \mathbb{R}^3$, show that

$$\mathbf{x} = \sum_{j=1}^{3} (\mathbf{x} \cdot \hat{\mathbf{e}}_j) \mathbf{e}_j.$$

(iii) Show that $\hat{\mathbf{e}}_1, \hat{\mathbf{e}}_2, \hat{\mathbf{e}}_3$ form a basis and that, if $\mathbf{x} \in \mathbb{R}^3$,

$$\mathbf{x} = \sum_{j=1}^{3} (\mathbf{x} \cdot \mathbf{e}_j) \hat{\mathbf{e}}_j.$$

(iv) Show that $\hat{\hat{\mathbf{e}}}_j = \mathbf{e}_j$.
(v) Show that

$$\hat{\mathbf{e}}_1 = \frac{\mathbf{e}_2 \times \mathbf{e}_3}{(\mathbf{e}_1 \times \mathbf{e}_2) \cdot \mathbf{e}_3}$$

and write down similar expressions for $\hat{\mathbf{e}}_2$ and $\hat{\mathbf{e}}_3$. Find an expression for

$$\big((\mathbf{e}_3 \times \mathbf{e}_1) \times (\mathbf{e}_1 \times \mathbf{e}_2)\big) \cdot (\mathbf{e}_2 \times \mathbf{e}_3)$$

in terms of $(\mathbf{e}_1 \times \mathbf{e}_2) \cdot \mathbf{e}_3$.

(vi) Find $\hat{\mathbf{e}}_1, \hat{\mathbf{e}}_2, \hat{\mathbf{e}}_3$ in the special case when $\mathbf{e}_1, \mathbf{e}_2, \mathbf{e}_3$ are orthonormal and $\det E > 0$.
[The basis $\hat{\mathbf{e}}_1, \hat{\mathbf{e}}_2, \hat{\mathbf{e}}_3$ is sometimes called the dual basis of $\mathbf{e}_1, \mathbf{e}_2, \mathbf{e}_3$, but we shall introduce a much more general notion of dual basis elsewhere in this book. The 'dual basis' considered here is very useful in crystallography.]

**Exercise 9.5.12** Let $\mathbf{r} = (x_1, x_2, x_3)$ be the usual position vector and write $r = \|\mathbf{r}\|$. If $\mathbf{a}$ is a constant vector, find the divergence of the following functions.
  (i) $r^n \mathbf{a}$ (for $\mathbf{r} \neq \mathbf{0}$).
  (ii) $r^n (\mathbf{a} \times \mathbf{r})$ (for $\mathbf{r} \neq \mathbf{0}$).
  (iii) $(\mathbf{a} \times \mathbf{r}) \times \mathbf{a}$.
  Find the curl of $r\mathbf{a}$ and $r^2 \mathbf{r} \times \mathbf{a}$ when $\mathbf{r} \neq \mathbf{0}$. Find the grad of $\mathbf{a} \cdot \mathbf{r}$.
  If $\psi$ is a smooth function of position, show that

$$\nabla \cdot (\mathbf{r} \times \nabla \psi) = 0.$$

Suppose that $f : (0, \infty) \to \mathbb{R}$ is smooth and $\phi(\mathbf{r}) = f(r)$. Show that

$$\nabla \phi = \frac{f'(r)}{r} \mathbf{r} \quad \text{and} \quad \nabla^2 \phi = \frac{1}{r^2} \frac{d}{dr} \big(r^2 f'(r)\big).$$

Deduce that, if $\nabla^2 \phi = 0$ on $\mathbb{R}^3 \setminus \{\mathbf{0}\}$, then

$$\phi(\mathbf{r}) = A + \frac{B}{r}$$

for some constants $A$ and $B$.

**Exercise 9.5.13** Let $\mathbf{r} = (x_1, x_2, x_3)$ be the usual position vector and write $r = \|\mathbf{r}\|$. Suppose that $f, g : (0, \infty) \to \mathbb{R}$ are smooth and

$$R_{ij}(\mathbf{x}) = f(r)x_i x_j + g(r)\delta_{ij}$$

on $\mathbb{R}^3 \setminus \{\mathbf{0}\}$. Explain why $R_{ij}$ is a Cartesian tensor of order 2. Find

$$\frac{\partial R_{ij}}{\partial x_i} \quad \text{and} \quad \epsilon_{ijk}\frac{\partial R_{kl}}{\partial x_j}$$

in terms of $f$ and $g$ and their derivatives. Show that, if both expressions vanish identically, $f(r) = Ar^{-5}$ for some constant $A$.

# 10

# More on tensors

## 10.1 Some tensorial theorems

This section deals mainly with tensors having special properties. We first look at isotropic tensors. These are tensors like $\delta_{ij}$ and $\epsilon_{ijk}$ which remain unchanged under rotation of our coordinate system. They are important because many physical systems (for example, empty space) look the same when we rotate them. We would expect such systems to be described by isotropic tensors.

**Theorem 10.1.1** (*i*) *Every Cartesian tensor of order 0 is isotropic.*
(*ii*) *The zero tensor is the only Cartesian tensor of order 1 which is isotropic.*
(*iii*) *The isotropic Cartesian tensors of order 2 have the form $\lambda\delta_{ij}$ with $\lambda$ a real number.*
(*iv*) *The isotropic Cartesian tensors of order 3 have the form $\lambda\epsilon_{ijk}$ with $\lambda$ a real number.*

*Proof* (i) Automatic.
(ii) Suppose that $a_i$ is an isotropic Cartesian tensor of order 1.
If we take

$$L = \begin{pmatrix} 1 & 0 & 0 \\ 0 & 0 & 1 \\ 0 & -1 & 0 \end{pmatrix},$$

then $LL^T = I$ and $\det L = 1$, so $L \in SO(\mathbb{R}^3)$. (The reader should describe $L$ geometrically.) Thus

$$a_3 = a_3' = l_{3i}a_i = -a_2 \quad \text{and} \quad a_2 = a_2' = l_{2i}a_i = a_3.$$

It follows that $a_2 = a_3 = 0$ and, by symmetry among the indices, we must also have $a_1 = 0$.
(iii) Suppose that $a_{ij}$ is an isotropic Cartesian tensor of order 2.
If we take $L$ as in part (ii), we obtain

$$a_{22} = a_{22}' = l_{2i}l_{2j}a_{ij} = a_{33}$$

so, by symmetry among the indices, we must have $a_{11} = a_{22} = a_{33}$. We also have

$$a_{12} = a_{12}' = l_{1i}l_{2j}a_{ij} = -a_{13} \quad \text{and} \quad a_{13} = a_{13}' = l_{1i}l_{3j}a_{ij} = a_{12}.$$

It follows that $a_{12} = a_{13} = 0$ and, by symmetry among the indices,

$$a_{12} = a_{13} = a_{21} = a_{23} = a_{31} = a_{32} = 0.$$

Thus $a_{ij} = \lambda \delta_{ij}$.

(iv) Suppose that $a_{ijk}$ is an isotropic Cartesian tensor of order 3. If we take $L$ as in part (ii) we obtain

$$a_{123} = a'_{123} = l_{1i} l_{2j} l_{3k} a_{ijk} = -a_{132}$$
$$a_{122} = a'_{122} = l_{1i} l_{2j} l_{2k} a_{ijk} = a_{133}$$
$$a_{211} = a'_{211} = l_{2i} l_{1j} l_{1k} a_{ijk} = -a_{311}$$
$$a_{222} = a'_{122} = l_{1i} l_{2j} l_{3k} a_{ijk} = -a_{333}$$
$$a_{333} = a'_{333} = l_{3i} l_{3j} l_{3k} a_{ijk} = a_{222}.$$

Using symmetry among the indices, we can also see that

$$a_{122} = a_{133} = -a_{122},$$

so $a_{122} = 0$. Symmetry among the indices now gives $a_{ijk} = \lambda \epsilon_{ijk}$ as required.     □

Matters become more complicated when the order is greater than 3.

**Exercise 10.1.2** *Show that, if $\alpha$, $\beta$ and $\gamma$ are real, then*

$$\alpha \delta_{ij} \delta_{kl} + \beta \delta_{ik} \delta_{jl} + \gamma \delta_{il} \delta_{jk} \qquad\qquad ★$$

*is an isotropic Cartesian tensor.*

*Show that, if*

$$\alpha \delta_{ij} \delta_{kl} + \beta \delta_{ik} \delta_{jl} + \gamma \delta_{il} \delta jk = 0,$$

*then $\alpha = \beta = \gamma = 0$.*

*Exercise 10.5.9 sketches a proof that the expression in ★ is the most general isotropic tensor of order 4, but requires the reader to work quite hard.*

*Use the statement that ★ is, indeed, the most general isotropic tensor of order 4 to produce a proof of the the Levi-Civita identity*

$$\epsilon_{ijk} \epsilon_{ist} = \delta_{js} \delta_{kt} - \delta_{ks} \delta_{jt}.$$

Recall the tensorial equation

$$p_{ij} = c_{ijkm} e_{km}$$

governing the relation between the stress tensor $e_{km}$ and the strain tensor $p_{ij}$ in elasticity. If we deal with a material like steel, which looks the same in all directions, we must have $c_{ijkm}$ isotropic.

By the results stated in Exercise 10.1.2, we have

$$c_{ijkm} = \alpha \delta_{ij} \delta_{km} + \beta \delta_{ik} \delta_{jm} + \gamma \delta_{im} \delta_{jk}$$

and so

$$p_{ij} = 3\alpha\delta_{ij}e_{kk} + \beta e_{ij} + \gamma e_{ji}$$

for some constants $\alpha$, $\beta$ and $\gamma$. In fact, $e_{ij} = e_{ji}$, so the equation reduces to

$$p_{ij} = \lambda\delta_{ij}e_{kk} + 2\mu e_{ij}$$

showing that the elastic behaviour of an isotropic material depends on two constants $\lambda$ and $\mu$.

Sometimes we get the 'opposite of isotropy' and a problem is best considered with a very particular set of coordinate axes. The wide-awake reader will not be surprised to see the appearance of a *symmetric Cartesian tensor of order* 2, that is to say, a tensor $a_{ij}$, like the stress tensor in the paragraph above, with $a_{ij} = a_{ji}$. Our first task is to show that symmetry is a tensorial property.

**Exercise 10.1.3** *Suppose that $a_{ij}$ is a Cartesian tensor. If $a_{ij} = a_{ji}$ in one coordinate system $S$, show that $a'_{ij} = a'_{ji}$ in any rotated coordinate system $S'$.*

**Exercise 10.1.4** *According to the theory of magnetostriction, the mechanical stress is a second order symmetric Cartesian tensor $\sigma_{ij}$ induced by the magnetic field $B_i$ according to the rule*

$$\sigma_{ij} = a_{ijk}B_k,$$

*where $a_{ijk}$ is a third order Cartesian tensor which depends only on the material. Show that $\sigma_{ij} = 0$ if the material is isotropic.*

A $3 \times 3$ matrix and a Cartesian tensor of order 2 are very different beasts, so we must exercise caution in moving between these two sorts of objects. However, our results on symmetric matrices give us useful information about symmetric tensors.

**Theorem 10.1.5** *If $a_{ij}$ is a symmetric Cartesian tensor, there exists a rotated coordinate system $S'$ in which*

$$\begin{pmatrix} a'_{11} & a'_{12} & a'_{13} \\ a'_{21} & a'_{22} & a'_{23} \\ a'_{31} & a'_{32} & a'_{33} \end{pmatrix} = \begin{pmatrix} \lambda & 0 & 0 \\ 0 & \mu & 0 \\ 0 & 0 & \nu \end{pmatrix}.$$

*Proof* Let us write down the *matrix*

$$A = \begin{pmatrix} a_{11} & a_{12} & a_{13} \\ a_{21} & a_{22} & a_{23} \\ a_{31} & a_{32} & a_{33} \end{pmatrix}.$$

By Theorem 8.1.6, we can find a $3 \times 3$ *matrix* $M \in O(\mathbb{R}^3)$ such that

$$MAM^T = D,$$

where $D$ is the *matrix* given by

$$D = \begin{pmatrix} \lambda & 0 & 0 \\ 0 & \mu & 0 \\ 0 & 0 & \nu \end{pmatrix}$$

for some $\lambda$, $\mu$, $\nu \in \mathbb{R}$.

If $M \in SO(\mathbb{R}^3)$, we set $L = M$. If $M \notin SO(\mathbb{R}^3)$, we set $L = -M$. In either case, $L \in SO(\mathbb{R}^3)$ and

$$LAL^T = D.$$

Let

$$L = \begin{pmatrix} l_{11} & l_{12} & l_{13} \\ l_{21} & l_{22} & l_{23} \\ l_{31} & l_{32} & l_{33} \end{pmatrix}.$$

If we consider the coordinate rotation associated with $l_{ij}$, we obtain the *tensor relation*

$$a'_{ij} = l_{ir}l_{js}a_{rs}$$

with

$$\begin{pmatrix} a'_{11} & a'_{12} & a'_{13} \\ a'_{21} & a'_{22} & a'_{23} \\ a'_{31} & a'_{32} & a'_{33} \end{pmatrix} = \begin{pmatrix} \lambda & 0 & 0 \\ 0 & \mu & 0 \\ 0 & 0 & \nu \end{pmatrix}$$

as required.                                                                      □

*Remark 1* If you want to use the summation convention, it is a very bad idea to replace $\lambda$, $\mu$ and $\nu$ by $\lambda_1$, $\lambda_2$ and $\lambda_3$.

*Remark 2* Although we are working with tensors, it is usual to say that if $a_{ij}u_j = \lambda u_i$ with $u_i \neq 0$, then **u** is an eigenvector and $\lambda$ an eigenvalue of the tensor $a_{ij}$. If we work with symmetric tensors, we often refer to *principal axes* instead of eigenvectors.

We can also consider *antisymmetric Cartesian tensors*, that is to say, tensors $a_{ij}$ with $a_{ij} = -a_{ji}$.

**Exercise 10.1.6** (*i*) *Suppose that $a_{ij}$ is a Cartesian tensor. If $a_{ij} = -a_{ji}$ in one coordinate system, show that $a'_{ij} = -a'_{ji}$ in any rotated coordinate system.*

(*ii*) *By looking at $\frac{1}{2}(b_{ij} + b_{ji})$ and $\frac{1}{2}(b_{ij} - b_{ji})$, or otherwise, show that any Cartesian tensor of order 2 is the sum of a symmetric and an antisymmetric tensor.*

(*iii*) *Show that, if we consider the vector space of Cartesian tensors of order 2 (see Exercise 9.3.8), then the symmetric tensors form a subspace of dimension 6 and the antisymmetric tensors form a subspace of dimension 3.*

If we look at the antisymmetric matrix

$$\begin{pmatrix} 0 & \sigma_3 & -\sigma_2 \\ -\sigma_3 & 0 & \sigma_1 \\ \sigma_2 & -\sigma_1 & 0 \end{pmatrix}$$

long enough, the following result may present itself.

**Lemma 10.1.7** *If $a_{ik}$ is an antisymmetric tensor and $\omega_j = \frac{1}{2}\epsilon_{ijk}a_{ik}$, then*

$$a_{ik} = \epsilon_{ijk}\omega_j.$$

*Proof* By the Levi-Civita identity,

$$\epsilon_{ijk}\omega_j = \tfrac{1}{2}\epsilon_{ijk}\epsilon_{rjs}a_{rs} = \tfrac{1}{2}\epsilon_{jki}\epsilon_{jsr}a_{rs} = \tfrac{1}{2}(\delta_{ks}\delta_{ir} - \delta_{kr}\delta_{is})a_{rs}$$
$$= \tfrac{1}{2}(a_{ik} - a_{ki}) = a_{ik}$$

as required. □

**Exercise 10.1.8** *Using the minimum of computation, identify the eigenvectors of the tensor $\epsilon_{ijk}\omega_j$ where $\boldsymbol{\omega}$ is a non-zero vector. (Note that we are working in $\mathbb{R}^3$.)*

## 10.2 A (very) little mechanics

We illustrate the use of the vector product by looking at the behaviour of a collection of particles in Newtonian mechanics. We suppose that the $\alpha$th particle has position $\mathbf{x}_\alpha$ and mass $m_\alpha$ for $1 \le \alpha \le N$. We do not apply the summation convention to the integers $\alpha$ and $\beta$. The $\alpha$th particle is subject to a force $\mathbf{F}_{\alpha,\beta}$ due to the $\beta$th particle $[\beta \ne \alpha]$ and an external force $\mathbf{F}_\alpha$. Newton's laws tell us that forces are vectors and

$$m_\alpha\ddot{\mathbf{x}}_\alpha = \mathbf{F}_\alpha + \sum_{\beta \ne \alpha}\mathbf{F}_{\alpha,\beta}.$$

In addition since 'action and reaction are equal and opposite'

$$\mathbf{F}_{\alpha,\beta} = -\mathbf{F}_{\beta,\alpha}.$$

These laws hold whether we consider a galaxy of stars with the stars as particles or a falling raindrop with the constituent molecules as particles.

In order to get some idea of the nature of the motion, we introduce a new vector called the *centre of mass*

$$\mathbf{x}_G = M^{-1}\sum_\alpha m_\alpha\mathbf{x}_\alpha,$$

where $M = \sum_\alpha m_\alpha$ is the total mass of the system, and observe that

$$M\ddot{\mathbf{x}}_G = \sum_\alpha m_\alpha\ddot{\mathbf{x}}_\alpha$$

$$= \sum_\alpha\left(\mathbf{F}_\alpha + \sum_{\beta \ne \alpha}\mathbf{F}_{\alpha,\beta}\right) = \sum_\alpha\mathbf{F}_\alpha + \sum_\alpha\sum_{\beta \ne \alpha}\mathbf{F}_{\alpha,\beta}$$

$$= \sum_\alpha\mathbf{F}_\alpha + \sum_{1 \le \beta < \alpha \le N}(\mathbf{F}_{\alpha,\beta} + \mathbf{F}_{\beta,\alpha})$$

$$= \sum_\alpha\mathbf{F}_\alpha + \sum_{1 \le \beta < \alpha \le N}\mathbf{0} = \mathbf{F}$$

where $\mathbf{F} = \sum_\alpha \mathbf{F}_\alpha$. Thus the centre of mass behaves like a particle of mass $M$ (the total mass of the system) under a force $\mathbf{F}$ (the total vector sum of the external forces on the system). This represents one of the most important results in mechanics, since it allows us, for example, when considering the orbit of the Earth (a body with unknown and complicated internal forces) round the Sun (whose internal forces are likewise unknown) to take the two bodies as point masses. We call

$$M\dot{\mathbf{x}}_G = \sum_\alpha m_\alpha \dot{\mathbf{x}}_\alpha$$

the *momentum* of the system.

If we make the additional assumption that the force exerted by one particle on the other acts along the line of centres (that is to say,

$$\mathbf{F}_{\alpha,\beta} = \lambda_{\alpha,\beta}(\mathbf{x}_\alpha - \mathbf{x}_\beta),$$

for some scalar $\lambda_{\alpha,\beta}$), then we can find another 'global equation' which does not involve internal forces by using the vector product. We define the *angular momentum* $\mathbf{H}$ of the system about the origin by the formula

$$\mathbf{H} = \sum_\alpha m_\alpha \mathbf{x}_\alpha \times \dot{\mathbf{x}}_\alpha.$$

Our assumptions that the forces are opposite and act along the line of centres tell us that

$$\mathbf{x}_\alpha \times \mathbf{F}_{\alpha,\beta} + \mathbf{x}_\beta \times \mathbf{F}_{\beta,\alpha} = (\mathbf{x}_\alpha - \mathbf{x}_\beta) \times \mathbf{F}_{\alpha,\beta} = \lambda_{\alpha,\beta}(\mathbf{x}_\alpha - \mathbf{x}_\beta) \times (\mathbf{x}_\alpha - \mathbf{x}_\beta) = \mathbf{0}$$

and so

$$\dot{\mathbf{H}} = \sum_\alpha m_\alpha \frac{d}{dt}\mathbf{x}_\alpha \times \dot{\mathbf{x}}_\alpha = \sum_\alpha m_\alpha(\dot{\mathbf{x}}_\alpha \times \dot{\mathbf{x}}_\alpha + \mathbf{x}_\alpha \times \ddot{\mathbf{x}}_\alpha)$$

$$= \sum_\alpha m_\alpha \mathbf{x}_\alpha \times \ddot{\mathbf{x}}_\alpha = \sum_\alpha \mathbf{x}_\alpha \times \left( \mathbf{F}_\alpha + \sum_{\beta \neq \alpha} \mathbf{F}_{\alpha,\beta} \right)$$

$$= \sum_\alpha \mathbf{x}_\alpha \times \mathbf{F}_\alpha + \sum_{1 \leq \beta < \alpha \leq N} (\mathbf{x}_\alpha \times \mathbf{F}_{\alpha,\beta} + \mathbf{x}_\beta \times \mathbf{F}_{\beta,\alpha})$$

$$= \sum_\alpha \mathbf{x}_\alpha \times \mathbf{F}_\alpha + \sum_{1 \leq \beta < \alpha \leq N} \mathbf{0} = \mathbf{G},$$

where $\mathbf{G} = \sum_\alpha \mathbf{x}_\alpha \times \mathbf{F}_\alpha$ is called the *total couple* of the external forces.

Thus, in the absence of external forces, the angular momentum, like the momentum, remains unchanged.

**Exercise 10.2.1** *It is interesting to consider what happens if we look at matters relative to the centre of mass. If, using the notation of this section, we write $\mathbf{r}_\alpha = \mathbf{x}_\alpha - \mathbf{x}_G$, show that*

$$\mathbf{H} = M\mathbf{x}_G \times \dot{\mathbf{x}}_G + \sum_\alpha m_\alpha \mathbf{r}_\alpha \times \dot{\mathbf{r}}_\alpha.$$

*Describe this result in words.*

**Exercise 10.2.2** *Suppose that particle $\alpha$ experiences an external force*

$$\mathbf{F}_\alpha = -km_\alpha \dot{\mathbf{x}}.$$

*Show that the angular momentum $\mathbf{H} = \mathbf{H}_0 e^{-kt}$, where $\mathbf{H}_0$ is constant.*

For the rest of the section, we look at what happens when we rotate a rigid body (fixed so that the point at the origin does not move) round a fixed axis through the origin.

It is important to understand that $l_{ij}$ itself *is not a tensor*. If we look at the relation

$$x_i' = l_{ij} x_j,$$

we can think of $l_{ij}$ standing with one foot $i$ in the rotated coordinate system and with the other foot $j$ in the original coordinate system. On the other hand, we can think of rotation about the origin within a *single* coordinate system as the map which takes the vector $\mathbf{y}$ to the vector $\mathbf{x}$ according to the rule

$$x_i = k_{ij} y_j.$$

By the quotient rule, $k_{ij}$ is a second order tensor.

If the rotation is about a fixed axis, we may choose a coordinate system in which the array associated with the tensor takes the form

$$(k_{ij}) = \begin{pmatrix} 1 & 0 & 0 \\ 0 & \cos\theta & \sin\theta \\ 0 & -\sin\theta & \cos\theta \end{pmatrix}.$$

If we consider $k_{ij}$ as changing with time (but with fixed axis of rotation) then, in our chosen coordinate system,

$$\dot{k}_{ij} = \begin{pmatrix} 0 & 0 & 0 \\ 0 & -\sin\theta & \cos\theta \\ 0 & -\cos\theta & -\sin\theta \end{pmatrix} \dot{\theta}.$$

If we make a further simplification by choosing our coordinate system in such a way that $\theta = 0$ at the time that we are considering, then

$$(\dot{k}_{ij}) = \begin{pmatrix} 0 & 0 & 0 \\ 0 & 0 & 1 \\ 0 & -1 & 0 \end{pmatrix} \dot{\theta},$$

so

$$\dot{k}_{ij} = \epsilon_{irj} \omega_r$$

where $\omega$ is the vector which (in our chosen system) corresponds to the array $(\omega_1, \omega_2, \omega_3) = (\dot{\theta}, 0, 0)$.

Since a tensor equation which is true in one system is true in all systems, we have shown the existence of a vector $\omega$ such that $\dot{k}_{ij} = \epsilon_{irj} \omega_r$. In particular, if

$$x_i = k_{ij} y_j,$$

then

$$\dot{x}_i = \dot{k}_{ij} y_j + k_{ij} \dot{y}_j = \epsilon_{irj} y_j \omega_r + k_{ij} \dot{y}_j.$$

We now specialise this rather general result. Since we are dealing with a rigid body fixed at the origin, $\dot{y}_j = 0$ and so $\dot{x}_i = \epsilon_{irj} \omega_r x_j$, that is to say,

$$\dot{\mathbf{x}} = \boldsymbol{\omega} \times \mathbf{y}.$$

Finally, we remark that we can choose to observe our system at the instant when $\mathbf{y} = \mathbf{x}$ and so

$$\dot{\mathbf{x}} = \boldsymbol{\omega} \times \mathbf{x}.$$

Thus, if our collection of particles form a rigid system revolving round the origin, the $\alpha$th particle in position $\mathbf{x}_\alpha$ has velocity $\boldsymbol{\omega} \times \mathbf{x}_\alpha$. If the reader feels that we have spent too much time proving the obvious or, alternatively, that we have failed to provide a proper proof of anything, she may simply take the previous sentence as our *definition* of the equation of motion of a point in a rigid body fixed at the origin revolving round an axis. The vector $\boldsymbol{\omega}$ is called the angular velocity of the body.

By definition, the angular momentum of our body is given by

$$\mathbf{H} = \sum_\alpha m_\alpha \mathbf{x}_\alpha \times \dot{\mathbf{x}}_\alpha = \sum_\alpha m_\alpha \mathbf{x}_\alpha \times (\boldsymbol{\omega} \times \mathbf{x}_\alpha) = \sum_\alpha m_\alpha \big( (\mathbf{x}_\alpha \cdot \mathbf{x}_\alpha) \boldsymbol{\omega} - (\mathbf{x}_\alpha \cdot \boldsymbol{\omega}) \mathbf{x}_\alpha \big).$$

We can rewrite this equation in tensor form to get

$$H_i = \sum_\alpha m_\alpha (x_{\alpha k} x_{\alpha k} \omega_i - x_{\alpha j} \omega_j x_{\alpha i}) = I_{ij} \omega_j,$$

where

$$I_{ij} = \sum_\alpha m_\alpha (x_{\alpha k} x_{\alpha k} \delta_{ij} - x_{\alpha i} x_{\alpha j}).$$

(Note that, as throughout this section, the $i$, $j$ and $k$ are subject to the summation convention, but $\alpha$ is not.)

We call $I_{ij}$ the *inertia tensor* of the rigid body. Passing from the discrete case of a finite set of point masses to the continuum case of a lump of matter with density $\rho$, we obtain the expression

$$I_{ij} = \int (x_k x_k \delta_{ij} - x_i x_j) \rho(\mathbf{x}) \, dV(\mathbf{x})$$

for the inertia tensor for such a body. The notion of angular momentum $\mathbf{H}$ and the total couple $\mathbf{G}$ can be extended in a similar way to give us

$$H_i = I_{ij} \omega_j, \quad \dot{H}_i = G_i$$

and so

$$I_{ij} \dot{\omega}_j = G_i.$$

**Exercise 10.2.3** **[The parallel axis theorem]** *Suppose that one lump of matter with centre of mass at the origin has density $\rho(\mathbf{x})$, total mass $M$, and moment of inertia $I_{ij}$ whilst a second lump of matter has density $\rho(\mathbf{x} - \mathbf{a})$ and moment of inertia $J_{ij}$. Prove that*

$$J_{ij} = I_{ij} + M(a_k a_k \delta_{ij} - a_i a_j).$$

If the reader takes some heavy object (a chair, say) and attempts to spin it, she will find that it is much easier to spin it in certain ways than in others. The inertia tensor tells us why.

Of course, we need to decide what we mean by 'easy to spin', but a little thought suggests that we should mean that applying a couple $\mathbf{G}$ produces rotation about the axis defined by $\mathbf{G}$, that is to say, we want $\dot{\omega}$ to be a multiple of $\mathbf{G}$ and so

$$I_{ij} w_j = \lambda w_i$$

with $\mathbf{w} = \dot{\omega}$ and $\mathbf{G} = \lambda^{-1} \mathbf{w}$ for some non-zero scalar $\lambda$ and some non-zero vector $\mathbf{w}$. (In other words, $\mathbf{w}$ is an eigenvector with eigenvalue $\lambda$.)

Observe that

$$I_{ji} = \int (x_k x_k \delta_{ji} - x_j x_i) \rho(\mathbf{x}) \, dV(\mathbf{x}) = \int (x_k x_k \delta_{ij} - x_i x_j) \rho(\mathbf{x}) \, dV(\mathbf{x}) = I_{ij},$$

so $I_{ij}$ is a symmetric Cartesian tensor of order 2. It follows, by Theorem 10.1.5, that we can find a coordinate system in which the array associated with the inertia tensor takes the simple form

$$\begin{pmatrix} A & 0 & 0 \\ 0 & B & 0 \\ 0 & 0 & C \end{pmatrix}.$$

**Exercise 10.2.4** *Show that, if we work in the coordinate system of the previous sentence,*

$$A = \iiint (y^2 + z^2) \rho(x, y, z) \, dx \, dy \, dz \geq 0$$

*and write down similar formulae for $B$ and $C$.*

If $A$, $B$ and $C$ are all unequal, we see that we can 'easily spin' the body about each of the three coordinate axes of our specially chosen system and about no other axis. If $B = C$ but $A \neq B$, we can 'easily spin' the body about the axis corresponding to $A$ and about any axis perpendicular to this (passing through the origin).

The reader who thinks that Cartesian tensors are a long winded way of stating the obvious should ask herself why it is obvious (as the discussion above shows) that, if we can 'easily spin' a rigid body about two axes through a fixed point, we can 'easily spin' the body about the axis perpendicular to both.

Rugby is a thugs' game played by gentlemen, soccer is a gentlemen's game played by thugs and Australian rules football is an Australian game played by Australians. The soccer ball is essentially spherical (so, by symmetry, $A = B = C$), but the balls used in rugby, Australian rules football and American football have $A > B = C$. When watching games with $A > B = C$, we are watching an inertia tensor in flight.

### 10.3 Left-hand, right-hand

We say that a $3 \times 3 \times \cdots \times 3$ array $a_{ij\ldots m}$ is a Cartesian tensor if it transforms according to the rule

$$a'_{ij\ldots m} = l_{ir} l_{js} \ldots l_{mu} a_{rs\ldots u}$$

for a rotation of the coordinate system corresponding to $L \in SO(\mathbb{R}^3)$. I claimed that this was a useful idea because we expect the laws of physics to be unchanged under rotation.

Why not use some other test? In particular, why not allow $L \in O(\mathbb{R}^3)$? Let us call $a_{ij\ldots m}$ an 'extended Cartesian tensor' (this is non-standard terminology) if it transforms according to the rule

$$a'_{ij\ldots m} = l_{ir} l_{js} \ldots l_{mu} a_{rs\ldots u}$$

for a transformation of the coordinate system corresponding to $L \in O(\mathbb{R}^3)$ (so that the coordinate axes remain orthogonal). If we do so, we get a rather nasty surprise.

**Lemma 10.3.1** *The array $\epsilon_{ijk}$ is not an extended Cartesian tensor.*

*Proof* The argument of Lemma 9.4.1 shows that

$$l_{ir} l_{js} l_{kt} \epsilon_{rst} = \epsilon_{ijk} \det L = -\epsilon_{ijk}$$

if $L \in O(\mathbb{R}^3) \setminus SO(\mathbb{R}^3)$.                                                             $\square$

**Exercise 10.3.2** (*i*) *Show that $\delta_{ij}$ is an extended Cartesian tensor.*

(*ii*) *Show that there are no non-zero isotropic extended Cartesian tensors of order 3.*

(*iii*) *If* **a**, **b** *and* **c** *are extended Cartesian tensors of order 1 (so Cartesian tensors of order 1) and* $(\mathbf{a} \times \mathbf{b}) \cdot \mathbf{c} \neq 0$, *show that* $(\mathbf{a} \times \mathbf{b}) \cdot \mathbf{c}$ *is not an extended Cartesian tensor.*

(*iv*) *If* **a**, **b** *are extended Cartesian tensors of order 1 show that $a_i b_i$ is an extended Cartesian tensor of order 0.*

(*v*) *If* **a** *and* **b** *are extended Cartesian tensors of order 1 (so Cartesian tensors of order 1) and* $\mathbf{a} \times \mathbf{b} \neq \mathbf{0}$, *show that* $\mathbf{a} \times \mathbf{b}$ *is not an extended Cartesian tensor.*

Exercise 10.3.2 (iii) can be expressed more vividly. If we form the apparent scalar $(\mathbf{a} \times \mathbf{b}) \cdot \mathbf{c} \neq 0$ and then look in the mirror, the result will have changed sign! (Of course, if we remember the ideas discussed in Section 4.1, we may be less flabbergasted at seeing a signed volume.)

Do the laws of nature remain the same when we look in a mirror? Tartaric acid derived from wine lees rotates the plane of polarisation when polarised light passes through it. Tartaric acid obtained by chemical synthesis does not. In 1849 Pasteur observed that when chemically derived tartaric acid crystallised it produced two types of crystals which (like your left-hand and right-hand) were mirror images but not rotational images. He sorted the crystals according to their handedness and discovered that a solution of 'left-handed' crystals rotated the plane of polarisation one way and a solution of 'right-handed' crystals rotated it the other way. A mixture of the two solutions produced no rotation. This extraordinary result excited much scepticism and Pasteur was invited to demonstrate his result in the presence of Biot, one of the grand old men of French science. After Pasteur had done the separation of the crystals in his presence, Biot performed the rest of the experiment himself. Pasteur recalled that 'When the result became clear he seized me by the hand and said "My dear boy, I have loved science so much during my life that this touches my very heart."'[1]

Pasteur's discovery showed that, in some sense life is 'handed'. In 1952, Weyl wrote

... the deeper chemical constitution of our human body shows that we have a screw, a screw that is turning the same way in every one of us. Thus our body contains the dextro-rotatory form of glucose and the laevo-rotatory form of fructose. A horrid manifestation of this genotypical asymmetry is a metabolic disease called phenylketonuria, leading to insanity, that man contracts when a small quantity of laevo-phenylalanine is added to his food, while the dextro-form has no such deleterious effects.

(Weyl *Symmetry* [33], Chapter 1)

In 1953, Crick and Watson showed that DNA had the structure of a double helix. We now know that the instructions for making an Earthly living thing are encoded on this 'handed' double helix. If life is discovered elsewhere in the solar system, one of our first questions will be whether it is 'handed' and, if so, whether it has the same handedness as us.

A first look at Maxwell's equations for electromagnetism

$$\nabla \cdot \mathbf{D} = \rho,$$

$$\nabla \cdot \mathbf{B} = 0,$$

$$\nabla \times \mathbf{E} = -\frac{\partial \mathbf{B}}{\partial t},$$

$$\nabla \times \mathbf{H} = \mathbf{j} + \frac{\partial \mathbf{D}}{\partial t},$$

$$\mathbf{D} = \epsilon \mathbf{E}, \quad \mathbf{B} = \mu \mathbf{H}, \quad \mathbf{j} = \sigma \mathbf{E},$$

seem to show handedness (or, to use the technical term, exhibit *chirality*), since they involve the vector product. However, suppose that we were in indirect communication with beings in a different part of the universe and we tried to use Maxwell's equations to establish the difference between left- and right-handedness. We would talk about the magnetic field **B** and explain that it is the force experienced by a unit 'north pole', but we would be unable to tell our correspondents which of the two possible poles we mean. Maxwell's electromagnetic theory is mirror symmetric if we change the pole naming convention when we reflect.

Of course, experiment trumps philosophy, so it may turn out that the universe is handed, though, even then, most people would prefer to ascribe the observed left- or right-handedness to chance.

---

[1] 'Mon cher enfant, j'ai tant aimé les sciences dans ma vie que cela fait battre le coeur.' See, for example, [14].

Be that as it may, many physical theories including Newtonian mechanics and Maxwell's electromagnetic theory are mirror symmetric. How should we deal with the failure of some Cartesian tensors to 'transform correctly' under reflection? A common sense solution is simply to check that our equations remain correct under reflection. In order to ensure that terrestrial mathematicians can talk to each other about things like vector products we introduce the 'right-hand convention' or 'right-hand rule'. 'If the forefinger of the right-hand points in the direction of the vector $\mathbf{a}$ and the middle finger in the direction of $\mathbf{b}$, then the thumb points in the direction of $\mathbf{a} \times \mathbf{b}$.' If the reader is a pure mathematician or lacks digital dexterity, she may prefer to note that this is equivalent to using a 'right-handed coordinate system' so that when the right hand is held easily with the forefinger in the direction of the vector $(1, 0, 0)$ and the middle finger in the direction of $(0, 1, 0)$ then the thumb points in the direction of $(0, 0, 1)$.

## 10.4 General tensors

Some physicists refer to Cartesian tensors which change sign under reflection as 'pseudo-tensors', so, if $\mathbf{a}$, $\mathbf{b}$ and $\mathbf{c}$ are linearly independent extended Cartesian tensors of order 1, then $\mathbf{a} \times \mathbf{b}$, is a 'pseudo-vector' and $\mathbf{a} \cdot (\mathbf{b} \times \mathbf{c})$ is a pseudo-scalar. Since Maxwell, Gibbs and Heaviside considered $\mathbf{a} \times \mathbf{b}$ to be a vector, the author thinks this is an unfortunate choice of phrase. It seems more reasonable to say that a physical vector is an object which transforms in a certain way for certain choices of coordinate transformations. As we change our choice of transformations, so we change our notion of a physical vector.

What happens if we choose the most general collection of invertible linear transformations? The answer is very interesting, but requires the introduction of a new notation. In this section we write our position vectors as

$$\mathbf{x} = \begin{pmatrix} x^1 \\ x^2 \\ x^3 \end{pmatrix},$$

that is to say, we use *upper indices* and consider $\mathbf{x}$ as a column vector. We write the matrix of an invertible linear map as $l^i_j$ so that $l^i_j$ is the entry in the $i$th (upper index) row and $j$th (lower index) column.

If we make a change of coordinate system from our initial system $S$ to a new system $\bar{S}$ (note the use of an upper bar rather than a dash), we have a relation

$$\bar{x}^i = l^i_j x^j,$$

where we sum over the $j$ which occurs once as an upper index and once as a lower index and $l^i_j$ is a $3 \times 3$ invertible matrix. We call any column $\mathbf{a} = (a^1, a^2, a^3)^T$ with three elements a *contravariant vector* if it transforms according to the rule

$$\bar{a}^i = l^i_j a^j.$$

Notice that the position vector is automatically a contravariant vector.

The reader should have no difficulty with the next exercise.

**Exercise 10.4.1** *(i) If* u *and* v *are contravariant vectors,* $\lambda$, $\mu \in \mathbb{R}$ *and we set* $w^i = \lambda u^i + \mu v^i$, *then* w *is a contravariant vector.*

*(ii) If a contravariant vector* u$(t)$ *varies in a smooth manner with time, then*

$$\dot{u}(t) = (\dot{u}^1(t), \dot{u}^2(t), \dot{u}^3(t))^T$$

*is a contravariant vector.*

However, matters become more complicated when we seek an analogue of Lemma 9.1.1.

**Lemma 10.4.2** *If* u *is a contravariant vector, then, with the notation just introduced,*

$$u^i = m^i_j \bar{u}^j$$

*where, in matrix notation,* $M = (L^T)^{-1}$.

Note that, in our upper and lower index notation, the definition of $M$ is equivalent to

$$m^i_k l^k_j = l^i_k m^k_j = \delta^i_j,$$

where our new Kronecker delta $\delta^i_j$ corresponds to a $3 \times 3$ array with 1s on the diagonal and 0s off the diagonal.

*Proof of Lemma 10.4.2* Observe that

$$m^i_j \bar{u}^j = m^i_j l^j_k u^k = \delta^i_k u^k = u^i,$$

as required. $\qquad\square$

**Exercise 10.4.3** *Show that* $m^i_j = l^i_j$ *if and only if* $L \in O(\mathbb{R}^3)$.

The new aspect of matters just revealed comes to the fore when we look for an analogue of Example 9.1.3.

**Lemma 10.4.4** *If* $\phi : \mathbb{R}^3 \to \mathbb{R}$ *is smooth, then*

$$\left( \frac{\partial \phi}{\partial x^1}, \frac{\partial \phi}{\partial x^2}, \frac{\partial \phi}{\partial x^3} \right)$$

*transforms according to the rule*

$$\frac{\partial \phi}{\partial \bar{x}^i} = m^j_i \frac{\partial \phi}{\partial x^j}.$$

*Proof* By the chain rule

$$\frac{\partial \phi}{\partial \bar{x}^i} = \frac{\partial \phi}{\partial x^j} \frac{\partial x^j}{\partial \bar{x}^i} = \frac{\partial \phi}{\partial x^j} \frac{\partial m^j_k \bar{x}^k}{\partial \bar{x}^i} = \frac{\partial \phi}{\partial x^j} m^j_k \delta^k_i = m^j_i \frac{\partial \phi}{\partial x^j}$$

as stated. $\qquad\square$

We get round this difficulty by introducing a new kind of vector.[2] We call any *row* $\mathbf{a} = (a_1, a_2, a_3)$ with three elements a *covariant vector* if it transforms according to the rule

$$\bar{a}_i = m_i^j a_j.$$

With this definition, Lemma 10.4.4 may be restated as follows.

**Lemma 10.4.5** *If $\phi : \mathbb{R}^3 \to \mathbb{R}$ is smooth, then, if we write*

$$u_i = \frac{\partial \phi}{\partial x^i},$$

$\mathbf{u}$ *is a covariant vector.*

To go much further than this would lead us too far afield, but the reader will probably observe that we now get three sorts of tensors of order 2, $R_{ij}$, $S_j^i$ and $T^{ij}$ with the transformation rules

$$\bar{R}_{ij} = l_i^u l_j^v R_{uv}, \quad \bar{S}_j^i = m_u^i l_j^v S_v^u, \quad \bar{T}^{ij} = m_u^i m_v^j T^{uv}.$$

**Exercise 10.4.6**  (*i*) *Show that $\delta_i^j$ is second order tensor (in the sense of this section) but $\delta_{ij}$ is not.*

(*ii*) *Write down the appropriate transformation rules for the fourth order tensor $a_{ij}^{kn}$.*

(*iii*) *Let $a_{ij}^{kn} = \delta_i^k \delta_j^n$. Check that $a_{ij}^{kn}$ is a fourth order tensor. Show that $a_{ij}^{in}$ is a second order tensor but $a_{ii}^{kn}$ and $a_{ij}^{nn}$ are not.*

(*iv*) *Write down the appropriate transformation rules for the fourth order tensor $a_{ijk}^n$. Show that $a_{ijn}^n$ is a second order tensor.*

Exercise 10.4.6 indicates that our indexing convention for general tensors should run as follows. 'No index should appear more than twice. If an index appears twice it should appear once as an upper and once as a lower suffix and we should then sum over that index.'

At this point the reader is entitled to say that 'This is all very pretty, but does it lead anywhere?' So far as this book goes, all it does is to give a fleeting notion of covariance and contravariance which pervade much of modern mathematics. (We get another glimpse in Exercise 11.4.6.)

In a wider context, tensors have their roots in the study by Gauss and Riemann of surfaces as 'objects in themselves' rather than 'objects embedded in a larger space'. Ricci-Curbastro and his pupil Levi-Civita developed what they called 'absolute differential calculus' and would now be called 'tensor calculus', but it was not until Einstein discovered that tensor calculus provided the correct language for the development of General Relativity that the importance of these ideas was generally understood.

---

[2] 'When *I* use a word,' Humpty Dumpty said in a rather scornful tone, 'it means just what I choose it to mean – neither more nor less.'
　'The question is,' said Alice, 'whether you *can* make words mean so many different things.'
　'The question is,' said Humpty Dumpty, 'who is to be Master – that's all.'
　　　　　　　　　　　　　　　　(Lewis Carroll *Through the Looking Glass, and What Alice Found There* [9])

Later, when people looked to see whether the same ideas could be useful in classical physics, they discovered that this was indeed the case, but it was often more appropriate to use what we now call Cartesian tensors.

## 10.5 Further exercises

**Exercise 10.5.1** Suppose that $\mathbf{E}(\mathbf{x}, t)$ and $\mathbf{B}(\mathbf{x}, t)$ satisfy Maxwell's equations in vacuum

$$\nabla \cdot \mathbf{D} = 0, \quad \nabla \cdot \mathbf{B} = 0,$$

$$\nabla \times \mathbf{E} = -\frac{\partial \mathbf{B}}{\partial t}, \quad \nabla \times \mathbf{H} = \frac{\partial \mathbf{D}}{\partial t},$$

$$\mathbf{D} = \epsilon_0 \mathbf{E}, \quad \mathbf{B} = \mu_0 \mathbf{H}.$$

Show that

$$\frac{\partial^2 E_i}{\partial x_j \partial x_j} - c^{-2} \frac{\partial^2 E_i}{\partial t^2} = 0 \quad \text{and} \quad \frac{\partial^2 B_i}{\partial x_j \partial x_j} - c^{-2} \frac{\partial^2 B_i}{\partial t^2} = 0,$$

where $c^2 = (\epsilon_0 \mu_0)^{-1}$. Show, by substitution, that the equations of the first paragraph are satisfied by

$$\mathbf{E} = \mathbf{e} \cos(\omega t - \mathbf{k} \cdot \mathbf{x} + \phi) \quad \text{and} \quad \mathbf{B} = \mathbf{b} \cos(\omega t - \mathbf{k} \cdot \mathbf{x} + \phi)$$

where $\phi$, $\mathbf{k}$ and $\mathbf{e}$ are freely chosen constants and $\omega$ and $\mathbf{b}$ are to be determined.[3]

**Exercise 10.5.2** Consider Maxwell's equations for a uniform conducting medium

$$\nabla \cdot \mathbf{D} = \rho, \quad \nabla \cdot \mathbf{B} = 0,$$

$$\nabla \times \mathbf{E} = -\frac{\partial \mathbf{B}}{\partial t}, \quad \nabla \times \mathbf{H} = \mathbf{j} + \frac{\partial \mathbf{D}}{\partial t},$$

$$\mathbf{D} = \epsilon \mathbf{E}, \quad \mathbf{B} = \mu \mathbf{H}, \quad \mathbf{j} = \sigma \mathbf{E}.$$

(Here $\epsilon$, $\mu$, $\sigma > 0$.) Show that the charge $\rho$ decays exponentially to zero at all points (more precisely $\rho(\mathbf{x}, t) = e^{-\kappa t} \rho(\mathbf{x}, 0)$ for some $\kappa > 0$ to be found).
[In order to go further, we would need to introduce the ideas of boundaries and boundary conditions.]

**Exercise 10.5.3** Lorentz showed that a charged particle of mass $m$ and charge $q$ moving in an electromagnetic field experiences a force

$$\mathbf{F} = q(\mathbf{E} + \dot{\mathbf{r}} \times \mathbf{B}),$$

where $\mathbf{r}$ is the position vector of the particle, $\mathbf{E}$ is the electric field and $\mathbf{B}$ is the magnetic field. In this exercise we shall assume that $\mathbf{E}$ and $\mathbf{B}$ are constant in space and time and that they are non-zero.

---

[3] 'The precise formulation of the time-space laws was the work of Maxwell. Imagine his feelings when the differential equations he had formulated proved to him that electromagnetic fields spread in the form of polarised waves and with the speed of light! To few men in the world has such an experience been vouchsafed.' (Einstein writing in the journal *Science* [16]. The article is on the web.)

Suppose that $\mathbf{E} \cdot \mathbf{B} = 0$. By taking an appropriate set of orthogonal coordinates, writing the equations of motion for these coordinates and solving the resulting differential equations (you will get one simple equation and a pair of simultaneous equations), show that the motion of the particle consists, in general, of circular movement in a plane perpendicular to $\mathbf{B}$ about a centre which moves with constant velocity $\mathbf{u} + \mathbf{v}$, where $\mathbf{u}$ is a multiple of $\mathbf{E} \times \mathbf{B}$ depending on the initial conditions and $\mathbf{v}$ is a multiple of $\mathbf{B}$ with $\|\mathbf{v}\| = \|\mathbf{E}\|/\|\mathbf{B}\|$. What is the period of the circular motion?

Someone tells you that the result above cannot be right because, as the magnetic field gets smaller, the speed of the centre of the circular motion gets larger. (If we keep $\mathbf{E}$ fixed, $\|\mathbf{E}\|/\|\mathbf{B}\| \to \infty$ as $\|\mathbf{B}\| \to 0$.) Answer your interlocutor's doubts.

If we remove the condition $\mathbf{E} \cdot \mathbf{B} = 0$, show that the motion is similar, but now the particle experiences a constant acceleration in the direction of $\mathbf{B}$.

**Exercise 10.5.4** (i) If $T_{ij}$ is an antisymmetric Cartesian tensor of order 2, show that we can find vectors $v_i$ and $u_i$ such that $T_{ij} = u_i v_j - u_j v_i$. Are the vectors $u_i$ and $v_i$ unique? Give reasons.

(ii) Show that it is not always possible, given $U_{ij}$, a symmetric Cartesian tensor of order 2, to find vectors $r_i$ and $s_i$ such that $U_{ij} = r_i s_j + r_j s_i$.

**Exercise 10.5.5** Let $v_i$ be a non-zero vector. By finding $a$, $b$, $c_i$ and $d_{ij}$ explicitly, show that any symmetric Cartesian tensor $t_{ij}$ of order 2 can be written uniquely as

$$t_{ij} = a\delta_{ij} + bv_i v_j + (c_i v_j + c_j v_i) + d_{ij},$$

where $a$ and $b$ are scalars, $c_i$ is a vector and $d_{ij}$ is a symmetric Cartesian tensor of order 2 such that

$$c_i v_i = 0, \quad d_{ii} = 0, \quad d_{ij} v_j = 0.$$

**Exercise 10.5.6** The relation between the stress $\sigma_{ij}$ and strain $e_{ij}$ for an isotropic medium is

$$\sigma_{ij} = \lambda e_{kk}\delta_{ij} + 2\mu e_{ij}.$$

Use the fact that $e_{ij}$ is symmetric to show that $\sigma_{ij}$ is symmetric. Show that the two tensors $\sigma_{ij}$ and $e_{ij}$ have the same principal axes.

Show that the stored elastic energy density $E = \frac{1}{2}\sigma_{ij}e_{ij}$ is non-negative for all $e_{ij}$ if and only if $\mu \geq 0$ and $\lambda \geq -\frac{2}{3}\mu$.

If $\lambda \neq -\frac{2}{3}\mu$, show that

$$e_{ij} = p\delta_{ij} + d_{ij},$$

where $p$ is a scalar and $d_{ij}$ is a traceless tensor (that is to say $d_{ii} = 0$) both to be determined explicitly in terms of $\sigma_{ij}$. Find $\sigma_{ij}$ in terms of $p$ and $d_{ij}$.

**Exercise 10.5.7** A homogeneous, but anisotropic, crystal has the conductivity tensor

$$\sigma_{ij} = \alpha\delta_{ij} + \gamma n_i n_j,$$

where $\alpha$ and $\gamma$ are real constants and **n** is a fixed unit vector. The electric current **J** is given by the equation

$$J_i = \sigma_{ij} E_j,$$

where **E** is the electric field.

(i) Show that there is a plane passing through the origin such that, if **E** lies in the plane, then $\mathbf{J} = \lambda \mathbf{E}$ for some $\lambda$ to be specified.

(ii) If $\alpha \neq 0$ and $\alpha \neq -\gamma$, show that

$$\mathbf{E} \neq \mathbf{0} \Rightarrow \mathbf{J} \neq \mathbf{0}.$$

(iii) If $D_{ij} = \epsilon_{ijk} n_k$, find the value of $\gamma$ which gives $\sigma_{ij} D_{jk} D_{km} = -\sigma_{im}$.

**Exercise 10.5.8** (In parts (i) to (iii) of this question you may use the results of Theorem 10.1.1, but should make no reference to formulae like ★ in Exercise 10.1.2.) Suppose that $T_{ijkm}$ is an isotropic Cartesian tensor of order 4.

(i) Show that $\epsilon_{ijk} T_{ijkm} = 0$.

(ii) Show that $\delta_{ij} T_{ijkm} = \alpha \delta_{km}$ for some scalar $\alpha$.

(iii) Show that $\epsilon_{iju} T_{ijkm} = \beta \epsilon_{kmu}$ for some scalar $\beta$.

Verify these results in the case $T_{ijkm} = \lambda \delta_{ij} \delta_{km} + \mu \delta_{ik} \delta_{jm} + \nu \delta_{im} \delta_{jk}$, finding $\alpha$ and $\beta$ in terms of $\lambda$, $\mu$ and $\nu$.

**Exercise 10.5.9** [**The most general isotropic Cartesian tensor of order 4**] In this question $A$, $B$, $C$, $D \in \{1, 2, 3, 4\}$ and we *do not* apply the summation convention to them.

Suppose that $a_{ijkm}$ is an isotropic Cartesian tensor.

(i) By considering the rotation associated with

$$L = \begin{pmatrix} 1 & 0 & 0 \\ 0 & 0 & 1 \\ 0 & -1 & 0 \end{pmatrix},$$

or otherwise, show that $a_{ABCD} = 0$ unless $A = B = C = D$ or two pairs of indices are equal (for example, $A = C$, $B = D$). Show also that $a_{1122} = a_{2211}$.

(ii) Suppose that

$$L = \begin{pmatrix} 0 & 1 & 0 \\ 0 & 0 & 1 \\ 1 & 0 & 0 \end{pmatrix}.$$

Show algebraically that $L \in SO(\mathbb{R}^3)$. Identify the associated rotation geometrically. By using the relation

$$a_{ijkm} = l_{ir} l_{js} l_{kt} l_{mu} a_{rstu}$$

and symmetry among the indices, show that

$$a_{AAAA} = \kappa, \quad a_{AABB} = \lambda, \quad a_{ABAB} = \mu, \quad a_{ABBA} = \nu$$

for all $A$, $B \in \{1, 2, 3, 4\}$ with $A \neq B$ and some real $\kappa$, $\lambda$, $\mu$, $\nu$.

(iii) Deduce that

$$a_{ijkm} = \lambda\delta_{ij}\delta_{km} + \mu\delta_{ik}\delta_{jm} + \nu\delta_{im}\delta_{jk} + (\kappa - \lambda - \mu - \nu)\upsilon_{ijkm},$$

where $\upsilon_{ABCD} = 1$ if $A = B = C = D$ and $\upsilon_{ABCD} = 0$ otherwise.

(iv) Explain why $(\kappa - \lambda - \mu - \nu)\upsilon_{ijkm}$ must be a tensor of order 4 and

$$(\kappa - \lambda - \mu - \nu)\upsilon_{ijkm}x_i x_j x_k x_m$$

must be a scalar. Show that

$$\upsilon_{ijkm}x_i x_j x_k x_m = x_1^4 + x_2^4 + x_3^4 + x_4^4$$

is not invariant under rotation of axes and so cannot be a scalar. Conclude that

$$\kappa - \lambda - \mu - \nu = 0$$

and

$$a_{ijkm} = \lambda\delta_{ij}\delta_{km} + \mu\delta_{ik}\delta_{jm} + \nu\delta_{im}\delta_{jk}.$$

**Exercise 10.5.10** Use the fact that the most general isotropic tensor of order 4 takes the form

$$a_{ijkm} = \lambda\delta_{ij}\delta_{km} + \mu\delta_{ik}\delta_{jm} + \nu\delta_{im}\delta_{jk}$$

to evaluate

$$b_{ijkm} = \int_{r\leq a} x_i x_j \frac{\partial^2}{\partial x_k \partial x_m}\left(\frac{1}{r}\right) dV,$$

where $\mathbf{x}$ is the position vector and $r = \|\mathbf{x}\|$.

**Exercise 10.5.11** (This exercise presents an alternative proof of the 'master identity' of Theorem 9.4.2 (iii).)

Show that, if a Cartesian tensor $T_{ijkrst}$ of order 6 is isotropic and has the symmetry properties

$$T_{ijkrst} = T_{kijrst} = -T_{jikrst},$$

then $T_{ijkrst}$ is a scalar multiple of $\epsilon_{ijk}\epsilon_{rst}$.

Deduce that

$$\epsilon_{ijk}\epsilon_{rst} = \delta_{ir}\delta_{js}\delta_{kt} + \delta_{it}\delta_{jr}\delta_{ks} + \delta_{is}\delta_{jt}\delta_{kr} - \delta_{ir}\delta_{ks}\delta_{jt} - \delta_{it}\delta_{kr}\delta_{js} - \delta_{is}\delta_{kt}\delta_{jr}.$$

**Exercise 10.5.12** Show that $\epsilon_{ijk}\delta_{lm}$ is an isotropic tensor. Show that $\epsilon_{ijk}\delta_{lm}$, $\epsilon_{jkl}\delta_{im}$ and $\epsilon_{kli}\delta_{jm}$ are linearly independent.

Now admire the equation

$$\epsilon_{ijk}\delta_{lm} - \epsilon_{jkl}\delta_{im} + \epsilon_{kli}\delta_{jm} - \epsilon_{lij}\delta_{km} = 0$$

and prove it by using ideas from the previous question, or otherwise.

[Clearly this is about as far as we can get without some new ideas. Such ideas do exist (the magic word is syzygy) but they are beyond the scope of this book and the competence of its author.]

**Exercise 10.5.13** Consider two particles of mass $m_\alpha$ at $\mathbf{x}_\alpha$ [$\alpha = 1, 2$] subject to forces $\mathbf{F}_1 = -\mathbf{F}_2 = \mathbf{f}(\mathbf{x}_1 - \mathbf{x}_2)$. Show directly that their centre of mass $\bar{\mathbf{x}}$ has constant velocity. If we write $\mathbf{s} = \mathbf{x}_1 - \mathbf{x}_2$, show that

$$\mathbf{x}_1 = \bar{\mathbf{x}} + \lambda_1 \mathbf{s} \quad \text{and} \quad \mathbf{x}_2 = \bar{\mathbf{x}} + \lambda_2 \mathbf{s},$$

where $\lambda_1$ and $\lambda_2$ are to be determined. Show that

$$\mu\ddot{\mathbf{s}} = \mathbf{f}(\mathbf{s}),$$

where $\mu$ is to be determined.

(i) Suppose that $\mathbf{f}(\mathbf{s}) = -k\mathbf{s}$. If the particles are initially at rest at distance $d$ apart, calculate how long it takes before they collide.

(ii) Suppose that $\mathbf{f}(\mathbf{s}) = -k\|\mathbf{s}\|^{-3}\mathbf{s}$. If the particles are initially at rest at distance $d$ apart, calculate how long it takes before they collide.
[Hint: Recall that $dv/dt = (dx/dt)(dv/dx)$.]

**Exercise 10.5.14** We consider the system of particles and forces described in the first paragraph of Section 10.2. Recall that we define the total momentum and angular momentum for a given origin $O$ by

$$\mathbf{P} = \sum_\alpha m_\alpha \dot{\mathbf{x}}_\alpha \quad \text{and} \quad \mathbf{L} = \sum_\alpha m_\alpha \mathbf{x}_\alpha \times \dot{\mathbf{x}}_\alpha.$$

If we choose a new origin $O'$ so that particle $\alpha$ is at $\mathbf{x}'_\alpha = \mathbf{x}_\alpha - \mathbf{b}$, show that the new total momentum and angular momentum satisfy

$$\mathbf{P}' = \mathbf{P}, \quad \mathbf{L}' = \mathbf{L} - \mathbf{b} \times \mathbf{P}.$$

Compute $\mathbf{L}' \cdot \mathbf{P}'$ and $\mathbf{L}' \times \mathbf{P}'$ in terms of $\mathbf{P}$ and $\mathbf{L}$.

Show that, if $\mathbf{P} \neq \mathbf{0}$, we can choose $\mathbf{b}$ in a manner to be specified so that $\mathbf{L}' = \lambda\mathbf{P}'$ for some $\lambda > 0$.

**Exercise 10.5.15** [Conservation of energy] We consider the system of particles and forces described in the first paragraph of Section 10.2. Suppose that there are no external forces, that is to say, $\mathbf{F}_\alpha = \mathbf{0}$ for all $\alpha$, and that the internal forces are derived from a potential $\phi_{\alpha,\beta} = \phi_{\beta,\alpha}$, that is to say,

$$F_{\alpha i} = -\sum_{\beta \neq \alpha} \frac{\partial \phi_{\alpha,\beta}}{\partial x_{\alpha i}}\left(\|\mathbf{x}_\alpha - \mathbf{x}_\beta\|\right).$$

Show that, if we write

$$U = \sum_{\alpha > \beta} \phi_{\alpha,\beta}\left(\|\mathbf{x}_\alpha - \mathbf{x}_\beta\|\right)$$

(so we might consider $U$ as the total potential energy of the system) and

$$T = \sum_\alpha \frac{1}{2} m_\alpha \|\dot{\mathbf{x}}_\alpha\|^2$$

(so $T$ corresponds to kinetic energy), then $\dot{T} + \dot{U} = 0$. Deduce that $T + U$ does not change with time.

**Exercise 10.5.16** Consider the system of the previous question, but now specify

$$\phi_{\alpha,\beta}(\mathbf{x}) = -Gm_\alpha m_\beta \|\mathbf{x}\|^{-1}$$

(so we are describing motion under gravity). If we take

$$J = \frac{1}{2} \sum_\alpha m_\alpha \|\mathbf{x}_\alpha\|^2,$$

show that

$$\frac{d^2 J}{dt^2} = 2T + U.$$

Hence show that, if the total energy $T + U > 0$, then the system must be unbounded both in the future and in the past.

**Exercise 10.5.17** Consider a uniform solid sphere of mass $M$ and radius $a$ with centre the origin. Explain why its inertia tensor $I_{ij}$ satisfies the equation $I_{ij} = K\delta_{ij}$ for some $K$ depending on $a$ and $M$ and, by computing $I_{ii}$, show that, in fact,

$$I_{ij} = \frac{3}{5} M a^2 \delta_{ij}.$$

(If you are interested, you can compute $I_{11}$ directly. This is not hard, but I hope you will agree that the method of the question is easier.)

If the centre of the sphere is moved to $(b, 0, 0)$, find the new inertia tensor (relative to the origin).

What are eigenvectors and eigenvalues of the new inertia tensor?

**Exercise 10.5.18** It is clear that an important question to ask about a symmetric tensor $a_{ij}$ of order 2 is whether all its eigenvalues are strictly positive.

(i) Consider the cubic

$$f(t) = t^3 - b_2 t^2 + b_1 t - b_0$$

with the $b_j$ real and strictly positive. Show that the real roots of $f$ must be strictly positive.

(ii) By looking at $g(t) = (t - \lambda_1)(t - \lambda_2)(t - \lambda_3)$, or otherwise, show that the real numbers $\lambda_1, \lambda_2, \lambda_3$ are strictly positive if and only if $\lambda_1 + \lambda_2 + \lambda_3 > 0$, $\lambda_1\lambda_2 + \lambda_2\lambda_3 + \lambda_3\lambda_1 > 0$ and $\lambda_1\lambda_2\lambda_3 > 0$.

(iii) Show that the eigenvalues of $a_{ij}$ are strictly positive if and only if

$$a_{rr} > 0,$$
$$a_{rr}a_{ss} - a_{rs}a_{rs} > 0,$$
$$a_{rr}a_{ss}a_{tt} - 3a_{rs}a_{rs}a_{tt} + 2a_{rs}a_{st}a_{tr} > 0.$$

**Exercise 10.5.19** A Cartesian tensor is said to have *cubic symmetry* if its components are unchanged by rotations through $\pi/2$ about each of the three coordinate axes. What is the most general tensor of order 2 having cubic symmetry? Give reasons.

Consider a cube of uniform density and side $2a$ with centre at the origin with sides aligned parallel to the coordinate axes. Find its inertia tensor.

If we now move a vertex to the origin, keeping the sides aligned parallel to the coordinate axes, find the new inertia tensor. You should use a coordinate system (to be specified) in which the associated array is a diagonal matrix.

**Exercise 10.5.20** A rigid *thin* plate $D$ has density $\rho(\mathbf{x})$ per unit area so that its inertia tensor is

$$M_{ij} = \int_D (x_k x_k \delta_{ij} - x_i x_j)\rho \, dS.$$

Show that one eigenvector is perpendicular to the plate and write down an integral expression for the corresponding eigenvalue $\lambda$.

If the other two eigenvalues are $\mu$ and $\nu$, show that $\lambda = \mu + \nu$.

Find $\lambda$, $\mu$ and $\nu$ for a circular disc of uniform density, of radius $a$ and mass $m$ having its centre at the origin.

**Exercise 10.5.21** Show that the total kinetic energy of a rigid body with inertia tensor $I_{ij}$ spinning about an axis through the origin with angular velocity $\omega_j$ is $\frac{1}{2}\omega_i I_{ij}\omega_j$. If $\|\boldsymbol{\omega}\|$ is fixed, how would you minimise the kinetic energy?

**Exercise 10.5.22** It may be helpful to look at Exercise 3.6.5 to put the next two exercises in context.

(i) Let $\Gamma$ be the set of all $2 \times 2$ matrices of the form $A = a_0 I + a_1 K$ with $a_r \in \mathbb{R}$ and

$$I = \begin{pmatrix} 1 & 0 \\ 0 & 1 \end{pmatrix}, \quad K = \begin{pmatrix} 0 & 1 \\ -1 & 0 \end{pmatrix}.$$

Show that $\Gamma$ is a vector space over $\mathbb{R}$ and find its dimension.

Show that $K^2 = -I$. Deduce that $\Gamma$ is closed under multiplication (that is to say, $A$, $B \in \Gamma \Rightarrow AB \in \Gamma$).

(ii) Let $\Omega$ be the set of all $2 \times 2$ matrices of the form $A = a_0 I + a_1 J + a_2 K + a_3 L$ with $a_r \in \mathbb{R}$ and

$$I = \begin{pmatrix} 1 & 0 \\ 0 & 1 \end{pmatrix}, \quad J = \begin{pmatrix} i & 0 \\ 0 & -i \end{pmatrix}, \quad K = \begin{pmatrix} 0 & 1 \\ -1 & 0 \end{pmatrix}, \quad L = \begin{pmatrix} 0 & i \\ i & 0 \end{pmatrix},$$

where, as usual, $i$ is a square root of $-1$.

Show that $\Omega$ is a vector space over $\mathbb{R}$ and find its dimension.

Show that $JKL = -I$ and prove that $\Omega$ is closed under multiplication (that is to say, $A, B \in \Omega \Rightarrow AB \in \Omega$).

**Exercise 10.5.23** For many years, Hamilton tried to find a generalisation of the complex numbers $a_0 + a_1 i$. Whilst walking along the Royal Canal near Maynooth he had a flash of inspiration and carved

$$ijk = -1$$

in the stone of the nearest bridge.[4]

His idea was to consider the system $\mathcal{Q}$ of 'numbers' of the form

$$a = a_0 + a_1 i + a_2 j + a_3 k$$

manipulated following the 'standard rules of algebra'[5] and the rule $i^2 = j^2 = k^2 = ijk = -1$. Explain why Exercise 10.5.22 (ii) shows that such a system exists. (Hamilton used a geometric argument.)

(i) Use the rules just given to show that

$$ij = -ji = k, \quad jk = -kj = i, \quad ki = -ik = j.$$

(ii) Let us write

$$(a_0 + a_1 i + a_2 j + a_3 k)^* = a_0 - a_1 i - a_2 j - a_3 k$$

and

$$\|a_0 + a_1 i + a_2 j + a_3 k\| = (a_0^2 + a_1^2 + a_2^2 + a_3^2)^{1/2}.$$

Show that

$$\mathbf{aa}^* = \|\mathbf{a}\|^2$$

and deduce that, if $\mathbf{a} \neq 0$, there exists a $\mathbf{b}$ such that

$$\mathbf{ab} = \mathbf{ba} = 1.$$

(iii) Show, in as much detail as you consider appropriate, that $(\mathcal{Q}, +, \times)$ satisfies the same laws of arithmetic as $\mathbb{R}$ except that multiplication does not commute.[6]

(iv) Hamilton called the elements of $\mathcal{Q}$ *quaternions*. Show that, if $\mathbf{a}$ and $\mathbf{b}$ are quaternions, then

$$\|\mathbf{ab}\| = \|\mathbf{a}\| \|\mathbf{b}\|.$$

---

[4] The equation has disappeared, but the Maynooth Mathematics Department celebrates the discovery with an annual picnic at the site. The farmer on whose land they picnic views the occasion with bemused benevolence.

[5] If the reader objects to this formulation, she is free to translate back into the language of Exercise 10.5.22 (ii).

[6] That is to say, $\mathcal{Q}$ satisfies all the conditions of Definition 13.2.1 (supplemented by $1 \times a = a$ in (vii) and $a^{-1} \times a = 1$ in (viii)) except for (v) which fails in certain cases.

Deduce the following result of Euler concerning *integers*. If $A = a_0^2 + a_1^2 + a_2^2 + a_3^2$ and $B = b_0^2 + b_1^2 + b_2^2 + b_3^2$ with $a_j$, $b_j \in \mathbb{Z}$, then there exist $c_j \in \mathbb{Z}$ such that

$$AB = c_0^2 + c_1^2 + c_2^2 + c_3^2.$$

In other words, the product of two sums of four squares is itself the sum of four squares.

(v) Let $\mathbf{a} = (a_1, a_2, a_3) \in \mathbb{R}^3$ and $\mathbf{b} = (b_1, b_2, b_3) \in \mathbb{R}^3$. If $\mathbf{c} = \mathbf{a} \times \mathbf{b}$, the *vector product* of $\mathbf{a}$ and $\mathbf{b}$, and we write $\mathbf{c} = (c_1, c_2, c_3)$, show that

$$(a_1 i + a_2 j + a_3 k)(b_1 i + b_2 j + b_3 k) = -\mathbf{a} \cdot \mathbf{b} + c_1 i + c_2 j + c_3 k.$$

The use of vectors in physics began when (to cries of anguish from holders of the pure quaternionic faith) Maxwell, Gibbs and Heaviside extracted the *vector part* $a_1 i + a_2 j + a_3 k$ from the quaternion $a_0 + a_1 i + a_2 j + a_3 k$ and observed that the vector part had an associated inner product and vector product.[7]

---

[7] Quaternions still have their particular uses. A friend of mine made a (small) fortune by applying them to lighting effects in computer games.

# Part II

## General vector spaces

# 11

# Spaces of linear maps

## 11.1 A look at $\mathcal{L}(U, V)$

In the first part of this book we looked at $n$-dimensional vector spaces as generalisations of two and three dimensional spaces. In this part we look at $n$-dimensional vector spaces with an eye to generalisation to infinite dimensional spaces.

We looked mainly at linear maps of a vector space into itself (the so-called *endomorphisms*). For various reasons, some of which we consider in the next section, these are often the most interesting, but it is worth looking at the more general case.

Let us start with the case when $U$ and $V$ are finite dimensional. We begin by generalising Definition 6.1.1.

**Definition 11.1.1** *Let $U$ and $V$ be vector spaces over $\mathbb{F}$ with bases $\mathbf{e}_1, \mathbf{e}_2, \ldots, \mathbf{e}_n$ and $\mathbf{f}_1, \mathbf{f}_2, \ldots, \mathbf{f}_m$. If $\alpha : U \to V$ is linear, we say that $\alpha$ has matrix $A = (a_{ij})_{1 \leq i \leq m}^{1 \leq j \leq n}$ with respect to the given bases if*

$$\alpha(\mathbf{e}_j) = \sum_{i=1}^{m} a_{ij} \mathbf{f}_i.$$

The next few exercises are essentially revision.

**Exercise 11.1.2** *Let $U$ and $V$ be vector spaces over $\mathbb{F}$ with bases $\mathbf{e}_1, \mathbf{e}_2, \ldots, \mathbf{e}_n$ and $\mathbf{f}_1, \mathbf{f}_2, \ldots, \mathbf{f}_m$. Show that, if $\alpha, \beta : U \to V$ are linear and have matrices $A$ and $B$ with respect to the given bases, then $\alpha + \beta$ has matrix $A + B$ and, if $\lambda \in \mathbb{F}$, $\lambda\alpha$ has matrix $\lambda A$.*

**Exercise 11.1.3** *Let $U$, $V$ and $W$ be vector spaces over $\mathbb{F}$ with bases $\mathbf{e}_1, \mathbf{e}_2, \ldots, \mathbf{e}_n$, $\mathbf{f}_1, \mathbf{f}_2, \ldots, \mathbf{f}_m$ and $\mathbf{g}_1, \mathbf{g}_2, \ldots, \mathbf{g}_p$. If $\alpha : V \to W$ and $\beta : U \to V$ are linear and have matrices $A$ and $B$ with respect to the appropriate bases, show that $\alpha\beta$ has matrix $AB$ with respect to the appropriate bases.*

**Exercise 11.1.4** *(i) Let $U$, $V$ and $W$ be (not necessarily finite dimensional) vector spaces over $\mathbb{F}$ and let $\alpha, \beta \in \mathcal{L}(U, V)$, $\gamma \in \mathcal{L}(W, V)$. Show that*

$$(\alpha + \beta)\gamma = \alpha\gamma + \beta\gamma.$$

*(ii) Use (i) to show that, if A and B are m × n matrices over $\mathbb{F}$ and C is an n × p matrix over $\mathbb{F}$, we have*

$$(A + B)C = AC + BC.$$

*(iii) Show that (i) follows from (ii) if U, V and W are finite dimensional.*

*(iv) State the results corresponding to (i) and (ii) when we replace the equation in (i) by*

$$\gamma(\alpha + \beta) = \gamma\alpha + \gamma\beta.$$

*(Be careful to make $\gamma$ a linear map between correct vector spaces.)*

We need a more general change of basis formula.

**Exercise 11.1.5** *Let U and V be finite dimensional vector spaces over $\mathbb{F}$ and suppose that $\alpha \in \mathcal{L}(U, V)$ has matrix A with respect to bases $\mathbf{e}_1, \mathbf{e}_2, \ldots, \mathbf{e}_n$ for U and $\mathbf{f}_1, \mathbf{f}_2, \ldots, \mathbf{f}_m$ for V. If $\alpha$ has matrix B with respect to bases $\mathbf{e}'_1, \mathbf{e}'_2, \ldots, \mathbf{e}'_n$ for U and $\mathbf{f}'_1, \mathbf{f}'_2, \ldots, \mathbf{f}'_m$ for V, then*

$$B = P^{-1}AQ,$$

*where P is an m × m invertible matrix and Q is an n × n invertible matrix.*

*If we write $P = (p_{ij})$ and $Q = (q_{rs})$, then*

$$\mathbf{f}'_j = \sum_{i=1}^{m} p_{ij}\mathbf{f}_i \qquad [1 \le j \le m]$$

*and*

$$\mathbf{e}'_s = \sum_{r=1}^{n} q_{rs}\mathbf{e}_r \qquad [1 \le s \le n].$$

**Lemma 11.1.6** *Let U and V be finite dimensional vector spaces over $\mathbb{F}$ and suppose that $\alpha \in \mathcal{L}(U, V)$. Then we can find bases $\mathbf{e}_1, \mathbf{e}_2, \ldots, \mathbf{e}_n$ for U and $\mathbf{f}_1, \mathbf{f}_2, \ldots, \mathbf{f}_m$ for V with respect to which $\alpha$ has matrix $C = (c_{ij})$ such that $c_{ii} = 1$ for $1 \le i \le r$ and $c_{ij} = 0$ otherwise (for some $0 \le r \le \min\{n, m\}$).*

*Proof* By Lemma 5.5.7, we can find bases $\mathbf{e}_1, \mathbf{e}_2, \ldots, \mathbf{e}_n$ for U and $\mathbf{f}_1, \mathbf{f}_2, \ldots, \mathbf{f}_m$ for V such that

$$\alpha(\mathbf{e}_j) = \begin{cases} \mathbf{f}_j & \text{for } 1 \le j \le r \\ \mathbf{0} & \text{otherwise.} \end{cases}$$

If we use these bases, then the matrix has the required form. $\square$

The change of basis formula has the following corollary.

**Lemma 11.1.7** *If A is an m × n matrix we can find an m × m invertible matrix P and an n × n invertible matrix Q such that*

$$P^{-1}AQ = C,$$

where $C = (c_{ij})$, such that $c_{ii} = 1$ for $1 \le i \le r$ and $c_{ij} = 0$ otherwise for some $0 \le r \le$ min$(m, n)$.

*Proof* Left to the reader. □

A particularly interesting case occurs when we consider a linear map $\alpha : U \to U$, but give $U$ two *different* bases.

**Lemma 11.1.8** (*i*) *Let $U$ be a finite dimensional vector space over $\mathbb{F}$ and suppose that $\alpha \in \mathcal{L}(U, U)$. Then we can find bases $\mathbf{e}_1, \mathbf{e}_2, \ldots, \mathbf{e}_n$ and $\mathbf{f}_1, \mathbf{f}_2, \ldots, \mathbf{f}_n$ for $U$ with respect to which $\alpha$ has matrix $C = (c_{ij})$ with $c_{ii} = 1$ for $1 \le i \le r$ and $c_{ij} = 0$ otherwise for some $0 \le r \le n$.*

(*ii*) *If $A$ is an $n \times n$ matrix we can find $n \times n$ invertible matrices $P$ and $Q$ such that*

$$P^{-1}AQ = C$$

*where $C = (c_{ij})$ is such that $c_{ii} = 1$ for $1 \le i \le r$ and $c_{ij} = 0$ otherwise for some $0 \le r \le n$.*

*Proof* Immediate. □

The reader should compare this result with Theorem 1.3.2 and may like to look at Exercise 3.6.8.

**Exercise 11.1.9** *The object of this exercise is to look at one of the main themes of this book in a unified manner. The reader will need to recall the notions of equivalence relation and equivalence class (see Exercise 6.8.34) and the notion of a subgroup (see Definition 5.3.15). We work with matrices over $\mathbb{F}$.*

(*i*) *Let $G$ be a subgroup of $GL(\mathbb{F}^n)$, the group of $n \times n$ invertible matrices, and let $X$ be a non-empty collection of $n \times m$ matrices such that*

$$P \in G, \ A \in X \Rightarrow PA \in X.$$

*If we write $A \sim_1 B$ whenever $A, B \in X$ and there exists a $P \in G$ with $B = PA$, show that $\sim_1$ is an equivalence relation on $X$.*

(*ii*) *Let $G$ be a subgroup of $GL(\mathbb{F}^n)$, $H$ a subgroup of $GL(\mathbb{F}^m)$ and let $X$ be a non-empty collection of $n \times m$ matrices such that*

$$P \in G, \ Q \in H, \ A \in X \Rightarrow P^{-1}AQ \in X.$$

*If we write $A \sim_2 B$ whenever $A, B \in X$ and there exist $P \in G, \ Q \in H$ with $B = P^{-1}AQ$, show that $\sim_2$ is an equivalence relation on $X$.*

(*iii*) *Suppose that $n = m$, $X$ is the set of $n \times n$ matrices and $H = G = GL(\mathbb{F}^n)$. Show that there are precisely $n + 1$ equivalence classes for $\sim_2$.*

(*iv*) *Let $G$ be a subgroup of $GL(\mathbb{F}^n)$ and let $X$ be a non-empty collection of $n \times n$ matrices such that*

$$P \in G, \ A \in X \Rightarrow P^{-1}AP \in X.$$

*If we write $A \sim_3 B$ whenever $A, B \in X$ and there exists $P \in G$ with $B = PAP^{-1}$, show that $\sim_3$ is an equivalence relation on $X$.*

(v) *Suppose $\mathbb{F} = \mathbb{R}$, $X$ is the collection of real symmetric $n \times n$ matrices and $G = O(\mathbb{F}^n)$, the group of orthogonal $n \times n$ matrices. Show that there are infinitely many equivalence classes for $\sim_3$ and give a criterion in terms of characteristic equations for two members of $X$ to be equivalent.*

(vi) *Let $G$ be a subgroup of $GL(\mathbb{R}^n)$ and let $X$ be a non-empty collection of $n \times n$ matrices such that*

$$P \in G, \ A \in X \Rightarrow P^T AP \in X.$$

*If we write $A \sim_4 B$ whenever $A, B \in X$ and there exists $P \in G$ with $B = P^T AP$, show that $\sim_4$ is an equivalence relation on $X$.*

(vii) *Suppose that $X$ is the collection of real symmetric $n \times n$ matrices and $G = GL(\mathbb{R}^n)$. Show that there are only finitely many equivalence classes for $\sim_4$. (We shall identify them precisely in Section 16.2.)*

*If we think along the lines of this exercise, the various 'diagonalisation theorems' and 'Jordan normal form theorems' (see, for example, our earlier Theorem 6.4.3 and the later Section 12.4) in this book may be thought of as identifying typical elements of equivalence classes for different equivalence relations.*

Algebraists are very fond of quotienting. If the reader has met the notion of the quotient of a group (or of a topological space or any similar object), she will find the rest of the section rather easy. If she has not, then she may find the discussion rather strange.[1] She should take comfort from the fact that, although a few of the exercises will involve quotients of vector spaces, I will not make use of the concept outside them.

**Definition 11.1.10** *Let $V$ be a vector space over $\mathbb{F}$ with a subspace $W$. We write*

$$[\mathbf{x}] = \{\mathbf{v} \in V : \mathbf{x} - \mathbf{v} \in W\}.$$

*We denote the set of such $[\mathbf{x}]$ with $\mathbf{x} \in V$ by $V/W$.*

**Lemma 11.1.11** *With the notation of Definition 11.1.10,*

$$[\mathbf{x}] = [\mathbf{y}] \Leftrightarrow \mathbf{x} - \mathbf{y} \in W.$$

*Proof* The reader who is happy with the notion of equivalence class will be able to construct her own proof. If not, we give the proof that

$$[\mathbf{x}] = [\mathbf{y}] \Rightarrow \mathbf{x} - \mathbf{y} \in W.$$

If $[\mathbf{x}] = [\mathbf{y}]$, then, since $\mathbf{0} \in W$, we have $\mathbf{x} \in [\mathbf{x}] = [\mathbf{y}]$ and so

$$\mathbf{y} - \mathbf{x} \in W.$$

---

[1] I would not choose quotients of vector spaces as a first exposure to quotients.

Thus, using the fact that $W$ is a subspace,

$$\mathbf{z} \in [\mathbf{y}] \Rightarrow \mathbf{y} - \mathbf{z} \in W \Rightarrow \mathbf{x} - \mathbf{z} = (\mathbf{y} - \mathbf{z}) - (\mathbf{y} - \mathbf{x}) \in W \Rightarrow \mathbf{z} \in [\mathbf{x}].$$

Thus $[\mathbf{y}] \subseteq [\mathbf{x}]$. Similarly $[\mathbf{x}] \subseteq [\mathbf{y}]$ and so $[\mathbf{x}] = [\mathbf{y}]$.

The proof that

$$\mathbf{x} - \mathbf{y} \in W \Rightarrow [\mathbf{x}] = [\mathbf{y}]$$

is left as a recommended exercise for the reader. $\qquad\square$

**Theorem 11.1.12** *Let $V$ be a vector space over $\mathbb{F}$ with a subspace $W$. Then $V/W$ can be made into a vector space by adopting the definitions*

$$[\mathbf{x}] + [\mathbf{y}] = [\mathbf{x} + \mathbf{y}] \quad and \quad \lambda[\mathbf{x}] = [\lambda\mathbf{x}].$$

*Proof* The key point to check is that the putative definitions do indeed make sense. Observe that

$$\begin{aligned}
[\mathbf{x}'] = [\mathbf{x}], \ [\mathbf{y}'] = [\mathbf{x}] &\Rightarrow \mathbf{x}' - \mathbf{x}, \ \mathbf{y}' - \mathbf{y} \in W \\
&\Rightarrow (\mathbf{x}' + \mathbf{y}') - (\mathbf{x} + \mathbf{y}) = (\mathbf{x}' - \mathbf{x}) + (\mathbf{y}' - \mathbf{y}) \in W \\
&\Rightarrow [\mathbf{x}' + \mathbf{y}'] = [\mathbf{x} + \mathbf{y}],
\end{aligned}$$

so that our definition of addition is unambiguous. The proof that our definition of scalar multiplication is unambiguous is left as a recommended exercise for the reader.

It is easy to check that the axioms for a vector space hold. The following verifications are typical:

$$[\mathbf{x}] + [\mathbf{0}] = [\mathbf{x} + \mathbf{0}] = [\mathbf{x}]$$

$$(\lambda + \mu)[\mathbf{x}] = [(\lambda + \mu)\mathbf{x}] = [\lambda\mathbf{x} + \mu\mathbf{x}] = \lambda[\mathbf{x}] + \mu[\mathbf{x}].$$

We leave it to the reader to check as many further axioms as she pleases. $\qquad\square$

If the reader has met quotients elsewhere in algebra, she will expect an 'isomorphism theorem' and, indeed, there is such a theorem following the standard pattern. To bring out the analogy we recall the following synonyms.[2] If $\alpha \in \mathcal{L}(U, V)$, we write

$$\operatorname{im}\alpha = \alpha(U) = \{\alpha\mathbf{u} : \mathbf{u} \in U\},$$

$$\ker\alpha = \alpha^{-1}(\mathbf{0}) = \{\mathbf{u} \in U : \alpha\mathbf{u} = \mathbf{0}\}.$$

Generalising an earlier definition, we say that $\operatorname{im}\alpha$ is the *image* or *image space* of $\alpha$ and that $\ker\alpha$ is the *kernel* or *null-space* of $\alpha$.

We also adopt the standard practice of writing

$$[\mathbf{x}] = \mathbf{x} + W \quad and \quad [\mathbf{0}] = \mathbf{0} + W = W.$$

---

[2] If the reader objects to my practice, here and elsewhere, of using more than one name for the same thing, she should reflect that all the names I use are standard and she will need to recognise them when they are used by other authors.

This is a very suggestive notation, but the reader is warned that

$$0(\mathbf{x} + W) = 0[\mathbf{x}] = [0\mathbf{x}] = [\mathbf{0}] = W.$$

Should she become confused by the new notation, she should revert to the old.

**Theorem 11.1.13** [**Vector space isomorphism theorem**] *Suppose that U and V are vector spaces over $\mathbb{F}$ and $\alpha : U \to V$ is linear. Then $U/\ker \alpha$ is isomorphic to $\operatorname{im} \alpha$.*

*Proof* We know that $\ker \alpha$ is a subspace of $U$. Further,

$$\mathbf{u}_1 + \ker \alpha = \mathbf{u}_2 + \ker \alpha \Rightarrow \mathbf{u}_1 - \mathbf{u}_2 \in \ker \alpha \Rightarrow \alpha(\mathbf{u}_1 - \mathbf{u}_2) = \mathbf{0} \Rightarrow \alpha(\mathbf{u}_1) = \alpha(\mathbf{u}_2).$$

Thus we may define $\tilde{\alpha} : U/\ker \alpha \to \operatorname{im} \alpha$ unambiguously by

$$\tilde{\alpha}(\mathbf{u} + \ker \alpha) = \alpha(\mathbf{u}).$$

We observe that

$$\tilde{\alpha}\big(\lambda_1(\mathbf{u}_1 + \ker \alpha) + \lambda_2(\mathbf{u}_2 + \ker \alpha)\big) = \tilde{\alpha}\big((\lambda_1\mathbf{u}_1 + \lambda_2\mathbf{u}_2) + \ker \alpha\big)$$
$$= \alpha(\lambda_1\mathbf{u}_1 + \lambda_2\mathbf{u}_2) = \lambda_1\alpha(\mathbf{u}_1) + \lambda_2\alpha(\mathbf{u}_2)$$
$$= \lambda_1\tilde{\alpha}(\mathbf{u}_1 + \ker \alpha) + \lambda_2\tilde{\alpha}(\mathbf{u}_2 + \ker \alpha)$$

and so $\tilde{\alpha}$ is linear.

Since our spaces may be infinite dimensional, we must verify both that $\tilde{\alpha}$ is surjective and that it is injective. Both verifications are easy. Since

$$\alpha(\mathbf{u}) = \tilde{\alpha}(\mathbf{u} + \ker \alpha),$$

it follows that $\tilde{\alpha}$ is surjective. Since $\tilde{\alpha}$ is linear and

$$\tilde{\alpha}(\mathbf{u} + \ker \alpha) = \mathbf{0} \Rightarrow \alpha\mathbf{u} = \mathbf{0} \Rightarrow \mathbf{u} \in \ker \alpha \Rightarrow \mathbf{u} + \ker \alpha = \mathbf{0} + \ker \alpha,$$

it follows that $\tilde{\alpha}$ is injective. Thus $\tilde{\alpha}$ is an isomorphism and we are done. $\qquad \square$

The dimension of a quotient space behaves as we would wish.

**Lemma 11.1.14** *If V is a finite dimensional space with a subspace W, then*

$$\dim V = \dim W + \dim V/W.$$

*Proof* Observe first that, if $\mathbf{u}_1, \mathbf{u}_2, \ldots, \mathbf{u}_m$ span $V$, then $\mathbf{u}_1 + W, \mathbf{u}_2 + W, \ldots, \mathbf{u}_m + W$ span $V/W$. Thus $V/W$ is finite dimensional.

Let $\mathbf{e}_1, \mathbf{e}_2, \ldots, \mathbf{e}_k$ form a basis for $W$ and $\mathbf{e}_{k+1} + W, \mathbf{e}_{k+2} + W, \ldots, \mathbf{e}_n + W$ form a basis for $V/W$. We claim that $\mathbf{e}_1, \mathbf{e}_2, \ldots, \mathbf{e}_n$ form a basis for $V$.

We start by showing that we have a spanning set. If $\mathbf{v} \in V$, then, since $\mathbf{e}_{k+1} + W$, $\mathbf{e}_{k+2} + W, \ldots, \mathbf{e}_n + W$ span $V/W$, we can find $\lambda_{k+1}, \lambda_{k+2}, \ldots, \lambda_n \in \mathbb{F}$ such that

$$\mathbf{v} + W = \sum_{j=k+1}^n \lambda_j(\mathbf{e}_j + W) = \left( \sum_{j=k+1}^n \lambda_j\mathbf{e}_j \right) + W.$$

We now have

$$\mathbf{v} - \sum_{j=k+1}^{n} \lambda_j \mathbf{e}_j \in W,$$

so we can find $\lambda_1, \lambda_2, \ldots, \lambda_k \in \mathbb{F}$ such that

$$\mathbf{v} - \sum_{j=k+1}^{n} \lambda_j \mathbf{e}_j = \sum_{j=1}^{k} \lambda_j \mathbf{e}_j$$

and so

$$\mathbf{v} = \sum_{j=1}^{n} \lambda_j \mathbf{e}_j.$$

Thus we have a spanning set.

Next we want to show linear independence. To this end, suppose that $\lambda_1, \lambda_2, \ldots, \lambda_n \in \mathbb{F}$ and

$$\sum_{j=1}^{n} \lambda_j \mathbf{e}_j = \mathbf{0}.$$

Since $\sum_{j=1}^{k} \lambda_j \mathbf{e}_j \in W$,

$$\sum_{j=k+1}^{n} \lambda_j (\mathbf{e}_j + W) = \left( \sum_{j=1}^{n} \lambda_j \mathbf{e}_j \right) + W = \mathbf{0} + W,$$

so, since $\mathbf{e}_{k+1} + W, \mathbf{e}_{k+2} + W, \ldots, \mathbf{e}_n + W$, are linearly independent, $\lambda_j = 0$ for $k + 1 \leq j \leq n$. We now have

$$\sum_{j=1}^{k} \lambda_j \mathbf{e}_j = \mathbf{0}$$

and so $\lambda_j = 0$ for $1 \leq j \leq k$. Thus $\lambda_j = 0$ for all $1 \leq j \leq n$ and we are done.

We now know that

$$\dim W + \dim V/W = k + (n - k) = n = \dim V$$

as required. $\qquad\square$

**Exercise 11.1.15** *Use Theorem 11.1.13 and Lemma 11.1.14 to obtain another proof of the rank-nullity theorem (Theorem 5.5.4).*

The reader should not be surprised by the many different contexts in which we meet the rank-nullity theorem. If you walk round the countryside, you will see many views of the highest hills. Our original proof and Exercise 11.1.15 show two different aspects of the same theorem.

**Exercise 11.1.16** *Suppose that we have a sequence of finite dimensional vector spaces $C_j$ with $C_{n+1} = C_0 = \{0\}$ and linear maps $\alpha_j : C_j \to C_{j-1}$ as shown in the next line*

$$C_{n+1} \overset{\alpha_{n+1}}{\to} C_n \overset{\alpha_n}{\to} C_{n-1} \overset{\alpha_{n-1}}{\to} C_{n-2} \overset{\alpha_{n-2}}{\to} \ldots \overset{\alpha_2}{\to} C_1 \overset{\alpha_1}{\to} C_0$$

*such that $\alpha_{j-1}\alpha_j = 0 \ [n \geq j \geq 2]$. Let $Z_j = \alpha_j^{-1}(0)$, $B_{j-1} = \alpha_j(C_j)$, and take $H_j = B_j/Z_j \ [n \geq j \geq 1]$. Show that*

$$\sum_{j=1}^{n}(-1)^j \dim C_j = -\sum_{j=1}^{n}(-1)^j \dim H_j.$$

## 11.2 A look at $\mathcal{L}(U, U)$

In the next few chapters we study the special spaces $\mathcal{L}(U, U)$ and $\mathcal{L}(U, \mathbb{F})$. (Recall that elements of $\mathcal{L}(U, U)$ are often called *endomorphisms*. Invertible endomorphisms are called *automorphisms*. Although we use the notation $\mathcal{L}(U, U)$, many authors use the notation $\mathcal{E}(U)$ for the vector space of endomorphisms.) The reader may ask why we do not simply study the more general space $\mathcal{L}(U, V)$ where $U$ and $V$ are any vector spaces and then specialise by setting $V = U$ or $V = \mathbb{F}$.

The special treatment of $\mathcal{L}(U, U)$ is easy to justify. If $\alpha$, $\beta \in \mathcal{L}(U, U)$, then $\alpha\beta \in \mathcal{L}(U, U)$. We get an algebraic structure which is much more intricate than that of $\mathcal{L}(U, V)$ in general. The next exercise is included for the reader's amusement.

**Exercise 11.2.1** *Explain why $\mathcal{P}$, the collection of real polynomials, can be made into a vector space over $\mathbb{R}$ by using the standard pointwise addition and scalar multiplication.*

*(i) Let $h \in \mathbb{R}$. Check that the following maps belong to $\mathcal{L}(\mathcal{P}, \mathcal{P})$:*

$$D \text{ given by } (DP)(t) = P'(t),$$

$$M \text{ given by } (MP)(t) = tP(t),$$

$$E_h \text{ given by } E_h P = P(t + h).$$

*(ii) Identify the map $DM - MD$.*

*(iii) Suppose that $\alpha_0, \alpha_1, \alpha_2, \ldots$ are elements of $\mathcal{L}(\mathcal{P}, \mathcal{P})$ with the property that, for each $P \in \mathcal{P}$, we can find an $N(P)$ such that $\alpha_j(P) = 0$ for all $j > N(P)$. Show that, if we set*

$$\sum_{j=0}^{\infty} \alpha_j P = \sum_{j=0}^{N(P)} \alpha_j P,$$

*then $\sum_{j=0}^{\infty} \alpha_j$ is a well defined element of $\mathcal{L}(\mathcal{P}, \mathcal{P})$.*

*(iv) Show that the sequence $\alpha_j = D^j$ has the property stated in the first sentence of (iii). Does the sequence $\alpha_j = M^j$? Does the sequence $\alpha_j = E^j$? Does the sequence $\alpha_j = E_h - \iota$? Give reasons.*

(*v*) *Suppose that* $\alpha \in \mathcal{L}(\mathcal{P}, \mathcal{P})$ *has the property that, for each* $P \in \mathcal{P}$, *we can find an* $N(P)$ *such that* $\alpha^j(P) = 0$ *for all* $j > N(P)$. *Show that we can define*

$$\exp \alpha = \sum_{j=0}^{\infty} \frac{1}{j!} \alpha^j$$

*and*

$$\log(\iota - \alpha) = \sum_{j=1}^{\infty} \frac{1}{j} \alpha^j.$$

*Explain why we can define* $\log(\exp \alpha)$. *By considering coefficients in standard power series, or otherwise, show that* $\log(\exp \alpha) = \alpha$.

  (*vi*) *Show that*

$$\exp hD = E_h.$$

*Deduce from* (*v*) *that, writing* $\Delta_h = E_h - \iota$, *we have*

$$hD = \sum_{j=1}^{\infty} \frac{(-1)^{j+1}}{j} \Delta_h^j.$$

  (*vii*) *Let* $\lambda \in \mathbb{R}$. *Show that*

$$(\iota - \lambda D) \sum_{j=0}^{\infty} \lambda^r D^r P = P$$

*for each* $P \in \mathcal{P}$ *and deduce that*

$$(\iota - \lambda D) \sum_{j=0}^{\infty} \lambda^r D^r = \iota = \sum_{j=0}^{\infty} \lambda^r D^r (\iota - \lambda D).$$

  *Solve the equation*

$$(\iota - \lambda D)P = Q$$

*with* $P$ *and* $Q$ *polynomials.*

  *Find a solution to the ordinary differential equation*

$$f'(x) - f(x) = x^2.$$

*We are taught that the solution of such an equation must have an arbitrary constant. The method given here does not produce one. What has happened to it?*[3]
*[The ideas set out in this exercise go back to Boole in his books* A Treatise on Differential Equations [5] *and* A Treatise on the Calculus of Finite Differences [6].*]*

---

[3]   Of course, a really clear thinking mathematician would not be a puzzled for an instant. If you are not puzzled for an instant, try to imagine why someone else might be puzzled and how you would explain things to them.

As we remarked earlier, mathematicians are very interested in iteration and, if $\alpha \in \mathcal{L}(U, U)$, we can iterate the action of $\alpha$ on $U$ (in other words, apply powers of $\alpha$). The following result is interesting in itself and provides the opportunity to revisit the rank-nullity theorem (Theorem 5.5.4). The reader should make sure that she can state and prove the rank-nullity theorem before proceeding.

**Theorem 11.2.2** *Suppose that $U$ is a finite vector space of dimension $n$ over $\mathbb{F}$. Then there exists an $m \leq n$ such that*

$$\text{rank } \alpha^k = \text{rank } \alpha^m$$

*for $k \geq m$ and $\text{rank } \alpha^k > \text{rank } \alpha^m$ for $0 \leq k < m$. If $m > 0$ we have*

$$n > \text{rank } \alpha > \text{rank } \alpha^2 > \ldots > \text{rank } \alpha^m.$$

*Further*

$$n - \text{rank } \alpha \geq \text{rank } \alpha - \text{rank } \alpha^2 \geq \ldots \geq \text{rank } \alpha^{m-1} - \text{rank } \alpha^m.$$

*Proof* Automatically $U \supseteq \alpha U$, so applying $\alpha^j$ to both sides of the inclusion we get $\alpha^j U \supseteq \alpha^{j+1} U$ for all $j \geq 0$. (This result will be used repeatedly in our proof.) By the observation in our first sentence,

$$\text{rank } \alpha^j = \dim \alpha^j U \geq \dim \alpha^{j+1} U = \text{rank } \alpha^{j+1}$$

and

$$n \geq \text{rank } \alpha \geq \text{rank } \alpha^2 \geq \ldots.$$

A strictly decreasing sequence of positive integers whose first element is $n$ cannot contain more than $n + 1$ terms, so there must exist a $0 \leq k \leq n$ with $\text{rank } \alpha^k = \text{rank } \alpha^{k+1}$. Let $m$ be the least such $k$.

Since $\alpha^m U \supseteq \alpha^{m+1} U$ and

$$\dim \alpha^m U = \text{rank } \alpha^m = \text{rank } \alpha^{m+1} = \dim \alpha^{m+1} U,$$

we have $\alpha^m U = \alpha^{m+1} U$. Applying $\alpha^j$ to both sides of the equality, we get $\alpha^{m+j} U = \alpha^{m+j+1} U$ for all $j \geq 0$. Thus $\text{rank } \alpha^k = \text{rank } \alpha^m$ for all $k \geq m$.

Applying the rank-nullity theorem to the restriction $\alpha|_{\alpha^j U} : \alpha^j U \to \alpha^j U$ of $\alpha$ to $\alpha^j U$, we get

$$\text{rank } \alpha^j = \dim \alpha^j U = \text{rank } \alpha|_{\alpha^j U} + \text{nullity } \alpha|_{\alpha^j U}$$
$$= \dim \alpha^j U + \dim N_{j+1} = \text{rank } \alpha^{j+1} + \dim N_{j+1},$$

where

$$N_{j+1} = \{\mathbf{u} \in \alpha^j U \ : \ \alpha \mathbf{u} = \mathbf{0}\} = \alpha^j U \cap \alpha^{-1}(\mathbf{0}).$$

Thus

$$\text{rank } \alpha^j - \text{rank } \alpha^{j+1} = \dim N_{j+1}.$$

Since $\alpha^j U \supseteq \alpha^{j+1} U$, it follows that $N_j \supseteq N_{j+1}$ and $\dim N_j \geq \dim N_{j+1}$ for all $j$. Thus

$$\operatorname{rank} \alpha^j - \operatorname{rank} \alpha^{j+1} = \dim N_{j+1} \geq \dim N_{j+2} = \operatorname{rank} \alpha^{j+1} - \operatorname{rank} \alpha^{j+2}.$$

$\square$

We give a further development of these ideas in Exercise 12.1.7.

**Exercise 11.2.3** *Suppose that $U$ is a vector space over $\mathbb{F}$ of dimension n. Use Theorem 11.2.2 to show that, if $\alpha$ is a nilpotent endomorphism (that is to say, $\alpha^m = 0$ for some m), then $\alpha^n = 0$.*

*Prove that, if $\alpha$ has rank r and $\alpha^m = 0$, then $r \leq n(1 - m^{-1})$.*

**Exercise 11.2.4** *Show that, given any sequence of integers*

$$n = s_0 > s_1 > s_2 > \ldots > s_m \geq 0$$

*satisfying the condition*

$$s_0 - s_1 \geq s_1 - s_2 \geq s_2 - s_3 \geq \ldots \geq s_{m-1} - s_m > 0$$

*and a vector space $U$ over $\mathbb{F}$ of dimension n, we can find a linear map $\alpha : U \to U$ such that*

$$\operatorname{rank} \alpha^j = \begin{cases} s_j & \text{if } 0 \leq j \leq m, \\ s_m & \text{otherwise.} \end{cases}$$

*[We look at the matter in a slightly different way in Exercise 12.5.2.]*

**Exercise 11.2.5** *Suppose that $V$ is a vector space over $\mathbb{F}$ of even dimension 2n. If $\alpha \in \mathcal{L}(U, U)$ has rank $2n - 2$ and $\alpha^n = 0$, what can you say about the rank of $\alpha^k$ for $2 \leq k \leq n - 1$? Give reasons for your answer.*

## 11.3 Duals (almost) without using bases

For the rest of this chapter we look at the *dual space* of $U$, that is to say, at $U' = \mathcal{L}(U, \mathbb{F})$. The elements of $U'$ are called *functionals* or *linear functionals*. So long as we only look at finite dimensional spaces, it is not easy to justify paying particular attention to linear functionals, but many important mathematical objects are linear functionals for infinite dimensional spaces.

**Exercise 11.3.1** *Let $C^\infty(\mathbb{R})$ be the space of infinitely differentiable functions $f : \mathbb{R} \to \mathbb{R}$. Show that $C^\infty(\mathbb{R})$ is a vector space over $\mathbb{R}$ under pointwise addition and scalar multiplication.*

*Show that the following definitions give linear functionals for $C^\infty(\mathbb{R})$. Here $a \in \mathbb{R}$.*

*(i) $\delta_a f = f(a)$.*

*(ii) $\delta'_a f = -f'(a)$. (The minus sign is introduced for consistency with more advanced work on the topic of 'distributions'.)*

*(iii) $Jf = \int_0^1 f(x) \, dx$.*

Because of the connection with infinite dimensional spaces, we try to develop as much of the theory of dual spaces as possible without using bases, but we hit a snag right at the beginning of the topic.

**Definition 11.3.2** (*This is a* non-standard definition *and is not used by other authors.*) *We shall say that a vector space $U$ over $\mathbb{F}$ is* separated by its dual *if, given $\mathbf{u} \in U$ with $\mathbf{u} \neq \mathbf{0}$, we can find a $T \in U'$ such that $T\mathbf{u} \neq 0$.*

Given any *particular* vector space, it is *always* easy to show that it is separated by its dual.[4] However, when we try to prove that *every* vector space is separated by its dual, we discover (and the reader is welcome to try for herself) that *the axioms for a general vector space do not provide enough information to enable us to construct an appropriate $T$* using the rules of reasoning appropriate to an elementary course.[5]

If we have a basis, everything becomes easy.

**Lemma 11.3.3** *Every finite dimensional space is separated by its dual.*

*Proof* Suppose that $\mathbf{u} \in U$ and $\mathbf{u} \neq \mathbf{0}$. Since $U$ is finite dimensional and $\mathbf{u}$ is non-zero, we can find a basis $\mathbf{e}_1 = \mathbf{u}, \mathbf{e}_2, \ldots, \mathbf{e}_n$ for $U$. If we set

$$T\left(\sum_{j=1}^{n} x_j \mathbf{e}_j\right) = x_1 \qquad [x_j \in \mathbb{F}],$$

then $T$ is a well defined function from $U$ to $\mathbb{F}$. Further, if $x_j, y_j, \lambda, \mu \in \mathbb{F}$, then

$$T\left(\lambda\left(\sum_{j=1}^{n} x_j \mathbf{e}_j\right) + \mu\left(\sum_{j=1}^{n} x_j \mathbf{e}_j\right)\right) = T\left(\sum_{j=1}^{n}(\lambda x_j + \mu y_j)\mathbf{e}_j\right)$$

$$= \lambda x_1 + \mu y_1$$

$$= \lambda T\left(\sum_{j=1}^{n} x_j \mathbf{e}_j\right) + \mu T\left(\sum_{j=1}^{n} y_j \mathbf{e}_j\right),$$

so $T$ is linear and $T \in U'$. Since $T\mathbf{u} = T\mathbf{e}_1 = 1 \neq 0$, we are done. $\qquad\square$

From now on until the end of the section, we will see what can be done without bases, assuming that our spaces are separated by their duals. We shall write a generic element of $U$ as $\mathbf{u}$ and a generic element of $U'$ as $\mathbf{u}'$.

We begin by proving a result linking a vector space with the dual of its dual.

**Lemma 11.3.4** *Let $U$ be a vector space over $\mathbb{F}$ separated by its dual. Then the map $\Theta : U \to U''$ given by*

$$(\Theta\mathbf{u})\mathbf{u}' = \mathbf{u}'(\mathbf{u})$$

---

[4] This statement does leave open exactly what I mean by 'particular'.
[5] More specifically, we require some form of the so-called axiom of choice.

*for all* $\mathbf{u} \in U$ *and* $\mathbf{u}' \in U'$ *is an injective linear map.*

*Remark.* Suppose that we wish to find a map $\Theta : U \to U''$. Since a function is defined by its effect, we need to know the value of $\Theta\mathbf{u}$ for all $\mathbf{u} \in U$. Now $\Theta\mathbf{u} \in U''$, so $\Theta\mathbf{u}$ is itself a function on $U'$. Since a function is defined by its effect, we need to know the value of $(\Theta\mathbf{u})\mathbf{u}'$ for all $\mathbf{u}' \in U'$. We observe that $(\Theta\mathbf{u})\mathbf{u}'$ depends on $\mathbf{u}$ and $\mathbf{u}'$. The simplest way of combining $\mathbf{u}$ and $\mathbf{u}'$ is $\mathbf{u}'(\mathbf{u})$, so we *try* the definition $(\Theta\mathbf{u})\mathbf{u}' = \mathbf{u}'(\mathbf{u})$. Either it will produce something interesting, or it will not, and the simplest way forward is just to see what happens.

*Proof of Lemma 11.3.4* We first need to show that $\Theta\mathbf{u} : U' \to \mathbb{F}$ is linear. To this end, observe that, if $\mathbf{u}'_1, \mathbf{u}'_2 \in U'$ and $\lambda_1, \lambda_2 \in \mathbb{F}$, then

$$\Theta\mathbf{u}(\lambda_1\mathbf{u}'_1 + \lambda_2\mathbf{u}'_2) = (\lambda_1\mathbf{u}'_1 + \lambda_2\mathbf{u}'_2)\mathbf{u} \qquad \text{(by definition)}$$
$$= \lambda_1\mathbf{u}'_1(\mathbf{u}) + \lambda_2\mathbf{u}'_2(\mathbf{u}) \qquad \text{(by definition)}$$
$$= \lambda_1\Theta\mathbf{u}(\mathbf{u}'_1) + \lambda_2\Theta\mathbf{u}(\mathbf{u}'_2) \qquad \text{(by definition)}$$

so $\Theta\mathbf{u} \in U''$.

Now we need to show that $\Theta : U \to U''$ is linear. In other words, we need to show that, whenever $\mathbf{u}_1, \mathbf{u}_2 \in U'$ and $\lambda_1, \lambda_2 \in \mathbb{F}$,

$$\Theta(\lambda_1\mathbf{u}_1 + \lambda_2\mathbf{u}_2) = \lambda_1\Theta\mathbf{u}_1 + \lambda_2\Theta\mathbf{u}_2.$$

The two sides of the equation lie in $U''$ and so are functions. In order to show that two functions are equal, we need to show that they have the same effect. Thus we need to show that

$$\big(\Theta(\lambda_1\mathbf{u}_1 + \lambda_2\mathbf{u}_2)\big)\mathbf{u}' = \big(\lambda_1\Theta\mathbf{u}_1 + \lambda_2\Theta\mathbf{u}_2\big)\mathbf{u}'$$

for all $\mathbf{u}' \in U'$. Since

$$\big(\Theta(\lambda_1\mathbf{u}_1 + \lambda_2\mathbf{u}_2)\big)\mathbf{u}' = \mathbf{u}'(\lambda_1\mathbf{u}_1 + \lambda_2\mathbf{u}_2) \qquad \text{(by definition)}$$
$$= \lambda_1\mathbf{u}'(\mathbf{u}_1) + \lambda_2\mathbf{u}'(\mathbf{u}_2) \qquad \text{(by linearity)}$$
$$= \lambda_1\Theta(\mathbf{u}_1)\mathbf{u}' + \lambda_2\Theta(\mathbf{u}_2)\mathbf{u}' \qquad \text{(by definition)}$$
$$= \big(\lambda_1\Theta\mathbf{u}_1 + \lambda_2\Theta\mathbf{u}_2\big)\mathbf{u}' \qquad \text{(by definition)}$$

the required result is indeed true and $\Theta$ is linear.

Finally, we need to show that $\Theta$ is injective. Since $\Theta$ is linear, it suffices (as we observed in Exercise 5.3.11) to show that

$$\Theta(\mathbf{u}) = \mathbf{0} \Rightarrow \mathbf{u} = \mathbf{0}.$$

In order to prove the implication, we observe that the two sides of the initial equation lie in $U''$ and two functions are equal if they have the same effect. Thus

$$\Theta(\mathbf{u}) = \mathbf{0} \Rightarrow \Theta(\mathbf{u})\mathbf{u}' = \mathbf{0}\mathbf{u}' \quad \text{for all } \mathbf{u}' \in U'$$
$$\Rightarrow \mathbf{u}'(\mathbf{u}) = \mathbf{0} \quad \text{for all } \mathbf{u}' \in U'$$
$$\Rightarrow \mathbf{u} = \mathbf{0}$$

as required. In order to prove the last implication we needed the fact that $U'$ separates $U$. This is the only point at which that hypothesis was used. $\qquad\square$

In Euclid's development of geometry there occurs a theorem[6] that generations of school-masters called the 'Pons Asinorum' (Bridge of Asses) partly because it is illustrated by a diagram that looks like a bridge, but mainly because the weaker students tended to get stuck there. Lemma 11.3.4 is a Pons Asinorum for budding mathematicians. It deals, not simply with functions, but with functions of functions, and represents a step up in abstraction. The author can only suggest that the reader writes out the argument repeatedly until she sees that it is merely a collection of trivial verifications and that the nature of the verifications is dictated by the nature of the result to be proved.

Note that we have no reason to suppose, in general, that the map $\Theta$ is surjective.

Our next lemma is another result of the same 'function of a function' nature as Lemma 11.3.4. We shall refer to such proofs as 'paper tigers', since they are fearsome in appearance, but easily folded away by any student prepared to face them with calm resolve.

**Lemma 11.3.5** *Let $U$ and $V$ be vector spaces over $\mathbb{F}$.*

*(i) If $\alpha \in \mathcal{L}(U, V)$, we can define a map $\alpha' \in \mathcal{L}(V', U')$ by the condition*

$$\alpha'(\mathbf{v}')(\mathbf{u}) = \mathbf{v}'(\alpha\mathbf{u}).$$

*(ii) If we now define $\Phi : \mathcal{L}(U, V) \to \mathcal{L}(V', U')$ by $\Phi(\alpha) = \alpha'$, then $\Phi$ is linear.*

*(iii) If, in addition, $V$ is separated by $V'$, then $\Phi$ is injective.*

We call $\alpha'$ the *dual map* of $\alpha$.

*Remark.* Suppose that we wish to find a map $\alpha' : V' \to U'$ corresponding to $\alpha \in \mathcal{L}(U, V)$. Since a function is defined by its effect, we need to know the value of $\alpha'\mathbf{v}'$ for all $\mathbf{v}' \in V'$. Now $\alpha'\mathbf{v}' \in U'$, so $\alpha'\mathbf{v}'$ is itself a function on $U$. Since a function is defined by its effect, we need to know the value of $\alpha'\mathbf{v}'(\mathbf{u})$ for all $\mathbf{u} \in U$. We observe that $\alpha'\mathbf{v}'(\mathbf{u})$ depends on $\alpha$, $\mathbf{v}'$ and $\mathbf{u}$. The simplest way of combining these elements is as $\mathbf{v}'\alpha(\mathbf{u})$, so we *try* the definition $\alpha'(\mathbf{v}')(\mathbf{u}) = \mathbf{v}'(\alpha\mathbf{u})$. Either it will produce something interesting, or it will not, and the simplest way forward is just to see what happens.

The reader may ask why we do not *try* to produce an $\tilde{\alpha} : U' \to V'$ in the same way. The answer is that she is welcome (indeed, strongly encouraged) to *try* but, although many people must have *tried*, no one has yet (to my knowledge) come up with anything satisfactory.

---

[6] The base angles of an isosceles triangle are equal.

*Proof of Lemma 11.3.5* By inspection, $\alpha'(\mathbf{v}')(\mathbf{u})$ is well defined. We must show that $\alpha'(\mathbf{v}')$ : $U \to \mathbb{F}$ is linear. To do this, observe that, if $\mathbf{u}_1, \mathbf{u}_2 \in U'$ and $\lambda_1, \lambda_2 \in \mathbb{F}$, then

$$
\begin{aligned}
\alpha'(\mathbf{v}')(\lambda_1\mathbf{u}_1 + \lambda_2\mathbf{u}_2) &= \mathbf{v}'\big(\alpha(\lambda_1\mathbf{u}_1 + \lambda_2\mathbf{u}_2)\big) && \text{(by definition)} \\
&= \mathbf{v}'(\lambda_1\alpha\mathbf{u}_1 + \lambda_2\alpha\mathbf{u}_2) && \text{(by linearity)} \\
&= \lambda_1\mathbf{v}'(\alpha\mathbf{u}_1) + \lambda_2\mathbf{v}'(\alpha\mathbf{u}_2) && \text{(by linearity)} \\
&= \lambda_1\alpha'(\mathbf{v}')\mathbf{u}_1 + \lambda_2\alpha'(\mathbf{v}')\mathbf{u}_2 && \text{(by definition).}
\end{aligned}
$$

We now know that $\alpha'$ maps $V'$ to $U'$. We want to show that $\alpha'$ is linear. To this end, observe that if $\mathbf{v}'_1, \mathbf{v}'_2 \in V'$ and $\lambda_1, \lambda_2 \in \mathbb{F}$, then

$$
\begin{aligned}
\alpha'(\lambda_1\mathbf{v}'_1 + \lambda_2\mathbf{v}'_2)\mathbf{u} &= (\lambda_1\mathbf{v}'_1 + \lambda_2\mathbf{v}'_2)\alpha(\mathbf{u}) && \text{(by definition)} \\
&= \lambda_1\mathbf{v}'_1(\alpha\mathbf{u}) + \lambda_2\mathbf{v}'_2(\alpha\mathbf{u}) && \text{(by definition)} \\
&= \lambda_1\alpha'(\mathbf{v}'_1)\mathbf{u} + \lambda_2\alpha'(\mathbf{v}'_2)\mathbf{u} && \text{(by definition)} \\
&= \big(\lambda_1\alpha'(\mathbf{v}'_1) + \lambda_2\alpha'(\mathbf{v}'_2)\big)\mathbf{u} && \text{(by definition)}
\end{aligned}
$$

for all $\mathbf{u} \in U$. By the definition of equality for functions,

$$
\alpha'(\lambda_1\mathbf{v}'_1 + \lambda_2\mathbf{v}'_2) = \lambda_1\alpha'(\mathbf{v}'_1) + \lambda_2\alpha'(\mathbf{v}'_2)
$$

and $\alpha'$ is linear as required.

(ii) We are now dealing with a function of a function of a function. In order to establish that

$$
\Phi(\lambda_1\alpha_1 + \lambda_2\alpha_2) = \lambda_1\Phi(\alpha_1) + \lambda_2\Phi(\alpha_2)
$$

when $\alpha_1, \alpha_2 \in \mathcal{L}(U, V)$ and $\lambda_1, \lambda_2 \in \mathbb{F}$, we must establish that

$$
\Phi(\lambda_1\alpha_1 + \lambda_2\alpha_2)(\mathbf{v}') = \big(\lambda_1\Phi(\alpha_1) + \lambda_2\Phi(\alpha_2)\big)(\mathbf{v}')
$$

for all $\mathbf{v}' \in V$ and, in order to establish this, we must show that

$$
\big(\Phi(\lambda_1\alpha_1 + \lambda_2\alpha_2)(\mathbf{v}')\big)\mathbf{u} = \big((\lambda_1\Phi(\alpha_1) + \lambda_2\Phi(\alpha_2))\mathbf{v}'\big)\mathbf{u}
$$

for all $\mathbf{u} \in U$.

As usual, this just a question of following things through. Observe that

$$
\begin{aligned}
\big(\Phi(\lambda_1\alpha_1 + \lambda_2\alpha_2)(\mathbf{v}')\big)(\mathbf{u}) &= \big((\lambda_1\alpha_1 + \lambda_2\alpha_2)'(\mathbf{v}')\big)(\mathbf{u}) && \text{(by definition)} \\
&= \mathbf{v}'\big((\lambda_1\alpha_1 + \lambda_2\alpha_2)\mathbf{u}\big) && \text{(by definition)} \\
&= \mathbf{v}'(\lambda_1\alpha_1\mathbf{u} + \lambda_2\alpha_2\mathbf{u}) && \text{(by linearity)} \\
&= \lambda_1\mathbf{v}'(\alpha_1\mathbf{u}) + \lambda_2\mathbf{v}'(\alpha_2\mathbf{u}) && \text{(by linearity)} \\
&= \lambda_1(\alpha'_1\mathbf{v}')\mathbf{u} + \lambda_2(\alpha'_2\mathbf{v}')\mathbf{u} && \text{(by definition)} \\
&= \big(\lambda_1(\alpha'_1\mathbf{v}') + \lambda_2(\alpha'_2\mathbf{v}')\big)\mathbf{u} && \text{(by definition)} \\
&= \big((\lambda_1\alpha'_1 + \lambda_2\alpha'_2)(\mathbf{v}')\big)\mathbf{u} && \text{(by definition)} \\
&= \big((\lambda_1\Phi(\alpha_1) + \lambda_2\Phi(\alpha_2))(\mathbf{v}')\big)\mathbf{u} && \text{(by definition)}
\end{aligned}
$$

as required.

(iii) Finally, if $V'$ separates $V$, then

$$\Phi(\alpha) = 0 \Rightarrow \alpha' = 0 \Rightarrow \alpha'(\mathbf{v}') = 0 \quad \text{for all } \mathbf{v}' \in V'$$
$$\Rightarrow \alpha'(\mathbf{v}')\mathbf{u} = \mathbf{0} \quad \text{for all } \mathbf{v}' \in V' \text{ and all } \mathbf{u} \in U$$
$$\Rightarrow \mathbf{v}'(\alpha\mathbf{u}) = \mathbf{0} \quad \text{for all } \mathbf{v}' \in V' \text{ and all } \mathbf{u} \in U$$
$$\Rightarrow \alpha\mathbf{u} = \mathbf{0} \quad \text{for all } \mathbf{u} \in U \Rightarrow \alpha = 0.$$

Thus $\Phi$ is injective                                                            □

We call $\alpha'$ the *dual map* of $\alpha$.

**Exercise 11.3.6** *Write down the reasons for each implication in the proof of part (iii) of Lemma 11.3.5.*

Here are a further couple of paper tigers.

**Exercise 11.3.7** *Suppose that $U$, $V$ and $W$ are vector spaces over $\mathbb{F}$.*
   *(i) If $\alpha \in \mathcal{L}(U, V)$ and $\beta \in \mathcal{L}(V, W)$, show that $(\beta\alpha)' = \alpha'\beta'$.*
   *(ii) Consider the identity map $\iota_U : U \to U$. Show that $\iota'_U : U' \to U$ is the identity map $\iota_{U'} : U' \to U'$. (We shall follow standard practice by simply writing $\iota' = \iota$.)*
   *(iii) If $\alpha \in \mathcal{L}(U, V)$ is invertible, show that $\alpha'$ is invertible and $(\alpha')^{-1} = (\alpha^{-1})'$.*

**Exercise 11.3.8** *Suppose that $V''$ is separated by $V'$. Show that, with the notation of the two previous lemmas,*

$$\alpha''(\Theta\mathbf{u}) = \alpha\mathbf{u}$$

*for all $\mathbf{u} \in U$ and $\alpha \in \mathcal{L}(U, V)$.*

The remainder of this section is not meant to be taken very seriously. We show that, if we deal with infinite dimensional spaces, the dual $U'$ of a space may be 'very much bigger' than the space $U$.

**Exercise 11.3.9** *Consider the space $s$ of all sequences*

$$\mathbf{a} = (a_1, a_2, \ldots)$$

*with $a_j \in \mathbb{R}$. (In more sophisticated language, $s = \mathbb{N}^{\mathbb{R}}$.) We know that, if we use pointwise addition and scalar multiplication, so that*

$$\mathbf{a} + \mathbf{b} = (a_1 + b_1, a_2 + b_2, \ldots) \quad and \quad \lambda\mathbf{a} = (\lambda a_1, \lambda a_2, \ldots),$$

*then $s$ is a vector space.*
   *(i) Show that, if $c_{00}$ is the set of $\mathbf{a}$ with only finitely many $a_j$ non-zero, then $c_{00}$ is a subspace of $s$.*

(ii) *Let* $\mathbf{e}_j$ *be the sequence whose jth term is 1 and whose other terms are all* 0. *If* $T \in c_{00}$ *and* $T\mathbf{e}_j = a_j$, *show that*

$$T\mathbf{x} = \sum_{j=1}^{\infty} a_j x_j$$

*for all* $\mathbf{x} \in c_{00}$. (*Note that, contrary to appearances, we are only looking at a sum of a* finite *set of terms.*)

(iii) *If* $\mathbf{a} \in s$, *show that the rule*

$$T_{\mathbf{a}}\mathbf{x} = \sum_{j=1}^{\infty} a_j x_j$$

*gives a well defined map* $T_{\mathbf{a}} : c_{00} \to \mathbb{R}$. *Show that* $T_{\mathbf{a}} \in c'_{00}$. *Deduce that* $c'_{00}$ *separates* $c_{00}$.

(iv) *Show that, using the notation of* (iii), *if* $\mathbf{a}, \mathbf{b} \in s$ *and* $\lambda \in \mathbb{R}$, *then*

$$T_{\mathbf{a}+\mathbf{b}} = T_{\mathbf{a}} + T_{\mathbf{b}}, \quad T_{\lambda\mathbf{a}} = \lambda T_{\mathbf{a}} \quad and \quad T_{\mathbf{a}} = 0 \Leftrightarrow \mathbf{a} = \mathbf{0}.$$

*Conclude that* $c'_{00}$ *is isomorphic to* $s$.

The space $s = c'_{00}$ certainly looks much bigger than $c_{00}$. To show that this is actually the case we need ideas from the study of countability and, in particular, Cantor's diagonal argument. (The reader who has not met these topics should skip the next exercise.)

**Exercise 11.3.10** (i) *Show, if you have not already done so, that the vectors* $\mathbf{e}_j$ *defined in Exercise 11.3.9* (ii) *span* $c_{00}$. *In other words, show that any* $\mathbf{x} \in c_{00}$ *can be written as a finite sum*

$$\mathbf{x} = \sum_{j=1}^{N} \lambda_j \mathbf{e}_j \quad with \ N \geq 0 \ and \ \lambda_j \in \mathbb{R}.$$

[*Thus* $c_{00}$ *has a countable spanning set. In the rest of the question we show that* $s$ *and so* $c'_{00}$ *does not have a countable spanning set.*]

(ii) *Suppose that* $\mathbf{f}_1, \mathbf{f}_2, \ldots, \mathbf{f}_n \in s$. *If* $m \geq 1$, *show that we can find* $b_m, b_{m+1}, b_{m+2}, \ldots,$ $b_{m+n+1}$ *such that if* $\mathbf{a} \in s$ *satisfies the condition* $a_r = b_r$ *for* $m \leq r \leq m+n+1$, *then the equation*

$$\sum_{j=1}^{n} \lambda_j \mathbf{f}_j = \mathbf{a}$$

*has no solution with* $\lambda_j \in \mathbb{R}$ $[1 \leq j \leq n]$.

(iii) *Suppose that* $\mathbf{f}_j \in s$ $[j \geq 1]$. *Show that there exists an* $\mathbf{a} \in s$ *such that the equation*

$$\sum_{j=1}^{n} \lambda_j \mathbf{f}_j = \mathbf{a}$$

*has no solution with* $\lambda_j \in \mathbb{R}$ $[1 \leq j \leq n]$ *for any* $n \geq 1$. *Deduce that* $c'_{00}$ *is not isomorphic to* $c_{00}$.

Exercises 11.3.9 and 11.3.10 suggest that the duals of infinite dimensional vector spaces may be very large indeed. Most studies of infinite dimensional vector spaces assume the existence of a distance given by a norm and deal with functionals which are *continuous* with respect to that norm.

### 11.4 Duals using bases

If we restrict $U$ to be finite dimensional, life becomes a lot simpler.

**Lemma 11.4.1** (*i*) *If* $U$ *is a vector space over* $\mathbb{F}$ *with basis* $\mathbf{e}_1, \mathbf{e}_2, \ldots, \mathbf{e}_n$, *then we can find unique* $\hat{\mathbf{e}}_1, \hat{\mathbf{e}}_2, \ldots, \hat{\mathbf{e}}_n \in U'$ *satisfying the equations*

$$\hat{\mathbf{e}}_i(\mathbf{e}_j) = \delta_{ij} \quad for\ 1 \leq i, j \leq n.$$

(*ii*) *The vectors* $\hat{\mathbf{e}}_1, \hat{\mathbf{e}}_2, \ldots, \hat{\mathbf{e}}_n$, *defined in* (*i*) *form a basis for* $U'$.

(*iii*) *The dual of a finite dimensional vector space is finite dimensional with the same dimension as the initial space.*

We call $\hat{\mathbf{e}}_1, \hat{\mathbf{e}}_2, \ldots, \hat{\mathbf{e}}_n$ the *dual basis* corresponding to $\mathbf{e}_1, \mathbf{e}_2, \ldots, \mathbf{e}_n$.

*Proof* (i) Left to the reader. (Look at Lemma 11.3.3 if necessary.)

(ii) To show that the $\hat{\mathbf{e}}_j$ are linearly independent, observe that, if

$$\sum_{j=1}^{n} \lambda_j \hat{\mathbf{e}}_j = 0,$$

then

$$0 = \left( \sum_{j=1}^{n} \lambda_j \hat{\mathbf{e}}_j \right) \mathbf{e}_k = \sum_{j=1}^{n} \lambda_j \hat{\mathbf{e}}_j(\mathbf{e}_k) = \sum_{j=1}^{n} \lambda_j \delta_{jk} = \lambda_k$$

for each $1 \leq k \leq n$.

To show that the $\hat{\mathbf{e}}_j$ span, suppose that $\mathbf{u}' \in U'$. Then

$$\left( \mathbf{u}' - \sum_{j=1}^{n} \mathbf{u}'(\mathbf{e}_j)\hat{\mathbf{e}}_j \right) \mathbf{e}_k = \mathbf{u}'(\mathbf{e}_k) - \sum_{j=1}^{n} \mathbf{u}'(\mathbf{e}_j)\delta_{jk}$$

$$= \mathbf{u}'(\mathbf{e}_k) - \mathbf{u}'(\mathbf{e}_k) = 0$$

so, using linearity,

$$\left( \mathbf{u}' - \sum_{j=1}^{n} (\mathbf{u}'(\mathbf{e}_j))\hat{\mathbf{e}}_j \right) \sum_{k=1}^{n} x_k \mathbf{e}_k = 0$$

for all $x_k \in \mathbb{F}$. Thus

$$\left( \mathbf{u}' - \sum_{j=1}^{n} \left( \mathbf{u}'(\mathbf{e}_j) \right) \hat{\mathbf{e}}_j \right) \mathbf{x} = \mathbf{0}$$

for all $\mathbf{x} \in U$ and so

$$\mathbf{u}' = \sum_{j=1}^{n} \left( \mathbf{u}'(\mathbf{e}_j) \right) \hat{\mathbf{e}}_j.$$

We have shown that the $\hat{\mathbf{e}}_j$ span and thus we have a basis.

(iii) Follows at once from (ii). $\qquad\square$

**Exercise 11.4.2** *Consider the vector space $\mathcal{P}_n$ of real polynomials $P$*

$$P(t) = \sum_{j=0}^{n} a_j t^j$$

*(where $t \in [a, b]$) of degree at most $n$. If $x_0, x_1, \ldots, x_n$ are distinct points of $[a, b]$ show that, if we set*

$$e_j(t) = \prod_{k \neq j} \frac{t - x_k}{x_j - x_k},$$

*then $e_0, e_1, \ldots, e_n$ form a basis for $\mathcal{P}_n$. Evaluate $\hat{e}_j P$ where $P \in \mathcal{P}_n$.*

We can now strengthen Lemma 11.3.4 in the finite dimensional case.

**Theorem 11.4.3** *Let $U$ be a finite dimensional vector space over $\mathbb{F}$. Then the map $\Theta :$ $U \to U''$ given by*

$$(\Theta \mathbf{u})\mathbf{u}' = \mathbf{u}'(\mathbf{u})$$

*for all $\mathbf{u} \in U$ and $\mathbf{u}' \in U'$ is an isomorphism.*

*Proof* Lemmas 11.3.4 and 11.3.3 tell us that $\Theta$ is an injective linear map. Lemma 11.4.1 tells us that

$$\dim U = \dim U' = \dim U''$$

and we know (for example, by the rank-nullity theorem) that an injective linear map between spaces of the same dimension is surjective, so bijective and so an isomorphism. $\qquad\square$

The reader may mutter under her breath that we have not proved anything special, since all vector spaces of the same dimension are isomorphic. However, the isomorphism given by $\Phi$ in Theorem 11.4.3 is a *natural isomorphism* in the sense that we do not have to make any arbitrary choices to specify it. The isomorphism of two general vector spaces of dimension $n$ depends on choosing a basis and *different choices* of bases produce *different isomorphisms*.

Because of the *natural isomorphism* we usually *identify* $U''$ with $U$ by writing $\mathbf{u} = \Theta(\mathbf{u})$. With this convention,

$$\mathbf{u}(\mathbf{u}') = \mathbf{u}'(\mathbf{u})$$

for all $\mathbf{u} \in U$, $\mathbf{u}' \in U'$. We then have

$$U = U'' = U'''' = \dots \quad \text{and} \quad U' = U''' = \dots.$$

**Exercise 11.4.4** *Let $U$ be a vector space over $\mathbb{F}$ with basis $\mathbf{e}_1$, $\mathbf{e}_2$, ..., $\mathbf{e}_n$. If we identify $U''$ with $U$ in the standard manner, find the dual basis of the dual basis, that is to say, find the vectors identified with $\hat{\hat{\mathbf{e}}}_1$, $\hat{\hat{\mathbf{e}}}_2$, ..., $\hat{\hat{\mathbf{e}}}_n$.*

**Exercise 11.4.5** *Consider a vector space $U$ over $\mathbb{F}$ with basis $\mathbf{e}_1$ and $\mathbf{e}_2$. Let $\hat{\mathbf{e}}_1$, $\hat{\mathbf{e}}_2$ be the dual basis of $U'$.*

*In each of the following cases, you are given a basis $\mathbf{f}_1$ and $\mathbf{f}_2$ for $U$ and asked to find the corresponding dual basis $\hat{\mathbf{f}}_1$, $\hat{\mathbf{f}}_2$ in terms of $\hat{\mathbf{e}}_1$, $\hat{\mathbf{e}}_2$. You are then asked to find the matrix $Q$ with respect to the bases $\mathbf{e}_1$, $\mathbf{e}_2$ of $U$ and $\hat{\mathbf{e}}_1$, $\hat{\mathbf{e}}_2$ of $U'$ for the linear map $\alpha : U \to U'$ with $\gamma \mathbf{e}_j = \hat{\mathbf{f}}_j$. (If you skip at once to (iv), look back briefly at what your general formula gives in the particular cases.)*

(i) $\mathbf{f}_1 = \mathbf{e}_2$, $\mathbf{f}_2 = \mathbf{e}_1$.
(ii) $\mathbf{f}_1 = 2\mathbf{e}_1$, $\mathbf{f}_2 = \mathbf{e}_2$.
(iii) $\mathbf{f}_1 = \mathbf{e}_1 + \mathbf{e}_2$, $\mathbf{f}_2 = \mathbf{e}_2$.
(iv) $\mathbf{f}_1 = a\mathbf{e}_1 + b\mathbf{e}_2$, $\mathbf{f}_2 = c\mathbf{e}_1 + d\mathbf{e}_2$ with $ad - bc \neq 0$.

**Exercise 11.4.6 [Change of basis and contravariance]** *Consider a vector space $U$ over $\mathbb{F}$ with two bases $\mathbf{e}_1$, $\mathbf{e}_2$, ..., $\mathbf{e}_n$ and $\mathbf{f}_1$, $\mathbf{f}_2$, ..., $\mathbf{f}_n$. If $L = (l_{ij})$ and $K = (k_{rs})$ are the $n \times n$ matrices defined by*

$$\mathbf{f}_j = \sum_{i=1}^{n} l_{ij} \mathbf{e}_i,$$

$$\hat{\mathbf{f}}_s = \sum_{r=1}^{n} k_{rs} \hat{\mathbf{e}}_r,$$

*show that $K = (L^T)^{-1}$. (The reappearance of the formula from Lemma 10.4.2 is no coincidence.)*

Lemma 11.3.5 strengthens in the expected way when $U$ and $V$ are finite dimensional.

**Theorem 11.4.7** *Let $U$ and $V$ be finite dimensional vector spaces over $\mathbb{F}$.*
(i) *If $\alpha \in \mathcal{L}(U, V)$, then we can define a map $\alpha' \in \mathcal{L}(V', U')$ by the condition*

$$\alpha'(\mathbf{v}')(\mathbf{u}) = \mathbf{v}'(\alpha \mathbf{u}).$$

(ii) *If we now define $\Phi : \mathcal{L}(U, V) \to \mathcal{L}(V', U')$ by $\Phi(\alpha) = \alpha'$, then $\Phi$ is an isomorphism.*
(iii) *If we identify $U''$ and $U$ and $V''$ and $V$ in the standard manner, then $\alpha'' = \alpha$ for all $\alpha \in \mathcal{L}(U, V)$.*

*Proof* (i) This is Lemma 11.3.5 (i).

(ii) Lemma 11.3.5 (ii) tells us that $\Phi$ is linear and injective. But

$$\dim \mathcal{L}(U, V) = \dim U \times \dim V = \dim U' \times \dim V' = \dim \mathcal{L}(V', U'),$$

so $\Phi$ is an isomorphism.

(iii) We have

$$(\alpha'' \mathbf{u})\mathbf{v}' = \mathbf{u}(\alpha' \mathbf{v}') = (\alpha' \mathbf{v}')\mathbf{u} = \mathbf{v}' \alpha \mathbf{u} = (\alpha \mathbf{u})\mathbf{v}'$$

for all $\mathbf{v}' \in V$ and $\mathbf{u} \in U$. Thus

$$\alpha'' \mathbf{u} = \alpha \mathbf{u}$$

for all $\mathbf{u} \in U$ and so $\alpha'' = \alpha$. (Compare Exercise 11.3.8.) $\qquad \square$

If we use bases, we can link the map $\alpha \mapsto \alpha'$ with a familiar matrix operation.

**Lemma 11.4.8** *If $U$ and $V$ are finite dimensional vector spaces over $\mathbb{F}$ and $\alpha \in \mathcal{L}(U, V)$ has matrix $A$ with respect to given bases of $U$ and $V$, then $\alpha'$ has matrix $A^T$ with respect to the dual bases.*

*Proof* Let $\mathbf{e}_1, \mathbf{e}_2, \ldots, \mathbf{e}_n$ be a basis for $U$ and $\mathbf{f}_1, \mathbf{f}_2, \ldots, \mathbf{f}_m$ a basis for $V$. Let the corresponding dual bases be $\hat{\mathbf{e}}_1, \hat{\mathbf{e}}_2, \ldots, \hat{\mathbf{e}}_n$ and $\hat{\mathbf{f}}_1, \hat{\mathbf{f}}_2, \ldots, \hat{\mathbf{f}}_m$. If $\alpha$ has matrix $(a_{ij})$ with respect to the given bases for $U$ and $V$ and $\alpha'$ has matrix $(c_{rs})$ with respect to the dual bases, then, by definition,

$$
\begin{aligned}
c_{rs} &= \sum_{k=1}^{n} c_{ks} \delta_{rk} = \sum_{k=1}^{n} c_{ks} \hat{\mathbf{e}}_k \mathbf{e}_r \\
&= \left( \sum_{k=1}^{n} c_{ks} \hat{\mathbf{e}}_k \right) \mathbf{e}_r = \alpha'(\hat{\mathbf{f}}_s) \mathbf{e}_r \\
&= \hat{\mathbf{f}}_s(\alpha \mathbf{e}_r) = \hat{\mathbf{f}}_s \left( \sum_{l=1}^{m} a_{lr} \mathbf{f}_l \right) \\
&= \sum_{l=1}^{m} a_{lr} \delta_{sl} = a_{sr}
\end{aligned}
$$

for all $1 \le r \le n$, $1 \le s \le m$. $\qquad \square$

**Exercise 11.4.9** *Use results on the map $\alpha \mapsto \alpha'$ and Exercise 11.3.7 (ii) to recover the familiar results*

$$(A + B)^T = A^T + B^T, \quad (\lambda A)^T = \lambda A^T, \quad A^{TT} = A, \quad I^T = I$$

*for appropriate matrices.*

Use Exercise 11.3.7 to prove that $(AB)^T = B^T A^T$ for appropriate matrices. Show, similarly, that if $A$ is an $n \times n$ invertible matrix then $A^T$ is invertible and $(A^T)^{-1} = (A^{-1})^T$.

The reader may ask why we do not prove the result of Exercise 11.3.7 (at least for finite dimensional spaces) by using direct calculation to obtain Exercise 11.4.9 and then obtaining the result on maps from the result on matrices. An algebraist would reply that this would tell us that (i) was true but not *why* it was true.

However, we shall not be overzealous in our pursuit of algebraic purity.

**Exercise 11.4.10** *Suppose that $U$ is a finite dimensional vector space over $\mathbb{F}$. If $\alpha \in \mathcal{L}(U, U)$, use the matrix representation to show that $\det \alpha = \det \alpha'$.*

*Hence show that $\det(\iota\iota - \alpha) = \det(\iota\iota - \alpha')$ and deduce that $\alpha$ and $\alpha'$ have the same eigenvalues. (See also Exercise 11.4.19.)*

*Use the result $\det(\iota\iota - \alpha) = \det(\iota\iota - \alpha')$ to show that $\operatorname{Tr} \alpha = \operatorname{Tr} \alpha'$. Deduce the same result directly from the matrix representation of $\alpha$.*

We now introduce the notion of an annihilator.

**Definition 11.4.11** *If $W$ is a subspace of a vector space $U$ over $\mathbb{F}$, we define the* annihilator *of $W^0$ of $W$ by taking*

$$W^0 = \{\mathbf{u}' \in U' : \mathbf{u}'\mathbf{w} = 0 \text{ for all } \mathbf{w} \in W\}.$$

**Exercise 11.4.12** *Show that, with the notation of Definition 11.4.11, $W^0$ is a subspace of $U'$.*

**Lemma 11.4.13** *If $W$ is a subspace of a finite dimensional vector space $U$ over $\mathbb{F}$, then*

$$\dim U = \dim W + \dim W^0.$$

*Proof* Since $W$ is a subspace of a finite dimensional space, it has a basis $\mathbf{e}_1, \mathbf{e}_2, \ldots, \mathbf{e}_m$, say. Extend this to a basis $\mathbf{e}_1, \mathbf{e}_2, \ldots, \mathbf{e}_n$ of $U$ and consider the dual basis $\hat{\mathbf{e}}_1, \hat{\mathbf{e}}_2, \ldots, \hat{\mathbf{e}}_n$.

We have

$$\sum_{j=1}^{n} y_j \hat{\mathbf{e}}_j \in W^0 \Rightarrow \left(\sum_{j=1}^{n} y_j \hat{\mathbf{e}}_j\right) \mathbf{w} = 0 \quad \text{for all } \mathbf{w} \in W$$

$$\Rightarrow \left(\sum_{j=1}^{n} y_j \hat{\mathbf{e}}_j\right) \mathbf{e}_k = 0 \quad \text{for all } 1 \le k \le m$$

$$\Rightarrow y_k = 0 \quad \text{for all } 1 \le k \le m.$$

On the other hand, if $\mathbf{w} \in W$, then we have $\mathbf{w} = \sum_{j=1}^{m} x_j \mathbf{e}_j$ for some $x_j \in \mathbb{F}$ so (if $y_r \in \mathbb{F}$)

$$\left(\sum_{r=m+1}^{n} y_r \hat{\mathbf{e}}_r\right)\left(\sum_{j=1}^{m} x_j \mathbf{e}_j\right) = \sum_{r=m+1}^{n} \sum_{j=1}^{m} y_r x_j \delta_{rj} = \sum_{r=m+1}^{n} \sum_{j=1}^{m} 0 = 0.$$

Thus

$$W^0 = \left\{ \sum_{r=m+1}^{n} y_r \hat{\mathbf{e}}_r \; : \; y_r \in \mathbb{F} \right\}$$

and

$$\dim W + \dim W^0 = m + (n - m) = n = \dim U$$

as stated. □

We have a nice corollary.

**Lemma 11.4.14** *Let $W$ be a subspace of a finite dimensional vector space $U$ over $\mathbb{F}$. If we identify $U''$ and $U$ in the standard manner, then $W^{00} = W$.*

*Proof* Observe that, if $\mathbf{w} \in W$, then

$$\mathbf{w}(\mathbf{u}') = \mathbf{u}'(\mathbf{w}) = 0$$

for all $\mathbf{u}' \in W^0$ and so $\mathbf{w} \in W^{00}$. Thus

$$W^{00} \supseteq W.$$

However,

$$\dim W^{00} = \dim U' - \dim W^0 = \dim U - \dim W^0 = \dim W$$

so $W^{00} = W$. □

**Exercise 11.4.15** *(An alternative proof of Lemma 11.4.14.) By using the bases $\mathbf{e}_j$ and $\hat{\mathbf{e}}_k$ of the proof of Lemma 11.4.13, identify $W^{00}$ directly.*

The next lemma gives a connection between null-spaces and annihilators.

**Lemma 11.4.16** *Suppose that $U$ and $V$ are vector spaces over $\mathbb{F}$ and $\alpha \in \mathcal{L}(U, V)$. Then*

$$(\alpha')^{-1}(\mathbf{0}) = (\alpha U)^0.$$

*Proof* Observe that

$$\mathbf{v}' \in (\alpha')^{-1}(\mathbf{0}) \Leftrightarrow \alpha' \mathbf{v}' = \mathbf{0}$$
$$\Leftrightarrow (\alpha' \mathbf{v}')\mathbf{u} = \mathbf{0} \quad \text{for all } \mathbf{u} \in U$$
$$\Leftrightarrow \mathbf{v}'(\alpha \mathbf{u}) = \mathbf{0} \quad \text{for all } \mathbf{u} \in U$$
$$\Leftrightarrow \mathbf{v}' \in (\alpha U)^0,$$

so $(\alpha')^{-1}(\mathbf{0}) = (\alpha U)^0$. □

**Lemma 11.4.17** *Suppose that $U$ and $V$ are finite dimensional spaces over $\mathbb{F}$ and $\alpha \in \mathcal{L}(U, V)$. Then, making the standard identification of $U''$ with $U$ and $V''$ with $V$, we have*

$$\alpha' V' = \left(\alpha^{-1}(\mathbf{0})\right)^0.$$

*Proof* Applying Lemma 11.4.16 to $\alpha' \in \mathcal{L}(V', U')$, we obtain

$$\alpha^{-1}(\mathbf{0}) = (\alpha'')^{-1}(\mathbf{0}) = (\alpha'V')^0$$

so, taking the annihilator of both sides, we obtain

$$\left(\alpha^{-1}(\mathbf{0})\right)^0 = (\alpha'V')^{00} = \alpha'V'$$

as stated.                                                                  □

We can summarise the results of the last two lemmas (when $U$ and $V$ are finite dimensional) in the formulae

$$\ker \alpha' = (\operatorname{im} \alpha)^0 \quad \text{and} \quad \operatorname{im} \alpha' = (\ker \alpha)^0.$$

We can now obtain a computation free proof of the fact that the row rank of a matrix equals its column rank.

**Lemma 11.4.18** *(i) If $U$ and $V$ are finite dimensional spaces over $\mathbb{F}$ and $\alpha \in \mathcal{L}(U, V)$, then*

$$\dim \operatorname{im} \alpha = \dim \operatorname{im} \alpha'.$$

*(ii) If $A$ is an $m \times n$ matrix, then the dimension of the space spanned by the rows of $A$ is equal to the dimension of the space spanned by the columns of $A$.*

*Proof* (i) Using the rank-nullity theorem we have

$$\dim \operatorname{im} \alpha = \dim U - \ker \alpha = \dim(\ker \alpha)^0 = \dim \operatorname{im} \alpha'.$$

(ii) Let $\alpha$ be the linear map having matrix $A$ with respect to the standard basis $\mathbf{e}_j$ (so $\mathbf{e}_j$ is the column vector with 1 in the $j$th place and 0 in all other places). Since $A\mathbf{e}_j$ is the $j$th column of $A$

$$\operatorname{span} \text{columns of } A = \operatorname{im} \alpha.$$

Similarly

$$\operatorname{span} \text{columns of } A^T = \operatorname{im} \alpha',$$

so

$$\dim(\operatorname{span} \text{rows of } A) = \dim(\operatorname{span} \text{columns of } A^T) = \dim \operatorname{im} \alpha'$$
$$= \dim \operatorname{im} \alpha = \dim(\operatorname{span} \text{columns of } A)$$

as stated.                                                                  □

**Exercise 11.4.19** *Suppose that $U$ is a finite dimensional space over $\mathbb{F}$ and $\alpha \in \mathcal{L}(U, U)$. By applying one of the results just obtained to $\lambda\iota - \alpha$, show that*

$$\dim\{\mathbf{u} \in U \ : \ \alpha\mathbf{u} = \lambda\mathbf{u}\} = \dim\{\mathbf{u}' \in U' \ : \ \alpha'\mathbf{u} = \lambda\mathbf{u}'\}.$$

*Interpret your result.*

**Exercise 11.4.20** *Let $V$ be a finite dimensional space over $\mathbb{F}$. If $\alpha$ is an endomorphism of $V$ and $U$ is a subspace of $V$, show that $\alpha'(\alpha U)^0$ is a subspace of $U^0$. Give examples (with $U \neq \{0\}$, $V$ and $\alpha \neq 0$, $\iota$) when $\alpha'(\alpha U)^0 = U^0$ and when $\alpha'(\alpha U)^0 \neq U^0$.*

## 11.5 Further exercises

**Exercise 11.5.1** We work over $\mathbb{F}$. We take $A$ to be an $m \times n$ matrix, $\mathcal{B}$ to be the set of all $n \times m$ matrices and $I_p$ to be the $p \times p$ identity matrix. Prove the following results by working with the corresponding linear maps.

(i) The equation $AB = I_m$ has a solution $B \in \mathcal{B}$ if and only if $m \leq n$ and rank $A = m$.

(ii) The equation $AB = I_m$ has a unique solution $B \in \mathcal{B}$ if and only if $m = n$ and rank $A = m$.

(iii) State and prove similar results for the equation $BA = I_n$.

Prove these results by using our earlier work on simultaneous linear equations.

**Exercise 11.5.2** If $\alpha$ is a singular endomorphism (that is to say, a non-invertible endomorphism) of the finite dimensional vector space $V \neq \{0\}$, show that we can find a non-zero endomorphism $\beta$ such that $\beta^2 = \beta\alpha = 0$.

**Exercise 11.5.3** Let $V$ be a finite dimensional vector space over $\mathbb{C}$ and let $U$ be a non-trivial subspace of $V$ (i.e. a subspace which is neither $\{0\}$ nor $V$). Without assuming any other results about linear mappings, prove that there is a linear mapping of $V$ onto $U$.

Are the following statements (a) always true, (b) sometimes true but not always true, (c) never true? Justify your answers in each case.

(i) There is a linear mapping of $U$ onto $V$ (that is to say, a surjective linear map).

(ii) There is a linear mapping $\alpha : V \to V$ such that $\alpha U = U$ and $\alpha \mathbf{v} = \mathbf{0}$ if $\mathbf{v} \in V \setminus U$.

(iii) Let $U_1, U_2$ be non-trivial subspaces of $V$ such that $U_1 \cap U_2 = \{0\}$ and let $\alpha_1, \alpha_2$ be linear mappings of $V$ into $V$. Then there is a linear mapping $\alpha : V \to V$ such that

$$\alpha \mathbf{v} = \begin{cases} \alpha_1 \mathbf{v} & \text{if } \mathbf{v} \in U_1, \\ \alpha_2 \mathbf{v} & \text{if } \mathbf{v} \in U_2. \end{cases}$$

(iv) Let $U_1, U_2, \alpha_1, \alpha_2$ be as in part (iii), and let $U_3, \alpha_3$ be similarly defined with $U_1 \cap U_3 = U_2 \cap U_3 = \{0\}$. Then there is a linear mapping $\alpha : V \to V$ such that

$$\alpha \mathbf{v} = \begin{cases} \alpha_1 \mathbf{v} & \text{if } \mathbf{v} \in U_1, \\ \alpha_2 \mathbf{v} & \text{if } \mathbf{v} \in U_2, \\ \alpha_3 \mathbf{v} & \text{if } \mathbf{v} \in U_3. \end{cases}$$

**Exercise 11.5.4** Let $\alpha : U \to V$ and $\beta : V \to W$ be maps between finite dimensional spaces, and suppose that $\ker \beta = \operatorname{im} \alpha$. Show that bases may be chosen for $U$, $V$ and $W$ with respect to which $\alpha$ and $\beta$ have matrices

$$\begin{pmatrix} I_r & 0 \\ 0 & 0 \end{pmatrix} \quad \text{and} \quad \begin{pmatrix} 0 & 0 \\ 0 & I_{n-r} \end{pmatrix}.$$

**Exercise 11.5.5** (i) If $V$ is an infinite dimensional space with a finite dimensional subspace $W$, show that $V/W$ is infinite dimensional.

(ii) Let $n \geq 0$. Give an example of an infinite dimensional space $V$ with a subspace $W$ such that

$$\dim V/W = n.$$

(iii) Give an example of an infinite dimensional space $V$ with an infinite dimensional subspace $W$ such that $V/W$ is infinite dimensional.

**Exercise 11.5.6** Suppose that $W$ is a finite dimensional vector space over $\mathbb{F}$ with subspaces $U$ and $V$. Show that $U \cap V$ and

$$U + V = \{\mathbf{u} + \mathbf{v} : \mathbf{u} \in U, \ \mathbf{v} \in V\}$$

are subspaces of $W$. Show that $(U + V)/U$ is isomorphic to $V/(U \cap V)$.

**Exercise 11.5.7** Let $U$, $V$, $W$, $X$ be finite dimensional spaces, let $\alpha \in \mathcal{L}(U, V)$ have rank $r$ and let $\beta \in \mathcal{L}(W, X)$ have rank $s$. Show that $\Gamma(\theta) = \beta \theta \alpha$ defines a linear map from $\mathcal{L}(V, W)$ to $\mathcal{L}(U, X)$ and find its rank.

**Exercise 11.5.8** Let $U$ and $V$ be finite dimensional vector spaces over $\mathbb{F}$. If $\alpha : U \to V$ is linear, show that there is a linear map $\beta : V \to U$ such that $\alpha \beta \alpha = \alpha$. Show that the linear maps $\beta$ such that $\alpha \beta \alpha$ is a scalar multiple of $\alpha$ (that is to say, $\alpha \beta \alpha = \lambda \alpha$ for some $\lambda \in \mathbb{F}$) form a subspace of $\mathcal{L}(U, V)$ and find its dimension in terms of the dimensions of $U$ and $V$ and the rank of $\alpha$.

**Exercise 11.5.9** Let $U$ and $V$ be finite dimensional spaces over $\mathbb{F}$ and let $\theta : U \to V$ be a linear map.

(i) Show that $\theta$ is injective if and only if, given any finite dimensional vector space $W$ over $\mathbb{F}$ and given any linear map $\alpha : U \to W$, there is a linear map $\hat{\alpha} : V \to W$ such that $\alpha = \theta \hat{\alpha}$.

(ii) Show that $\theta$ is surjective if and only if, given any finite dimensional vector space $W$ over $\mathbb{F}$ and given any linear map $\beta : W \to V$, there is a linear map $\hat{\beta} : W \to U$ such that $\beta = \hat{\beta} \theta$.

**Exercise 11.5.10** Let $\alpha_1, \alpha_2, \ldots, \alpha_k$ be endomorphisms of an $n$-dimensional vector space $V$. Show that

$$\dim(\alpha_1 \alpha_2 \ldots \alpha_n V) \geq \sum_{i=1}^{k} \dim(\alpha_j V) - n(k - 1).$$

Hence show that, if $\alpha$ is an endomorphism of $V$,

$$\dim V + \dim \alpha^2 V \geq 2 \dim \alpha V.$$

Show, more generally, that

$$\tfrac{1}{2}(\dim \alpha^j V + \dim \alpha^{j+2} V) \geq \dim \alpha^{j+1} V.$$

**Exercise 11.5.11** (Variation on a theme.) If $U$ is a finite dimensional vector space over $\mathbb{F}$ and $\alpha$, $\beta$, $\gamma \in \mathcal{L}(U, U)$ show that

$$\operatorname{rank} \alpha + \operatorname{rank} \alpha \beta \gamma \geq \operatorname{rank} \alpha \beta + \operatorname{rank} \beta \gamma.$$

[Exercise 5.7.8 gives a proof involving matrices, but you should provide a proof in the style of this chapter.]

**Exercise 11.5.12** Suppose that $V$ is a vector space over $\mathbb{F}$ with subspaces $U_1, U_2, \ldots, U_m$. Show that

$$(U_1 + U_2 + \cdots + U_m)^0 = U_1^0 \cap U_2^0 \cap \ldots \cap U_k^0.$$

If $V$ is finite dimensional, show that

$$(U_1 \cap U_2 \cap \ldots \cap U_k)^0 = U_1^0 + U_2^0 + \cdots + U_m^0.$$

**Exercise 11.5.13** Let $V$ be a vector space over $\mathbb{F}$. We make $V \times V$ into a vector space in the usual manner by setting

$$\lambda(\mathbf{x}, \mathbf{y}) = (\lambda \mathbf{x}, \lambda \mathbf{y}) \quad \text{and} \quad (\mathbf{x}, \mathbf{y}) + (\mathbf{u}, \mathbf{v}) = (\mathbf{x} + \mathbf{u}, \mathbf{y} + \mathbf{v}).$$

Prove the following results.

(i) The equation

$$(\alpha \phi)(\mathbf{x}, \mathbf{y}) = \tfrac{1}{2}\big(\phi(\mathbf{x}, \mathbf{y}) - \phi(\mathbf{y}, \mathbf{x})\big)$$

for all $\mathbf{x}, \mathbf{y} \in V$, $\phi \in \mathcal{L}(V \times V, \mathbb{F})$ defines an endomorphism $\alpha$ of $\mathcal{L}(V \times V, \mathbb{F})$.

(ii) $\alpha$ is a projection (that is to say, $\alpha^2 = \alpha$).

(iii) $\alpha$ has rank $\tfrac{1}{2}n(n-1)$ and nullity $\tfrac{1}{2}n(n+1)$.

(iv) If $\theta$ is an endomorphism of $V$, the equation

$$(\tilde{\theta}\phi)(\mathbf{x}, \mathbf{y}) = \phi(\theta \mathbf{x}, \theta \mathbf{y})$$

for all $\mathbf{x}, \mathbf{y} \in V$, $\phi \in \mathcal{L}(V \times V, \mathbb{F})$ defines an endomorphism $\tilde{\theta}$ of $\mathcal{L}(V \times V, \mathbb{F})$.

(v) With the notation above $\tilde{\theta}\alpha = \alpha\tilde{\theta}$.

(vi) State and sketch proofs for the corresponding results for $\beta$ given by

$$(\beta \phi)(\mathbf{x}, \mathbf{y}) = \tfrac{1}{2}\big(\phi(\mathbf{x}, \mathbf{y}) + \phi(\mathbf{y}, \mathbf{x})\big).$$

What is $\alpha + \beta$?

**Exercise 11.5.14** Are the following statements true or false. Give proofs or counterexamples. In each case $U$ and $V$ are vector spaces over $\mathbb{F}$ and $\alpha : U \to V$ is linear.

(i) If $U$ and $V$ are finite dimensional, then $\alpha : U \to V$ is injective if and only if the image $\alpha(\mathbf{e}_1), \alpha(\mathbf{e}_2), \ldots, \alpha(\mathbf{e}_n)$ of every finite linearly independent set $\mathbf{e}_1, \mathbf{e}_2, \ldots, \mathbf{e}_n$ is linearly independent.

(ii) If $U$ and $V$ are possibly infinite dimensional, then $\alpha : U \to V$ is injective if and only if the image $\alpha(\mathbf{e}_1), \alpha(\mathbf{e}_2), \ldots, \alpha(\mathbf{e}_n)$ of every finite linearly independent set $\mathbf{e}_1, \mathbf{e}_2, \ldots,$ $\mathbf{e}_n$ is linearly independent.

(iii) If $U$ and $V$ are finite dimensional, then the dual map $\alpha' : V' \to U'$ is surjective if and only if $\alpha : U \to V$ is injective.

(iv) If $U$ and $V$ are finite dimensional, then the dual map $\alpha' : V' \to U'$ is injective if and only if $\alpha : U \to V$ is surjective.

**Exercise 11.5.15** In the following diagram of finite dimensional spaces over $\mathbb{F}$ and linear maps

$$
\begin{array}{ccccc}
U_1 & \xrightarrow{\phi_1} & V_1 & \xrightarrow{\psi_1} & W_1 \\
\alpha\downarrow & & \beta\downarrow & & \gamma\downarrow \\
U_2 & \xrightarrow{\phi_2} & V_2 & \xrightarrow{\psi_2} & W_2
\end{array}
$$

$\phi_1$ and $\phi_2$ are injective, $\psi_1$ and $\psi_2$ are surjective, $\psi_i^{-1}(\mathbf{0}) = \phi_i(U_i)$ $(i = 1, 2)$ and the two squares commute (that is to say $\phi_2\alpha = \beta\phi_1$ and $\psi_2\beta = \gamma\psi_1$). If $\alpha$ and $\gamma$ are both injective, prove that $\beta$ is injective. (Start by asking what follows if $\beta\mathbf{v}_1 = \mathbf{0}$. You will find yourself at the first link in a long chain of reasoning where, at each stage, there is exactly one deduction you can make. The reasoning involved is called 'diagram chasing' and most mathematicians find it strangely addictive.)

By considering the duals of all the maps involved, prove that if $\alpha$ and $\gamma$ are surjective, then so is $\beta$.

The proof suggested in the previous paragraph depends on the spaces being finite dimensional. Produce an alternative diagram chasing proof.

**Exercise 11.5.16** (It may be helpful to have done the previous question.) Suppose that we are given vector spaces $U, V_1, V_2, W$ over $\mathbb{F}$ and linear maps $\phi_1 : U \to V_1, \phi_2 : U \to V_2,$ $\psi_1 : V_1 \to W, \psi_2 : V_2 \to W,$ and $\beta : V_1 \to V_2$. Suppose that the following four conditions hold.

(i) $\phi_i^{-1}(\{\mathbf{0}\}) = \{\mathbf{0}\}$ for $i = 1, 2$.
(ii) $\psi_i(V_i) = W$ for $i = 1, 2$.
(iii) $\psi_i^{-1}(\{\mathbf{0}\}) = \phi_1(U_i)$ for $i = 1, 2$.
(iv) $\phi_1\alpha = \phi_2$ and $\alpha\psi_2 = \psi_1$.

Prove that $\beta : V_1 \to V_2$ is an isomorphism. You may assume that the spaces are finite dimensional if you wish.

**Exercise 11.5.17** In the diagram below $A$, $B$, $A_1$, $B_1$, $C_1$, $B_2$, $C_2$ are vector spaces over $\mathbb{F}$ and $\alpha$, $\alpha_1$, $\beta_1$, $\beta_2$, $\phi$, $\psi$, $\psi_1$, $\eta_1$ are linear maps between the spaces indicated such that

(a) $\psi\alpha = \alpha_1\phi$, (b) $\eta_1\beta_1 = \beta_2\psi_1$, (c) $\phi$ is surjective, (d) $\eta_1$ is injective.

$$
\begin{array}{ccc}
A & \xrightarrow{\alpha} & B \\
\downarrow{\phi} & & \downarrow{\psi} \\
A_1 \xrightarrow{\alpha_1} & B_1 \xrightarrow{\beta_1} & C_1 \\
& \downarrow{\psi_1} & \downarrow{\eta_1} \\
& B_2 \xrightarrow{\beta_2} & C_2
\end{array}
$$

Prove the following results.

(i) If the null-space of $\beta_1$ is contained in the image of $\alpha_1$, then the null-space of $\psi_1$ is contained in the image of $\psi$.

(ii) If $\psi_1\psi$ is a zero map, then so is $\beta_1\alpha_1$.

**Exercise 11.5.18** We work in the space $M_n(\mathbb{R})$ of $n \times n$ real matrices. Recall the definition of a trace given, for example, in Exercise 6.8.20.

If we write $t(A) = \mathrm{Tr}(A)$, show the following.

(i) $t : M_n(\mathbb{R}) \to \mathbb{R}$ is linear.

(ii) $t(P^{-1}AP) = t(A)$ whenever $A$, $P \in M_n(\mathbb{R})$ and $P$ is invertible.

(iii) $t(I) = n$.

Show, conversely, that, if $t$ satisfies these conditions, then $t(A) = \mathrm{Tr}\, A$.

**Exercise 11.5.19** Consider $M_n$ the vector space of $n \times n$ matrices over $\mathbb{F}$ with the usual matrix addition and scalar multiplication. If $f$ is an element of the dual $M_n'$, show that

$$f(XY) = f(YX)$$

for all $X$, $Y \in M_n$ if and only if

$$f(X) = \lambda\, \mathrm{Tr}\, X$$

for all $X \in M_n$ and some fixed $\lambda \in \mathbb{F}$.

Deduce that, if $A \in M_n$ is the sum of matrices of the form $[X, Y] = XY - YX$, then $\mathrm{Tr}\, A = 0$. Show, conversely, that, if $\mathrm{Tr}\, A = 0$, then $A$ is the sum of matrices of the form $[X, Y]$. (In Exercise 12.6.24 we shall see that a stronger result holds.)

**Exercise 11.5.20** Let $M_{p,q}$ be the usual vector space of $p \times q$ matrices over $\mathbb{F}$. Let $A \in M_{n,m}$ be fixed. If $B \in M_{m,n}$ we write

$$\tau_A B = \mathrm{Tr}\, AB.$$

Show that $\tau$ is a linear map from $M_{m,n}$ to $\mathbb{F}$. If we set $\Theta(A) = \tau_A$ show that $\Theta : M_{n,m} \to M_{m,n}'$ is an isomorphism.

**Exercise 11.5.21 [The Binet–Cauchy formula]** (This is included as a mathematical curiosity. It may be helpful to experiment with small matrices.) Let $B$ be an $m \times n$ matrix and $C$ an $n \times m$ matrix over $\mathbb{F}$. If $m \leq n$ the $m \times m$ matrix formed from $B$ by using the $i_r$th column of $B$ as the $r$th column of the new matrix is called $B^{i_1 i_2 \ldots i_m}$. The $m \times m$ matrix formed from $C$ by using the $i_r$th row of $C$ as the $r$th row of the new matrix is called $C_{i_1 i_2 \ldots i_m}$. The Binet–Cauchy formula states that

$$\det BC = \sum \det B^{i_1 i_2 \ldots i_m} \det C_{i_1 i_2 \ldots i_m}$$

where the sum is over all $i_1, i_2, \ldots, i_m$ with

$$1 \leq i_1 < i_2 < \ldots < i_m.$$

(i) By considering row operations on $B$ and column operations on $C$, show that the full Binet–Cauchy formula will follow from the special case when $B$ has first row $\mathbf{b}_1 = (1, 0, 0, \ldots, 0)$ and $C$ has first column $\mathbf{c}_1 = (1, 0, 0, \ldots, 0)^T$.

(ii) By using (i), or otherwise, show that if the Binet–Cauchy formula holds when $m = p, n = q - 1$ and when $m = p - 1, n = q - 1$, then it holds for $m = p, n = q$ [$2 \leq p \leq q - 1$]. Deduce that the Binet–Cauchy formula holds for all $1 \leq m \leq n$.

(iii) What can you say about $\det AB$ if $m > n$?

(iv) Prove the Binet–Cauchy identity

$$\left( \sum_{i=1}^{n} a_i b_i \right) \left( \sum_{j=1}^{n} c_j d_j \right) = \left( \sum_{i=1}^{n} a_i d_i \right) \left( \sum_{j=1}^{n} b_j c_j \right) + \sum_{1 \leq i < j \leq n} (a_i b_j - a_j b_i)(c_i d_j - c_j d_i)$$

for all $a_i, b_i, c_i, d_i \in \mathbb{F}$ and deduce Lagrange's identity

$$\left( \sum_{i=1}^{n} a_i^2 \right) \left( \sum_{j=1}^{n} c_j^2 \right) = \left( \sum_{i=1}^{n} a_i c_i \right)^2 + \sum_{1 \leq i < j \leq n} (a_i c_j - a_j c_i)^2.$$

(v) Use Lagrange's identity to prove the Cauchy–Schwarz inequality in $\mathbb{R}^n$, identifying the cases of equality.

[Cauchy and Binet were the first to prove the result which we would now write $\det AB = \det A \det B$. As with many other topics in mathematics, Cauchy's work marked the beginning of the modern era.]

**Exercise 11.5.22** (Requires elementary group theory.) Show that the set of matrices

$$\begin{pmatrix} 0 & 0 \\ 0 & x \end{pmatrix}$$

with $x$ real and non-zero forms a group under matrix multiplication.

Let $V$ be a vector space over $\mathbb{F}$ and let $G$ be a set of linear maps $\alpha : V \to V$ which forms a group under composition. Show that *either* every $\alpha \in G$ is invertible *or* no $\alpha \in G$ is invertible.

Show that all $\alpha \in G$ have the same image space $E = \alpha(V)$ and the same null-space $\alpha^{-1}(\mathbf{0})$.

For each $\alpha \in G$ define $T(\alpha) : E \to E$ by

$$T(\alpha)(\mathbf{x}) = \alpha(\mathbf{x}).$$

Show that $T$ is group isomorphism between $G$ and $\tilde{G}$ a group of invertible linear mappings on $E$.

Give an example to show that $\tilde{G}$ need not contain all the invertible linear mappings on $E$.

**Exercise 11.5.23** Consider the set $\mathbb{F}^X$ of functions $f : X \to \mathbb{F}$. We have seen in Lemma 5.2.6 how to make $\mathbb{F}^X$ into a vector space $(\mathbb{F}^X, \times, +, \mathbb{F})$ over $\mathbb{F}$.

Suppose that we make $\mathbb{F}^X$ into a vector space $V_\square = (\mathbb{F}^X, \boxtimes, \boxplus, \mathbb{F})$ over $\mathbb{F}$. Show that the *point evaluation* functions $\phi_x : V_\square \to \mathbb{F}$, defined by $\phi_x(f) = f(x)$, are linear maps for each $x \in X$ if and only if $\lambda \boxtimes f = \lambda \times f$ and $f \boxplus g = f + g$ for all $\lambda \in \mathbb{F}$ and $f, g \in \mathbb{F}^X$. (More succinctly, the standard vector space structure on $\mathbb{F}^X$ is the only one for which point evaluations are linear maps.)

[Your answer may be shorter than the statement of the exercise.]

**Exercise 11.5.24** Suppose that $X$ is a subset of a vector space $U$ over $\mathbb{F}$. Show that

$$X^0 = \{\mathbf{u}' \in U' : \mathbf{u}'\mathbf{x} = 0 \quad \text{for all } \mathbf{x} \in X\}$$

is a subspace of $U'$.

If $U$ is finite dimensional and we identify $U''$ and $U$ in the usual manner, show that $X^{00} = \text{span } X$.

**Exercise 11.5.25** The object of this question is to show that the mapping $\Theta : U \to U''$ defined in Lemma 11.3.4 is not bijective for all vector spaces $U$. I owe this example to Imre Leader and the reader is warned that the argument is at a slightly higher level of sophistication than the rest of the book. We take $c_{00}$ and $s$ as in Exercise 11.3.9. We write $\Theta_U = \Theta$ to make the space $U$ explicit.

(i) Show that, if $\mathbf{a} \in c_{00}$ and $(\Theta_{c_{00}}\mathbf{a})\mathbf{b} = 0$ for all $\mathbf{b} \in c_{00}$, then $\mathbf{a} = \mathbf{0}$.

(ii) Consider the vector space $V = s/c_{00}$. By looking at $(1, 1, 1, \ldots)$, or otherwise, show that $V$ is not zero dimensional. (That is to say, $V \neq \{\mathbf{0}\}$.)

(iii) If $V'$ is zero dimensional, explain why $V''$ is zero dimensional and so $\Theta_V$ is not injective.

(iv) If $V'$ is not zero dimensional, pick a non-zero $T \in V'$. Define $\tilde{T} : s \to \mathbb{R}$ by

$$\tilde{T}\mathbf{a} = T(\mathbf{a} + c_{00}).$$

Show that $T \in s'$ and $T \neq 0$, but $T\mathbf{b} = 0$ for all $\mathbf{b} \in c_{00}$. Deduce that $\Theta_{c_{00}}$ is not injective.

# 12

# Polynomials in $\mathcal{L}(U, U)$

## 12.1 Direct sums

In this section we develop some of the ideas from Section 5.4 which the reader may wish to reread. We start with some useful definitions.

**Definition 12.1.1** *Let $U$ be a vector space over $\mathbb{F}$ with subspaces $U_j$ $[1 \le j \le m]$.*
*(i) We say that $U$ is the* sum *of the subspaces $U_j$ and write*

$$U = U_1 + U_2 + \cdots + U_m$$

*if*

$$U = \{\mathbf{u}_1 + \mathbf{u}_2 + \cdots + \mathbf{u}_m \, : \, \mathbf{u}_j \in U_j\}.$$

*(ii) We say that $U$ is the* direct sum *of the subspaces $U_j$ and write*

$$U = U_1 \oplus U_2 \oplus \ldots \oplus U_m$$

*if $U = U_1 + U_2 + \cdots + U_m$ and, in addition, the equation*

$$\mathbf{0} = \mathbf{v}_1 + \mathbf{v}_2 + \cdots + \mathbf{v}_m$$

*with $\mathbf{v}_j \in U_j$ implies*

$$\mathbf{v}_1 = \mathbf{v}_2 = \ldots = \mathbf{v}_m = \mathbf{0}.$$

[*We discuss a related idea in Exercise 12.6.4.*]

Before starting the discussion that follows, the reader should recall the useful result about $\dim(V + W)$ which we proved in Lemma 5.4.10.

**Exercise 12.1.2** *Let $U$ be a vector space over $\mathbb{F}$ with subspaces $U_j$ $[1 \le j \le m]$. Show that $U = U_1 \oplus U_2 \oplus \ldots \oplus U_m$ if and only if the equation*

$$\mathbf{u} = \mathbf{u}_1 + \mathbf{u}_2 + \cdots + \mathbf{u}_m$$

*has exactly one solution with $\mathbf{u}_j \in U_j$ for each $\mathbf{u} \in U$.*

The following exercise is easy, but the result is useful.

**Exercise 12.1.3** *Let $U$ be a vector space over $\mathbb{F}$ which is the direct sum of subspaces $U_j$ $[1 \leq j \leq m]$.*

*(i) If $U_j$ has a basis $\mathbf{e}_{jk}$ with $1 \leq k \leq n(j)$ for $1 \leq j \leq m$, show that the vectors $\mathbf{e}_{jk}$ $[1 \leq k \leq n(j), 1 \leq j \leq m]$ form a basis for $U$.*

*(ii) Show that $U$ is finite dimensional if and only if all the $U_j$ are.*

*(iii) If $U$ is finite dimensional, show that*

$$\dim U = \dim U_1 + \dim U_2 + \cdots + \dim U_m.$$

**Exercise 12.1.4** *Let $U$ and $W$ be subspaces of a finite dimensional vector space $V$. Show that there exist subspaces $A$, $B$ and $C$ of $V$ such that*

$$U = A \oplus B, \quad W = B \oplus C, \quad U + W = U \oplus C = W \oplus A.$$

*Show that $B$ is specified uniquely by the conditions just given. Is the same true for $A$ and $C$? Give a proof or counterexample.*

**Lemma 12.1.5** *Let $U$ be a vector space over $\mathbb{F}$ which is the direct sum of subspaces $U_j$ $[1 \leq j \leq m]$. If $U$ is finite dimensional, we can find linear maps $\pi_j : U \to U_j$ such that*

$$\mathbf{u} = \pi_1\mathbf{u} + \pi_2\mathbf{u} + \cdots + \pi_m\mathbf{u}.$$

*Automatically $\pi_j|_{U_j} = \iota|_{U_j}$ (i.e. $\pi_j\mathbf{u} = \mathbf{u}$ whenever $\mathbf{u} \in U_j$).*

*Proof* Let $U_j$ have a basis $\mathbf{e}_{jk}$ with $1 \leq k \leq n(j)$. Then, as remarked in Exercise 12.1.3, the vectors $\mathbf{e}_{jk}$ $[1 \leq k \leq n(j), 1 \leq j \leq m]$ form a basis for $U$. It follows that, if $\mathbf{u} \in U$, there are unique $\lambda_{jk} \in \mathbb{F}$ such that

$$\mathbf{u} = \sum_{j=1}^{m} \sum_{k=1}^{n(j)} \lambda_{jk} \mathbf{e}_{jk}$$

and we may define $\pi_j\mathbf{u} \in U_j$ by

$$\pi_j\mathbf{u} = \sum_{k=1}^{n(j)} \lambda_{jk} \mathbf{e}_{jk}.$$

We note that

$$\pi_j \left( \lambda \sum_{r=1}^{m} \sum_{k=1}^{n(r)} \lambda_{rk}\mathbf{e}_{rk} + \mu \sum_{r=1}^{m} \sum_{k=1}^{n(r)} \mu_{rk}\mathbf{e}_{rk} \right) = \pi_j \left( \sum_{r=1}^{m} \sum_{k=1}^{n(r)} (\lambda\lambda_{rk} + \mu\mu_{rk})\mathbf{e}_{rk} \right)$$

$$= \sum_{k=1}^{n(j)} (\lambda\lambda_{jk} + \mu\mu_{jk})\mathbf{e}_{jk}$$

$$= \lambda \sum_{k=1}^{n(j)} \lambda_{jk}\mathbf{e}_{jk} + \mu \sum_{k=1}^{n(j)} \mu_{jk}\mathbf{e}_{jk}$$

$$= \lambda\pi_j \left( \sum_{r=1}^{m} \sum_{k=1}^{n(r)} \lambda_{rk}\mathbf{e}_{rk} \right) + \mu\pi_j \left( \sum_{r=1}^{m} \sum_{k=1}^{n(r)} \mu_{rk}\mathbf{e}_{rk} \right).$$

Thus $\pi_j : U \to U_j$ is linear. The equality

$$\mathbf{u} = \pi_1 \mathbf{u} + \pi_2 \mathbf{u} + \cdots + \pi_m \mathbf{u}$$

follows directly from our definition of $\pi_j$.

The final remark follows from the fact that we have a direct sum. $\qquad\square$

**Exercise 12.1.6** *Suppose that $V$ is a finite dimensional vector space over $\mathbb{F}$ with subspaces $V_1$ and $V_2$. Which of the following possibilities can occur? Prove your answers.*

*(i) $V_1 + V_2 = V$, but $\dim V_1 + \dim V_2 > \dim V$.*

*(ii) $V_1 + V_2 = V$, but $\dim V_1 + \dim V_2 < \dim V$.*

*(iii) $\dim V_1 + \dim V_2 = \dim V$, but $V_1 + V_2 \neq V$.*

The next exercise develops the ideas of Theorem 11.2.2.

**Exercise 12.1.7** *A subspace $V$ of a vector space $U$ over $\mathbb{F}$ is said to be an* invariant subspace *of an $\alpha \in \mathcal{L}(V, V)$ if $V \subseteq \alpha V$. We say that $V$ is a* maximal invariant subspace *of $\alpha$ if $V$ is an invariant space of $\alpha$ and, whenever $W$ is an invariant subspace of $\alpha$ with $W \supseteq V$, it follows that $W = V$.*

*Show that, if $U$ is finite dimensional, and $\alpha$ is an endomorphism of $U$, the following statements are true.*

*(i) There is a non-negative integer $m$ such that $\alpha^m U$ is the unique maximal invariant subspace of $\alpha$.*

*(ii) If we write $M = \alpha^m U$ and $N = (\alpha^m)^{-1}(\mathbf{0})$, then $U = M \oplus N$.*

*(iii) $\alpha(M) \subseteq M$, $\alpha(N) \subseteq N$.*

*(iv) If we define $\beta : M \to M$ by $\beta\mathbf{u} = \alpha\mathbf{u}$ for $\mathbf{u} \in M$ and $\gamma : N \to N$ by $\gamma\mathbf{u} = \alpha\mathbf{u}$ for $\mathbf{u} \in N$, then $\beta$ is an automorphism and $\gamma$ is nilpotent (that is to say, $\gamma^r = 0$ for some $r \geq 1$).*

*(v) Suppose that $\tilde{M}$ and $\tilde{N}$ are subspaces of $U$ such that $V = \tilde{M} \oplus \tilde{N}$, $\tilde{\beta}$ is an automorphism on $\tilde{M}$ and $\tilde{\gamma} : M \to M$ is a nilpotent linear map. If*

$$\alpha(\mathbf{a} + \mathbf{b}) = \tilde{\beta}\mathbf{a} + \tilde{\gamma}\mathbf{b}$$

*for all $\mathbf{a} \in \tilde{M}$, $\mathbf{b} \in \tilde{N}$, show that $\tilde{M} = M$, $\tilde{N} = N$, $\tilde{\beta} = \beta$ and $\tilde{\gamma} = \gamma$.*

*(vi) If $A$ is an $n \times n$ matrix over $\mathbb{F}$, show that we can find an invertible $n \times n$ matrix $P$, an invertible $r \times r$ matrix $B$ and a nilpotent $n - r \times n - r$ matrix $C$ such that*

$$P^{-1}AP = \begin{pmatrix} B & 0 \\ 0 & C \end{pmatrix}.$$

As the previous exercise indicates, we are often interested in decomposing a space into the direct sum of two subspaces.

**Definition 12.1.8** *Let $U$ be a vector space over $\mathbb{F}$ with subspaces $U_1$ and $U_2$. If $U$ is the direct sum of $U_1$ and $U_2$, then we say that $U_2$ is a* complementary subspace *of $U_1$.*

It is very important to remember that complementary subspaces are not unique in general. The following simple example should be kept constantly in mind.

**Example 12.1.9** *Consider $\mathbb{F}^2$ as a row vector space over $\mathbb{F}$. If*

$$U_1 = \{(x,0),\ x \in \mathbb{F}\}, \quad U_2 = \{(0,y),\ y \in \mathbb{F}\}, \quad U_3 = \{(t,t),\ t \in \mathbb{F}\},$$

*then the $U_j$ are subspaces and both $U_2$ and $U_3$ are complementary subspaces of $U_1$.*

We give some variations on this theme in Exercise 12.6.29. When we consider inner product spaces, we shall look at the notion of an 'orthogonal complement' (see Lemma 14.3.6).

We shall need the following result.

**Lemma 12.1.10** *Every subspace $V$ of a finite dimensional vector space $U$ has a complementary subspace.*

*Proof* Since $V$ is the subspace of a finite dimensional space, it is itself finite dimensional and has a basis $\mathbf{e}_1, \mathbf{e}_2, \ldots, \mathbf{e}_k$, say, which can be extended to basis $\mathbf{e}_1, \mathbf{e}_2, \ldots, \mathbf{e}_n$ of $U$. If we take

$$W = \mathrm{span}\{\mathbf{e}_{k+1},\ \mathbf{e}_{k+2},\ \ldots,\ \mathbf{e}_n\},$$

then $W$ is a complementary subspace of $V$.                                  $\square$

**Exercise 12.1.11** *Let $V$ be a subspace of a finite dimensional vector space $U$. Show that $V$ has a unique complementary subspace if and only if $V = \{\mathbf{0}\}$ or $V = U$.*

We use the ideas of this section to solve a favourite problem of the Tripos examiners in the 1970s.

**Example 12.1.12** *Consider the vector space $M_n(\mathbb{R})$ of real $n \times n$ matrices with the usual definitions of addition and multiplication by scalars. If $\theta : M_n(\mathbb{R}) \to M_n(\mathbb{R})$ is given by $\theta(A) = A^T$, show that $\theta$ is linear and that*

$$M_n(\mathbb{R}) = \{A \in M_n(\mathbb{R}) : \theta(A) = A\} \oplus \{A \in M_n(\mathbb{R}) : \theta(A) = -A\}.$$

*Hence, or otherwise, find $\det \theta$.*

*Solution.* Suppose that $A = (a_{ij}) \in M_n(\mathbb{R})$, $B = (b_{ij}) \in M_n(\mathbb{R})$ and $\lambda, \mu \in \mathbb{R}$. If $C = (\lambda A + \mu B)^T$ and we write $C = (c_{ij})$, then

$$c_{ij} = \lambda a_{ji} + \mu b_{ji},$$

so $\theta(\lambda A + \mu B) = \lambda \theta(A) + \mu \theta(B)$. Thus $\theta$ is linear.

If we write

$$U = \{A \in M_n(\mathbb{R}) : \theta(A) = A\}, \quad V = \{A \in M_n(\mathbb{R}) : \theta(A) = -A\},$$

then $U$ is the null-space of $\iota - \theta$ and $V$ is the null-space of $\iota + \theta$, so $U$ and $V$ are subspaces of $M_n(\mathbb{R})$.

To see that $U + V = M_n(\mathbb{R})$, observe that, if $A \in M_n(\mathbb{R})$, then

$$A = 2^{-1}(A + A^T) + 2^{-1}(A - A^T)$$

and $2^{-1}(A + A^T) \in U, 2^{-1}(A - A^T) \in V$. To see that $U \cap V = \{0\}$ observe that

$$A \in U \cap V \Rightarrow A^T = A, \quad A^T = -A \Rightarrow A = -A \Rightarrow A = 0.$$

Next observe that, if $1 \le s < r \le n$ and $F(r, s)$ is the matrix $(\delta_{ir}\delta_{js} - \delta_{is}\delta_{jr})$ (that is to say, with entry 1 in the $r$, $s$th place, entry $-1$ in the $s$, $r$th place and 0 in all other places), then $F_{r,s} \in V$. Further, if $A = (a_{ij}) \in V$,

$$A = \sum_{1 \le s < r \le n} \lambda_{r,s} F_{r,s} \Leftrightarrow \lambda_{r,s} = a_{r,s} \quad \text{for all } 1 \le s < r \le n.$$

Thus the set of matrices $F(r, s)$ with $1 \le s < r \le n$ form a basis for $V$. This shows that $\dim V = n(n - 1)/2$.

A similar argument shows that the matrices $E(r, s)$ given by $(\delta_{ir}\delta_{js} + \delta_{is}\delta_{jr})$ when $r \ne s$ and $(\delta_{ir}\delta_{jr})$ when $r = s$ $[1 \le s \le r \le n]$ form a basis for $U$ which thus has dimension $n(n + 1)/2$.

If we give $M_n(\mathbb{R})$ the basis consisting of the $E(r, s)$ $[1 \le s \le r \le n]$ and $F(r, s)$ $[1 \le s < r \le n]$, then, with respect to this basis, $\theta$ has an $n^2 \times n^2$ diagonal matrix with $\dim U$ diagonal entries taking the value 1 and $\dim V$ diagonal entries taking the value $-1$. Thus

$$\det \theta = (-1)^{\dim V} = (-1)^{n(n-1)/2}.$$

$\square$

**Exercise 12.1.13** *We continue with the notation of Example 12.1.12.*

*(i) Suppose that we form a basis for $M_n(\mathbb{R})$ by taking the union of bases of U and V (but not necessarily those in the solution). What can you say about the corresponding matrix of $\theta$?*

*(ii) Show that the value of $\det \theta$ depends on the value of n modulo 4. State and prove the appropriate rule for obtaining $\det \theta$ from the value of n modulo 4.*

**Exercise 12.1.14** *Consider the real vector space $C(\mathbb{R})$ of continuous functions $f : \mathbb{R} \to \mathbb{R}$ with the usual pointwise definitions of addition and multiplication by a scalar. If*

$$U = \{f \in C(\mathbb{R}) : f(x) = f(-x) \text{ for all } x \in \mathbb{R}\},$$
$$V = \{f \in C(\mathbb{R}) : f(x) = -f(-x) \text{ for all } x \in \mathbb{R}, \},$$

*show that U and V are complementary subspaces.*

The following exercise introduces some very useful ideas.

**Exercise 12.1.15 [Projection]** *Prove that the following three conditions on an endomorphism $\alpha$ of a finite dimensional vector space V are equivalent.*

*(i) $\alpha^2 = \alpha$.*

*(ii) V can be expressed as a direct sum $U \oplus W$ of subspaces in such a way that $\alpha|_U$ is the identity mapping of U and $\alpha|_W$ is the zero mapping of W.*

*(iii) A basis of V can be chosen so that all the non-zero elements of the matrix representing $\alpha$ lie on the main diagonal and take the value 1.*

*[You may find it helpful to use the identity $\iota = \alpha + (\iota - \alpha)$.]*

*An endomorphism of V satisfying any (and hence all) of the above conditions is called a* projection.[1]

*Consider the following linear maps $\alpha_j : \mathbb{R}^2 \to \mathbb{R}^2$ (we use row vectors). Which of them are projections and why?*

$$\alpha_1(x, y) = (x, 0), \quad \alpha_2(x, y) = (0, x), \quad \alpha_3(x, y) = (y, x), \quad \alpha_4(x, y) = (x + y, 0),$$
$$\alpha_5(x, y) = (x + y, x + y), \quad \alpha_6(x, y) = \left(\tfrac{1}{2}(x + y), \tfrac{1}{2}(x + y)\right).$$

**Exercise 12.1.16** *If $\alpha$ is an endomorphism of a finite dimensional space V, show that $\alpha$ is a projection if and only if $\iota - \alpha$ is.*

**Exercise 12.1.17** *Suppose that $\alpha, \beta \in \mathcal{L}(V, V)$ are both projections of V. Prove that, if $\alpha\beta = \beta\alpha$, then $\alpha\beta$ is also a projection of V. Show that the converse is false by giving examples of projections $\alpha, \beta$ such that (a) $\alpha\beta$ is a projection, but $\beta\alpha$ is not, and (b) $\alpha\beta$ and $\beta\alpha$ are both projections, but $\alpha\beta \neq \beta\alpha$.*

**Exercise 12.1.18** *Suppose that $\alpha, \beta \in \mathcal{L}(V, V)$ are both projections of V.*

*(i) By considering what happens if we multiply by $\alpha$ on the left and what happens if we multiply by $\alpha$ on the right, show that*

$$\alpha\beta = -\beta\alpha \Rightarrow \alpha\beta = \beta\alpha = 0.$$

*(ii) Show that $\alpha + \beta$ is a projection if and only if $\alpha\beta = \beta\alpha = 0$.*
*(iii) Show that $\alpha - \beta$ is a projection if and only if $\alpha\beta = \beta\alpha = \beta$.*

**Exercise 12.1.19** *Let V be a finite dimensional vector space over $\mathbb{F}$ and $\alpha$ an endomorphism. Show that $\alpha$ is diagonalisable if and only if there exist distinct $\lambda_j \in \mathbb{F}$ and projections $\pi_j$ such that $\pi_k \pi_j = 0$ when $k \neq j$,*

$$\iota = \pi_1 + \pi_2 + \cdots + \pi_m \quad and \quad \alpha = \lambda_1 \pi_1 + \lambda_2 \pi_2 + \cdots + \lambda_m \pi_m.$$

## 12.2 The Cayley–Hamilton theorem

We start with a couple of observations.

**Exercise 12.2.1** *Show, by exhibiting a basis, that the space $M_n(\mathbb{F})$ of $n \times n$ matrices (with the standard structure of vector space over $\mathbb{F}$) has dimension $n^2$. Deduce that we can find $a_j \in \mathbb{C}$, not all zero, such that $\sum_{j=0}^{n^2} a_j A^j = 0$. Conclude that there is a non-trivial polynomial P of degree at most $n^2$ such that $P(A) = 0$.*

In Exercise 12.6.13 we show that Exercise 12.2.1 can be used to give a quick proof, without using determinants, that, if $U$ is a finite dimensional vector space over $\mathbb{C}$, every $\alpha \in \mathcal{L}(U, U)$ has an eigenvalue.

---

[1] We discuss *orthogonal projection* in Exercise 14.3.14.

**Exercise 12.2.2** (i) *If D is an $n \times n$ diagonal matrix over $\mathbb{F}$ with $j$th diagonal entry $\lambda_j$, write down the characteristic polynomial*

$$\chi_D(t) = \det(tI - D)$$

*as the product of linear factors.*

*If we write $\chi_D(t) = \sum_{k=0}^{n} b_k t^k$ with $b_k \in \mathbb{F}$ [$t \in \mathbb{F}$], show that*

$$\sum_{k=0}^{n} b_k D^k = 0.$$

*More briefly, we say that $\chi_D(D) = 0$.*

*(ii) If A is an $n \times n$ diagonalisable matrix over $\mathbb{F}$ with characteristic polynomial*

$$\chi_A(t) = \det(tI - A) = \sum_{k=0}^{n} c_k t^k,$$

*show that $\chi_A(A) = 0$, that is to say,*

$$\sum_{k=0}^{n} c_k A^k = 0.$$

**Exercise 12.2.3** *Recall that the trace of an $n \times n$ matrix $A = (a_{ij})$ is given by $\mathrm{Tr}\, A = \sum_{j=1}^{n} a_{jj}$. In this exercise A and B will be $n \times n$ matrices over $\mathbb{F}$.*

*(i) If B is invertible, show that $\mathrm{Tr}\, B^{-1} A B = \mathrm{Tr}\, A$.*

*(ii) If I is the identity matrix, show that $Q_A(t) = \mathrm{Tr}(tI - A)$ is a (rather simple) polynomial in t. If B is invertible, show that $Q_{B^{-1}AB} = Q_A$.*

*(iii) Show that, if $n \geq 2$, there exists an A with $Q_A(A) \neq 0$. What happens if $n = 1$?*

Exercise 12.2.2 suggests that the following result might be true.

**Theorem 12.2.4 [Cayley–Hamilton over $\mathbb{C}$]** *If U is a vector space of dimension n over $\mathbb{C}$ and $\alpha : U \to U$ is linear, then, writing*

$$\chi_\alpha(t) = \sum_{k=0}^{n} a_k t^k = \det(t\iota - \alpha),$$

*we have $\sum_{k=0}^{n} a_k \alpha^k = 0$ or, more briefly, $\chi_\alpha(\alpha) = 0$.*

We sometimes say that '$\alpha$ satisfies its own characteristic equation'.

Exercise 12.2.3 tells us that any attempt to prove Theorem 12.2.4 by ignoring the difference between a scalar and a linear map (or matrix) and 'just setting $t = A$' is bound to fail.

Exercise 12.2.2 tells us that Theorem 12.2.4 is true when $\alpha$ is diagonalisable, but we know that not every linear map is diagonalisable, even over $\mathbb{C}$.

However, we can apply the following useful substitute for diagonalisation.

**Theorem 12.2.5** *If $V$ is a finite dimensional vector space over $\mathbb{C}$ and $\alpha : V \to V$ is linear, we can find a basis for $V$ with respect to which $\alpha$ has a upper triangular matrix $A$ (that is to say, a matrix $A = (a_{ij})$ with $a_{ij} = 0$ for $i > j$).*

*Proof* We use induction on the dimension $m$ of $V$. Since every $1 \times 1$ matrix is upper triangular, the result is true when $m = 1$. Suppose that the result is true when $m = n - 1$ and that $V$ has dimension $n$.

Since we work over $\mathbb{C}$, the linear map $\alpha$ must have at least one eigenvalue $\lambda_1$ with a corresponding eigenvector $\mathbf{e}_1$. Let $W$ be a complementary subspace for $\mathrm{span}\{\mathbf{e}_1\}$. By Lemma 12.1.5, we can find linear maps $\tau : V \to \mathrm{span}\{\mathbf{e}_1\}$ and $\pi : V \to W$ such that

$$\mathbf{u} = \tau\mathbf{u} + \pi\mathbf{u}.$$

Now $(\pi\alpha)|_W$ is a linear map from $W$ to $W$ and $W$ has dimension $n - 1$ so, by the inductive hypothesis, we can find a basis $\mathbf{e}_2, \mathbf{e}_3, \ldots, \mathbf{e}_n$ with respect to which $(\pi\alpha)|_W$ has an upper triangular matrix. The statement that $(\pi\alpha)|_W$ has an upper triangular matrix means that

$$(\pi\alpha)\mathbf{e}_j \in \mathrm{span}\{\mathbf{e}_2, \mathbf{e}_3, \ldots, \mathbf{e}_j\} \qquad\qquad \bigstar$$

for $2 \leq j \leq n$.

Since $W$ is a complementary space of $\mathrm{span}\{\mathbf{e}_1\}$, it follows that $\mathbf{e}_1, \mathbf{e}_2, \ldots, \mathbf{e}_n$ form a basis of $V$. But $\bigstar$ tells us that

$$\alpha\mathbf{e}_j \in \mathrm{span}\{\mathbf{e}_1, \mathbf{e}_2, \ldots, \mathbf{e}_j\}$$

for $2 \leq j \leq n$ and the statement

$$\alpha\mathbf{e}_1 \in \mathrm{span}\{\mathbf{e}_1\}$$

is automatic. Thus $\alpha$ has upper triangular matrix with respect to the given matrix and the induction is complete $\qquad\qquad\qquad\qquad\qquad\qquad\qquad\qquad\qquad\qquad\qquad\qquad\quad\square$

A slightly different proof using quotient spaces is outlined in Exercise 12.6.21 (If we deal with inner product spaces, then, as we shall see later, Theorem 12.2.5 can be improved to give Theorem 15.2.1.)

**Exercise 12.2.6** *By considering roots of the characteristic polynomial, or otherwise, show, by example, that the result corresponding to Theorem 12.2.5 is false if $V$ is a finite dimensional vector space of dimension greater than 1 over $\mathbb{R}$. What can we say if $\dim V = 1$?*

**Exercise 12.2.7** *(i) Let $r$ be a strictly positive integer. Use Theorem 12.2.5 to show that, if we work over $\mathbb{C}$ and $\alpha : U \to U$ is an endomorphism of a finite dimensional space $U$, then $\alpha^r$ has an eigenvalue $\mu$ if and only if $\alpha$ has an eigenvalue $\lambda$ with $\lambda^r = \mu$.*

*(ii) State and prove an appropriate corresponding result if we allow $r$ to take any integer value.*

*(iii) Does the result of (i) remain true if we work over $\mathbb{R}$? Give a proof or counterexample. [Compare the treatment via characteristic polynomials in Exercise 6.8.15.]*

Now that we have Theorem 12.2.5 in place, we can quickly prove the Cayley–Hamilton theorem for $\mathbb{C}$.

*Proof of Theorem 12.2.4* By Theorem 12.2.5, we can find a basis $\mathbf{e}_1, \mathbf{e}_2, \ldots, \mathbf{e}_n$ for $U$ with respect to which $\alpha$ has matrix $A = (a_{ij})$ where $a_{ij} = 0$ if $i > j$. Setting $\lambda_j = a_{jj}$ we see, at once that,

$$\chi_\alpha(t) = \det(t\iota - \alpha) = \det(tI - A) = \prod_{j=1}^{n}(t - \lambda_j).$$

Next observe that

$$\alpha \mathbf{e}_j \in \text{span}\{\mathbf{e}_1, \mathbf{e}_2, \ldots, \mathbf{e}_j\}$$

and

$$(\alpha - \lambda_j \iota)\mathbf{e}_j = \mathbf{0},$$

so

$$(\alpha - \lambda_j \iota)\mathbf{e}_j \in \text{span}\{\mathbf{e}_1, \mathbf{e}_2, \ldots, \mathbf{e}_{j-1}\}$$

and, if $k \neq j$,

$$(\alpha - \lambda_k \iota)\mathbf{e}_j \in \text{span}\{\mathbf{e}_1, \mathbf{e}_2, \ldots, \mathbf{e}_j\}.$$

Thus

$$(\alpha - \lambda_j \iota)\big(\text{span}\{\mathbf{e}_1, \mathbf{e}_2, \ldots, \mathbf{e}_j\}\big) \subseteq \text{span}\{\mathbf{e}_1, \mathbf{e}_2, \ldots, \mathbf{e}_{j-1}\}$$

and, using induction on $n - j$,

$$(\alpha - \lambda_j \iota)(\alpha - \lambda_{j+1}\iota)\ldots(\alpha - \lambda_n \iota)U \subseteq \text{span}\{\mathbf{e}_1, \mathbf{e}_2, \ldots, \mathbf{e}_{j-1}\}.$$

Taking $j = n$, we obtain

$$(\alpha - \lambda_1 \iota)(\alpha - \lambda_2 \iota)\ldots(\alpha - \lambda_n \iota)U = \{\mathbf{0}\}$$

and $\chi_\alpha(\alpha) = 0$ as required. $\qquad\square$

**Exercise 12.2.8** (*i*) *Prove directly by matrix multiplication that*

$$\begin{pmatrix} 0 & a_{12} & a_{13} \\ 0 & a_{22} & a_{23} \\ 0 & 0 & a_{33} \end{pmatrix} \begin{pmatrix} b_{11} & b_{12} & b_{13} \\ 0 & 0 & b_{23} \\ 0 & 0 & b_{33} \end{pmatrix} \begin{pmatrix} c_{11} & c_{12} & c_{13} \\ 0 & c_{22} & c_{23} \\ 0 & 0 & 0 \end{pmatrix} = \begin{pmatrix} 0 & 0 & 0 \\ 0 & 0 & 0 \\ 0 & 0 & 0 \end{pmatrix}.$$

*State and prove a general theorem along these lines.*

(*ii*) *Is the product*

$$\begin{pmatrix} c_{11} & c_{12} & c_{13} \\ 0 & c_{22} & c_{23} \\ 0 & 0 & 0 \end{pmatrix} \begin{pmatrix} b_{11} & b_{12} & b_{13} \\ 0 & 0 & b_{23} \\ 0 & 0 & b_{33} \end{pmatrix} \begin{pmatrix} 0 & a_{12} & a_{13} \\ 0 & a_{22} & a_{23} \\ 0 & 0 & a_{33} \end{pmatrix}$$

*necessarily zero?*

**Exercise 12.2.9** *Let U be a vector space. Suppose that* $\alpha$, $\beta \in \mathcal{L}(U, U)$ *and* $\alpha\beta = \beta\alpha$. *Show that, if P and Q are polynomials, then* $P(\alpha)Q(\beta) = Q(\beta)P(\alpha)$.

The Cayley–Hamilton theorem for $\mathbb{C}$ implies a Cayley–Hamilton theorem for $\mathbb{R}$.

**Theorem 12.2.10 [Cayley–Hamilton over** $\mathbb{R}$**]** *If U is a vector space of dimension n over* $\mathbb{R}$ *and* $\alpha : U \to U$ *is linear, then, writing*

$$\chi_\alpha(t) = \sum_{k=0}^{n} a_k t^k = \det(t\iota - \alpha),$$

*we have* $\sum_{k=0}^{n} a_k \alpha^k = 0$ *or, more briefly,* $\chi_\alpha(\alpha) = 0$.

*Proof* By the correspondence between matrices and linear maps, it suffices to prove the corresponding result for matrices. In other words, we need to show that, if $A$ is an $n \times n$ real matrix, then, writing $\chi_A(t) = \det(tI - A)$, we have $\chi_A(A) = 0$.

But, if $A$ is an $n \times n$ real matrix, then $A$ may also be considered as an $n \times n$ complex matrix and the Cayley–Hamilton theorem for $\mathbb{C}$ tells us that $\chi_A(A) = 0$. $\qquad\square$

Exercise 12.6.14 sets out a proof of Theorem 12.2.10 which does not depend on the Cayley–Hamilton theorem for $\mathbb{C}$. We discuss this matter further on page 332.

**Exercise 12.2.11** *Let A be an* $n \times n$ *matrix over* $\mathbb{F}$ *with* $\det A \neq 0$. *Explain why*

$$\det(tI - A) = \sum_{j=0}^{n} a_j t^j$$

*with* $a_0 \neq 0$. *Show that*

$$A^{-1} = -a_0^{-1} \sum_{j=1}^{n} a_j A^{j-1}.$$

*Is this likely to be a good way of computing* $A^{-1}$ *and why?*

In case the reader feels that the Cayley–Hamilton theorem is trivial, she should note that Cayley merely verified it for $3 \times 3$ matrices and stated his conviction that the result would be true for the general case. It took twenty years before Frobenius came up with the first proof.

**Exercise 12.2.12** *Suppose that U is a vector space over* $\mathbb{F}$ *of dimension n. Use the Cayley–Hamilton theorem to show that, if* $\alpha$ *is a nilpotent endomorphism (that is to say,* $\alpha^m = 0$ *for some m), then* $\alpha^n = 0$. *(Of course there are many other ways to prove this of which the most natural is, perhaps, that of Exercise 11.2.3.)*

## 12.3 Minimal polynomials

As we saw in our study of the Cayley–Hamilton theorem and elsewhere in this book, the study of endomorphisms (that is to say, members of $\mathcal{L}(U, U)$) for a finite dimensional vector space over $\mathbb{C}$ is much easier when all the roots of the characteristic polynomial are unequal. For the rest of this chapter we shall be concerned with what happens when some of the roots are equal.

The reader may object that the 'typical' endomorphism has all the roots of its characteristic polynomial distinct (we shall discuss this further in Theorem 15.2.3) and it is not worth considering non-typical cases. This is a little too close to the argument 'the typical number is non-zero so we need not bother to worry about dividing by zero' for comfort. The reader will recall that, when we discussed differential and difference equations in Section 6.4, we discovered that the case when several of the roots were equal was particularly interesting and this phenomenon may be expected to recur.

Here are some examples where the roots of the characteristic polynomial are not distinct. We shall see that, in some sense, they are typical.

**Exercise 12.3.1** (i) (Revision) Find the characteristic polynomials of the following matrices.

$$A_1 = \begin{pmatrix} 0 & 0 \\ 0 & 0 \end{pmatrix}, \quad A_2 = \begin{pmatrix} 0 & 1 \\ 0 & 0 \end{pmatrix}.$$

Show that there does not exist a non-singular matrix $B$ with $A_1 = BA_2B^{-1}$.

(ii) Find the characteristic polynomials of the following matrices.

$$A_3 = \begin{pmatrix} 0 & 0 & 0 \\ 0 & 0 & 0 \\ 0 & 0 & 0 \end{pmatrix}, \quad A_4 = \begin{pmatrix} 0 & 1 & 0 \\ 0 & 0 & 0 \\ 0 & 0 & 0 \end{pmatrix}, \quad A_5 = \begin{pmatrix} 0 & 1 & 0 \\ 0 & 0 & 1 \\ 0 & 0 & 0 \end{pmatrix}.$$

Show that there does not exist a non-singular matrix $B$ with $A_i = BA_jB^{-1}$ [$3 \le i < j \le 5$].

(iii) Find the characteristic polynomials of the following matrices.

$$A_6 = \begin{pmatrix} 0 & 1 & 0 & 0 \\ 0 & 0 & 0 & 0 \\ 0 & 0 & 0 & 0 \\ 0 & 0 & 0 & 0 \end{pmatrix}, \quad A_7 = \begin{pmatrix} 0 & 1 & 0 & 0 \\ 0 & 0 & 0 & 0 \\ 0 & 0 & 0 & 1 \\ 0 & 0 & 0 & 0 \end{pmatrix}.$$

Show that there does not exist a non-singular matrix $B$ with $A_6 = BA_7B^{-1}$. Write down three further matrices $A_j$ with the same characteristic polynomial such that there does not exist a non-singular matrix $B$ with $A_i = BA_jB^{-1}$ [$6 \le i < j \le 10$] explaining why this is the case.

We now get down to business.

**Theorem 12.3.2** *If $U$ is a vector space of dimension n over $\mathbb{F}$ and $\alpha : U \to U$ a linear map, then there is a unique monic[2] polynomial $Q_\alpha$ of smallest degree such that $Q_\alpha(\alpha) = 0$.*

---

[2] That is to say, having leading coefficient 1.

*Further, if $P$ is any polynomial with $P(\alpha) = 0$, then $P(t) = S(t)Q_\alpha(t)$ for some polynomial $S$.*

More briefly we say that there is a unique monic polynomial $Q_\alpha$ of smallest degree which *annihilates* $\alpha$. We call $Q_\alpha$ the *minimal polynomial* of $\alpha$ and observe that $Q$ divides any polynomial $P$ which annihilates $\alpha$.

The proof takes a form which may be familiar to the reader from elsewhere (for example, from the study of Euclid's algorithm[3]).

*Proof of Theorem 12.3.2* Consider the collection $\mathcal{P}$ of polynomials $P$ with $P(\alpha) = 0$. We know, from the Cayley–Hamilton theorem (or by the simpler argument of Exercise 12.2.1), that $\mathcal{P} \setminus \{0\}$ is non-empty. Thus $\mathcal{P} \setminus \{0\}$ contains a polynomial of smallest degree and, by multiplying by a constant, a monic polynomial of smallest degree. If $Q_1$ and $Q_2$ are two monic polynomials in $\mathcal{P} \setminus \{0\}$ of smallest degree, then $Q_1 - Q_2 \in \mathcal{P}$ and $Q_1 - Q_2$ has strictly smaller degree than $Q_1$. It follows that $Q_1 - Q_2 = 0$ and $Q_1 = Q_2$ as required. We write $Q_\alpha$ for the unique monic polynomial of smallest degree in $\mathcal{P}$.

Suppose that $P(\alpha) = 0$. We know, by long division, that

$$P(t) = S(t)Q_\alpha(t) + R(t)$$

where $R$ and $S$ are polynomial and the degree of the 'remainder' $R$ is strictly smaller than the degree of $Q_\alpha$. We have

$$R(\alpha) = P(\alpha) - S(\alpha)Q_\alpha(\alpha) = 0 - 0 = 0$$

so $R \in \mathcal{P}$ and, by minimality, $R = 0$. Thus $P(t) = S(t)Q_\alpha(t)$ as required. $\qquad\square$

**Exercise 12.3.3** (i) *Making the usual switch between linear maps and matrices, find the minimal polynomials for each of the $A_j$ in Exercise 12.3.1.*

(ii) *Find the characteristic and minimal polynomials for*

$$A = \begin{pmatrix} 1 & 1 & 0 & 0 \\ 0 & 1 & 0 & 0 \\ 0 & 0 & 2 & 1 \\ 0 & 0 & 0 & 2 \end{pmatrix} \quad and \quad B = \begin{pmatrix} 1 & 0 & 0 & 0 \\ 1 & 1 & 0 & 0 \\ 0 & 0 & 2 & 0 \\ 0 & 0 & 0 & 2 \end{pmatrix}.$$

The minimal polynomial becomes a powerful tool when combined with the following observation.

**Lemma 12.3.4** *Suppose that $\lambda_1, \lambda_2, \ldots, \lambda_r$ are distinct elements of $\mathbb{F}$. Then there exist $q_j \in \mathbb{F}$ with*

$$1 = \sum_{j=1}^{r} q_j \prod_{i \neq j}(t - \lambda_i).$$

[3] Look at Exercise 6.8.32 if this is unfamiliar.

*Proof* Let us write

$$q_j = \prod_{i \neq j} (\lambda_j - \lambda_i)^{-1}$$

and

$$R(t) = \left( \sum_{j=1}^{r} q_j \prod_{i \neq j} (t - \lambda_i) \right) - 1.$$

Then $R$ is a polynomial of degree at most $r - 1$ which vanishes at the $r$ points $\lambda_j$. Thus $R$ is identically zero (since a polynomial of degree $k \geq 1$ can have at most $k$ roots) and the result follows $\qquad \square$

**Theorem 12.3.5** **[Diagonalisability theorem]** *Suppose that $U$ is a finite dimensional vector over $\mathbb{F}$. Then a linear map $\alpha : U \to U$ is diagonalisable if and only if its minimal polynomial factorises completely into linear factors and no factor is repeated.*

*Proof* If $D$ is an $n \times n$ diagonal matrix whose diagonal entries take the distinct values $\lambda_1$, $\lambda_2, \ldots, \lambda_r$, then $\prod_{j=1}^{r} (D - \lambda_j I) = 0$, so the minimal polynomial of $D$ can contain no repeated factors. The necessity part of the proof is immediate.

We now prove sufficiency. Suppose that the minimal polynomial of $\alpha$ is $\prod_{i=1}^{r} (t - \lambda_i)$. By Lemma 12.3.4, we can find $q_j \in \mathbb{F}$ such that

$$1 = \sum_{j=1}^{r} q_j \prod_{i \neq j} (t - \lambda_i)$$

and so, writing

$$\theta_j = q_j \prod_{i \neq j} (\alpha - \lambda_i \iota),$$

we have

$$\iota = \theta_1 + \theta_2 + \cdots + \theta_m \quad \text{and} \quad (\alpha - \lambda_j \iota)\theta_j = 0 \quad \text{for } 1 \leq j \leq m.$$

It follows at once that, if $\mathbf{u} \in U$ and we write $\mathbf{u}_j = \theta_j \mathbf{u}$, we have

$$\mathbf{u} = \mathbf{u}_1 + \mathbf{u}_2 + \cdots + \mathbf{u}_m \quad \text{and} \quad \alpha \mathbf{u}_j = \lambda_j \mathbf{u}_j \quad \text{for } 1 \leq j \leq m.$$

Observe that, if we write

$$U_j = \{ \mathbf{v} : \alpha \mathbf{v} = \lambda_j \mathbf{v} \},$$

then $U_j$ is a subspace of $U$ (we call $U_j$ the *eigenspace* corresponding to $\lambda_j$) and we have shown that

$$U = U_1 + U_2 + \cdots + U_m.$$

If $\mathbf{u}_j \in U_j$ and

$$\sum_{p=1}^{m} \mathbf{u}_j = \mathbf{0},$$

then, using the same idea as in the proof of Theorem 6.3.3, we have

$$\mathbf{0} = \prod_{i \neq j} (\alpha - \lambda_j \iota)\mathbf{0} = \prod_{i \neq j} (\alpha - \lambda_j \iota) \sum_{p=1}^{m} \mathbf{u}_p$$

$$= \sum_{p=1}^{m} \prod_{i \neq j} (\lambda_p - \lambda_i)\mathbf{u}_p = \prod_{p \neq j} (\lambda_p - \lambda_j)\mathbf{u}_j$$

and so $\mathbf{u}_j = \mathbf{0}$ for each $1 \leq j \leq m$. Thus

$$U = U_1 \oplus U_2 \oplus \ldots \oplus U_m.$$

If we take a basis for each subspace $U_j$ and combine them to form a basis for $U$, then we will have a basis of eigenvectors for $U$. Thus $\alpha$ is diagonalisable. $\qquad \square$

Exercise 12.6.33 indicates an alternative, less constructive, proof.

**Exercise 12.3.6** (*i*) *If $D$ is an $n \times n$ diagonal matrix whose diagonal entries take the distinct values $\lambda_1, \lambda_2, \ldots, \lambda_r$, show that the minimal polynomial of $D$ is $\prod_{i=1}^{r}(t - \lambda_i)$.*

(*ii*) *Let $U$ be a finite dimensional vector space and $\alpha : U \to U$ a diagonalisable linear map. If $\lambda$ is an eigenvalue of $\alpha$, explain, with proof, how we can find the dimension of the eigenspace*

$$U_\lambda = \{\mathbf{u} \in U \,:\, \alpha\mathbf{u} = \lambda\mathbf{u}\}$$

*from the characteristic polynomial of $\alpha$.*

**Exercise 12.3.7** (*i*) *If a polynomial $P$ has a repeated root, show that $P$ and its derivative $P'$ have a non-trivial common factor. Is the converse true? Give a proof or counterexample.*

(*ii*) *If $A$ is an $n \times n$ matrix over $\mathbb{C}$ with $A^m = I$ for some integer $m \geq 1$ show that $A$ is diagonalisable. If $A$ is a real symmetric matrix, show that $A^2 = I$.*

We can push these ideas a little further by extending Lemma 12.3.4.

**Lemma 12.3.8** *If $m(1), m(2), \ldots, m(r)$ are strictly positive integers and $\lambda_1, \lambda_2, \ldots, \lambda_r$ are distinct elements of $\mathbb{F}$, show that there exist polynomials $Q_j$ with*

$$1 = \sum_{j=1}^{r} Q_j(t) \prod_{i \neq j} (t - \lambda_i)^{m(i)}.$$

Since the proof uses the same ideas by which we established the existence and properties of the minimal polynomial (see Theorem 12.3.2), we set it out as an exercise for the reader.

**Exercise 12.3.9** *Consider the collection $A$ of polynomials with coefficients in $\mathbb{F}$. Let $\mathcal{P}$ be a non-empty subset of $A$ such that*

$$P, Q \in \mathcal{P}, \quad \lambda, \mu \in \mathbb{F} \Rightarrow \lambda P + \mu Q \in \mathcal{P} \quad and \quad P \in \mathcal{P}, Q \in A \Rightarrow P \times Q \in \mathcal{P}.$$

*(In this exercise, $P \times Q(t) = P(t)Q(t)$.)*

(i) *Show that either $\mathcal{P} = \{0\}$ or $\mathcal{P}$ contains a monic polynomial $P_0$ of smallest degree. In the second case, show that $P_0$ divides every $P \in \mathcal{P}$.*

(ii) *Suppose that $P_1, P_2, \ldots, P_r$ are non-zero polynomials. By considering*

$$\mathcal{P} = \left\{ \sum_{j=1}^{r} T_j \times P_j : T_j \in A \right\},$$

*show that we can find $Q_j \in A$ such that, writing*

$$P_0 = \sum_{j=1}^{r} Q_j \times P_j,$$

*we know that $P_0$ is a monic polynomial dividing each $Q_j$. By considering the given expression for $P_0$, show that any polynomial dividing each $P_j$ also divides $P_0$. (Thus $P_0$ is the 'greatest common divisor of the $P_j$'.)*

(iii) *Prove Lemma 12.3.4.*

We also need a natural definition.

**Definition 12.3.10** *Suppose that $U$ is a vector space over $\mathbb{F}$ with subspaces $U_j$ such that*

$$U = U_1 \oplus U_2 \oplus \ldots \oplus U_r.$$

*If $\alpha_j : U_j \to U_j$ is linear, we define $\alpha_1 \oplus \alpha_2 \oplus \ldots \oplus \alpha_r$ as a function from $U$ to $U$ by*

$$(\alpha_1 \oplus \alpha_2 \oplus \cdots \oplus \alpha_r)(\mathbf{u}_1 + \mathbf{u}_2 + \cdots + \mathbf{u}_r) = \alpha_1 \mathbf{u}_1 + \alpha_2 \mathbf{u}_2 + \cdots + \alpha_r \mathbf{u}_r.$$

**Exercise 12.3.11** *Explain why, with the notation and assumptions of Definition 12.3.10, $\alpha_1 \oplus \alpha_2 \oplus \ldots \oplus \alpha_r$ is well defined. Show that $\alpha_1 \oplus \alpha_2 \oplus \cdots \oplus \alpha_r$ is linear.*

We can now state our extension of Theorem 12.3.5.

**Theorem 12.3.12** *Suppose that $U$ is a finite dimensional vector space over $\mathbb{F}$. If the linear map $\alpha : U \to U$ has minimal polynomial*

$$Q(t) = \prod_{j=1}^{r} (t - \lambda_j)^{m(j)},$$

*where $m(1), m(2), \ldots, m(r)$ are strictly positive integers and $\lambda_1, \lambda_2, \ldots, \lambda_r$ are distinct elements of $\mathbb{F}$, then we can find subspaces $U_j$ and linear maps $\alpha_j : U_j \to U_j$ such that $\alpha_j$ has minimal polynomial $(t - \lambda_j)^{m(j)}$,*

$$U = U_1 \oplus U_2 \oplus \ldots \oplus U_r \quad and \quad \alpha = \alpha_1 \oplus \alpha_2 \oplus \ldots \oplus \alpha_r.$$

The proof is so close to that of Theorem 12.3.2 that we set it out as another exercise for the reader.

**Exercise 12.3.13** *Suppose that $U$ and $\alpha$ satisfy the hypotheses of Theorem 12.3.12. By Lemma 12.3.4, we can find polynomials $Q_j$ with*

$$1 = \sum_{j=1}^{r} Q_j(t) \prod_{i \neq j} (t - \lambda_i)^{m(i)}. \qquad\qquad \bigstar$$

*Set*

$$U_k = \{\mathbf{u} \in U \ : \ (\alpha - \lambda_k \iota)^{m(k)} \mathbf{u} = \mathbf{0}\}.$$

*(i) Show, using $\bigstar$, that*

$$U = U_1 + U_2 + \cdots + U_r.$$

*(ii) Show, using $\bigstar$, that, if $\mathbf{u} \in U_j$, then*

$$Q_j(\alpha) \prod_{i \neq j} (\alpha - \lambda_i \iota)^{m(i)} \mathbf{u} = \mathbf{0} \Rightarrow \mathbf{u} = \mathbf{0}$$

*and deduce that*

$$\prod_{i \neq j} (\alpha - \lambda_i \iota)^{m(i)} \mathbf{u} = \mathbf{0} \Rightarrow \mathbf{u} = \mathbf{0}.$$

*Hence, or otherwise, show that*

$$U = U_1 \oplus U_2 \oplus \ldots \oplus U_r.$$

*(iii) Show that $\alpha U_j \subseteq U_j$, so that we can define a linear map $\alpha_j : U_j \to U_j$ by taking $\alpha_j(\mathbf{u}) = \alpha\mathbf{u}$ for all $\mathbf{u} \in U_j$. Show that*

$$\alpha = \alpha_1 \oplus \alpha_2 \oplus \ldots \oplus \alpha_r.$$

*(iv) Show that $\alpha_j$ has minimal polynomial $(t - \lambda_j)^{p(j)}$ for some $p(j) \leq m(j)$. Show that $\alpha$ has minimal polynomial dividing $\prod_{j=1}^{r} (t - \lambda_j)^{m(j)}$ and deduce that $p(j) = m(j)$.*

**Exercise 12.3.14** *Suppose that $U$ and $V$ are subspaces of a finite dimensional vector space $W$ with $U \oplus V = W$. If $\alpha \in \mathcal{L}(U, U)$ and $\beta \in \mathcal{L}(V, V)$, show, by choosing an appropriate basis for $W$, that $\det(\alpha \oplus \beta) = \det \alpha \det \beta$. Find, with proof, the characteristic and minimal polynomials of $\alpha \oplus \beta$ in terms of the characteristic and minimal polynomials of $\alpha$ and $\beta$.*

**Exercise 12.3.15** *We work over $\mathbb{C}$.*

*(i) Explain why the following statement is false. Given monic polynomials $P$, $Q$ and $S$ with*

$$P(t) = S(t)Q(t),$$

*we can find a matrix with characteristic polynomial $P$ and minimal polynomial $Q$.*

(*ii*) *Write down an* $n \times n$ *matrix* $A$ *with characteristic polynomial* $t^n$ *and minimal polynomial* $t^m$ $[n \geq m \geq 1]$. *What are the characteristic and minimal polynomials of* $A - \lambda I$?

(*iii*) *Show that, given monic polynomials* $P$, $Q$, $S$ *and* $R$ *with*

$$P(t) = S(t)Q(t)$$

*such that* $Q(z) = 0 \Rightarrow S(z) = 0$, *we can find a matrix with characteristic polynomial* $P$ *and minimal polynomial* $Q$.

## 12.4 The Jordan normal form

If $\alpha : U \to U$ is a linear map such that $\alpha^m = 0$ for some $m \geq 0$, we say that $\alpha$ is *nilpotent*.

If we work over $\mathbb{C}$, Theorem 12.3.12 implies the following result.

**Theorem 12.4.1** *Suppose that* $U$ *is a finite dimensional vector space over* $\mathbb{C}$. *Then, given any linear map* $\alpha : U \to U$, *we can find* $r \geq 1$, $\lambda_j \in \mathbb{C}$, *subspaces* $U_j$ *and nilpotent linear maps* $\beta_j : U_j \to U_j$ *such that, writing* $\iota_j$ *for the identity map on* $U_j$ *we have*

$$U = U_1 \oplus U_2 \oplus \ldots \oplus U_r,$$
$$\alpha = (\beta_1 + \lambda_1 \iota_1) \oplus (\beta_2 + \lambda_2 \iota_2) \oplus \ldots \oplus (\beta_r + \lambda_r \iota_r).$$

*Proof* Since we work over $\mathbb{C}$, the minimal polynomial $Q$ of $\alpha$ will certainly factorise in the manner required by Theorem 12.3.12. Setting $\beta_j = \alpha_j - \lambda_j \iota_j$ we have the required result. $\square$

Thus, in some sense, the study of $\mathcal{L}(U, U)$ for finite dimensional vector spaces $U$ over $\mathbb{C}$ reduces to the study of nilpotent linear maps. We shall see that the study of nilpotent linear maps can be reduced to the study of a particular type of nilpotent linear map.

**Lemma 12.4.2** (*i*) *If* $U$ *is a vector space over* $\mathbb{F}$, $\alpha$ *is a nilpotent linear map on* $U$ *and* $\mathbf{e}$ *satisfies* $\alpha^m \mathbf{e} \neq \mathbf{0}$, *then* $\mathbf{e}$, $\alpha \mathbf{e}$, $\ldots$, $\alpha^m \mathbf{e}$ *are linearly independent.*

(*ii*) *If* $U$ *is a vector space of dimension* $n$ *over* $\mathbb{F}$, $\alpha$ *is a nilpotent linear map on* $U$ *and* $\mathbf{e}$ *satisfies* $\alpha^{n-1} \mathbf{e} \neq \mathbf{0}$, *then* $\mathbf{e}$, $\alpha \mathbf{e}$, $\ldots$, $\alpha^{n-1} \mathbf{e}$ *form a basis for* $U$.

*Proof* (i) If $\mathbf{e}$, $\alpha \mathbf{e}$, $\ldots \alpha^m \mathbf{e}$ are not independent, then we must be able to find a $j$ with $1 \leq j \leq m$, a $\lambda_j \neq 0$ and $\lambda_{j+1}, \lambda_{j+2}, \ldots, \lambda_m$ such that

$$\lambda_j \alpha^j \mathbf{e} + \lambda_{j+1} \alpha^{j+1} \mathbf{e} + \cdots + \lambda_m \alpha^m \mathbf{e} = \mathbf{0}.$$

Since $\alpha$ is nilpotent there must be an $N \geq m$ with $\alpha^{N+1} \mathbf{e} = \mathbf{0}$ but $\alpha^N \mathbf{e} \neq \mathbf{0}$. We observe that

$$\begin{aligned}
\mathbf{0} &= \alpha^{N-j} \mathbf{0} \\
&= \alpha^{N-j}(\lambda_j \alpha^j \mathbf{e} + \lambda_{j+1} \alpha^{j+1} \mathbf{e} + \cdots + \lambda_m \alpha^m \mathbf{e}) \\
&= \lambda_j \alpha^N \mathbf{e},
\end{aligned}$$

and so $\lambda_j = 0$ contradicting our initial assumption. The required result follows by reductio ad absurdum.

(ii) Any linearly independent set with $n$ elements is a basis. $\qquad\square$

**Exercise 12.4.3** *If the conditions of Lemma 12.4.2 (ii) hold, write down the matrix of $\alpha$ with respect to the given basis.*

**Exercise 12.4.4** *Use Lemma 12.4.2 (i) to provide yet another proof of the statement that, if $\alpha$ is a nilpotent linear map on a vector space of dimension $n$, then $\alpha^n = 0$.*

We now come to the central theorem of the section. The general view, with which this author concurs, is that it is more important to understand what it says than how it is proved.

**Theorem 12.4.5** *Suppose that $U$ is a finite dimensional vector space over $\mathbb{F}$ and $\alpha : U \to U$ is a nilpotent linear map. Then we can find subspaces $U_j$ and nilpotent linear maps $\alpha_j : U_j \to U_j$ such that $\alpha_j^{\dim U_j - 1} \neq 0$,*

$$U = U_1 \oplus U_2 \oplus \ldots \oplus U_r, \quad \text{and} \quad \alpha = \alpha_1 \oplus \alpha_2 \oplus \ldots \oplus \alpha_r.$$

The proof given here is in a form due to Tao. Although it would be a very long time before the average mathematician could come up with the idea behind this proof,[4] once the idea is grasped, the proof is not too hard.

We make a temporary and non-standard definition which will not be used elsewhere.[5]

**Definition 12.4.6** *Suppose that $U$ is a finite dimensional vector space over $\mathbb{F}$ and $\alpha : U \to U$ is a nilpotent linear map. If*

$$E = \{\mathbf{e}_1, \mathbf{e}_2, \ldots, \mathbf{e}_m\}$$

*is a finite subset of $U$ not containing $\mathbf{0}$, we say that $E$ generates the set*

$$\text{gen } E = \{\alpha^k \mathbf{e}_i : k \geq 0, 1 \leq i \leq m\}.$$

*If gen $E$ spans $U$, we say that $E$ is sufficiently large.*

**Exercise 12.4.7** *Why is gen $E$ finite?*

We set out the proof of Theorem 12.4.5 in the following lemma of which part (ii) is the key step.

**Lemma 12.4.8** *Suppose that $U$ is a vector space over $\mathbb{F}$ of dimension $n$ and $\alpha : U \to U$ is a nilpotent linear map.*

*(i) There exists a sufficiently large set.*

*(ii) If $E$ is a sufficiently large set and gen $E$ contains more than $n$ elements, we can find a sufficiently large set $F$ such that gen $F$ contains strictly fewer elements.*

*(iii) There exists a sufficiently large set $E$ such that gen $E$ has exactly $n$ elements.*

*(iv) The conclusion of Theorem 12.4.5 holds.*

---

[4] Fitzgerald says to Hemingway 'The rich are different from us' and Hemingway replies 'Yes they have more money'.
[5] So the reader should not use it outside this context and should always give the definition within this context.

*Proof* (i) Any basis for $E$ will be a sufficiently large set.

(ii) Suppose that

$$E = \{\mathbf{e}_1, \mathbf{e}_2, \ldots, \mathbf{e}_m\}$$

is a sufficiently large set, but gen $E > n$. We define $N_i$ by the condition $\alpha^{N_i}\mathbf{e}_i \neq \mathbf{0}$ but $\alpha^{N_i+1}\mathbf{e}_i = \mathbf{0}$.

Since gen $E > n$, gen $E$ cannot be linearly independent and so we can find $\lambda_{ik}$, not all zero, such that

$$\sum_{i=1}^{m}\sum_{k=0}^{N_i} \lambda_{ik}\alpha^k\mathbf{e}_i = \mathbf{0}.$$

Rearranging, we obtain

$$\sum_{1\leq i\leq m} P_i(\alpha)\mathbf{e}_i$$

where the $P_i$ are polynomials of degree at most $N_i$ and not all the $P_i$ are zero. By Lemma 12.4.2 (ii), this means that at least two of the $P_i$ are non-zero.

Factorising out the highest power of $\alpha$ possible, we have

$$\alpha^l \sum_{1\leq i\leq m} Q_i(\alpha)\mathbf{e}_i = \mathbf{0}$$

where the $Q_i$ are polynomials of degree at most $N_i - l$, $N(i) \geq l$ whenever $Q_i$ is non-zero, at least two of the $Q_i$ are non-zero and at least one of the $Q_i$ has non-zero constant term. Multiplying by a scalar and renumbering if necessary we may suppose that $Q_1$ has constant term 1 and $Q_2$ is non-zero. Then

$$\alpha^l \left(\mathbf{e}_1 + \alpha R(\alpha)\mathbf{e}_1 + \sum_{2\leq i\leq m} Q_i(\alpha)\mathbf{e}_i\right) = \mathbf{0}$$

where $R$ is a polynomial.

There are now three possibilities.

(A) If $l = 0$ and $\alpha R(\alpha)\mathbf{e}_1 = \mathbf{0}$, then

$$\mathbf{e}_1 \in \text{span gen}\{\mathbf{e}_2, \mathbf{e}_3, \ldots, \mathbf{e}_m\},$$

so we can take $F = \{\mathbf{e}_2, \mathbf{e}_3, \ldots, \mathbf{e}_m\}$.

(B) If $l = 0$ and $\alpha R(\alpha)\mathbf{e}_1 \neq \mathbf{0}$, then

$$\mathbf{e}_1 \in \text{span gen}\{\alpha\mathbf{e}_1, \mathbf{e}_2, \ldots, \mathbf{e}_m\},$$

so we can take $F = \{\alpha\mathbf{e}_1, \mathbf{e}_2, \ldots, \mathbf{e}_m\}$.

(C) If $l \geq 1$, we set $\mathbf{f} = \mathbf{e}_1 + \alpha R(\alpha)\mathbf{e}_1 + \sum_{1\leq i\leq m} Q_i(\alpha)\mathbf{e}_i$ and observe that

$$\mathbf{e}_1 \in \text{span gen}\{\mathbf{f}, \mathbf{e}_2, \ldots, \mathbf{e}_m\},$$

so the set $F = \{\mathbf{f}, \mathbf{e}_2, \ldots, \mathbf{e}_m\}$ is sufficiently large. Since $\alpha^l \mathbf{f} = \mathbf{0}$, $\alpha^{N_1} \mathbf{e}_1 \neq \mathbf{0}$ and $l \leq N_1$, gen $F$ contains strictly fewer elements than gen $E$.

It may be useful to note note the general resemblance of the argument in (ii) to Gaussian elimination and the Steinitz replacement lemma.

(iii) Use (i) and then apply (ii) repeatedly.

(iv) By (iii) we can find a set of non-zero vectors

$$E = \{\mathbf{e}_1, \mathbf{e}_2, \ldots, \mathbf{e}_m\}$$

such that gen $E$ has $n$ elements and spans $U$. It follows that gen $E$ is a basis for $U$. If we set

$$U_i = \text{span gen}\{\mathbf{e}_i\}$$

and define $\alpha_i : U_i \to U_i$ by $\alpha_i \mathbf{u} = \alpha \mathbf{u}$ whenever $\mathbf{u} \in U$, the conclusions of Theorem 12.4.5 follow at once. $\qquad\square$

Combining Theorem 12.4.5 with Theorem 12.4.1, we obtain a version of the Jordan normal form theorem.

**Theorem 12.4.9** *Suppose that* $U$ *is a finite dimensional vector space over* $\mathbb{C}$. *Then, given any linear map* $\alpha : U \to U$, *we can find* $r \geq 1$, $\lambda_j \in \mathbb{C}$, *subspaces* $U_j$ *and linear maps* $\beta_j : U_j \to U_j$ *such that, writing* $\iota_j$ *for the identity map on* $U_j$, *we have*

$$U = U_1 \oplus U_2 \oplus \ldots \oplus U_r,$$

$$\alpha = (\beta_1 + \lambda_1 \iota_1) \oplus (\beta_2 + \lambda_2 \iota_2) \oplus \ldots \oplus (\beta + \lambda_r \iota_r),$$

$$\beta_j^{\dim U_j - 1} \neq 0 \quad \text{and} \quad \beta_j^{\dim U_j} = 0 \quad \text{for } 1 \leq j \leq r.$$

*Proof* Left to the reader. $\qquad\square$

Using Lemma 12.4.2 and remembering the change of basis formula, we obtain the standard version of our theorem.

**Theorem 12.4.10 [The Jordan normal form]** *We work over* $\mathbb{C}$. *We shall write* $J_k(\lambda)$ *for the* $k \times k$ *matrix*

$$J_k(\lambda) = \begin{pmatrix} \lambda & 1 & 0 & 0 & \ldots & 0 & 0 \\ 0 & \lambda & 1 & 0 & \ldots & 0 & 0 \\ 0 & 0 & \lambda & 1 & \ldots & 0 & 0 \\ \vdots & \vdots & \vdots & \vdots & & \vdots & \vdots \\ 0 & 0 & 0 & 0 & \ldots & \lambda & 1 \\ 0 & 0 & 0 & 0 & \ldots & 0 & \lambda \end{pmatrix}.$$

*If* $A$ *is any* $n \times n$ *matrix, we can find an invertible* $n \times n$ *matrix* $M$, *an integer* $r \geq 1$, *integers* $k_j \geq 1$ *and complex numbers* $\lambda_j$ *such that* $MAM^{-1}$ *is a matrix with the matrices*

$J_{k_j}(\lambda_j)$ *laid out along the diagonal and all other entries* 0. *Thus*

$$MAM^{-1} = \begin{pmatrix} J_{k_1}(\lambda_1) & & & & & \\ & J_{k_2}(\lambda_2) & & & & \\ & & J_{k_3}(\lambda_3) & & & \\ & & & \ddots & & \\ & & & & J_{k_{r-1}}(\lambda_{r-1}) & \\ & & & & & J_{k_r}(\lambda_r) \end{pmatrix}.$$

*Proof* Left to the reader. □

The $J_k(\lambda)$ are called *Jordan blocks*.

**Exercise 12.4.11** *Why is a diagonal matrix already in Jordan form?*

**Exercise 12.4.12** (*i*) *We adopt the notation of Theorem 12.4.10 and use column vectors. If* $\lambda \in \mathbb{C}$, *find the dimension of*

$$\{\mathbf{x} \in \mathbb{C}^n : (\lambda I - A)^k \mathbf{x} = \mathbf{0}\}$$

*in terms of the* $\lambda_j$ *and* $k_j$.

(*ii*) *Consider the matrix A of Theorem 12.4.10. If* $\tilde{M}$ *is an invertible* $n \times n$ *matrix,* $\tilde{r} \geq 1$, $\tilde{k}_j \geq 1$ $\tilde{\lambda}_j \in \mathbb{C}$ *and*

$$\tilde{M}A\tilde{M}^{-1} = \begin{pmatrix} J_{\tilde{k}_1}(\tilde{\lambda}_1) & & & & & \\ & J_{\tilde{k}_2}(\tilde{\lambda}_2) & & & & \\ & & J_{\tilde{k}_3}(\tilde{\lambda}_3) & & & \\ & & & \ddots & & \\ & & & & J_{\tilde{k}_{\tilde{r}-1}}(\tilde{\lambda}_{\tilde{r}-1}) & \\ & & & & & J_{\tilde{k}_{\tilde{r}}}(\tilde{\lambda}_{\tilde{r}}) \end{pmatrix},$$

*show that,* $\tilde{r} = r$ *and, possibly after renumbering,* $\tilde{\lambda}_j = \lambda_j$ *and* $\tilde{k}_j = k_j$ *for* $1 \leq j \leq r$.

Theorem 12.4.10 and the result of Exercise 12.4.12 are usually stated as follows. 'Every $n \times n$ complex matrix can be reduced by a similarity transformation to Jordan form. This form is unique up to rearrangements of the Jordan blocks.' The author is not sufficiently enamoured with the topic to spend time giving formal definitions of the various terms in this statement. Notice that we have shown that 'two complex matrices are similar if and only if they have the same Jordan form'. Exercise 6.8.25 shows that this last result remains useful when we look at real matrices.

**Exercise 12.4.13** *We work over* $\mathbb{C}$. *Let* $\mathcal{P}_n$ *be the usual vector space of polynomials of degree at most n in the variable z. Find the Jordan normal form for the endomorphisms T and S given by* $(Tp)(z) = p'(z)$ *and* $(Sp)(z) = zp'(z)$.

**Exercise 12.4.14** *Suppose that $U$ is a vector space of dimension $n$ over $\mathbb{F}$. If $\alpha \in \mathcal{L}(U, U)$, we say that $\lambda$ has* algebraic multiplicity *$m_a(\lambda)$ if $\lambda$ is a root of multiplicity $m$ of the characteristic polynomial $\chi_\alpha$ of $\alpha$. (That is to say, $(t - \lambda)^{m_a(\lambda)}$ is a factor of $\chi_\alpha(t)$, but $(t - \lambda)^{m_a(\lambda)+1}$ is not.) We say that $\lambda$ has* geometric multiplicity

$$m_g(\lambda) = \dim\{\mathbf{u} \, : \, (\alpha - \lambda\iota)(\mathbf{u}) = \mathbf{0}\}.$$

*(i) Show that, if $\lambda$ is a root of $\chi_\alpha$, then*

$$1 \le m_g(\lambda) \le m_a(\lambda).$$

*(ii) Show that if $\lambda_k \in \mathbb{F}$, $1 \le n_g(\lambda_k) \le n_a(\lambda_k)$ $[1 \le k \le r]$ and $\sum_{k=1}^{r} n_a(\lambda_k) = n$ we can find an $\alpha \in \mathcal{L}(U, U)$ with $m_g(\lambda_k) = n_g(\lambda_k)$ and $m_a(\lambda_k) = n_a(\lambda_k)$ for $1 \le k \le r$.*

*(iii) Suppose now that $\mathbb{F} = \mathbb{C}$. Show how to compute $m_a(\lambda)$ and $m_g(\lambda)$ from the Jordan normal form associated with $\alpha$.*

## 12.5 Applications

The Jordan normal form provides another method of studying the behaviour of $\alpha \in \mathcal{L}(U, U)$.

**Exercise 12.5.1** *Use the Jordan normal form to prove the Cayley–Hamilton theorem over $\mathbb{C}$. Explain why the use of Exercise 12.2.1 enables us to avoid circularity.*

**Exercise 12.5.2** *(i) Let $A$ be a matrix written in Jordan normal form. Find the rank of $A^j$ in terms of the terms of the numbers of Jordan blocks of certain types.*

*(ii) Use (i) to show that, if $U$ is a finite vector space of dimension $n$ over $\mathbb{C}$, and $\alpha \in \mathcal{L}(U, U)$, then the rank $r_j$ of $\alpha^j$ satisfies the conditions*

$$n = r_0 > r_1 > r_2 > \ldots > r_m = r_{m+1} = r_{m+2} = \ldots,$$

*for some $m \le n$ together with the condition*

$$r_0 - r_1 \ge r_1 - r_2 \ge r_2 - r_3 \ge \ldots \ge r_{m-1} - r_m.$$

*Show also that, if a sequence $s_j$ satisfies the condition*

$$n = s_0 > s_1 > s_2 > \ldots > s_m = s_{m+1} = s_{m+2} = \ldots$$

*for some $m \le n$ together with the condition*

$$s_0 - s_1 \ge s_1 - s_2 \ge s_2 - s_3 \ge \ldots \ge s_{m-1} - s_m > 0,$$

*then there exists an $\alpha \in \mathcal{L}(U, U)$ such that the rank of $\alpha^j$ is $s_j$.*
*[We thus have an alternative proof of Theorem 11.2.2 in the case when $\mathbb{F} = \mathbb{C}$. If the reader cares to go into the matter more closely, she will observe that we only need the Jordan form theorem for nilpotent matrices and that this result holds in $\mathbb{R}$. Thus the proof of Theorem 11.2.2 outlined here also works for $\mathbb{R}$.]*

It also enables us to extend the ideas of Section 6.4 which the reader should reread. She should do the following exercise in as much detail as she thinks appropriate.

**Exercise 12.5.3** *Write down the five essentially different Jordan forms of $4 \times 4$ nilpotent complex matrices. Call them $A_1$, $A_2$, $A_3$, $A_4$, $A_5$.*

*(i) For each $1 \le j \le 5$, write down the the general solution of*

$$\mathbf{x}'(t) = A_j\mathbf{x}(t)$$

*where $\mathbf{x}(t) = \big(x_1(t), x_2(t), x_3(t), x_4(t)\big)^T$, $\mathbf{x}'(t) = \big(x_1'(t), x_2'(t), x_3'(t), x_4'(t)\big)^T$ and we assume that the functions $x_j : \mathbb{R} \to \mathbb{C}$ are well behaved.*

*(ii) If $\lambda \in \mathbb{C}$, obtain the the general solution of*

$$\mathbf{y}'(t) = (\lambda I + A_j)\mathbf{x}(t)$$

*from general solutions to the problems in (i).*

*(iii) Suppose that $B$ is an $n \times n$ matrix in normal form. Write down the general solution of*

$$\mathbf{x}'(t) = B\mathbf{x}(t)$$

*in terms of the Jordan blocks.*

*(iv) Suppose that $A$ is an $n \times n$ matrix. Explain how to find the general solution of*

$$\mathbf{x}'(t) = A\mathbf{x}(t).$$

If we wish to solve the differential equation

$$x^{(n)}(t) + a_{n-1}x^{(n-1)}(t) + \cdots + a_0x(t) = 0 \qquad\qquad \bigstar$$

using the ideas of Exercise 12.5.3, then it is natural to set $x_j(t) = x^{(j)}(t)$ and rewrite the equation as

$$\mathbf{x}'(t) = A\mathbf{x}(t),$$

where

$$A = \begin{pmatrix} 0 & 1 & 0 & \dots & 0 & 0 \\ 0 & 0 & 1 & \dots & 0 & 0 \\ & \vdots & & & \vdots & \\ 0 & 0 & 0 & \dots & 1 & 0 \\ 0 & 0 & 0 & \dots & 0 & 1 \\ -a_0 & -a_1 & -a_2 & \dots & -a_{n-2} & -a_{n-1} \end{pmatrix}. \qquad\qquad \bigstar\bigstar$$

**Exercise 12.5.4** *Check that the rewriting is correct.*

If we now try to apply the results of Exercise 12.5.3, we run into an immediate difficulty. At first sight, it appears there could be many Jordan forms associated with $A$. Fortunately, we can show that there is only one possibility.

**Exercise 12.5.5** *Let $U$ be a vector space over $\mathbb{C}$ and $\alpha : U \to U$ be a linear map. Show that $\alpha$ can be associated with a Jordan form*

$$\begin{pmatrix} J_{k_1}(\lambda_1) & & & & & \\ & J_{k_2}(\lambda_2) & & & & \\ & & J_{k_3}(\lambda_3) & & & \\ & & & \ddots & & \\ & & & & J_{k_{r-1}}(\lambda_{r-1}) & \\ & & & & & J_{k_r}(\lambda_r) \end{pmatrix},$$

*with all the $\lambda_j$ distinct, if and only if the characteristic polynomial of $\alpha$ is also its minimal polynomial.*

**Exercise 12.5.6** *Let $A$ be the matrix given by ★★. Suppose that $a_0 \neq 0$ By looking at*

$$\left( \sum_{j=0}^{n-1} b_j A^j \right) \mathbf{e},$$

*where $\mathbf{e} = (1, 0, 0, \dots, 0)^T$, or otherwise, show that the minimal polynomial of $A$ must have degree at least $n$. Explain why this implies that the characteristic polynomial of $A$ is also its minimal polynomial. Using Exercise 12.5.5, deduce that there is a Jordan form associated with $A$ in which all the blocks $J_k(\lambda_k)$ have distinct $\lambda_k$.*

**Exercise 12.5.7** *(i) Find the general solution of ★ when the Jordan normal form associated with $A$ is a single Jordan block.*

*(ii) Find the general solution of ★ in terms of the structure of the Jordan form associated with $A$.*

We can generalise our previous work on difference equations (see for example Exercise 6.6.11 and the surrounding discussion) in the same way.

**Exercise 12.5.8** *(i) Prove the well known formula for binomial coefficients*

$$\binom{r}{k-1} + \binom{r}{k} = \binom{r+1}{k} \qquad [1 \le k \le r].$$

*(ii) If $k \ge 0$, we consider the $k+1$ two sided sequences*

$$\mathbf{u}_j = (\dots, u_j(-3), u_j(-2), u_j(-1), u_j(0), u_j(1), u_j(2), u_j(3), \dots)$$

with $-1 \le j \le k$. *(More exactly, consider the functions* $\mathbf{u}_j : \mathbb{Z} \to \mathbb{F}$ *with* $-1 \le j \le k$.*)*
*Consider the system of* $k + 1$ *equations*

$$u_r(-1) = 0$$
$$u_r(0) - u_{r-1}(0) = u_r(-1)$$
$$u_r(1) - u_{r-1}(1) = u_r(0)$$
$$u_r(2) - u_{r-1}(2) = u_r(1)$$
$$\vdots$$
$$u_r(k) - u_{r-1}(k) = u_r(k - 1)$$

*where* $r$ *ranges freely over* $\mathbb{Z}$.

*Show that* $u_r(0) = b_0$ *with* $b_0$ *constant. Show that* $u_r(1) = b_0 r + b_1$ *(with* $b_0$ *and* $b_1$
*constants). Find, with proof, the solution for all* $k \ge 0$.

*(ii) Let* $\lambda \in \mathbb{F}$ *and* $\lambda \ne 0$. *Find, with proof, the general solution of*

$$v_r(-1) = 0$$
$$v_r(0) - \lambda v_{r-1}(0) = v_r(-1)$$
$$v_r(1) - \lambda v_{r-1}(1) = v_r(0)$$
$$v_r(2) - \lambda v_{r-1}(2) = v_r(1)$$
$$\vdots$$
$$v_r(k) - \lambda v_{r-1}(k) = v_r(k - 1).$$

*(iii) Use the ideas of this section to find the general solution of*

$$u_r + \sum_{j=0}^{n-1} a_j u_{j-n+r} = 0$$

*(where* $a_0 \ne 0$ *and* $r$ *ranges freely over* $\mathbb{Z}$*) in terms of the roots of*

$$P(t) = t^n + \sum_{j=0}^{n-1} a_j t^j.$$

*(iv) When we worked on differential equations we did not impose the condition* $a_0 \ne 0$.
*Why do we impose it for linear difference equations but not for linear differential equations?*

Students are understandably worried by the prospect of having to find Jordan normal
forms. However, in real life, there will usually be a good reason for the failure of the roots
of the characteristic polynomial to be distinct and the nature of the original problem may
well give information about the Jordan form. (For example, one reason for suspecting that
Exercise 12.5.6 holds is that, otherwise, we would get rather implausible solutions to our
original differential equation.)

In an examination problem, the worst that might happen is that we are asked to find a Jordan normal form of an $n \times n$ matrix $A$ with $n \leq 4$ and, because $n$ is so small,[6] there are very few possibilities.

**Exercise 12.5.9** *Write down the six possible types of Jordan forms for a $3 \times 3$ matrix.* [*Hint: Consider the cases all characteristic roots the same, two characteristic roots the same, all characteristic roots distinct.*]

A natural procedure runs as follows.

(a) Think. (This is an examination question so there cannot be too much calculation involved.)
(b) Factorise the characteristic polynomial $\chi(t)$.
(c) Think.
(d) We can deal with non-repeated factors. We now look at each repeated factor $(t - \lambda)^m$.
(e) Think.
(f) Find the general solution of $(A - \lambda I)\mathbf{x} = \mathbf{0}$. Now find the general solution of $(A - \lambda I)\mathbf{x} = \mathbf{y}$ with $\mathbf{y}$ a general solution of $(A - \lambda I)\mathbf{y} = \mathbf{0}$ and so on. (But because the dimensions involved are small there will be not much 'so on'.)
(g) Think.

**Exercise 12.5.10** *Let $A$ be a $5 \times 5$ complex matrix with $A^4 = A^2 \neq A$. What are the possible minimal and characteristic polynomials of $A$? How many possible Jordan forms are there? Give reasons. (You are not asked to write down the Jordan forms explicitly. Two Jordan forms which can be transformed into each other by renumbering rows and columns should be considered identical.)*

**Exercise 12.5.11** *Find a Jordan normal form $J$ for the matrix*

$$M = \begin{pmatrix} 1 & 0 & 1 & 0 \\ 0 & 1 & 0 & 0 \\ 0 & -1 & 2 & 0 \\ 0 & 0 & 0 & 2 \end{pmatrix}.$$

*Determine both the characteristic and the minimal polynomial of $M$.*

*Find a basis of $\mathbb{C}^4$ with respect to which the linear map corresponding to $M$ for the standard basis has matrix $J$. Write down a matrix $P$ such that $P^{-1}MP = J$.*

## 12.6 Further exercises

**Exercise 12.6.1** Let $U$, $V$, $W$ and $X$ be finite dimensional vector spaces over $\mathbb{F}$. Suppose that $\alpha \in \mathcal{L}(U, V)$ and $\beta \in \mathcal{L}(V, W)$ are such that the image of $\alpha$ is the null-space of $\beta$.

If $\sigma \in \mathcal{L}(V, X)$ is such that $\sigma\alpha = 0$, show that there exists a $\tau \in \mathcal{L}(W, X)$ with $\tau\beta = \sigma$. Is $\tau$ necessarily unique? Give a proof or or a counterexample.

---

[6] If $n \geq 5$, then there is some sort of trick involved and direct calculation is foolish.

**Exercise 12.6.2** Suppose that $U_1, U_2, \ldots, U_n$ are subspaces of a vector space $V$ over $\mathbb{F}$. Show that $V = U_1 \oplus U_2 \oplus \ldots \oplus U_n$ if and only if

(i) $V = \sum_{j=1}^{n} U_j$

(ii) $U_i \cap \sum_{j \neq i} U_j = \{\mathbf{0}\}$ for each $i = 1, 2, \ldots, n$.

Let $\mathbb{F} = \mathbb{R}$ and $V = \mathbb{R}^3$. Show that there are distinct one dimensional subspaces $U_j$ such that $U_1 + U_2 + U_3 + U_4 = V$ and $U_i \cap (U_j + U_k) = \{\mathbf{0}\}$ whenever $i, j, k$ are distinct integers taken from $\{1, 2, 3, 4\}$, but $V$ is not the direct sum of the $U_i$.

**Exercise 12.6.3** Let $V$ and $W$ be finite dimensional vector spaces over $\mathbb{F}$, let $U$ be a subspace of $V$ and let $\alpha : V \to W$ be a surjective linear map. Which of the following statements are true and which may be false? Give proofs or counterexamples.

(i) There exists a linear map $\beta : V \to W$ such that $\beta(\mathbf{v}) = \alpha(\mathbf{v})$ if $\mathbf{v} \in U$, and $\beta(\mathbf{v}) = \mathbf{0}$ otherwise.

(ii) There exists a linear map $\gamma : W \to V$ such that $\alpha\gamma$ is the identity map on $W$.

(iii) If $X$ is a subspace of $V$ such that $V = U \oplus X$, then $W = \alpha U \oplus \alpha X$.

(iv) If $Y$ is a subspace of $V$ such that $W = \alpha U \oplus \alpha Y$, then $V = U \oplus Y$.

**Exercise 12.6.4** Suppose that $U$ and $V$ are vector spaces over $\mathbb{F}$. Show that $U \times V$ is a vector space over $\mathbb{F}$ if we define

$$(\mathbf{u}_1, \mathbf{v}_1) + (\mathbf{u}_2, \mathbf{v}_2) = (\mathbf{u}_1 + \mathbf{u}_2, \mathbf{v}_1 + \mathbf{v}_2) \quad \text{and} \quad \lambda(\mathbf{u}, \mathbf{v}) = (\lambda\mathbf{u}, \lambda\mathbf{v})$$

in the natural manner.

Let

$$\tilde{U} = \{(\mathbf{u}, \mathbf{0}) : \mathbf{u} \in U\} \quad \text{and} \quad \tilde{V} = \{(\mathbf{0}, \mathbf{v}) : \mathbf{v} \in V\}.$$

Show that there are natural isomorphisms[7] $\theta : U \to \tilde{U}$ and $\phi : V \to \tilde{V}$. Show that

$$\tilde{U} \oplus \tilde{V} = U \times V.$$

[Because of the results of this exercise, mathematicians denote the space $U \times V$, equipped with the addition and scalar multiplication given here, by $U \oplus V$. They call $U \oplus V$ the *exterior direct sum* (or the *external direct sum*).]

**Exercise 12.6.5** If $C(\mathbb{R})$ is the space of continuous functions $f : \mathbb{R} \to \mathbb{R}$ and

$$X = \{f \in C(\mathbb{R}) : f(x) = f(-x)\},$$

show that $X$ is subspace of $C(\mathbb{R})$ and find two subspaces $Y_1$ and $Y_2$ of $C(\mathbb{R})$ such that $C(\mathbb{R}) = X \oplus Y_1 = X \oplus Y_2$ but $Y_1 \cap Y_2 = \{0\}$.

Show, by exhibiting an isomorphism, that if $V$, $W_1$ and $W_2$ are subspaces of a vector space $U$ with $U = V \oplus W_1 = V \oplus W_2$, then $W_1$ is isomorphic to $W_2$. If $V$ is finite dimensional, is it necessarily true that $W_1 = W_2$? Give a proof or a counterexample.

---

[7] Take *natural* as a synonym for *defined without the use of bases*.

**Exercise 12.6.6** (A second bite at the cherry.)

(i) Let $U$ be a vector space over $\mathbb{F}$.

If $V$ and $W$ are subspaces of $U$ with $U = V \oplus W$ show that there is an isomorphism $\theta : W \to U/V$. Deduce that if $A, B, C$ are subspaces of a vector space $X$ over $\mathbb{F}$, then

$$A \oplus B = A \oplus C \Rightarrow B \cong C$$

(where, as usual, $B \cong C$ means that $B$ is isomorphic to $C$).

(ii) Let $\mathcal{P}$ be the standard real vector space of polynomials on $\mathbb{R}$ with real coefficients. Let

$$\mathcal{Q}_n = \{Q \in \mathcal{P} : Q(x) = x^n P(x) \quad \text{for some } P \in \mathcal{P}\}.$$

Show that $\mathcal{Q}_n$ is a subspace of $\mathcal{P}$ with $\mathcal{P} \cong \mathcal{Q}_n$ for all $n$.

If $A, B, C$ are subspaces of a vector space $X$ over $\mathbb{F}$ and $A \oplus B \cong A \oplus C$, does it follow that $B \cong C$? Give reasons for your answer.

**Exercise 12.6.7** Let $W_1$, $W_2$ and $W_3$ be subspaces of a finite dimensional vector space $V$. Which of the following statements are true and which are false? Give proofs or counterexamples as appropriate.

(i) If $V = W_1 \oplus W_2$, then $\dim W_3 = \dim(W_1 \cap W_3) + \dim(W_2 \cap W_3)$.

(ii) If $\dim W_1 + \dim W_2 + \dim W_3 = \dim V$ and

$$\dim(W_1 \cap W_2) = \dim(W_2 \cap W_3) = \dim(W_3 \cap W_1) = 0,$$

then $V = W_1 \oplus W_2 \oplus W_3$.

(iii) If $W_1 \cap W_2 \subseteq W_3$, then $W_3/(W_1 \cap W_2)$ is isomorphic with

$$(W_1 + W_2 + W_3)/(W_1 + W_2).$$

**Exercise 12.6.8** Let $P$ and $Q$ be real polynomials with no non-trivial common factor. If $M$ is an $n \times n$ real matrix and we write $A = f(M)$, $B = g(M)$, show that $\mathbf{x}$ is a solution of $AB\mathbf{x} = \mathbf{0}$ if and only if we can find $\mathbf{y}$ and $\mathbf{z}$ with $A\mathbf{y} = B\mathbf{z} = \mathbf{0}$ such that $\mathbf{x} = \mathbf{y} + \mathbf{z}$.

Is the same result true for general $n \times n$ real matrices $A$ and $B$? Give a proof or counterexample.

**Exercise 12.6.9** Let $V$ be a vector space of dimension $n$ over $\mathbb{F}$ and consider $\mathcal{L}(V, V)$ as a vector space in the usual way. If $\alpha \in \mathcal{L}(V, V)$ has rank $r$, show that

$$X = \{\beta \in \mathcal{L}(V, V) : \beta\alpha = 0\} \quad \text{and} \quad Y = \{\beta \in \mathcal{L}(V, V) : \alpha\beta = 0\}$$

are subspaces of $\mathcal{L}(V, V)$ and find their dimensions.

Suppose that $n \geq r \geq 1$. Is it always true that $X = Y$? Is it never true that $X = Y$? Give proofs or counterexamples.

**Exercise 12.6.10** Let $U$ and $V$ be finite dimensional vector spaces over $\mathbb{F}$ and let $\phi : U \to V$ be linear. We take $U_j$ to be subspace of $U$. Which of the following statements are true and which are false? Give proofs or counterexamples as appropriate.

(i) If $U_1 + U_2 = U_3$, then $\phi(U_1) + \phi(U_2) = \phi(U_3)$.

(ii) If $U_1 \oplus U_2 = U_3$, then $\phi(U_1) \oplus \phi(U_2) = \phi(U_3)$.

(iii) If $U_2 \subseteq U_3$ and $V_1$ is a subspace of $V$ such that $V_1 \oplus \phi(U_2) = \phi(U_3)$, then there exists a subspace $U_1$ of $U$ such that $U_1 \oplus U_2 = U_3$ and $\phi(U_1) = V_1$.

**Exercise 12.6.11** Are the following statements about a linear map $\alpha : \mathbb{F}^n \to \mathbb{F}^n$ true or false? Give proofs of counterexamples as appropriate.

(i) $\alpha$ is invertible if and only if its characteristic polynomial has non-zero constant coefficient.

(ii) $\alpha$ is invertible if and only if its minimal polynomial has non-zero constant coefficient.

(iii) $\alpha$ is invertible if and only if $\alpha^2$ is invertible.

**Exercise 12.6.12** Let $W$ be a subspace of a finite dimensional vector space $V$ over $\mathbb{F}$. Suppose that $\alpha$ is an endomorphism of $V$ such that $\alpha(W) \subseteq W$. Let $\beta = \alpha|_W$ be the restriction of $\alpha$ to $V$. Show that the minimal polynomial $m_\beta$ of $\beta$ divides the minimal polynomial $m_\alpha$ of $\alpha$.

Let $\mathbb{F} = \mathbb{R}$ and $V = \mathbb{R}^4$. Find an $\alpha$ and two subspaces $W_1$, $W_2$ of dimension 2 with $\alpha(W_j) \subseteq W_j$ such that, writing $\beta_j = \alpha|_{W_j}$, we have $m_{\beta_1} = m_\alpha$ and $m_{\beta_2}$ has degree at least 1, but $m_{\beta_2} \neq m_\alpha$.

**Exercise 12.6.13 [Eigenvalues without determinants]** Suppose that $U$ is a vector space of dimension $n$ over $\mathbb{C}$ and let $\alpha \in \mathcal{L}(U, U)$. Explain, without using any results which depend on determinants, why there is a monic polynomial $P$ such that $P(\alpha) = 0$.

Since we work over $\mathbb{C}$, we can write

$$P(t) = \prod_{j=1}^{N} (t - \mu_j)$$

for some $\mu_j \in \mathbb{C}$ $[1 \leq j \leq N]$. Explain carefully why there must be some $k$ with $1 \leq k \leq N$ such that $\mu_k \iota - \alpha$ is not invertible. Let us write $\mu = \mu_k$. Show that there is a non-zero vector $\mathbf{u}_k$ such that

$$\alpha \mathbf{u} = \mu \mathbf{u}.$$

[If the reader needs a hint, she should look at Exercise 12.2.1 and the rank-nullity theorem (Theorem 5.5.4). She should note that both theorems are proved without using determinants. Sheldon Axler wrote an article entitled '*Down with determinants*' ([3], available on the web) in which he proposed a determinant free treatment of linear algebra. Later he wrote a textbook *Linear Algebra Done Right* [4] to carry out the program.]

**Exercise 12.6.14 [A direct proof of Cayley–Hamilton over $\mathbb{R}$]** In this question we deal with the space $M_n(\mathbb{R})$ of real $n \times n$ matrices. (Any reader interested in the matter should note that the proof will work over any field.)

(i) Suppose that $B_r \in M_n(\mathbb{R})$ and

$$\sum_{r=0}^{R} B_r t^r = 0$$

for all $t \in \mathbb{R}$. By looking at matrix entries, or otherwise, show that $B_r = 0$ for all $0 \le r \le R$.

(ii) Suppose that $C_r, B, C \in M_n(\mathbb{R})$ and

$$(tI - C) \sum_{r=0}^{R} C_r t^r = B$$

for all $t \in \mathbb{R}$. By using (i), or otherwise, show that $C_R = 0$. Hence show that $C_r = 0$ for all $0 \le r \le R$ and conclude that $B = 0$.

(iii) Let $\chi_A(t) = \det(tI - A)$. Verify that

$$(t^k I - A^k) = (tI - A)(t^{k-1} I + t^{k-2} A + t^{k-3} A + \cdots + A^{k-1}),$$

and deduce that

$$\chi_A(t) I - \chi_A(A) = (tI - A) \sum_{j=0}^{n-1} B_j t^j$$

for some $B_j \in M_n(\mathbb{R})$.

(iv) Use the formula

$$(tI - A) \operatorname{Adj}(tI - A) = \det(tI - A) I,$$

from Section 4.5, to show that

$$\chi_A(t) I = (tI - A) \sum_{j=0}^{n-1} C_j t^j$$

for some $C_j \in M_n(\mathbb{R})$.

Conclude that

$$\chi_A(A) = (tI - A) \sum_{j=0}^{n-1} A_j t^j$$

for some $A_j \in M_n(\mathbb{R})$ and deduce the Cayley–Hamilton theorem in the form $\chi_A(A) = 0$.

**Exercise 12.6.15** Let $A$ be an $n \times n$ matrix over $\mathbb{F}$ with characteristic polynomial

$$\det(tI - A) = \sum_{j=0}^{n} b_j t^j.$$

If $A$ is non-singular and $n \ge 0$, show that

$$\operatorname{Adj} A = (-1)^{n+1} \sum_{j=1}^{n} b_j A^{j-1}.$$

Use a limiting argument to show that the result is true for all $A$.

**Exercise 12.6.16** Let $V$ be vector space over $\mathbb{F}$ of dimension $n$ and $T$ an endomorphism of $V$. If $\mathbf{x} \in V$, show that

$$U = \{P(T)\mathbf{x} \ : \ P \text{ a polynomial}\}$$

is a subspace of $V$. If $U = V$, we say that $\mathbf{x}$ is *cyclic* for $T$.

(i) If $T$ has a cyclic vector, show that the minimal and characteristic polynomials of $T$ coincide.

(ii) If $T$ has a cyclic vector and the eigenvectors of $T$ span $V$, show that $T$ has $n$ distinct eigenvalues.

(iii) If $T$ has $n$ distinct eigenvalues, show that $T$ has a cyclic vector. (If you need a hint, think for five more minutes and then look at Exercise 4.4.9.)

(iv) If $T^n = 0$ but $T^{n-1} \neq 0$, explain why we can find a vector $\mathbf{v}$ such that $T^{n-1}\mathbf{v} \neq \mathbf{0}$ and show that $\mathbf{v}$ is cyclic. What are the eigenvalues of $T$?

(v) If $T$ has a cyclic vector, show that an endomorphism $S$ commutes with $T$ if and only if $S = Q(T)$ for some polynomial $Q$.

(vi) Give an example of two commuting endomorphisms $S$ and $T$ such that there does not exist a polynomial $Q$ with $S = Q(T)$.

**Exercise 12.6.17** Let $V$ be a finite dimensional vector space over $\mathbb{F}$ and let $\Theta$ be a collection of endomorphisms of $V$. A subspace $U$ of $V$ is said to be *stable* under $\Theta$ if $\theta U \subseteq U$ for all $\theta \in \Theta$ and $\Theta$ is said to be *irreducible* if the only stable subspaces under $\Theta$ are $\{\mathbf{0}\}$ and $V$.

(i) Show that, if an endomorphism $\alpha$ commutes with every $\theta \in \Theta$, then $\ker \alpha$, $\operatorname{im} \alpha$ and the eigenspaces

$$E_\lambda = \{\mathbf{v} \in V \ : \ \alpha\mathbf{v} = \lambda\mathbf{v}\}$$

are all stable under $\Theta$.

(ii) If $\mathbb{F} = \mathbb{C}$, show that, if $\Theta$ is irreducible, the only endomorphisms which commute with every $\theta \in \Theta$ are the scalar multiples of the identity isomorphism $\iota$.

(iii) Suppose that $\mathbb{F} = \mathbb{R}$ and $V = \mathbb{R}^2$. By thinking geometrically, or otherwise, find an irreducible $\Theta$ such that it is not true that the only endomorphisms which commute with every $\theta \in \Theta$ are the scalar multiples of the identity isomorphism $\iota$.

**Exercise 12.6.18** Suppose that $V$ is a vector space over $\mathbb{F}$ and $\alpha$, $\beta$, $\gamma \in \mathcal{L}(V, V)$ are projections. Show that $\alpha + \beta + \gamma = \iota$ implies that

$$\alpha\beta = \beta\alpha = \gamma\beta = \beta\gamma = \gamma\alpha = \alpha\gamma = 0.$$

Deduce that if $\alpha$ and $\beta$ are projections, $\alpha + \beta$ is a projection if and only if $\alpha\beta = \beta\alpha = 0$.

**Exercise 12.6.19 [Simultaneous diagonalisation]** Suppose that $U$ is an $n$-dimensional vector space over $\mathbb{F}$ and $\alpha$ and $\beta$ are endomorphisms of $U$. The object of this question is to show that there exists a basis $\mathbf{e}_1, \mathbf{e}_2, \ldots, \mathbf{e}_n$ of $U$ such that each $\mathbf{e}_j$ is an eigenvector of both $\alpha$ and $\beta$ if and only if $\alpha$ and $\beta$ are separately diagonalisable (that is to say, have minimal polynomials all of whose roots lie in $\mathbb{F}$ and have no repeated roots) and $\alpha\beta = \beta\alpha$.

(i) (Easy.) Check that the condition is necessary.

(ii) From now on, we suppose that the stated condition holds. If $\lambda$ is an eigenvalue of $\beta$, write

$$E(\lambda) = \{\mathbf{e} \in U : \beta\mathbf{e} = \lambda\mathbf{e}\}.$$

Show that $E(\lambda)$ is a subspace of $U$ such that, if $\mathbf{e} \in E(\lambda)$, then $\alpha\mathbf{e} \in E(\lambda)$.

(iii) Consider the restriction map $\alpha|_{E(\lambda)} : E(\lambda) \to E(\lambda)$. By looking at the minimal polynomial of $\alpha|_{E(\lambda)}$, show that $E(\lambda)$ has a basis of eigenvectors of $\alpha$.

(iv) Use (iii) to show that there is a basis for $U$ consisting of vectors which are eigenvectors of both $\alpha$ and $\beta$.

(v) Is the following statement true? If $\alpha$ and $\beta$ are simultaneously diagonalisable (i.e. satisfy the conditions of (i)) and we write

$$E(\lambda) = \{\mathbf{e} \in U : \beta\mathbf{e} = \lambda\mathbf{e}\}, \quad F(\mu) = \{\mathbf{f} \in U : \alpha\mathbf{e} = \mu\mathbf{e}\}$$

then at least one of the following occurs: $F(\mu) \supseteq E(\lambda)$ or $E(\lambda) \supseteq F(\mu)$ or $E(\lambda) \cap F(\mu) = \{\mathbf{0}\}$. Give a proof or counterexample.

**Exercise 12.6.20** Suppose that $U$ is an $n$-dimensional vector space over $\mathbb{F}$ and $\alpha_1, \alpha_2, \ldots, \alpha_m$ are endomorphisms of $U$. Show that there exists a basis $\mathbf{e}_1, \mathbf{e}_2, \ldots, \mathbf{e}_n$ of $U$ such that each $\mathbf{e}_k$ is an eigenvector of all the $\alpha_j$ if and only if the $\alpha_j$ are separately diagonalisable and $\alpha_j\alpha_k = \alpha_k\alpha_j$ for all $1 \leq k, j \leq m$.

**Exercise 12.6.21** Suppose that $V$ is a finite dimensional space with a subspace $U$ and that $\alpha \in \mathcal{L}(V, V)$ has the property that $\alpha U \subseteq U$.

(i) Show that, if $\mathbf{v}_1 + U = \mathbf{v}_2 + U$, then $\alpha(\mathbf{v}_1) = \alpha(\mathbf{v}_2)$. Conclude that the map $\tilde{\alpha} : V/U \to U/V$ given by

$$\tilde{\alpha}(\mathbf{u} + U) = \alpha(\mathbf{u}) + U$$

is well defined. Show that $\tilde{\alpha}$ is linear.

(ii) Suppose that $\mathbf{e}_1, \mathbf{e}_2, \ldots, \mathbf{e}_k$ is a basis for $U$ and $\mathbf{e}_{k+1} + U, \mathbf{e}_{k+2} + U, \ldots, \mathbf{e}_n + U$ is a basis for $V/U$. Explain why $\mathbf{e}_1, \mathbf{e}_2, \ldots, \mathbf{e}_n$ is a basis for $V$. If $\alpha|_U : U \to U$ (the restriction of $\alpha$ to $U$) has matrix $B$ with respect to the basis $\mathbf{e}_1, \mathbf{e}_2, \ldots, \mathbf{e}_k$ of U and $\tilde{\alpha} : V/U \to V/U$ has matrix $C$ with respect to the basis $\mathbf{e}_{k+1} + U, \mathbf{e}_{k+2} + U, \ldots, \mathbf{e}_n + U$ for $V/U$, show that the matrix $A$ of $\alpha$ with respect to the basis $\mathbf{e}_1, \mathbf{e}_2, \ldots, \mathbf{e}_n$ can be written as

$$A = \begin{pmatrix} B & G \\ 0 & C \end{pmatrix}$$

where 0 is a matrix of the appropriate size consisting of zeros and $G$ is a matrix of the appropriate size.

(iii) Show that the characteristic polynomials satisfy the relation

$$\chi_\alpha(t) = \chi_{\alpha|U}(t)\chi_{\tilde{\alpha}}(t).$$

Does a similar result hold for minimal polynomials? Give reasons for your answer.

(iv) Explain why, if $V$ is a finite dimensional vector space over $\mathbb{C}$ and $\alpha \in \mathcal{L}(V, V)$, we can always find a one-dimensional subspace $U$ with $\alpha(U) \subseteq U$. Use this result and induction to prove Theorem 12.2.5.

[Of course, this proof is not very different from the proof in the text, but some people will prefer it.]

**Exercise 12.6.22** Suppose that $\alpha$ and $\beta$ are endomorphisms of a (not necessarily finite dimensional) vector space $U$ over $\mathbb{F}$. Show that, if

$$\alpha\beta - \beta\alpha = \iota, \qquad\qquad\qquad ★$$

then

$$\beta\alpha^m - \alpha^m\beta = m\alpha^{m-1}$$

for all integers $m \geq 0$.

By considering the minimal polynomial of $\alpha$, show that ★ cannot hold if $U$ is finite dimensional.

Let $\mathcal{P}$ be the vector space of all polynomials $p$ with coefficients in $\mathbb{F}$. If $\alpha$ is the differentiation map given by $(\alpha p)(t) = p'(t)$, find a $\beta$ such that ★ holds.

[Recall that $[\alpha, \beta] = \alpha\beta - \beta\alpha$ is called the *commutator* of $\alpha$ and $\beta$.]

**Exercise 12.6.23** Suppose that $\alpha$ and $\beta$ are endomorphisms of a vector space $U$ over $\mathbb{F}$ such that

$$\alpha\beta - \beta\alpha = \iota.$$

Suppose that $\mathbf{y} \in U$ is a non-zero vector such that $\alpha\mathbf{y} = \mathbf{0}$. Let $W$ be the subspace spanned by $\mathbf{y}, \beta\mathbf{y}, \beta^2\mathbf{y}, \ldots$. Show that $\alpha\beta\mathbf{y} \in W$ and find a simple expression for it. More generally show that $\alpha\beta^n\mathbf{y} \in W$ and find a simple formula for it.

By using your formula, or otherwise, show that $\mathbf{y}, \beta\mathbf{y}, \beta^2\mathbf{y}, \ldots, \beta^n\mathbf{y}$ are linearly independent for all $n$.

Find $U$, $\beta$ and $\mathbf{y}$ satisfying the conditions of the question.

**Exercise 12.6.24** Recall, or prove that, if we deal with finite dimensional spaces, the commutator of two endomorphisms has trace zero.

Suppose that $T$ is an endomorphism of a finite dimensional space over $\mathbb{F}$ with a basis $\mathbf{e}_1, \mathbf{e}_2, \ldots, \mathbf{e}_n$ such that

$$T\mathbf{e}_j \in \text{span}\{\mathbf{e}_1, \mathbf{e}_2, \ldots, \mathbf{e}_j\}.$$

Let $S$ be the endomorphism with $S\mathbf{e}_1 = \mathbf{0}$ and $S\mathbf{e}_j = \mathbf{e}_{j-1}$ for $2 \leq j \leq n$. Show that, if $\text{Tr}\, T = 0$, we can find an endomorphism such that $SR - RS = T$.

Deduce that, if $\gamma$ is an endomorphism of a finite dimensional vector space over $\mathbb{C}$, with $\text{Tr}\, \gamma = 0$ we can find endomorphisms $\alpha$ and $\beta$ such that

$$\gamma = \alpha\beta - \beta\alpha.$$

[Shoda proved that this result also holds for $\mathbb{R}$. Albert and Muckenhoupt showed that it holds for all fields. Their proof, which is perfectly readable by anyone who can do the exercise above, appears in the *Michigan Mathematical Journal* [1].]

**Exercise 12.6.25** Let us write $M_n$ for the set of $n \times n$ matrices over $\mathbb{F}$.

Suppose that $A \in M_n$ and $A$ has 0 as an eigenvalue. Show that we can find a non-singular matrix $P \in M_n$ such that $P^{-1}AP = B$ and $B$ has all entries in its first column zero. If $\tilde{B}$ is the $(n-1) \times (n-1)$ matrix obtained by deleting the first row and column from $B$, show that $\operatorname{Tr} \tilde{B}^k = \operatorname{Tr} B^k$ for all $k \geq 1$.

Let $C \in M_n$. By using the Cayley–Hamilton theorem, or otherwise, show that, if $\operatorname{Tr} C^k = 0$ for all $1 \leq k \leq n$, then $C$ has 0 as an eigenvalue. Deduce that $C = 0$.

Suppose that $F \in M_n$ and $\operatorname{Tr} F^k = 0$ for all $1 \leq k \leq n-1$. Does it follow that $F$ has 0 as an eigenvalue? Give a proof or counterexample.

**Exercise 12.6.26** We saw in Exercise 6.2.15 that, if $A$ and $B$ are $n \times n$ matrices over $\mathbb{F}$, then the characteristic polynomials of $AB$ and $BA$ are the same. By considering appropriate nilpotent matrices, or otherwise, show that $AB$ and $BA$ may have different minimal polynomials.

**Exercise 12.6.27** (i) Are the following statements about an $n \times n$ matrix $A$ true or false if we work in $\mathbb{R}$? Are they true or false if we work in $\mathbb{C}$? Give reasons for your answers.

(a) If $P$ is a polynomial and $\lambda$ is an eigenvalue of $P$, then $P(\lambda)$ is an eigenvalue of $P(A)$.

(b) If $P(A) = 0$ whenever $P$ is a polynomial with $P(\lambda) = 0$ for all eigenvalues $\lambda$ of $A$, then $A$ is diagonalisable.

(c) If $P(\lambda) = 0$ whenever $P$ is a polynomial and $P(A) = 0$, then $\lambda$ is an eigenvalue of $A$.

(ii) We work in $\mathbb{C}$. Let

$$
B = \begin{pmatrix} a & d & c & b \\ b & a & d & c \\ c & b & a & d \\ d & c & b & a \end{pmatrix} \quad \text{and} \quad A = \begin{pmatrix} 0 & 0 & 0 & 1 \\ 1 & 0 & 0 & 0 \\ 0 & 1 & 0 & 0 \\ 0 & 0 & 1 & 0 \end{pmatrix}.
$$

By computing powers of $A$, or otherwise, find a polynomial $P$ with $B = P(A)$ and find the eigenvalues of $B$. Compute $\det B$.

(iii) Generalise (ii) to $n \times n$ matrices and then compare the results with those of Exercise 6.8.29.

**Exercise 12.6.28** Let $V$ be a vector space over $\mathbb{F}$ with a basis $e_1, e_2, \ldots, e_n$. If $\sigma$ is a permutation of $1, 2, \ldots, n$ we define $\alpha_\sigma$ to be the unique endomorphism with

$$
\alpha_\sigma e_j = e_{\sigma j}
$$

for $1 \leq j \leq n$. If $\mathbb{F} = \mathbb{R}$, show that $\alpha_\sigma$ is diagonalisable if and only if $\sigma^2$ is the identity permutation. If $\mathbb{F} = \mathbb{C}$, show that $\alpha_\sigma$ is always diagonalisable.

**Exercise 12.6.29** Consider $V = \mathbb{F}^n$ with the usual inner product.

By using an appropriate orthonormal basis, or otherwise (there are lots of ways of doing this), show that, if $E$ is a subspace of $V$ and $\mathbf{a} \notin E$, then we can find a $\delta > 0$ such that

$$\|\mathbf{x} - \mathbf{a}\| < \delta \Rightarrow \mathbf{x} \notin E.$$

Show also that, if $E$ is a proper subspace of $V$ (so $E \neq V$), then, given any $\mathbf{v} \in V$ and any $\delta > 0$, we can find $\mathbf{a} \notin E$ with

$$\|\mathbf{v} - \mathbf{a}\| < \delta.$$

Deduce that, given any $1 \leq p \leq n$, we can find a sequence $\mathbf{a}_1, \mathbf{a}_2, \ldots$ of distinct vectors in $V$ such that any collection of $p$ members of the sequence is linearly independent, but no collection of $p + 1$ members is.

Since all vector spaces over $\mathbb{F}$ of the same dimension are isomorphic, the result holds for all finite dimensional vector spaces.[8]

**Exercise 12.6.30** (This continuation of Exercise 12.6.29 requires the notion of a Cauchy sequence in $\mathbb{R}^m$ and the knowledge of analysis which goes with it.) Consider $V = \mathbb{R}^m$ with the usual inner product.

If $E_1, E_2, \ldots$ are proper subspaces of $V$, show that we can find inductively $\mathbf{a}_n$ and $\delta_n > 0$, with $\mathbf{a}_0 = \mathbf{0}$ and $\delta_0 = 1$, satisfying the following conditions.

(i) $\|\mathbf{a}_n - \mathbf{a}_{n-1}\| < \delta_{n-1}/4$.
(ii) $\|\mathbf{x} - \mathbf{a}_n\| < \delta_n \Rightarrow \mathbf{x} \notin E_n$.
(iii) $\delta_n < \delta_{n-1}/4$.

Show that the $\mathbf{a}_n$ form a Cauchy sequence and deduce that there exists an $\mathbf{a} \in V$ with $\|\mathbf{a}_n - \mathbf{a}\| \to 0$. Show that $\|\mathbf{a}_k - \mathbf{a}\| \leq \delta_k/3$ for each $k \geq 1$ and deduce that $\mathbf{a} \notin V$.

Thus a finite dimensional vector space over $\mathbb{R}$ cannot be the countable union of proper subspaces. (The same argument works for $\mathbb{C}$.)

**Exercise 12.6.31** We work in a finite dimensional vector space $V$ over $\mathbb{F}$. Show that any two subspaces $U_1$, $U_2$ of the same dimension have a common complementary subspace. In other words, show that there is a subspace $W$ such that

$$U_1 \oplus W = U_2 \oplus W = V.$$

**Exercise 12.6.32** Let $U$ and $V$ be vector spaces over $\mathbb{F}$ of dimensions $m$ and $n$ respectively. Suppose that $X$ and $Y$ are subspaces of $U$ with $X \subseteq Y$, that $Z$ is a subspace of $V$ and that

$$\dim X = r, \quad \dim Y = s, \quad \dim Z = t.$$

Show that the set $\mathcal{L}_0$ of all $\alpha \in \mathcal{L}(U, V)$ such that $X \subseteq \ker \alpha$ and $\alpha(Y) \subseteq Z$ is a subspace of $\mathcal{L}(U, V)$ with dimension

$$mn + st - rt - sn.$$

---

[8] This is an ugly way of doing things, but, as we shall see in the next chapter (for example, in Exercise 13.4.1), we must use some 'non-algebraic' property of $\mathbb{R}$ and $\mathbb{C}$.

**Exercise 12.6.33** Here is another proof of the diagonalisation theorem (Theorem 12.3.5). Let $V$ be a finite dimensional vector space over $\mathbb{F}$. If $\alpha_j \in \mathcal{L}(V, V)$ show that the nullities satisfy

$$n(\alpha_1\alpha_2) \le n(\alpha_1) + n(\alpha_2)$$

and deduce that

$$n(\alpha_1\alpha_2 \cdots \alpha_k) \le n(\alpha_1) + n(\alpha_2) + \cdots + n(\alpha_k).$$

Hence show that, if $\alpha \in \mathcal{L}(V, V)$ satisfies $p(\alpha) = 0$ for some polynomial $p$ which factorises into distinct linear terms, then $\alpha$ is diagonalisable.

**Exercise 12.6.34** In our treatment of the Jordan normal form we worked over $\mathbb{C}$. In this question you may not use any result concerning $\mathbb{C}$, but you may use the theorem that every real polynomial factorises into linear and quadratic terms.

Let $\alpha$ be an endomorphism on a finite dimensional real vector space $V$. Explain briefly why there exists a real monic polynomial $m$ with $m(\alpha) = 0$ such that, if $f$ is a real polynomial with $f(\alpha) = 0$, then $m$ divides $f$.

Show that, if $k$ is a non-constant polynomial dividing $m$, there is a non-zero subspace $W$ of $V$ such that $k(W) = \{\mathbf{0}\}$. Deduce that $V$ has a subspace $U$ of dimension 1 or 2 such that $\alpha(U) \subseteq U$ (that is to say, $U$ is an $\alpha$-*invariant subspace*).

Let $\alpha$ be the endomorphism of $\mathbb{R}^4$ whose matrix with respect to the standard basis is

$$\begin{pmatrix} 0 & 1 & 0 & 0 \\ 0 & 0 & 1 & 0 \\ 0 & 0 & 0 & 1 \\ -1 & 0 & -2 & 0 \end{pmatrix}.$$

Show that $\mathbb{R}^4$ has an $\alpha$-invariant subspace of dimension 2 but no $\alpha$-invariant subspaces of dimension 1 or 3.

**Exercise 12.6.35** Suppose that $V$ is a finite dimensional space over $\mathbb{F}$ and $\alpha : V \to V$ is a linear map such that $\alpha^n = \iota$. Show that, if $V_1$ is a subspace with $\alpha V_1 \subseteq V_1$, then there is a subspace $V_2$ such that $V = V_1 \oplus V_2$ and $\alpha(V_2) \subseteq V_2$.
[Hint: Let $\pi$ be a projection with $\pi(V) \subseteq V_1$ and $\pi\mathbf{u} = \mathbf{u}$ for all $\mathbf{u} \in V_1$. Consider the map $\rho$ defined by $\rho(\mathbf{v}) = n^{-1}\sum_{j=0}^{n-1} \alpha^j \pi \alpha^{-j}(\mathbf{v})$.]

**Exercise 12.6.36** (i) Let $U$ be a vector space over $\mathbb{F}$ and let $\alpha : U \to U$ be a linear map such that $\alpha^m = 0$ for some $m$ (that is to say, a nilpotent map). Show that

$$(\iota - \alpha)(\iota + \alpha + \alpha^2 + \cdots + \alpha^{m-1}) = \iota$$

and deduce that $\iota - \alpha$ is invertible.

(ii) Let $J_m(\lambda)$ be an $m \times m$ Jordan block matrix over $\mathbb{F}$, that is to say, let

$$J_m(\lambda) = \begin{pmatrix} \lambda & 1 & 0 & 0 & \ldots & 0 & 0 \\ 0 & \lambda & 1 & 0 & \ldots & 0 & 0 \\ 0 & 0 & \lambda & 1 & \ldots & 0 & 0 \\ \vdots & \vdots & \vdots & \vdots & & \vdots & \vdots \\ 0 & 0 & 0 & 0 & \ldots & \lambda & 1 \\ 0 & 0 & 0 & 0 & \ldots & 0 & \lambda \end{pmatrix}.$$

Show that $J_m(\lambda)$ is invertible if and only if $\lambda \neq 0$. Write down $J_m(\lambda)^{-1}$ explicitly in the case when $\lambda \neq 0$.

(iii) Use the Jordan normal form theorem to show that, if $U$ is a vector space over $\mathbb{C}$ and $\alpha : U \to U$ is a linear map, then $\alpha$ is invertible if and only if the equation $\alpha\mathbf{x} = \mathbf{0}$ has a unique solution.

[Part (iii) is for amusement only. It would require a lot of hard work to remove any suggestion of circularity.]

**Exercise 12.6.37** Let $\mathcal{Q}_n$ be the space of all real polynomials in two variables

$$Q(x, y) = \sum_{j=0}^{n} \sum_{k=0}^{n} q_{jk} x^j y^k$$

of degree at most $n$ in each variable. Let

$$(\alpha Q)(x, y) = \left( \frac{\partial}{\partial x} + \frac{\partial}{\partial y} \right) Q(x, y), \quad (\beta Q)(x, y) = Q(x + 1, y + 1).$$

Show that $\alpha$ and $\beta$ are endomorphisms of $\mathcal{Q}_n$ and find the associated Jordan normal forms. [It *may* be helpful to look at simple cases, but *just* looking for patterns without asking the appropriate questions is probably not the best way of going about things.]

**Exercise 12.6.38** We work over $\mathbb{C}$. Let $A$ be an invertible $n \times n$ matrix. Show that $\lambda$ is an eigenvalue of $A$ if and only if $\lambda^{-1}$ is an eigenvalue of $A^{-1}$. What is the relationship between the algebraic and geometric multiplicities of $\lambda$ as an eigenvalue of $A$ and the algebraic and geometric multiplicities of $\lambda^{-1}$ as an eigenvalue of $A^{-1}$? Obtain the characteristic polynomial of $A^{-1}$ in terms of the characteristic polynomial of $A$. Obtain the minimal polynomial of $A^{-1}$ in terms of the minimal polynomial of $A$. Give reasons.

**Exercise 12.6.39** Given a matrix in Jordan normal form, explain how to write down the associated minimal polynomial without further calculation. Why does your method work?

Is it true that two $n \times n$ matrices over $\mathbb{C}$ with the same minimal polynomial must have the same rank? Give a proof or counterexample.

**Exercise 12.6.40** By first considering Jordan block matrices, or otherwise, show that every $n \times n$ complex matrix is conjugate to its transpose $A^T$ (in other words, there exists an invertible $n \times n$ matrix $P$ such that $A^T = PAP^{-1}$).

**Exercise 12.6.41** We work with $n \times n$ matrices over $\mathbb{C}$. We say that an $n \times n$ matrix $U$ is *unipotent* if $U - I$ is nilpotent.

(i) Show that $U$ is unipotent if and only if its only eigenvalue is 1.

(ii) If $A$ is an invertible $n \times n$ matrix, show, by considering the Jordan normal form, that there exists an invertible matrix $P$ such that

$$PAP^{-1} = D_0 + N$$

where $D_0$ is an invertible diagonal matrix, $N$ is an upper triangular matrix with zeros on the diagonal and $D_0 N = N D_0$.

(iii) If we now set $D = P^{-1} D_0 P$, show that $U = D^{-1} A$ is unipotent.

(iv) Conclude that any invertible matrix $A$ can be written in the form $A = DU$ where $D$ is diagonalisable, $U$ is unipotent and $DU = UD$.

(v) Is it true that every $n \times n$ matrix $A$ can be written in the form $A = DU$ where $D$ is diagonalisable, $U$ is unipotent and $DU = UD$? Is it true that, if an $n \times n$ matrix $A$ can be written in the form $A = DU$ where $D$ is diagonalisable, $U$ is unipotent and $DU = UD$, then $A$ is invertible? Give reasons for your answers.

**Exercise 12.6.42** Let $\alpha$, $\beta \in \mathcal{L}(U, U)$, where $U$ is a finite dimensional vector space over $\mathbb{C}$. Show that, if $\alpha\beta = \beta\alpha$, then we can triangularise the two endomorphisms simultaneously. In other words, we can find a basis $\mathbf{e}_1, \mathbf{e}_2, \ldots, \mathbf{e}_n$ such that

$$\alpha\mathbf{e}_r, \ \beta\mathbf{e}_r \in \text{span}\{\mathbf{e}_1, \ \mathbf{e}_2, \ \ldots, \mathbf{e}_r\}.$$

If we can triangularise $\alpha$, $\beta \in \mathcal{L}(U, U)$ simultaneously, does it follow that $\alpha\beta = \beta\alpha$? Give reasons for your answer.

# 13

# Vector spaces without distances

## 13.1 A little philosophy

There are at least two ways that the notion of a finite dimensional vector space over $\mathbb{R}$ or $\mathbb{C}$ can be generalised. The first is that of the analyst who considers infinite dimensional spaces. The second is that of the algebraist who considers finite dimensional vector spaces over more general objects than $\mathbb{R}$ or $\mathbb{C}$.

It appears that infinite dimensional vector spaces are not very interesting unless we add additional structure. This additional structure is provided by the notion of distance or metric. It is natural for analysts to invoke metric considerations when talking about finite dimensional spaces, since they expect to invoke metric considerations when talking about infinite dimensional spaces.

It is also natural for numerical analysts to talk about distances, since they need to measure the errors in their computations, and for physicists to talk about distances, since they need to measure the results of their experiments.

Algebraists dislike mixing up concepts in this way. They point out that many results in vector space theory from Desargues' theorem (see Exercise 13.4.9) to determinants do not depend on the existence of a distance and that it is likely that the most perspicacious way of viewing these results will not involve this extraneous notion. They will also point out that many generalisations of the idea of vector spaces (including the ones considered in this chapter) produce structures which do not support a linked metric.[1]

In Chapters 14 and 15 we shall plunge eagerly into the world of distances, but in this chapter we look at the world through the eyes of the algebraist. Although I hope the reader will think about the contents of this chapter, she should realise that it only scratches the surface of its subject.

## 13.2 Vector spaces over fields

The simplest generalisation of the real and complex number systems is the notion of a field, that is to say an object which behaves *algebraically* like those two systems. We formalise this idea by writing down an axiom system.

---

[1] They may support metrics, but these metrics will not reflect the algebraic structure.

329

**Definition 13.2.1** *A field* $(\mathbb{G}, +, \times)$ *is a set* $\mathbb{G}$ *containing elements* 0 *and* 1, *with* $0 \neq 1$, *for which the operations of addition and multiplication obey the following rules (we suppose that* $a, b, c \in \mathbb{G}$).

(i) $a + b = b + a$.

(ii) $(a + b) + c = a + (b + c)$.

(iii) $a + 0 = a$.

(iv) *Given* $a$, *we can find* $-a \in \mathbb{G}$ *with* $a + (-a) = 0$.

(v) $a \times b = b \times a$.

(vi) $(a \times b) \times c = a \times (b \times c)$.

(vii) $a \times 1 = a$.

(viii) *Given* $a \neq 0$, *we can find* $a^{-1} \in \mathbb{G}$ *with* $a \times a^{-1} = 1$.

(ix) $a \times (b + c) = a \times b + a \times c$.

We write $ab = a \times b$ and refer to the field $\mathbb{G}$ rather than to the field $(\mathbb{G}, +, \times)$. The axiom system is merely intended as background and we shall not spend time checking that every step is justified from the axioms.[2]

It is easy to see that the rational numbers $\mathbb{Q}$ form a field and that, if $p$ is a prime, the integers modulo $p$ give rise to a field which we call $\mathbb{Z}_p$.

**Exercise 13.2.2** (i) *Write down addition and multiplication tables for* $\mathbb{Z}_2$. *Check that* $x + x = 0$, *so* $x = -x$ *for all* $x \in \mathbb{Z}_2$.

(ii) *If* $\mathbb{G}$ *is a field which satisfies the condition* $1 + 1 \neq 0$, *show that* $x = -x \Rightarrow x = 0$.

**Exercise 13.2.3** *We work in a field* $\mathbb{G}$.

(i) *Show that, if* $cd = 0$, *then at least one of* $c$ *and* $d$ *must be zero.*

(ii) *Show that if* $a^2 = b^2$, *then* $a = b$ *or* $a = -b$ *(or both).*

(iii) *If* $\mathbb{G}$ *is a field with* $k$ *elements which satisfies the condition* $1 + 1 \neq 0$, *show that* $k$ *is odd and exactly* $(k + 1)/2$ *elements of* $\mathbb{G}$ *are squares.*

(iv) *How many elements of* $\mathbb{Z}_2$ *are squares?*

If $\mathbb{G}$ is a field, we define a vector space $U$ over $\mathbb{G}$ by repeating Definition 5.2.2 with $\mathbb{F}$ replaced by $\mathbb{G}$. All the material on solution of linear equations, determinants and dimension goes through essentially unchanged. However, in general, there is no analogue of our work on Euclidean distance and inner product.

One interesting new vector space that turns up is described in the next exercise.

**Exercise 13.2.4** *Check that* $\mathbb{R}$ *is a vector space over the field* $\mathbb{Q}$ *of rationals if we define vector addition* $\dot{+}$ *and scalar multiplication* $\dot{\times}$ *in terms of ordinary addition* $+$ *and multiplication* $\times$ *on* $\mathbb{R}$ *by*

$$x \dot{+} y = x + y, \quad \lambda \dot{\times} x = \lambda \times x$$

*for* $x, y \in \mathbb{R}, \lambda \in \mathbb{Q}$.

---

[2] But I shall not lie to the reader and, if she wants, she can check that everything is indeed deducible from the axioms. If you *are* going to think about the axioms, you may find it useful to observe that conditions (i) to (iv) say that $(\mathbb{G}, +)$ is an Abelian group with 0 as identity, that conditions (v) to (viii) say that $(\mathbb{G} \setminus \{0\}, \times)$ is an Abelian group with 1 as identity and that condition (ix) links addition and multiplication through the 'distributive law'.

The world is divided into those who, once they see what is going on in Exercise 13.2.4, smile a little and those who become very angry.[3] Once we are clear what the definition means, we replace $\dot{+}$ and $\dot{\times}$ by $+$ and $\times$.

The proof of the next lemma requires you to know about countability.

**Lemma 13.2.5** *The vector space* $\mathbb{R}$ *over* $\mathbb{Q}$ *is infinite dimensional.*

*Proof* If $\mathbf{e}_1, \mathbf{e}_2, \ldots, \mathbf{e}_n \in \mathbb{R}$, then, since $\mathbb{Q}$, and so $\mathbb{Q}^n$ is countable, it follows that

$$E = \left\{ \sum_{j=1}^{n} \lambda_j \mathbf{e}_j \; : \; \lambda_j \in \mathbb{Q} \right\}$$

is countable. Since $\mathbb{R}$ is uncountable, it follows that $\mathbb{R} \neq E$. Thus no finite set can span $\mathbb{R}$. $\qquad\qquad\square$

Linearly independent sets for the vector space just described turn up in number theory and related disciplines.

The smooth process of generalisation comes to a halt when we reach eigenvalues. It remains true that every root of the characteristic equation of an $n \times n$ matrix corresponds to an eigenvalue, but, in many fields, it is not true that every polynomial has a root.

**Exercise 13.2.6** (*i*) *Suppose that we work over a field* $\mathbb{G}$ *in which* $1 + 1 \neq 0$. *Let*

$$A = \begin{pmatrix} 0 & 1 \\ a & 0 \end{pmatrix}$$

*with* $a \neq 0$. *Show that there exists an invertible* $2 \times 2$ *matrix* $M$ *with* $MAM^{-1}$ *diagonal if and only if* $t^2 = a$ *has a solution.*

(*ii*) *What happens if* $a = 0$?

(*iii*) *Does a suitable* $M$ *exist if* $\mathbb{G} = \mathbb{Q}$ *and* $a = 2$?

(*iv*) *Suppose that* $\mathbb{G}$ *is a finite field with* $1 + 1 \neq 0$ (*for example,* $\mathbb{G} = \mathbb{Z}_p$ *with* $p \geq 3$). *Use Exercise 13.2.3 to show that we cannot diagonalise* $A$ *for all non-zero* $a$.

(*v*) *Suppose that we drop the condition in* (*i*) *and consider* $\mathbb{G} = \mathbb{Z}_2$. *Can we diagonalise*

$$A = \begin{pmatrix} 0 & 1 \\ 1 & 0 \end{pmatrix}?$$

*Give reasons.*

*Find an explicit invertible* $M$ *such that* $MAM^{-1}$ *is lower triangular.*

There are many other fields besides the ones just discussed.

**Exercise 13.2.7** *Consider the set* $\mathbb{Z}_2^2$. *Suppose that we define addition and multiplication for* $\mathbb{Z}_2^2$, *in terms of standard addition and multiplication for* $\mathbb{Z}_2$, *by*

$$(a, b) + (c, d) = (a + c, b + d),$$
$$(a, b) \times (c, d) = (ac + bd, ad + bc + bd).$$

---

[3] Not a good sign if you want to become a pure mathematician.

*Write out the addition and multiplication tables and show that* $\mathbb{Z}_2^2$ *is a field for these operations. How many of the elements are squares?*

*[Secretly, $(a, b) = a + b\omega$ with $\omega^2 = 1 + \omega$ and there is a theory which explains this choice, but fiddling about with possible multiplication tables would also reveal the existence of this field.]*

The existence of fields in which $2 = 0$ (where we define $2 = 1 + 1$) such as $\mathbb{Z}_2$ and the field described in Exercise 13.2.7 produces an interesting difficulty.

**Exercise 13.2.8** (*i*) *If* $\mathbb{G}$ *is a field in which* $2 \neq 0$, *show that every* $n \times n$ *matrix over* $\mathbb{G}$ *can be written in a unique way as the sum of a symmetric and an antisymmetric matrix. (That is to say, $A = B + C$ with $B^T = B$ and $C^T = -C$.)*

(*ii*) *Show that an* $n \times n$ *matrix over* $\mathbb{Z}_2$ *is antisymmetric if and only if it is symmetric. Give an example of a* $2 \times 2$ *matrix over* $\mathbb{Z}_2$ *which cannot be written as the sum of a symmetric and an antisymmetric matrix. Give an example of a* $2 \times 2$ *matrix over* $\mathbb{Z}_2$ *which can be written as the sum of a symmetric and an antisymmetric matrix in two distinct ways.*

Thus, if you wish to extend a result from the theory of vector spaces over $\mathbb{C}$ to a vector space $U$ over a field $\mathbb{G}$, you must ask yourself the following questions.

(1) Does every polynomial in $\mathbb{G}$ have a root? (If it does we say that $\mathbb{G}$ is *algebraically closed*.)
(2) Is there a useful analogue of Euclidean distance? If you are an analyst, you will then need to ask whether your metric is complete (that is to say, every Cauchy sequence converges).
(3) Is the space finite or infinite dimensional?
(4) Does $\mathbb{G}$ have any algebraic quirks? (The most important possibility is that $2 = 0$.)

Sometimes the result fails to transfer. It is not true that every endomorphism of finite dimensional vector spaces $V$ over $\mathbb{G}$ has an eigenvector (and so it is not true that the triangularisation result Theorem 12.2.5 holds) unless every polynomial with coefficients $\mathbb{G}$ has a root in $\mathbb{G}$. (We discuss this further in Exercise 13.4.3.)

Sometimes it is only the proof which fails to transfer. We used Theorem 12.2.5 to prove the Cayley–Hamilton theorem for $\mathbb{C}$, but we saw that the Cayley–Hamilton theorem remains true for $\mathbb{R}$ although the triangularisation result of Theorem 12.2.5 now fails. In fact, the alternative proof of the Cayley–Hamilton theorem given in Exercise 12.6.14 works for every field and so the Cayley–Hamilton theorem holds for every finite dimensional vector space $U$ over any field. (Since we do not know how to define determinants in the infinite dimensional case, we cannot even state such a theorem if $U$ is not finite dimensional.)

**Exercise 13.2.9** *Here is another surprise that lies in wait for us when we look at general fields.*

(i) *Show that, if we work in* $\mathbb{Z}_2$,

$$x^2 + x = 0$$

*for all x. Thus a polynomial with non-zero coefficients may be zero everywhere.*

(ii) *Show that, if we work in any finite field, we can find a polynomial with non-zero coefficients which is zero everywhere.*

The reader may ask why I do not prove all results for the most general fields to which they apply. This is to view the mathematician as a worker on a conveyor belt who can always find exactly the parts that she requires. It is more realistic to think of the mathematician as a tinkerer in a garden shed who rarely has the exact part she requires, but has to modify some other part to make it fit her machine. A theorem is not a monument, but a signpost and the proof of a theorem is often more important than its statement.

We conclude this section by showing how vector space theory gives us information about the structure of finite fields. Since this is a digression within a digression, I shall use phrases like 'subfield' and 'isomorphic as a field' without defining them. If you find that this makes the discussion incomprehensible, just skip the rest of the section.

**Lemma 13.2.10** *Let* $\mathbb{G}$ *be a finite field. We write*

$$\bar{k} = \underbrace{1 + 1 + \cdots + 1}_{k}$$

*for k a positive integer. (Thus* $\bar{0} = 0$ *and* $\bar{1} = 1$.)

(i) *There is a prime p such that*

$$\mathbb{H} = \{\bar{r} : 0 \leq r \leq p - 1\}$$

*is a subfield of* $\mathbb{G}$ *and* $\mathbb{H}$ *is isomorphic as a field to* $\mathbb{Z}_p$.

(iii) $\mathbb{G}$ *may be considered as a vector space over* $\mathbb{H}$.

(iv) *There is an* $n \geq 1$ *such that* $\mathbb{G}$ *has exactly* $p^n$ *elements.*

Thus we know, without further calculation, that there is no field with 22 elements. It can be shown that, if $p$ is a prime and $n$ a strictly positive integer, then there does, indeed, exist a field with exactly $p^n$ elements, but a proof of this would take us too far afield.

*Proof of Lemma 13.2.10* (i) Since $\mathbb{G}$ is finite, there must exist integers $u$ and $v$ with $0 \leq v < u$ such that $\bar{u} = \bar{v}$ and so, setting $w = u - v$, there exists an integer $w > 0$ such that $\bar{w} = 0$. Let $p$ be the least strictly positive integer such that $\bar{p} = 0$. We must have $p$ prime, since, if $1 \leq r \leq s \leq p$,

$$p = rs \Rightarrow 0 = \bar{r}\bar{s} \Rightarrow \bar{r} = 0 \quad \text{and/or} \quad \bar{s} = 0 \Rightarrow s = p.$$

(ii) Suppose that $r$ and $s$ are integers with $r \geq s \geq 0$ and $\bar{r} = \bar{s}$. We know that $r - s = kp + q$ for some integer $k \geq 0$ and some integer $q$ with $p - 1 \geq q \geq 0$, so

$$0 = \bar{r} - \bar{s} = \overline{r - s} = \overline{kp + q} = \bar{q},$$

so $q = 0$ and $r \equiv s$ modulo $p$. Conversely, if $r \equiv s$ modulo $p$, then an argument of a similar type shows that $\bar{r} = \bar{s}$. We have established a bijection $[r] \leftrightarrow \bar{r}$ which matches the element $[r]$ of $\mathbb{Z}_p$ corresponding to the integer $r$ to the element $\bar{r}$. Since $\overline{rs} = \bar{r}\bar{s}$ and $\overline{r + s} = \bar{r} + \bar{s}$, this bijection preserves addition and multiplication.

(iii) Just as in Exercise 13.2.4, any field $\mathbb{H}$ with subfield $\mathbb{K}$ may be considered as a vector space over $\mathbb{K}$ by defining vector addition to correspond to ordinary field addition for $\mathbb{H}$ and multiplication of a vector in $\mathbb{H}$ by a scalar in $\mathbb{K}$ to correspond to ordinary field multiplication for $\mathbb{H}$.

(iv) Consider $\mathbb{G}$ as a vector space over $\mathbb{H}$. Since $\mathbb{G}$ is finite, it has a finite spanning set (for example, $\mathbb{G}$ itself) and so is finite dimensional. Let $\mathbf{e}_1, \mathbf{e}_2, \ldots, \mathbf{e}_n$ be a basis. Then each element of $\mathbb{G}$ corresponds to exactly one expression

$$\sum_{j=1}^{n} \lambda_j \mathbf{e}_j \quad \text{with } \lambda_j \in \mathbb{H} \, [1 \le j \le n]$$

and so $\mathbb{G}$ has $p^n$ elements. $\qquad\square$

The theory of finite fields has applications in the theory of data transmission and storage. (We discussed 'secret sharing' in Section 5.6 and the next section discusses an error correcting code, but many applications require deeper results.)

### 13.3 Error correcting codes

In this section we use vector spaces over $\mathbb{Z}_2$ to discuss some simple error correcting codes. We start with some easy exercises to get the reader used to working with $\mathbb{Z}_2$.

**Exercise 13.3.1** *Do Exercise 13.2.2 if you have not already done it.*

*If $U$ is a vector space over the field $\mathbb{Z}_2$, show that a subset $W$ of $U$ is a subspace if and only if*

*(i)* $\mathbf{0} \in W$ *and*

*(ii)* $\mathbf{u}, \mathbf{v} \in W \Rightarrow \mathbf{u} + \mathbf{v} \in W$.

**Exercise 13.3.2** *Show that the following statements about a vector space $U$ over the field $\mathbb{Z}_2$ are equivalent.*

*(i) $U$ has dimension $n$.*

*(ii) $U$ is isomorphic to $\mathbb{Z}_2^n$ (that is to say, there is a linear map $\alpha : U \to \mathbb{Z}_2^n$ which is a bijection).*

*(iii) $U$ has $2^n$ elements.*

**Exercise 13.3.3** *If $\mathbf{p} \in \mathbb{Z}_2^n$, show that the mapping $\alpha : \mathbb{Z}_2^n \to \mathbb{Z}_2$ defined by*

$$\alpha(\mathbf{e}) = \sum_{j=1}^{n} p_j e_j$$

is linear. (*In the language of Sections 11.3 and 11.4, we have* $\alpha \in (\mathbb{Z}_2^n)'$ *the* dual space *of* $\mathbb{Z}_2^n$.)

Show, conversely, that, if $\alpha \in (\mathbb{Z}_2^n)'$, then we can find a $\mathbf{p} \in \mathbb{Z}_2^n$ such that

$$\alpha(\mathbf{e}) = \sum_{j=1}^{n} p_j e_j.$$

Early computers received their instructions through paper tape. Each line of paper tape had a pattern of holes which may be thought of as a 'word' $\mathbf{x} = (x_1, x_2, \ldots, x_8)$ with $x_j$ either taking the value 0 (no hole) or 1 (a hole). Although this was the fastest method of reading in instructions, mistakes could arise if, for example, a hole was mispunched or a speck of dust interfered with the optical reader. Because of this, $x_8$ was used as a check digit defined by the relation

$$x_1 + x_2 + \cdots + x_8 \equiv 0 \pmod{2}.$$

The input device would check that this relation held for each line. If the relation failed for a single line the computer would reject the entire program.

Hamming had access to an early electronic computer, but was low down in the priority list of users. He would submit his programs encoded on paper tape to run over the weekend, but often he would have his tape returned on Monday because the machine had detected an error in the tape. 'If the machine can detect an error' he asked himself 'why can the machine not correct it?' and he came up with the following idea.

Hamming's scheme used seven of the available places so his words had the form $\mathbf{c} = (c_1, c_2, \ldots, c_7) \in \{0, 1\}^7$. The codewords[4] $\mathbf{c}$ are chosen to satisfy the following three conditions, modulo 2,

$$c_1 + c_3 + c_5 + c_7 \equiv 0$$
$$c_2 + c_3 + c_6 + c_7 \equiv 0$$
$$c_4 + c_5 + c_6 + c_7 \equiv 0.$$

By inspection, we may choose $c_3, c_5, c_6$ and $c_7$ freely and then $c_1, c_2$ and $c_4$ are completely determined.

Suppose that we receive the string $\mathbf{x} \in \mathbb{F}_2^7$. We form the *syndrome* $(z_1, z_2, z_4) \in \mathbb{F}_2^3$ given by

$$z_1 \equiv x_1 + x_3 + x_5 + x_7$$
$$z_2 \equiv x_2 + x_3 + x_6 + x_7$$
$$z_4 \equiv x_4 + x_5 + x_6 + x_7$$

where our arithmetic is modulo 2. If $\mathbf{x}$ is a codeword, then $(z_1, z_2, z_4) = (0, 0, 0)$. If one error has occurred then the place in which $\mathbf{x}$ differs from $\mathbf{c}$ is given by $z_1 + 2z_2 + 4z_4$ (using ordinary addition, not addition modulo 2).

---

[4] There is no suggestion of secrecy here or elsewhere in this section. A code is simply a collection of codewords and a codeword is simply a permitted pattern of zeros and ones. Notice that our codewords are *row* vectors.

**Exercise 13.3.4** *Construct a couple of examples of Hamming codewords* **c** *and change them in one place. Check that the statement just made holds for your examples.*

**Exercise 13.3.5** *Suppose that we use eight hole tape with the standard paper tape code and the probability that an error occurs at a particular place on the tape (i.e. a hole occurs where it should not or fails to occur where it should) is $10^{-4}$, errors occurring independently of each other. A program requires about 10 000 lines of tape (each line containing eight places) using the paper tape code. Using the Poisson approximation, direct calculation (possible with a hand calculator, but really no advance on the Poisson method), or otherwise, show that the probability that the tape will be accepted as error free by the decoder is less than 0.04%.*

*Suppose now that we use the Hamming scheme (making no use of the last place in each line). Explain why the program requires about 17 500 lines of tape but that any particular line will be accepted as error free and correctly decoded with probability about $1 - (21 \times 10^{-8})$ and the probability that the entire program will be accepted as error free and be correctly decoded is better than 99.6%.*

Hamming's scheme is easy to implement. It took a little time for his company to realise what he had done,[5] but they were soon trying to patent it. In retrospect, the idea of an error correcting code seems obvious (Hamming's scheme had actually been used as the basis of a Victorian party trick) but Hamming's idea opened up an entirely new field.[6]

Why does the Hamming scheme work? It is natural to look at strings of 0s and 1s as row vectors in the vector space $\mathbb{Z}_2^n$ over the field $\mathbb{Z}_2$. Let us write

$$A = \begin{pmatrix} 1 & 0 & 1 & 0 & 1 & 0 & 1 \\ 0 & 1 & 1 & 0 & 0 & 1 & 1 \\ 0 & 0 & 0 & 1 & 1 & 1 & 1 \end{pmatrix} \qquad \bigstar$$

and let $\mathbf{e}_i \in \mathbb{Z}_2^7$ be the row vector with 1 in the $i$th place and 0 everywhere else.

**Exercise 13.3.6** *We use the notation just introduced.*

*(i) Show that* **c** *is a Hamming codeword if and only if $A\mathbf{c}^T = \mathbf{0}$ (working in $\mathbb{Z}_2$).*

*(ii) If $j = a_1 + 2a_2 + 4a_3$ in ordinary arithmetic with $a_1$, $a_2$, $a_3 \in \{0, 1\}$ show that, working in $\mathbb{Z}_2$,*

$$A\mathbf{e}_j^T = \begin{pmatrix} a_1 \\ a_2 \\ a_3 \end{pmatrix}.$$

---

[5] Experienced engineers came away from working demonstrations muttering 'I still don't believe it'.

[6] When Fillipo Brunelleschi was in competition to build the dome for the Cathedral of Florence he refused to show his plans '... proposing instead ... that whosoever could make an egg stand upright on a flat piece of marble should build the cupola, since thus each man's intellect would be discerned. Taking an egg, therefore, all those craftsmen sought to make it stand upright, but not one could find the way. Whereupon Filippo, being told to make it stand, took it graciously, and, giving one end of it a blow on the flat piece of marble, made it stand upright. The craftsmen protested that they could have done the same; but Filippo answered, laughing, that they could also have raised the cupola, if they had seen his design.' Vasari *Lives of the Artists* [31].

*Deduce that, if* **c** *is a Hamming codeword,*

$$A(\mathbf{c} + \mathbf{e}_j)^T = \begin{pmatrix} a_1 \\ a_2 \\ a_3 \end{pmatrix}.$$

*Explain why the statement made in the paragraph preceding Exercise 13.3.4 holds.*

**Exercise 13.3.7** *We can make a few more observations. We take* $\mathbf{e}_i$ *as in the previous exercise.*

(i) *If* $1 \le i, j \le 7$ *and* $i \ne j$, *explain why* $A\mathbf{e}_i^T \ne A\mathbf{e}_j^T$ *and deduce that*

$$A(\mathbf{e}_i^T + \mathbf{e}_j^T) \ne \mathbf{0}^T. \qquad\qquad ★$$

*Conclude that, if we make two errors in transcribing a Hamming codeword, the result will not be a Hamming codeword. (Thus the Hamming system will* detect *two errors. However, you should look at the second part of the question before celebrating.)*

(ii) *If* $1 \le i, j \le 7$ *and* $i \ne j$, *show, by using* ★, *or otherwise, that*

$$A(\mathbf{e}_i^T + \mathbf{e}_j^T) = A\mathbf{e}_k^T$$

*for some* $k \ne i, j$. *Show that, if we make three errors in transcribing a Hamming codeword, the result will be a Hamming codeword. Deduce, or prove otherwise, that, if we make two errors, the Hamming system will indeed detect that an error has been made, but will always choose the wrong codeword.*

The Hamming code is a *parity check code*.

**Definition 13.3.8** *If $A$ is an $r \times n$ matrix with values in $\mathbb{Z}_2$ and $C$ consists of those row vectors $\mathbf{c} \in \mathbb{Z}_2^n$ with $A\mathbf{c}^T \in \mathbf{0}^T$, we say that $C$ is a* parity check code *with* parity check matrix $A$.

*If $W$ is a subspace of $\mathbb{Z}_2^n$ we say that $W$ is a* linear code *with codewords of length n.*

**Exercise 13.3.9** *We work in $\mathbb{Z}_2^n$. By using Exercise 13.3.3, or otherwise, show that $C \subseteq \mathbb{Z}_2^n$ is a parity code if and only if we can find $\alpha_j \in (\mathbb{Z}_2^n)'$ $[1 \le j \le r]$ such that*

$$\mathbf{c} \in C \Leftrightarrow \alpha_j \mathbf{c} = 0 \quad \text{for all } 1 \le j \le r.$$

**Exercise 13.3.10** *Show that any parity check code is a linear code.*

Our proof of the converse for Exercise 13.3.9 makes use of the idea of an annihilator. (See Definition 11.4.11. The cautious reader may check that everything runs smoothly when $\mathbb{F}$ is replaced by $\mathbb{Z}_2$, but the more impatient reader can accept my word.)

**Theorem 13.3.11** *Every linear code is a parity check code.*

*Proof* Let us write $V = \mathbb{Z}_2^n$ and let $U$ be a linear code, that is to say, a subspace of $V$. We look at the annihilator (called the *dual code* in coding theory)

$$W^0 = \{\alpha \in U' : \alpha(\mathbf{w}) = 0 \text{ for all } \mathbf{w} \in C\}.$$

Automatically $W^0$ is a subspace of $V'$ and Lemma 11.4.14 gives us $W = W^{00}$. Taking a basis $\alpha_1, \alpha_2, \ldots, \alpha_r$ for $W^0$ we have

$$\mathbf{u} \in U \Leftrightarrow \alpha\mathbf{u} = 0 \quad \text{for all } \alpha \in W^0$$
$$\Leftrightarrow \alpha_j\mathbf{u} = 0 \quad \text{for } 1 \leq j \leq r.$$

Thus, using Exercise 13.3.9, we see that $U$ is a parity check code.                     □

**Exercise 13.3.12** *Consider a parity check code $C$ of length $n$ given by an $n \times r$ parity check matrix $A$ of rank $r$. Show, by using some version of the rank-nullity theorem, or otherwise, that $C$ has dimension $n - r$ and so contains $2^{n-r}$ members.*

**Exercise 13.3.13** *Suppose that $C$ is a linear code of dimension $r$ with codewords of length $n$.*

*(i) Show, by thinking about elementary row and column operations, that (possibly after interchanging the order in which we write the elements of vectors) we can find a basis $\mathbf{e}(j)$ for $C$ in which $e_k(j) = \delta_{kj}$ for $1 \leq j, k \leq r$.*

*(ii) Find such a basis for the Hamming code.*

*(iii) Explain why the map $\alpha : \mathbb{Z}_2^r \to \mathbb{Z}_2^n$ given by*

$$\alpha(x_1, x_2, \ldots, x_r) = \sum_{j=1}^{r} x_j \mathbf{e}(j)$$

*is an injective linear map. Show that if we write*

$$\alpha\mathbf{x} = (\mathbf{x}, \beta\mathbf{x})$$

*then $\beta : \mathbb{Z}_2^r \to \mathbb{Z}_2^{n-r}$ is a linear map.*

*(iv) Consider the map $\beta$ defined in (iv). Is $\beta$ necessarily injective if $2r \geq n$? Is $\beta$ necessarily surjective if $n \geq 2r$? Give proofs or counterexamples.*

Looking at Exercise 13.3.5 again, we see that the Hamming code worked well because the probability of an error involving a particular hole (or to use the more modern terminology *bit*) was already fairly small.

**Exercise 13.3.14** *We consider the same setup as in Exercise 13.3.5. Suppose that the probability that an error occurs at a particular place is $10^{-1}$ and we use the Hamming scheme (making no use of the last place in each line). Show that the probability that the tape will be correctly read by the decoder is negligible.*

Hamming's scheme ceases to be useful when the probability of error is high and an enormous amount of work has been done to find codes that will work under these circumstances. (We give a simple example in Exercise 13.4.6.)

**Exercise 13.3.15** *Why is it impossible to recover the original message when the probability of an error in one bit is $1/2$ (independent of what happens to the other bits)? Why are we only interested in error probabilities $p$ with $p < 1/2$?*

A different problem arises if the probability of error is very low. Observe that an $n$ bit message is swollen to about $7n/4$ bits by Hamming encoding. Such a message takes $7/4$ times as long to transmit and its transmission costs $7/4$ times as much money. Can we use the low error rate to cut down the length of of the encoded message?

A simple generalisation of Hamming's scheme does just that.

**Exercise 13.3.16** *Suppose that $j$ is an integer with $1 \le j \le 2^n - 1$. Then, in ordinary arithmetic, $j$ has a binary expansion*

$$j = \sum_{i=1}^{n} a_{ij}(n)2^{i-1}$$

*where $a_{ij}(n) \in \{0, 1\}$. We define $A_n$ to be the $n \times (2^n - 1)$ matrix with entries $a_{ij}(n)$.*
*(i) Check that that $A_1 = (1)$,*

$$A_2 = \begin{pmatrix} 1 & 0 & 1 \\ 0 & 1 & 1 \end{pmatrix}$$

*and $A_3 = A$ where $A$ is the Hamming matrix defined in ★ on page 336. Write down $A_4$.*
*(ii) By looking at columns which contain a single 1, or otherwise, show that $A_n$ has rank $n$.*
*(iii) (This is not needed later.) Show that*

$$A_{n+1} = \begin{pmatrix} A_n & \mathbf{a}_n^T & A_n \\ \mathbf{c}_n & 1 & \mathbf{b}_n \end{pmatrix}$$

*with $\mathbf{a}_n$ the row vector of length $n$ with $0$ in each place, $\mathbf{b}_n$ the row vector of length $2^{n-1} - 1$ with $1$ in each place and $\mathbf{c}_n$ the row vector of length $2^{n-1} - 1$ with $0$ in each place.*

**Exercise 13.3.17** *Let $A_n$ be defined as in the previous exercise. Consider the parity check code $C_n$ of length $2^n - 1$ with parity check matrix $A_n$.*
*Show how, if a codeword is received with a single error, we can recover the original codeword in much the same way as we did with the original Hamming code. (See the paragraph preceding Exercise 13.3.4.)*
*How many elements does $C_n$ have?*

**Exercise 13.3.18** *I need to transmit a message with $5 \times 10^7$ bits (that is to say, $0$s and $1$s). If the probability of a transmission error is $10^{-7}$ for each bit (independent of each other), show that the probability that every bit is transmitted correctly is negligible.*
*If, on the other hand, I split the message up in to groups of $57$ bits, translate them into Hamming codewords of length $63$ (using Exercise 13.3.13 (iv) or some other technique) and then send the resulting message, show that the probability of the decoder failing to produce the correct result is negligible and we only needed to transmit a message about $63/57$ times as long to achieve the result.*

## 13.4 Further exercises

**Exercise 13.4.1** If $G$ is a finite field, show that the vector space $G^2$ over $G$ is the union of a finite collection of one dimensional subspaces. If $G$ is an infinite field, show that $G^2$ is not the union of a finite collection of one dimensional subspaces. If $G = \mathbb{R}$, show that $G^2$ is not the union of a countable collection of one dimensional subspaces.

**Exercise 13.4.2** (i) Let $G$ be a field such that the equation $x^2 = -1$ has no solution in $G$. Prove that, if $x$ and $y$ are elements of $G$ such that $x^2 + y^2 = 0$, then $x = y = 0$.

Prove that $G^2$ is made into a field by the operations

$$(x, y) + (z, w) = (x + y, z + w),$$
$$(x, y) \times (z, w) = (xz - yw, xw + yz).$$

(ii) Let $p$ be a prime of the form $4m + 3$. By using Fermat's little theorem (see Exercise 6.8.33), or otherwise, show that $-1$ is not a square modulo $p$. Deduce that there is a field with exactly $p^2$ elements.

**Exercise 13.4.3** [**Eigenvalues and algebraic closure**] In this question, $\mathbb{K}$ and $G$ are fields.

(i) Suppose that $\mathbb{K}$ is algebraically closed. If $V$ is a finite dimensional vector space over $\mathbb{K}$, show that every endomorphism $\alpha : V \to V$ has an eigenvalue (and so an eigenvector).

(ii) Suppose that every endomorphism $\alpha : G^n \to G^n$ has an eigenvalue (and so an eigenvector) for all $n \geq 1$. By considering matrices of the form

$$\begin{pmatrix} 0 & 1 & 0 & 0 & \ldots & 0 & 0 \\ 0 & 0 & 1 & 0 & \ldots & 0 & 0 \\ 0 & 0 & 0 & 1 & \ldots & 0 & 0 \\ \vdots & \vdots & \vdots & \vdots & \ddots & \vdots & \vdots \\ 0 & 0 & 0 & 0 & \ldots & 0 & 1 \\ -a_0 & -a_1 & -a_2 & -a_3 & \ldots & -a_{n-2} & -a_{n-1} \end{pmatrix},$$

or otherwise, show that every non-constant polynomial with coefficients in $G$ has a root in $G$.

(iii) Suppose that $G$ is a field such that every endomorphism $\alpha : V \to V$ of a finite dimensional vector space over $G$ has an eigenvalue. Explain why $G$ is algebraically closed.

(iv) Show that the analogue of Theorem 12.2.5 is true for all $V$ with $\mathbb{F}$ replaced by $G$ if and only if $G$ is algebraically closed.

**Exercise 13.4.4** (Only for those who share the author's taste for this sort of thing. You will need to know the meaning of 'countable', 'complete' and 'dense'.)

We know that field $\mathbb{R}$ is complete for the usual Euclidean distance, but is not algebraically closed. In this question we give an example of a subfield of $\mathbb{C}$ which is algebraically closed, but not complete for the usual Euclidean distance.

(i) Let $X \subseteq \mathbb{C}$. If $\mathcal{G}$ is the collection of subfields of $\mathbb{C}$ containing $X$, show that gen $X = \bigcap_{G \in \mathcal{G}} G$ is a subfield of $\mathbb{R}$ containing $X$.

(ii) If $X$ is countable, show that gen $X$ is countable.

(iii) If $Y \subseteq \mathbb{C}$, is countable, show that the set $\mathcal{P}_Y$ of polynomials with coefficients in $Y$ is countable and deduce that the set of roots of polynomials in $\mathcal{P}_Y$ is countable.

(iv) Deduce that, given a countable subfield $\mathbb{H}$ of $\mathbb{C}$, we can find a countable subfield $\mathbb{H}'$ of $\mathbb{C}$ containing all the roots of polynomials with coefficients in $\mathbb{H}$.

(v) Let $\mathbb{H}_0 = \mathbb{Q}$. By (iv), we can find countable subfields $\mathbb{H}_n$ of $\mathbb{C}$ such that $\mathbb{H}_n$ contains all the roots of polynomials with coefficients in $\mathbb{H}_{n-1}$ $[n \geq 1]$. Show that $\mathbb{H} = \cup_{n=0}^{\infty} \mathbb{H}_n$ is countable and algebraically closed.

(vi) Show that $\mathbb{H}_1$ is dense in $\mathbb{C}$ and so $\mathbb{H}$ is. On the other hand, $\mathbb{H}$ is countable so $\mathbb{H} \neq \mathbb{C}$. Deduce that $\mathbb{H}$ is not complete in the usual metric.

**Exercise 13.4.5 [The Hill cipher]** In a simple substitution code, letters are replaced by other letters, so, for example, $A$ is replaced by $C$ and $B$ by $Q$ and so on. Such secret codes are easy to break because of the statistical properties of English. (For example, we know that the most frequent letter in a long message will probably correspond to $E$.) One way round this is to substitute pairs of letters so that, for example, $AN$ becomes $RQ$ and $AM$ becomes $TC$. (Such a code is called a *digraph cipher*.) We could take this idea further and operate on $n$ letters at a time (obtaining a *polygraphic cipher*), but this requires an enormous codebook. It has been suggested that we could use matrices instead.

(i) Suppose that our alphabet has five elements which we write as 0, 1, 2, 3, 4. If we have a message $b_1 b_2 \ldots b_{2n}$, we form the vectors $\mathbf{b}_j = (b_{2j-1}, b_{2j})^T$ and consider $\mathbf{c}_j = A\mathbf{b}_j$ *where we do our arithmetic modulo 5 and $A$ is an invertible $2 \times 2$ matrix.* The encoded message is $c_1 c_2 \ldots c_{2n}$ where $\mathbf{c}_j = (c_{2j-1}, c_{2j})^T$.

If

$$A = \begin{pmatrix} 2 & 4 \\ 1 & 3 \end{pmatrix},$$

$n = 3$ and $b_1 b_2 b_3 b_4 b_5 b_6 = 340221$, show that $c_1 c_2 = 20$ and find $c_1 c_2 c_3 c_4 c_5 c_6$.

Find $A^{-1}$ and use it to recover $b_1 b_2 b_3 b_4 b_5 b_6$ from $c_1 c_2 c_3 c_4 c_5 c_6$.

In general, we want to work with an alphabet of $p$ elements (and so do arithmetic modulo $p$) where $p$ is a prime and to break up our message into vectors of length $n$. The next part of the question investigates how easy it is to find an invertible $n \times n$ matrix 'at random'.

(ii) We work in the vector space $\mathbb{Z}_p^n$. If $\mathbf{e}_1, \mathbf{e}_2, \ldots, \mathbf{e}_r$ are linearly independent and we choose a vector $\mathbf{u}$ at random,[7] show that the probability that

$$\mathbf{u} \notin \text{span}\{\mathbf{e}_1, \mathbf{e}_2, \ldots, \mathbf{e}_r\}$$

is $1 - p^{r-n}$. Hence show that, if we choose $n$ vectors from $\mathbb{Z}_p^n$ at random, the probability that they form a basis is $\prod_{r=1}^{n}(1 - p^{-r})$. Deduce that, if we work in $\mathbb{Z}_p$ and choose the entries of an $n \times n$ matrix $A$ at random, the probability that $A$ is invertible is $\prod_{r=1}^{n}(1 - p^{-r})$.

---

[7] We say that we choose 'at random' from a finite set $X$ if each element of $X$ has the same probability of being chosen. If we make several choices 'at random', we suppose the choices to be independent.

(iii) Show, by using calculus, that $\log(1 - x) \geq -x$ for $0 < x \leq 1$. Hence, or otherwise, show that

$$\prod_{r=1}^{n}(1 - p^{-r}) \geq \exp\left(-\frac{1}{p-1}\right).$$

Conclude that, if we choose a matrix $A$ at random, it is quite likely to be invertible. (If it is not, we just try again until we get one which is.)

(iv) We could take $p = 29$ (giving us 26 letters and 3 other signs). If $n$ is reasonably large and $A$ is chosen at random, it will not be possible to break the code by the simple statistical means described in the first paragraph of this question. However, the code is *not* secure by modern standards. Suppose that you know *both* a message $b_1b_2b_3 \ldots b_{mn}$ *and* its encoding $c_1c_2c_3 \ldots c_{mn}$. Describe a method which has a good chance of deducing $A$ and explain why it is likely to work (at least, if $m$ is substantially larger than $n$).

Although the Hill cipher by itself does not meet modern standards, it can be combined with modern codes to give an extra layer of security.

**Exercise 13.4.6** Recall, from Exercise 7.6.14, that a Hadamard matrix is a scalar multiple of an $n \times n$ orthogonal matrix with all entries $\pm 1$.

(i) Explain why, if $A$ is an $m \times m$ Hadamard matrix, then $m$ must be even and any two columns will differ in exactly $m/2$ places.

(ii) Explain why, if you are given a column of $4k$ entries with values $\pm 1$ and told that it comes from altering at most $k - 1$ entries in a column of a specified $4k \times 4k$ Hadamard matrix, you can identify the appropriate column.

(iii) Given a $4k \times 4k$ Hadamard matrix show how to produce a set of $4k$ codewords (strings of 0s and 1s) of length $4k$ such that you can identify the correct codeword provided that there are less than $k - 1$ errors.

**Exercise 13.4.7** In order to see how effective the Hadamard codes of the previous question actually are we need to make some simple calculus estimates.

(i) If we transmit $n$ bits and each bit has a probability $p$ of being mistransmitted, explain why the probability that we make exactly $r$ errors is

$$p_r = \binom{n}{r}p^r(1 - p)^{n-r}.$$

(ii) Show that, if $r \geq 2np$, then

$$\frac{p_r}{p_{r-1}} \leq \frac{1}{2}.$$

Deduce that, if $n \geq m \geq 2np$,

$$\text{Pr}(m \text{ errors or more}) = \sum_{m}^{n} p_r \leq 2p_m.$$

(iii) Consider a Hadamard code based on a $128 \times 128 = 2^7 \times 2^7$ Hadamard matrix. (Recall that Exercise 7.6.14 gave a simple procedure for constructing $2^n \times 2^n$ Hadamard matrices.) Suppose that $p = 1/20$. Show, using (ii) and Stirling's formula (or direct calculation if you do not know Stirling's formula, but do have a good scientific calculator), that the probability of having 31 errors or more when you send a codeword is negligible. (On the other hand, your message will be $128/7 \approx 18.3$ times longer in the coded form than in an unencoded form.)

**Exercise 13.4.8** (A simple remark.) Although $\mathbb{Z}_2^n$ does not carry distances linked to its structure as a vector space over $\mathbb{Z}_2$, it does have distances associated with it. Let

$$d_0(\mathbf{x}, \mathbf{y}) = \begin{cases} 0 & \text{if } \mathbf{x} = \mathbf{y} \\ 1 & \text{otherwise} \end{cases}$$

and $d_1(\mathbf{x}, \mathbf{y})$ equal the number of places in which $\mathbf{x}$ and $\mathbf{y}$ disagree. (We call $d_0$ the discrete metric and $d_1$ the Hamming metric.)

Prove the following results for $j = 0$ and $j = 1$. (Throughout, $\mathbf{x}$, $\mathbf{y}$ and $\mathbf{z}$ are general elements of $\mathbb{Z}_2^n$.)

(i) $d_j(\mathbf{x}, \mathbf{y}) \geq 0$ with equality if and only if $\mathbf{x} = \mathbf{y}$.

(ii) $d_j(\mathbf{x}, \mathbf{y}) = d_j(\mathbf{y}, \mathbf{x})$.

(iii) $d_j(\mathbf{x}, \mathbf{y}) + d_j(\mathbf{y}, \mathbf{z}) \geq d_j(\mathbf{x}, \mathbf{z})$.

Show also that $n d_0(\mathbf{x}, \mathbf{y}) \geq d_1(\mathbf{x}, \mathbf{y}) \geq d_0(\mathbf{x}, \mathbf{y})$.

**Exercise 13.4.9** Show that, if we use the appropriate notion of a line, Desargues' theorem (see Theorem 2.1.5) holds in any vector space over any field.

**Exercise 13.4.10** (i) Let $p$ be a prime. If $a_1, a_2, \ldots, a_{p+1} \in \mathbb{Z}_p$ show that we can find $1 = r \leq s \leq p + 1$ such that $\sum_{j=r}^{s} a_j = 0$.

(ii) Let $V$ be a finite dimensional vector space over a field $\mathbb{G}$ such that there exists a $w : V \times V \to \mathbb{G}$ with

(a) $w(\mathbf{x}, \mathbf{x}) = 0 \Rightarrow \mathbf{x} = \mathbf{0}$,

(b) $w(\mathbf{x}, \mathbf{y}) = w(\mathbf{x}, \mathbf{y})$,

(c) $w(\lambda \mathbf{x} + \mu \mathbf{y}, \mathbf{z}) = \lambda w(\mathbf{x}, \mathbf{z}) + \mu w(\mathbf{y}, \mathbf{z})$,

for all $\mathbf{x}, \mathbf{y}, \mathbf{z} \in V$ and $\lambda, \mu \in \mathbb{G}$.

By following the ideas of the Gram–Schmidt method, show that $V$ has a basis of vectors $\mathbf{e}_j$ with $w(\mathbf{e}_i, \mathbf{e}_j) = 0$ if $i \neq j$. (Note that you cannot assume that every element of $\mathbb{G}$ has a square root.)

(iii) If $\mathbb{G} = \mathbb{Z}_p$ and $V = \mathbb{Z}_p^{p+1}$, show that no function $w$ with the properties stated in (ii) can exist.

# 14

# Vector spaces with distances

## 14.1 Orthogonal polynomials

After looking at one direction that the algebraist might take, let us look at a direction that the analyst might take.

Earlier we introduced an inner product on $\mathbb{R}^n$ defined by

$$\langle \mathbf{x}, \mathbf{y} \rangle = \mathbf{x} \cdot \mathbf{y} = \sum_{j=1}^{n} x_j y_j$$

and obtained the properties given in Lemma 2.3.7. We now turn the procedure on its head and define an inner product by demanding that it obey the conclusions of Lemma 2.3.7.

**Definition 14.1.1** *If $U$ is a vector space over $\mathbb{R}$, we say that a function $M : U^2 \to \mathbb{R}$ is an* inner product *if, writing*

$$\langle \mathbf{x}, \mathbf{y} \rangle = M(\mathbf{x}, \mathbf{y}),$$

*the following results hold for all $\mathbf{x}$, $\mathbf{y}$, $\mathbf{w} \in U$ and all $\lambda$, $\mu \in \mathbb{R}$.*

*(i) $\langle \mathbf{x}, \mathbf{x} \rangle \geq 0$.*
*(ii) $\langle \mathbf{x}, \mathbf{x} \rangle = 0$ if and only if $\mathbf{x} = \mathbf{0}$.*
*(iii) $\langle \mathbf{y}, \mathbf{x} \rangle = \langle \mathbf{x}, \mathbf{y} \rangle$.*
*(iv) $\langle \mathbf{x}, (\mathbf{y} + \mathbf{w}) \rangle = \langle \mathbf{x}, \mathbf{y} \rangle + \langle \mathbf{x}, \mathbf{w} \rangle$.*
*(v) $\langle \lambda \mathbf{x}, \mathbf{y} \rangle = \lambda \langle \mathbf{x}, \mathbf{y} \rangle$.*
*We write $\|\mathbf{x}\|$ for the positive square root of $\langle \mathbf{x}, \mathbf{x} \rangle$.*

When we wish to recall that we are talking about *real* vector spaces, we talk about real inner products. We shall consider *complex* inner product spaces in Section 14.3. Exercise 14.1.3, which the reader should do, requires the following result from analysis.

**Lemma 14.1.2** *Let $a < b$. If $F : [a, b] \to \mathbb{R}$ is continuous, $F(x) \geq 0$ for all $x \in [a, b]$ and*

$$\int_a^b F(t) \, dt = 0,$$

*then $F = 0$.*

*Proof* If $F \neq 0$, then we can find an $x_0 \in [a, b]$ such that $F(x_0) > 0$. (Note that we might have $x_0 = a$ or $x_0 = b$.) By continuity, we can find a $\delta$ with $1/2 > \delta > 0$ such that $|F(x) - F(x_0)| \geq F(x_0)/2$ for all $x \in I = [x_0 - \delta, x_0 + \delta] \cap [a, b]$. Thus $F(x) \geq F(x_0)/2$ for all $x \in I$ and

$$\int_a^b F(t)\, dt \geq \int_I F(t)\, dt \geq \int_I \frac{F(x_0)}{2}\, dt \geq \frac{\delta F(x_0)}{2} > 0,$$

giving the required contradiction. □

**Exercise 14.1.3** *Let $a < b$. Show that the set $C([a, b])$ of continuous functions $f : [a, b] \to \mathbb{R}$ with the operations of pointwise addition and multiplication by a scalar (thus $(f + g)(x) = f(x) + g(x)$ and $(\lambda f)(x) = \lambda f(x)$) forms a vector space over $\mathbb{R}$. Verify that*

$$\langle f, g \rangle = \int_a^b f(x) g(x)\, dx$$

*defines an inner product for this space.*

We obtained most of our results on our original inner product by using Lemma 2.3.7 rather than the definition and these results must remain true for all real inner products.

**Exercise 14.1.4** *(i) Show that, if $U$ is a real inner product space, then the Cauchy–Schwarz inequality*

$$\langle \mathbf{x}, \mathbf{y} \rangle \leq \|\mathbf{x}\| \|\mathbf{y}\|$$

*holds for all $\mathbf{x}, \mathbf{y} \in U$. When do we have equality? Prove your answer.*

*(ii) Show that, if $U$ is a real inner product space, then the following results hold for all $\mathbf{x}, \mathbf{y} \in U$ and $\lambda \in \mathbb{R}$.*

*(a) $\|\mathbf{x}\| \geq 0$ with equality if and only if $\mathbf{x} = \mathbf{0}$.*
*(b) $\|\lambda \mathbf{x}\| = |\lambda| \|\mathbf{x}\|$.*
*(c) $\|\mathbf{x} + \mathbf{y}\| \leq \|\mathbf{x}\| + \|\mathbf{y}\|$.*
*[These results justify us in calling $\| \ \|$ the norm derived from the inner product.]*
*(iii) If $f, g \in C([a, b])$, show that*

$$\left( \int_a^b f(t) g(t)\, dt \right)^2 \leq \int_a^b f(t)^2\, dt \int_a^b g(t)^2\, dt.$$

Whenever we use a result about inner products, the reader can check that our earlier proofs, in Chapter 2 and elsewhere, extend to the more general situation.

Let us consider $C([-1, 1])$ in more detail. If we set $q_j(x) = x^j$, then, since a polynomial of degree $n$ can have at most $n$ roots,

$$\sum_{j=0}^n \lambda_j q_j = 0 \Rightarrow \sum_{j=0}^n \lambda_j x^j = 0 \quad \text{for all } x \in [-1, 1] \Rightarrow \lambda_0 = \lambda_1 = \ldots = \lambda_n = 0$$

and so $q_0, q_1, \ldots, q_n$ are linearly independent. Applying the Gram–Schmidt orthogonalisation method, we obtain a sequence of non-zero polynomials $p_0, p_1, \ldots, p_n, \ldots$ given by

$$p_n = q_n - \sum_{j=0}^{n-1} \frac{\langle q_n, p_j \rangle}{\|p_j\|^2} p_j.$$

**Exercise 14.1.5** (i) *Prove, by induction, or otherwise, that* $p_n$ *is a monic polynomial of degree exactly* n. *Compute* $p_0$, $p_1$ *and* $p_2$.

(ii) *Show that the collection* $\mathcal{P}_n$ *of polynomials of degree* n *or less is a subspace of* $C([-1, 1])$. *Show that* $\mathcal{P}_n$ *has dimension* $n + 1$ *and that, using the inner product introduced above, the functions* $\|p_j\|^{-1} p_j$ *with* $0 \le j \le n$ *form an orthonormal basis for* $\mathcal{P}_n$.

(iii) *Show that* $p_{n+1}$ *is the unique monic polynomial* P *of degree* $n + 1$ *such that* $\langle P, p \rangle = 0$ *whenever* p *is a polynomial of degree* n *or less*.

The polynomials $p_j$ are called the Legendre polynomials and turn up in several places in mathematics.[1] We give an alternative approach to Legendre polynomials in Exercise 14.4.3.

The next few lemmas lead up to a beautiful idea of Gauss.

**Lemma 14.1.6** *The Legendre polynomial* $p_n$ *has* n *distinct roots* $\theta_1, \theta_2, \ldots, \theta_n$. *All these roots are real and satisfy the condition* $-1 < \theta_j < 1$.

*Proof* Suppose that $p_n$ has $\theta_1, \theta_2, \ldots, \theta_k$ as roots of *odd order*[2] with

$$-1 < \theta_1 < \theta_2 < \ldots \theta_k < 1$$

and no other roots $\theta$ of odd order with $-1 < \theta < 1$.

If we set $Q(t) = \prod_{j=1}^{k}(t - \theta_j)$, then $Q$ is a polynomial of order $k$ and $Q(t)p_n(t)$ cannot change sign on $[-1, 1]$ (that is to say, either $p_n(t)Q(t) \ge 0$ for all $t \in [-1, 1]$ or $p_n(t)Q(t) \le 0$ for all $t \in [-1, 1]$). By Lemma 14.1.2, it follows that

$$\langle p_n, Q \rangle = \int_{-1}^{1} p_n(t)Q(t)\, dt \ne 0$$

and so, by Exercise 14.1.5 (iii), $k \ge n$. Since a polynomial of degree $n$ has at most $n$ roots, counting multiplicities, we have $k = n$. It follows that all the roots $\theta$ of $p_n$ are distinct, real and satisfy $-1 < \theta < 1$.                                                    $\square$

Suppose that we wish to estimate $\int_{-1}^{1} f(t)\, dt$ for some well behaved function $f$ from its values at points $t_1, t_2, \ldots, t_n$. The following result is clearly relevant.

**Lemma 14.1.7** *Suppose that we are given distinct points* $t_1, t_2, \ldots, t_n \in [-1, 1]$. *Then there are unique* $\lambda_1, \lambda_2, \ldots, \lambda_n \in \mathbb{R}$ *such that*

$$\int_{-1}^{1} P(t)\, dt = \sum_{j=1}^{n} \lambda_j P(t_j)$$

---

[1] Depending on context, *the Legendre polynomial of degree* j refers to $a_j p_j$, where the particular writer may take $a_j = 1$, $a_j = \|p_j\|^{-1}$, $a_j = p_j(1)^{-1}$ or make some other choice.

[2] Recall that $\alpha$ is a root of order $r$ of the polynomial $P$ if $(t - \alpha)^r$ is a factor of $P(t)$, but $(t - \alpha)^{r+1}$ is not.

*for every polynomial P of degree n − 1 or less.*

*Proof* Let us set

$$e_k(t) = \prod_{j \neq k} \frac{t - t_j}{t_k - t_j}.$$

Then $e_k$ is a polynomial of degree $n - 1$ with $e_k(t_j) = \delta_{kj}$. It follows that, if $P$ is a polynomial of degree at most $n - 1$ and we set

$$Q = P - \sum_{j=1}^{n} P(t_j)e_j,$$

then $Q$ is a polynomial of degree at most $n - 1$ which vanishes at the $n$ points $t_k$. Thus $Q = 0$ and

$$P(t) = \sum_{j=1}^{n} P(t_j)e_j(t).$$

Integrating, we obtain

$$\int_{-1}^{1} P(t)\,dt = \sum_{j=1}^{n} \lambda_j P(t_j)$$

with $\lambda_j = \int_{-1}^{1} e_j(t)\,dt$.

To prove uniqueness, suppose that

$$\int_{-1}^{1} P(t)\,dt = \sum_{j=1}^{n} \mu_j P(t_j)$$

for every polynomial of degree $n - 1$ or less. Taking $P = e_k$, we obtain

$$\mu_k = \sum_{j=1}^{n} \mu_j e_k(t_j) = \int_{-1}^{1} e_k(t)\,dt = \lambda_k$$

as required. $\qquad\qquad\square$

At first sight, it seems natural to take the $t_j$ equidistant, but Gauss suggested a different choice.

**Theorem 14.1.8 [Gaussian quadrature]** *If we choose the $t_j$ in Lemma 14.1.7 to be the roots of the Legendre polynomial $p_n$, then*

$$\int_{-1}^{1} P(t)\,dt = \sum_{j=1}^{n} \lambda_j P(t_j)$$

*for every polynomial P of degree 2n − 1 or less.*

*Proof* If $P$ has degree $2n - 1$ or less, then

$$P = Qp_n + R,$$

where $Q$ and $R$ are polynomials of degree $n - 1$ or less. Thus

$$\int_{-1}^{1} P(t)\,dt = \int_{-1}^{1} Q(t)p_n(t)\,dt + \int_{-1}^{1} R(t)\,dt = \langle Q, p_n \rangle + \int_{-1}^{1} R(t)\,dt$$

$$= 0 + \int_{-1}^{1} R(t)\,dt = \sum_{j=1}^{n} \lambda_j R(t_j)$$

$$= \sum_{j=1}^{n} \lambda_j \big( p_n(t_j)Q(t_j) + R(t_j) \big) = \sum_{j=1}^{n} \lambda_j P(t_j)$$

as required. $\qquad\square$

This is very impressive, but Gauss' method has a further advantage. If we use equidistant points, then it turns out that, when $n$ is large, $\sum_{j=1}^{n} |\lambda_j|$ becomes very large indeed and small changes in the values of the $f(t_j)$ may cause large changes in the value of $\sum_{j=1}^{n} \lambda_j f(t_j)$ making the sum useless as an estimate for the integral. This is not the case for Gaussian quadrature.

**Lemma 14.1.9** *If we choose the $t_k$ in Lemma 14.1.7 to be the roots of the Legendre polynomial $p_n$, then $\lambda_k > 0$ for each $k$ and so*

$$\sum_{j=1}^{n} |\lambda_j| = \sum_{j=1}^{n} \lambda_j = 2.$$

*Proof* If $1 \le k \le n$, set

$$Q_k(t) = \prod_{j \neq k} (t - t_j)^2.$$

We observe that $Q_k$ is an everywhere non-negative polynomial of degree $2n - 2$ and so

$$\lambda_k Q_k(t_k) = \sum_{j=1}^{n} \lambda_j Q_k(t_j) = \int_{-1}^{1} Q_k(t)\,dt > 0.$$

Thus $\lambda_k > 0$.

Taking $P = 1$ in the quadrature formula gives

$$2 = \int_{-1}^{1} 1\,dt = \sum_{j=1}^{n} \lambda_j.$$

$\qquad\square$

This gives us the following reassuring result.

**Theorem 14.1.10** *Suppose that* $f : [-1, 1] \to \mathbb{R}$ *is a continuous function such that there exists a polynomial $P$ of degree at most $2n - 1$ with*

$$|f(t) - P(t)| \le \epsilon \qquad \text{for all } t \in [-1, 1]$$

*for some $\epsilon > 0$. Then, if we choose the $t_j$ in Lemma 14.1.7 to be the roots of the Legendre polynomial $p_n$, we have*

$$\left| \int_{-1}^{1} f(t)\, dt - \sum_{j=1}^{n} \lambda_j f(t_j) \right| \le 4\epsilon.$$

*Proof* Observe that

$$\left| \int_{-1}^{1} f(t)\, dt - \sum_{j=1}^{n} \lambda_j f(t_j) \right|$$

$$= \left| \int_{-1}^{1} f(t)\, dt - \sum_{j=1}^{n} \lambda_j f(t_j) - \int_{-1}^{1} P(t)\, dt + \sum_{j=1}^{n} \lambda_j P(t_j) \right|$$

$$= \left| \int_{-1}^{1} \big( f(t) - P(t) \big)\, dt - \sum_{j=1}^{n} \lambda_j \big( f(t_j) - P(t_j) \big) \right|$$

$$\le \int_{-1}^{1} |f(t) - P(t)|\, dt + \sum_{j=1}^{n} |\lambda_j| |f(t_j) - P(t_j)|$$

$$\le 2\epsilon + 2\epsilon = 4\epsilon,$$

as stated. □

The following simple exercise may help illustrate the point at issue.

**Exercise 14.1.11** *Let $\epsilon > 0$, $n \ge 1$ and distinct points $t_1, t_2, \ldots, t_{2n} \in [-1, 1]$ be given. By Lemma 14.1.7, there are unique $\lambda_1, \lambda_2, \ldots, \lambda_{2n} \in \mathbb{R}$ such that*

$$\int_{-1}^{1} P(t)\, dt = \sum_{j=1}^{n} \lambda_j P(t_j) = \sum_{j=n+1}^{2n} \lambda_j P(t_j)$$

*for every polynomial $P$ of degree $n - 1$ or less.*

*(i) Explain why $\sum_{j=1}^{n} |\lambda_j| \ge 2$ and $\sum_{j=n+1}^{2n} |\lambda_j| \ge 2$.*

*(ii) If we set $\mu_j = (\epsilon^{-1} + 1)\lambda_j$ for $1 \le j \le n$ and $\mu_j = -\epsilon^{-1}\lambda_j$ for $n + 1 \le j \le 2n$, show that*

$$\int_{-1}^{1} P(t)\, dt = \sum_{j=1}^{2n} \mu_j P(t_j)$$

*for all polynomials P of degree n or less. Find a piecewise linear function f such that*

$$|f(t)| \leq \epsilon \quad \text{for all } t \in [-1, 1], \text{ but} \quad \sum_{j=1}^{2n} \mu_j f(t_j) \geq 2.$$

The following strongly recommended exercise acts as revision for our work on Legendre polynomials and shows that the ideas can be pushed much further.

**Exercise 14.1.12** *Let $r : [a, b] \to \mathbb{R}$ be an everywhere strictly positive continuous function. Set*

$$\langle f, g \rangle = \int_a^b f(x)g(x)r(x)\,dx.$$

*(i) Show that we have defined an inner product on $C([a, b])$.*

*(ii) Deduce that*

$$\left( \int_a^b f(x)g(x)r(x)\,dx \right)^2 \leq \left( \int_a^b f(x)^2 r(x)\,dx \right) \left( \int_a^b g(x)^2 r(x)\,dx \right)$$

*for all continuous functions $f, g : [-1, 1] \to \mathbb{R}$.*

*(iii) Show that we can find a monic polynomial $P_n$ of degree n such that*

$$\int_a^b P_n(x)Q(x)r(x)\,dx = 0$$

*for all polynomials Q of degree $n - 1$ or less.*

*(iv) Develop a method along the lines of Gaussian quadrature for the numerical estimation of*

$$\int_a^b f(x)r(x)\,dx$$

*when f is reasonably well behaved.*

The various kinds of polynomials $P_n$ that arise for different $r$ are called *orthogonal polynomials*. We give an interesting example in Exercise 14.4.4.

The subject of orthogonal polynomials is rich in elegant formulae.

**Example 14.1.13** **[Recurrence relations for orthogonal polynomials]** *Let $P_0$, $P_1$, $P_2$, . . . be a sequence of orthogonal polynomials of the type given in Exercise 14.1.12. Show that there exists a 'three term recurrence relation' of the form*

$$P_{n+1}(x) + (A_n - x)P_n(x) + B_n P_{n-1}(x) = 0$$

*for $n \geq 0$. Determine $A_n$ and $B_n$.*

*Solution* Since

$$\int_a^b \big( f(x)h(x) \big) g(x)r(x)\,dx = \int_a^b f(x)\big( h(x)g(x) \big)r(x)\,dx,$$

the inner product defined in Exercise 14.1.12 has a further property (which is not shared by inner products in general[3]) that $\langle fh, g \rangle = \langle f, gh \rangle$. We make use of this fact in what follows.

Let $q(x) = x$ and $Q_n = P_{n+1} - q P_n$. Since $P_{n+1}$ and $P_n$ are monic polynomials of degree $n + 1$ and $n$, $Q_n$ is polynomial of degree at most $n$ and

$$Q_n = \sum_{j=0}^{n} a_j P_j.$$

By orthogonality

$$\langle Q_n, P_k \rangle = \sum_{j=0}^{n} a_j \langle P_j, P_k \rangle = \sum_{j=0}^{n} a_j \delta_{jk} \| P_k \|^2 = a_k \| P_k \|^2.$$

Now $P_r$ is orthogonal to all polynomials of lower degree and so, if $k \le n - 2$,

$$\langle Q_n, P_k \rangle = \langle P_{n+1} - q P_n, P_k \rangle = \langle P_{n+1} - q P_n, P_k \rangle$$
$$= \langle P_{n+1}, P_k \rangle - \langle q P_n, P_k \rangle = -\langle P_n, q P_k \rangle = 0.$$

Thus $Q_n = A_n P_n + B_n P_{n-1}$ and our argument shows that

$$A_n = \frac{\langle P_n, q P_n \rangle}{\| P_n \|^2}, \quad B_n = \frac{\langle P_n, q P_{n-1} \rangle}{\| P_{n-1} \|^2}.$$

$\square$

For many important systems of orthogonal polynomials, $A_n$ and $B_n$ take a simple form and we obtain an efficient method for calculating the $P_n$ inductively (or, in more sophisticated language, recursively). We give examples in Exercises 14.4.4 and 14.4.5.

**Exercise 14.1.14** *By considering* $q P_{n-1} - P_n$, *or otherwise, show that, using the notation of the discussion above,* $B_n > 0$.

Once we start dealing with infinite dimensional inner product spaces, Bessel's inequality, which we met in Exercise 7.1.9, takes on a new importance.

**Theorem 14.1.15 [Bessel's inequality]** *Suppose that* $e_1, e_2, \ldots$ *is a sequence of orthonormal vectors (that is to say,* $\langle e_j, e_k \rangle = \delta_{jk}$ *for* $j, k \ge 1$*) in a real inner product space* $U$.
  *(i) If* $f \in U$, *then*

$$\left\| f - \sum_{j=1}^{n} a_j e_j \right\|$$

*attains a unique minimum when*

$$a_j = \hat{f}(j) = \langle f, e_j \rangle.$$

---

[3] Indeed, it does not make sense in general, since our definition of a general vector space does not allow us to multiply two vectors.

*We have*

$$\left\| \mathbf{f} - \sum_{j=1}^{n} \hat{\mathbf{f}}(j) \mathbf{e}_j \right\|^2 = \|\mathbf{f}\|^2 - \sum_{j=1}^{n} |\hat{\mathbf{f}}(j)|^2.$$

*(ii) If* $\mathbf{f} \in U$, *then* $\sum_{j=1}^{\infty} |\hat{\mathbf{f}}(j)|^2$ *converges and*

$$\sum_{j=1}^{\infty} |\hat{\mathbf{f}}(j)|^2 \le \|\mathbf{f}\|^2.$$

*(iii) We have*

$$\left\| \mathbf{f} - \sum_{j=1}^{n} \hat{\mathbf{f}}(j) \mathbf{e}_j \right\| \to 0 \quad as\ n \to \infty \Leftrightarrow \sum_{j=1}^{\infty} |\hat{\mathbf{f}}(j)|^2 = \|\mathbf{f}\|^2.$$

*Proof* (i) Following a familiar path, we have

$$\left\| \mathbf{f} - \sum_{j=1}^{n} a_j \mathbf{e}_j \right\|^2 = \left\langle \mathbf{f} - \sum_{j=1}^{n} a_j \mathbf{e}_j, \mathbf{f} - \sum_{j=1}^{n} a_j \mathbf{e}_j \right\rangle$$

$$= \|\mathbf{f}\|^2 - 2 \sum_{j=1}^{n} a_j \hat{\mathbf{f}}(j) + \sum_{j=1}^{n} a_j^2.$$

Thus

$$\left\| \mathbf{f} - \sum_{j=1}^{n} \hat{\mathbf{f}}(j) \mathbf{e}_j \right\|^2 = \|\mathbf{f}\|^2 - \sum_{j=1}^{n} |\hat{\mathbf{f}}(j)|^2$$

and

$$\left\| \mathbf{f} - \sum_{j=1}^{n} a_j \mathbf{e}_j \right\|^2 - \left\| \mathbf{f} - \sum_{j=1}^{n} \hat{\mathbf{f}}(j) \mathbf{e}_j \right\|^2 = \sum_{j=1}^{n} (a_j - \hat{\mathbf{f}}(j))^2.$$

The required results can now be read off.

(ii) We use the theorem[4] from analysis which tells us that an increasing sequence bounded above tends to a limit. Elementary analysis tell us that the limit is no greater than the upper bound.

(iii) This follows at once from (i). □

The reader may, or may not, need to be reminded that the type of convergence discussed in Theorem 14.1.15 is not the same as she is used to from elementary calculus.

---

[4] Or, in many treatments, axiom.

**Exercise 14.1.16** *Consider the continuous function $f_n : [-1, 1] \to \mathbb{R}$ defined by*

$$f_n(x) = \sum_{r=-2^n+1}^{r=2^n-1} 2^n \max(0, 1 - 2^{8n}|x - r2^{-n}|).$$

*Show that*

$$\int_1^1 f_n(x)^2 \, dx \to 0$$

*so $\|f_n\| \to 0$ in the norm given by the inner product of Exercise 14.1.3 as $n \to \infty$, but $f_n(u2^{-m}) \to \infty$ as $n \to \infty$, for all integers $u$ with $|u| \le 2^m - 1$ and all integers $m \ge 1$.*

## 14.2 Inner products and dual spaces

As I have already said, we can obtain any result we want on general finite dimensional real inner product spaces by imitating the proof for $\mathbb{R}^n$ with the standard inner product.

**Exercise 14.2.1** *If $U$ is a finite dimensional real vector space with inner product $\langle \, , \, \rangle$ show, by using the Gram–Schmidt orthogonalisation process, that $U$ has an orthonormal basis $\mathbf{e}_1, \mathbf{e}_2, \ldots, \mathbf{e}_n$ say. Show that the map $\theta : U \to \mathbb{R}^n$ given by*

$$\theta \left( \sum_{j=1}^n x_j \mathbf{e}_j \right) = (x_1, x_2, \ldots, x_n)^T$$

*for all $x_j \in \mathbb{R}$ is a well defined vector space isomorphism which preserves inner products in the sense that*

$$\theta \mathbf{u} \cdot \theta \mathbf{v} = \langle \mathbf{u}, \mathbf{v} \rangle$$

*for all $\mathbf{u}$, $\mathbf{v}$ where $\cdot$ is the dot product of Section 2.3.*

The reader need hardly be told that it is better mathematical style to prove results for general inner product spaces directly rather than to prove them for $\mathbb{R}^n$ with the dot product and then quote Exercise 14.2.1.

Bearing in mind that we shall do nothing basically new, it is, none the less, worth taking another look at the notion of an adjoint map. We start from the simple observation in three dimensional geometry that a plane through the origin can be described by the equation

$$\mathbf{x} \cdot \mathbf{b} = 0$$

where $\mathbf{b}$ is a non-zero vector. A natural generalisation to inner product spaces runs as follows.

**Lemma 14.2.2** *If $V$ is an $n - 1$ dimensional subspace of an $n$ dimensional real vector space $U$ with inner product $\langle \, , \, \rangle$, then we can find a $\mathbf{b} \ne \mathbf{0}$ such that*

$$V = \{\mathbf{x} \in U : \langle \mathbf{x}, \mathbf{b} \rangle = 0\}.$$

*Proof* Let $V$ have an orthonormal basis $\mathbf{e}_1, \mathbf{e}_2, \ldots, \mathbf{e}_{n-1}$. By using the Gram–Schmidt process, we can find $\mathbf{e}_n$ such that $\mathbf{e}_1, \mathbf{e}_2, \ldots, \mathbf{e}_n$ are orthonormal and so a basis for $U$. Setting $\mathbf{b} = \mathbf{e}_n$, we have the result. $\qquad\qquad\qquad\qquad\qquad\qquad\qquad\square$

**Exercise 14.2.3** *Suppose that $V$ is a subspace of a finite dimensional real vector space $U$ with inner product $\langle\ ,\ \rangle$ and derived norm $\|\ \|$. If $\mathbf{a} \in U$, show that there is a unique $\mathbf{b} \in V$ such that*

$$\|\mathbf{a} - \mathbf{b}\| \le \|\mathbf{a} - \mathbf{v}\|$$

*for all $\mathbf{v} \in V$.*

*Show, also, that $\mathbf{x} = \mathbf{b}$ is the unique vector with $\mathbf{x} \in V$ such that*

$$\langle \mathbf{a} - \mathbf{x}, \mathbf{v} \rangle = 0$$

*for all $\mathbf{v} \in V$.*

*[Hint: Choose an orthonormal basis for $V$ and use Bessel's inequality (Theorem 14.1.15).]*

Lemma 14.2.2 leads on to a finite dimensional version of a famous theorem of Riesz.

**Theorem 14.2.4  [Riesz representation theorem]**[5] *Let $U$ be a finite dimensional real vector space with inner product $\langle\ ,\ \rangle$. If $\alpha \in U'$, the dual space of $U$, then there exists a unique $\mathbf{a} \in U$ such that*

$$\alpha\mathbf{u} = \langle \mathbf{u}, \mathbf{a} \rangle$$

*for all $\mathbf{u} \in U$.*

*Proof Uniqueness.* If $\alpha\mathbf{u} = \langle \mathbf{u}, \mathbf{a}_1 \rangle = \langle \mathbf{u}, \mathbf{a}_2 \rangle$ for all $\mathbf{u} \in U$, then

$$\langle \mathbf{u}, \mathbf{a}_1 - \mathbf{a}_2 \rangle = 0$$

for all $\mathbf{u} \in U$ and, choosing $\mathbf{u} = \mathbf{a}_1 - \mathbf{a}_2$, we conclude, in the usual way, that $\mathbf{a}_1 - \mathbf{a}_2 = \mathbf{0}$. *Existence.* If $\alpha = 0$, then we set $\mathbf{a} = \mathbf{0}$. If not, then $\alpha$ has rank 1 (since $\alpha(U) = \mathbb{R}$) and, by the rank-nullity theorem, $\alpha$ has nullity $n - 1$. In other words,

$$\alpha^{-1}(\mathbf{0}) = \{\mathbf{u}\ :\ \alpha\mathbf{u} = 0\}$$

has dimension $n - 1$. By Lemma 14.2.2, we can find a $\mathbf{b} \neq \mathbf{0}$ such that

$$\alpha^{-1}(\mathbf{0}) = \{\mathbf{x} \in U\ :\ \langle\mathbf{x}, \mathbf{b}\rangle = 0\},$$

that is to say,

$$\alpha(\mathbf{x}) = 0 \Leftrightarrow \langle\mathbf{x}, \mathbf{b}\rangle = 0.$$

If we now set

$$\mathbf{a} = \|\mathbf{b}\|^{-2}\alpha(\mathbf{b})\mathbf{b},$$

---

[5] There are several Riesz representation theorems, but this is the only one we shall refer to. We should really use the longer form 'the Reisz *Hilbert space* representation theorem'.

we have

$$\alpha\mathbf{x} = 0 \Leftrightarrow \langle \mathbf{x}, \mathbf{a} \rangle = 0$$

and

$$\alpha\mathbf{a} = \|\mathbf{b}\|^{-2}\alpha(\mathbf{b})^2 = \langle \mathbf{a}, \mathbf{a} \rangle.$$

Now suppose that $\mathbf{u} \in U$. If we set

$$\mathbf{x} = \mathbf{u} - \frac{\alpha\mathbf{u}}{\alpha\mathbf{a}}\mathbf{a},$$

then $\alpha\mathbf{x} = 0$, so $\langle \mathbf{x}, \mathbf{a} \rangle = 0$, that is to say,

$$0 = \left\langle \mathbf{u} - \frac{\alpha\mathbf{u}}{\alpha\mathbf{a}}\mathbf{a}, \mathbf{a} \right\rangle = \langle \mathbf{u}, \mathbf{a} \rangle - \frac{\alpha\mathbf{a}}{\langle \mathbf{a}, \mathbf{a} \rangle}\alpha\mathbf{u} = \langle \mathbf{u}, \mathbf{a} \rangle - \alpha\mathbf{u}$$

and we are done. □

**Exercise 14.2.5** *Draw diagrams to illustrate the existence proof in two and three dimensions.*

**Exercise 14.2.6** *We can obtain a quicker (but less easily generalised) proof of the existence part of Theorem 14.2.4 as follows. Let $\mathbf{e}_1$, $\mathbf{e}_2$, ..., $\mathbf{e}_n$ be an orthonormal basis for a real inner product space $U$. If $\alpha \in U'$ set*

$$\mathbf{a} = \sum_{j=1}^{n} \alpha(\mathbf{e}_j)\mathbf{e}_j.$$

*Verify that*

$$\alpha\mathbf{u} = \langle \mathbf{u}, \mathbf{a} \rangle$$

*for all $\mathbf{u} \in U$.*

We note that Theorem 14.2.4 has a trivial converse.

**Exercise 14.2.7** *Let $U$ be a finite dimensional real vector space with inner product $\langle \,,\, \rangle$. If $\mathbf{a} \in U$, show that the equation*

$$\alpha\mathbf{u} = \langle \mathbf{u}, \mathbf{a} \rangle$$

*defines an $\alpha \in U'$.*

**Lemma 14.2.8** *Let $U$ be a finite dimensional real vector space with inner product $\langle \,,\, \rangle$. The equation*

$$\theta(\mathbf{a})\mathbf{u} = \langle \mathbf{u}, \mathbf{a} \rangle$$

*with $\mathbf{a}$, $\mathbf{u} \in U$ defines an isomorphism $\theta : U \to U'$.*

*Proof* Exercise 14.2.7 tells us that $\theta$ maps $U$ to $U'$. Next we observe that

$$\theta(\lambda\mathbf{a} + \mu\mathbf{b})\mathbf{u} = \langle\mathbf{u}, \lambda\mathbf{a} + \mu\mathbf{b}\rangle = \lambda\langle\mathbf{u}, \mathbf{a}\rangle + \mu\langle\mathbf{u}, \mathbf{b}\rangle$$
$$= \lambda\theta(\mathbf{a})\mathbf{u} + \mu\theta(\mathbf{b})\mathbf{u} = \big(\lambda\theta(\mathbf{a}) + \mu\theta(\mathbf{b})\big)\mathbf{u},$$

for all $\mathbf{u} \in U$. It follows that

$$\theta(\lambda\mathbf{a} + \mu\mathbf{b}) = \lambda\theta(\mathbf{a}) + \mu\theta(\mathbf{b})$$

for all $\mathbf{a}, \mathbf{b} \in U$ and $\lambda, \mu \in \mathbb{R}$. Thus $\theta$ is linear. Theorem 14.2.4 tells us that $\theta$ is bijective so we are done. $\square$

The sharp eyed reader will note that the function $\theta$ of Lemma 14.2.8 is, in fact, a *natural* (that is to say, *basis independent*) isomorphism and ask whether, just as we usually identify $U''$ with $U$ for ordinary finite dimensional vector spaces, so we should identify $U$ with $U'$ for finite dimensional real inner product spaces. The answer is that mathematicians whose work only involves finite dimensional real inner product spaces often make the identification, but those with more general interests do not. There are, I think, two reasons for this. The first is that the convention of identifying $U$ with $U'$ for finite dimensional real inner product spaces makes it hard to compare results on such spaces with more general vector spaces. The second is that the natural extension of our ideas to complex vector spaces does not produce an isomorphism.[6]

We now give a more abstract (but, in my view, more informative) proof than the one we gave in Lemma 7.2.1 of the existence of the adjoint.

**Lemma 14.2.9** *Let $U$ be a finite dimensional real vector space with inner product $\langle\ ,\ \rangle$. If $\alpha \in \mathcal{L}(U, U)$, there exists a unique $\alpha^* \in \mathcal{L}(U, U)$ such that*

$$\langle\alpha\mathbf{u}, \mathbf{v}\rangle = \langle\mathbf{u}, \alpha^*\mathbf{v}\rangle$$

*for all $\mathbf{u}, \mathbf{v} \in U$.*
*The map $\Phi : \mathcal{L}(U, U) \to \mathcal{L}(U, U)$ given by $\Phi\alpha = \alpha^*$ is an isomorphism.*

*Proof* Observe that the map $\beta_\mathbf{v} : U \to \mathbb{R}$ given by

$$\beta_\mathbf{v}\mathbf{u} = \langle\alpha\mathbf{u}, \mathbf{v}\rangle$$

is linear, so, by the Riesz representation theorem (Theorem 14.2.4), there exists a unique $\mathbf{a}_\mathbf{v} \in U$ such that

$$\beta_\mathbf{v}\mathbf{u} = \langle\mathbf{u}, \mathbf{a}_\mathbf{v}\rangle$$

for all $\mathbf{u} \in U$. Setting $\alpha^*\mathbf{v} = \mathbf{a}_\mathbf{v}$, we obtain a map $\alpha^* : U \to U$ such that

$$\langle\alpha\mathbf{u}, \mathbf{v}\rangle = \langle\mathbf{u}, \alpha^*\mathbf{v}\rangle.$$

---

[6] However, it does produce something very close to it. See Exercise 14.3.9.

The rest of the proof is routine. Observe that

$$\langle \mathbf{u}, \alpha^*(\lambda\mathbf{v} + \mu\mathbf{w})\rangle = \langle \alpha\mathbf{u}, \lambda\mathbf{v} + \mu\mathbf{w}\rangle$$
$$= \lambda\langle \alpha\mathbf{u}, \mathbf{v}\rangle + \mu\langle \alpha\mathbf{u}, \mathbf{w}\rangle$$
$$= \lambda\langle \mathbf{u}, \alpha^*\mathbf{v}\rangle + \mu\langle \alpha\mathbf{u}, \alpha^*\mathbf{w}\rangle$$
$$= \langle \mathbf{u}, \lambda\alpha^*\mathbf{v} + \mu\alpha^*\mathbf{w}\rangle$$

for all $\mathbf{u} \in U$, so

$$\alpha^*(\lambda\mathbf{v} + \mu\mathbf{w}) = \lambda\alpha^*\mathbf{v} + \mu\alpha^*\mathbf{w}$$

for all $\mathbf{v}, \mathbf{w} \in U$ and all $\lambda, \mu \in \mathbb{R}$. Thus $\alpha^*$ is linear.

A similar 'paper tiger proof', which is left as a strongly recommended exercise for the reader, shows that $\Phi$ is linear. We now remark that

$$\langle \alpha\mathbf{u}, \mathbf{v}\rangle = \langle \mathbf{u}, \alpha^*\mathbf{v}\rangle = \langle \alpha^*\mathbf{v}, \mathbf{u}\rangle$$
$$= \langle \mathbf{v}, \alpha^{**}\mathbf{u}\rangle = \langle \alpha^{**}\mathbf{u}, \mathbf{v}\rangle$$

for all $\mathbf{v} \in U$, so

$$\alpha\mathbf{u} = \alpha^{**}\mathbf{u}$$

for all $\mathbf{u} \in U$ whence

$$\alpha = \alpha^{**}$$

for all $\alpha \in \mathcal{L}(U, U)$. Thus $\Phi$ is bijective with $\Phi^{-1} = \Phi$ and we have shown that $\Phi$ is, indeed, an isomorphism. $\qquad\square$

**Exercise 14.2.10** *Supply the missing part of the proof of Lemma 14.2.9 by showing that, if $\alpha, \beta \in \mathcal{L}(U, U)$ and $\lambda, \mu \in \mathbb{F}$, then*

$$\langle \mathbf{u}, \Phi(\lambda\alpha + \mu\beta)\mathbf{v}\rangle = \langle \mathbf{u}, (\lambda\Phi(\alpha) + \mu\Phi(\beta))\mathbf{v}\rangle$$

*for all $\mathbf{u}, \mathbf{v} \in U$ and deducing that $\Phi$ is linear.*

We call the $\alpha^*$ obtained in Lemma 14.2.9 the *adjoint* of $\alpha$. The next lemma shows that this definition is consistent with our usage earlier in the book.

**Lemma 14.2.11** *Let $U$ be a finite dimensional real vector space with inner product $\langle\ ,\ \rangle$. Let $\mathbf{e}_1, \mathbf{e}_2, \ldots, \mathbf{e}_n$ be an orthonormal basis of $U$. If $\alpha \in \mathcal{L}(U, U)$ has matrix $A$ with respect to this basis, then $\alpha^*$ has matrix $A^T$ with respect to the same basis.*

*Proof* Let $A = (a_{ij})$ let $\alpha^*$ have matrix $B = (b_{ij})$. We then have

$$b_{ij} = \left\langle \mathbf{e}_i, \sum_{k=1}^{n} b_{kj}\mathbf{e}_k \right\rangle = \langle \mathbf{e}_i, \alpha^*\mathbf{e}_j\rangle = \langle \alpha\mathbf{e}_i, \mathbf{e}_j\rangle = \left\langle \sum_{k=1}^{n} a_{ki}\mathbf{e}_k.\mathbf{e}_j \right\rangle = a_{ji}$$

as required. $\qquad\square$

The reader will observe that it has taken us four or five pages to do what took us one or two pages at the beginning of Section 7.2. However, the notion of an adjoint map is sufficiently important for it to be worth looking at in various different ways and the longer journey has introduced a simple version of the Riesz representation theorem and given us practice in abstract algebra.

The reader may also observe that, although we have tried to make our proofs as basis free as possible, our proof of Lemma 14.2.2 made essential use of bases and our proof of Theorem 14.2.4 used the rank-nullity theorem. Exercise 15.5.7 shows that it is possible to reduce the use of bases substantially by using a little bit of analysis, but, ultimately, our version of Theorem 14.2.4 depends on the fact that we are working in finite dimensional spaces and therefore on the existence of a finite basis.

However, the notion of an adjoint appears in many infinite dimensional contexts. As a simple example, we note a result which plays an important role in the study of second order differential equations.

**Exercise 14.2.12** *Consider the space $\mathcal{D}$ of infinitely differentiable functions $f : [0, 1] \to \mathbb{R}$ with $f(0) = f(1) = 0$. Check that, if we set*

$$\langle f, g \rangle = \int_0^1 f(t)g(t)\, dt$$

*and use standard pointwise operations, $\mathcal{D}$ is an infinite dimensional real inner product space.*

*If $p \in \mathcal{D}$ is fixed and we write*

$$(\alpha f)(t) = \frac{d}{dt}\big(p(t)f'(t)\big),$$

*show that $\alpha : \mathcal{D} \to \mathbb{R}$ is linear and $\alpha^* = \alpha$. (In the language of this book $\alpha$ is a self-adjoint linear functional.)*

*Restate the result given in the first sentence of Example 14.1.13 in terms of self-adjoint linear functionals.*

It turns out that, provided our infinite dimensional inner product spaces are *Hilbert spaces*, the Riesz representation theorem has a basis independent proof and we can define adjoints in this more general context.

At the beginning of the twentieth century, it became clear that many ideas in the theory of differential equations and elsewhere could be better understood in the context of Hilbert spaces. In 1925, Heisenberg wrote a paper proposing a new quantum mechanics founded on the notion of observables. In his entertaining and instructive history, *Inward Bound* [27], Pais writes: 'If the early readers of Heisenberg's first paper on quantum mechanics had one thing in common with its author, it was an inadequate grasp of what was happening. The mathematics was unfamiliar, the physics opaque.' However, Born and Jordan realised that Heisenberg's key rule corresponded to matrix multiplication and introduced 'matrix mechanics'. Schrödinger produced another approach, 'wave mechanics' via partial differential equations. Dirac and then Von Neumann showed that these ideas could be unified using Hilbert spaces.[7]

The amount of time we shall spend studying self-adjoint (Hermitian) and normal linear maps reflects their importance in quantum mechanics, and our preference for basis free methods reflects the fact that the inner product spaces of quantum mechanics are infinite dimensional. The fact that we shall use complex inner product spaces (see the next section) reflects the needs of the physicist.

## 14.3 Complex inner product spaces

So far in this chapter we have looked at real inner product spaces because we wished to look at situations involving real numbers, but there is no obstacle to extending our ideas to complex vector spaces.

**Definition 14.3.1** *If $U$ is a vector space over $\mathbb{C}$, we say that a function $M : U^2 \to \mathbb{R}$ is an inner product if, writing*

$$\langle \mathbf{z}, \mathbf{w} \rangle = M(\mathbf{z}, \mathbf{w}),$$

*the following results hold for all $\mathbf{z}$, $\mathbf{w}$, $\mathbf{u} \in U$ and all $\lambda$, $\mu \in \mathbb{C}$.*

*(i) $\langle \mathbf{z}, \mathbf{z} \rangle$ is always real and positive.*
*(ii) $\langle \mathbf{z}, \mathbf{z} \rangle = 0$ if and only if $\mathbf{z} = \mathbf{0}$.*
*(iii) $\langle \lambda \mathbf{z}, \mathbf{w} \rangle = \lambda \langle \mathbf{z}, \mathbf{w} \rangle$.*
*(iv) $\langle \mathbf{z} + \mathbf{u}, \mathbf{w} \rangle = \langle \mathbf{z}, \mathbf{w} \rangle + \langle \mathbf{u}, \mathbf{w} \rangle$.*
*(v) $\langle \mathbf{w}, \mathbf{z} \rangle = \langle \mathbf{z}, \mathbf{w} \rangle^*$.*
*We write $\|\mathbf{z}\|$ for the positive square root of $\langle \mathbf{z}, \mathbf{w} \rangle$.*

The reader should compare Exercise 8.4.2. Once again, we note the particular form of rule (v). When we wish to recall that we are talking about complex vector spaces, we talk about complex inner products.

**Exercise 14.3.2** *Most physicists use a slightly different definition of an inner product. The physicist's inner product $\langle \ , \ \rangle_P$ obeys all the conditions of our inner product except that (iii) is replaced by*

*(iii)$_P$ $\langle \lambda \mathbf{z}, \mathbf{w} \rangle_P = \lambda^* \langle \mathbf{z}, \mathbf{w} \rangle_P$.*
*Show that $\langle \mathbf{z}, \lambda \mathbf{w} \rangle_P = \lambda \langle \mathbf{z}, \mathbf{w} \rangle_P$.*
*Show that, if $\langle \ , \ \rangle_P$ is a physicist's inner product, then*

$$\langle \mathbf{w}, \mathbf{z} \rangle_M = \langle \mathbf{w}, \mathbf{z} \rangle_P^*$$

*defines $\langle \ , \ \rangle_M$ as a mathematician's inner product.*

---

[7] Of course the physics was vastly more important and deeper than the mathematics, but the pre-existence of various mathematical theories related to Hilbert space theory made the task of the physicists substantially easier.

Hilbert studied specific spaces. Von Neumann introduced and named the general idea of a Hilbert space. There is an apocryphal story of Hilbert leaving a seminar and asking a colleague 'What is this Hilbert space that the youngsters are talking about?'.

*Derive a physicist's inner product from a mathematician's inner product.*

**Exercise 14.3.3** *Let* $a < b$. *Show that the set* $C([a, b])$ *of continuous functions* $f :$ $[a, b] \to \mathbb{C}$ *with the operations of pointwise addition and multiplication by a scalar (thus* $(f + g)(x) = f(x) + g(x)$ *and* $(\lambda f)(x) = \lambda f(x)$*) forms a vector space over* $\mathbb{C}$. *Verify that*

$$\langle f, g \rangle = \int_a^b f(x)g(x)^* \, dx$$

*defines an inner product for this space.*

Some of our results on orthogonal polynomials will not translate easily to the complex case. (For example, the proof of Lemma 14.1.6 depends on the order properties of $\mathbb{R}$.) With exceptions such as these, the reader will have no difficulty in stating and proving the results on complex inner products which correspond to our previous results on real inner products.

**Exercise 14.3.4** *Let us work in a complex inner product vector space* $U$. *Naturally we say that* $\mathbf{e}_1, \mathbf{e}_2, \ldots$ *are orthonormal if* $\langle \mathbf{e}_r, \mathbf{e}_s \rangle = \delta_{rs}$.

(i) *There is one point were we have to be careful. If* $\mathbf{e}_1, \mathbf{e}_2, \ldots, \mathbf{e}_n$ *are orthonormal and*

$$\mathbf{f} = \sum_{j=1}^n a_j \mathbf{e}_j$$

*show that* $a_j = \langle \mathbf{f}, \mathbf{e}_j \rangle$. *What is the value of* $\langle \mathbf{e}_j, \mathbf{f} \rangle$?

(ii) *Consider the system in Exercise 14.3.3 with* $[a, b] = [0, 1]$. *If* $e_n : [0, 1] \to \mathbb{C}$ *is given by* $e_n(t) = \exp(2\pi i n t)$ $[n \in \mathbb{Z}]$ *show that the* $e_n$ *are orthonormal.*

Here is the translation of Theorem 7.1.5 to the complex case.

**Exercise 14.3.5** *We work in a finite dimensional complex inner product space* $U$.

(i) *If* $\mathbf{e}_1, \mathbf{e}_2, \ldots, \mathbf{e}_k$ *are orthonormal and* $\mathbf{x} \in U$, *show that*

$$\mathbf{v} = \mathbf{x} - \sum_{j=1}^k \langle \mathbf{x}, \mathbf{e}_j \rangle \mathbf{e}_j$$

*is orthogonal to each of* $\mathbf{e}_1, \mathbf{e}_2, \ldots, \mathbf{e}_k$.

(ii) *If* $\mathbf{e}_1, \mathbf{e}_2, \ldots, \mathbf{e}_k$ *are orthonormal and* $\mathbf{x} \in U$, *show that either*

$$\mathbf{x} \in \text{span}\{\mathbf{e}_1, \mathbf{e}_2, \ldots, \mathbf{e}_k\}$$

*or the vector* $\mathbf{v}$ *defined in* (i) *is non-zero and, writing* $\mathbf{e}_{k+1} = \|\mathbf{v}\|^{-1}\mathbf{v}$, *we know that* $\mathbf{e}_1, \mathbf{e}_2,$ $\ldots, \mathbf{e}_{k+1}$ *are orthonormal and*

$$\mathbf{x} \in \text{span}\{\mathbf{e}_1, \mathbf{e}_2, \ldots, \mathbf{e}_{k+1}\}.$$

(iii) *If* $U$ *has dimension* $n$, $k \leq n$ *and* $\mathbf{e}_1, \mathbf{e}_2, \ldots, \mathbf{e}_k$ *are orthonormal vectors in* $U$, *show that we can find an orthonormal basis* $\mathbf{e}_1, \mathbf{e}_2, \ldots, \mathbf{e}_n$ *for* $U$.

Here is a typical result that works equally well in the real and complex cases.

**Lemma 14.3.6** *If $U$ is a real or complex inner product finite dimensional space with a subspace $V$, then*

$$V^{\perp} = \{\mathbf{u} \in U \, : \, \langle \mathbf{u}, \mathbf{v} \rangle = 0 \quad \text{for all } \mathbf{v} \in V\}$$

*is a complementary subspace of $U$.*

*Proof* We observe that $\langle \mathbf{0}, \mathbf{v} \rangle = 0$ for all $\mathbf{v}$, so that $\mathbf{0} \in V^{\perp}$. Further, if $\mathbf{u}_1, \mathbf{u}_2 \in V^{\perp}$ and $\lambda_1, \lambda_2 \in \mathbb{C}$, then

$$\langle \lambda_1 \mathbf{u}_1 + \lambda_2 \mathbf{u}_2, \mathbf{v} \rangle = \lambda_1 \langle \mathbf{u}_1, \mathbf{v} \rangle + \lambda_2 \langle \mathbf{u}_2, \mathbf{v} \rangle = \mathbf{0} + \mathbf{0} = \mathbf{0}$$

for all $\mathbf{v} \in V$ and so $\lambda_1 \mathbf{u}_1 + \lambda_2 \mathbf{u}_2 \in V^{\perp}$.

Since $U$ is finite dimensional, so is $V$ and $V$ must have an orthonormal basis $\mathbf{e}_1, \mathbf{e}_2, \ldots,$ $\mathbf{e}_m$. If $\mathbf{u} \in U$ and we write

$$\tau \mathbf{u} = \sum_{j=1}^{m} \langle \mathbf{u}, \mathbf{e}_j \rangle \mathbf{e}_j, \quad \pi = \iota - \tau,$$

then $\tau \mathbf{u} \in V$ and

$$\tau \mathbf{u} + \pi \mathbf{u} = \mathbf{u}.$$

Further

$$\langle \pi \mathbf{u}, \mathbf{e}_k \rangle = \left\langle \mathbf{u} - \sum_{j=1}^{m} \langle \mathbf{u}, \mathbf{e}_j \rangle \mathbf{e}_j, \mathbf{e}_k \right\rangle$$

$$= \langle \mathbf{u}, \mathbf{e}_k \rangle - \sum_{j=1}^{m} \langle \mathbf{u}, \mathbf{e}_j \rangle \langle \mathbf{e}_j, \mathbf{e}_k \rangle$$

$$= \langle \mathbf{u}, \mathbf{e}_k \rangle - \sum_{j=1}^{m} \langle \mathbf{u}, \mathbf{e}_j \rangle \delta_{jk} = 0$$

for $1 \leq k \leq m$. It follows that

$$\left\langle \pi \mathbf{u}, \sum_{j=1}^{m} \lambda_j \mathbf{e}_j \right\rangle = \sum_{j=1}^{m} \lambda_j^* \langle \pi \mathbf{u}, \mathbf{e}_j \rangle = 0$$

so $\langle \pi \mathbf{u}, \mathbf{v} \rangle = 0$ for all $\mathbf{v} \in V$ and $\pi \mathbf{u} \in V^{\perp}$. We have shown that

$$U = V + V^{\perp}.$$

If $\mathbf{u} \in V \cap V^{\perp}$, then, by the definition of $V^{\perp}$,

$$\|\mathbf{u}\|^2 = \langle \mathbf{u}, \mathbf{u} \rangle = 0,$$

so $\mathbf{u} = \mathbf{0}$. Thus

$$U = V \oplus V^{\perp}.$$

$\square$

We call the set $V^\perp$ of Lemma 14.3.6 the *orthogonal complement* (or *perpendicular complement*) of $V$. Note that, although $V$ has many complementary subspaces, it has only one orthogonal complement.

The next exercises parallel our earlier discussion of adjoint mappings for the real case. Even if the reader does not do all these exercises, she should make sure she understands what is going on in Exercise 14.3.9.

**Exercise 14.3.7** *If $V$ is an $n - 1$ dimensional subspace of an $n$ dimensional complex vector space $U$ with inner product $\langle\ ,\ \rangle$, show that we can find a $\mathbf{b} \neq \mathbf{0}$ such that*

$$V = \{\mathbf{x} \in U\ :\ \langle\mathbf{x}, \mathbf{b}\rangle = 0\}.$$

**Exercise 14.3.8 [Riesz representation theorem]** *Let $U$ be a finite dimensional complex vector space with inner product $\langle\ ,\ \rangle$. If $\alpha \in U'$, the dual space of $U$, show that there exists a unique $\mathbf{a} \in U$ such that*

$$\alpha\mathbf{u} = \langle\mathbf{u}, \mathbf{a}\rangle$$

*for all $\mathbf{u} \in U$.*

The complex version of Lemma 14.2.8 differs in a very interesting way from the real case.

**Exercise 14.3.9** *Let $U$ be a finite dimensional complex vector space with inner product $\langle\ ,\ \rangle$. Show that the equation*

$$\theta(\mathbf{a})\mathbf{u} = \langle\mathbf{u}, \mathbf{a}\rangle,$$

*with $\mathbf{a}, \mathbf{u} \in U$, defines an* anti-isomorphism $\theta : U \to U'$. *In other words, show that $\theta$ is a bijective map with*

$$\theta(\lambda\mathbf{a} + \mu\mathbf{b}) = \lambda^*\theta(\mathbf{a}) + \mu^*\theta(\mathbf{b})$$

*for all $\mathbf{a}, \mathbf{b} \in U$ and all $\lambda, \mu \in \mathbb{C}$.*

The fact that the mapping $\theta$ is not an isomorphism but an anti-isomorphism means that we cannot use it to identify $U$ and $U'$, even if we wish to.

**Exercise 14.3.10** *The natural first reaction to the result of Exercise 14.3.9 is that we have somehow made a mistake in defining $\theta$ and that a different definition would give an isomorphism. I very strongly recommend that you spend at least ten minutes trying to find such a definition.*

**Exercise 14.3.11** *Let $U$ be a finite dimensional complex vector space with inner product $\langle\ ,\ \rangle$. If $\alpha \in \mathcal{L}(U, U)$, show that there exists a unique $\alpha^* \in \mathcal{L}(U, U)$ such that*

$$\langle\alpha\mathbf{u}, \mathbf{v}\rangle = \langle\mathbf{u}, \alpha^*\mathbf{v}\rangle$$

*for all $\mathbf{u}, \mathbf{v} \in U$.*

*Show that the map $\Phi : \mathcal{L}(U, U) \to \mathcal{L}(U, U)$ given by $\Phi\alpha = \alpha^*$ is an anti-isomorphism.*

**Exercise 14.3.12** *Let $U$ be a finite dimensional real vector space with inner product $\langle \ , \ \rangle$. Let $\mathbf{e}_1, \mathbf{e}_2, \ldots, \mathbf{e}_n$ be an orthonormal basis of $U$. If $\alpha \in \mathcal{L}(U, U)$ has matrix $A$ with respect to this basis, show that $\alpha^*$ has matrix $A^*$ with respect to the same basis.*

The reader will see that the fact that $\Phi$ in Exercise 14.3.11 is anti-isomorphic should have come as no surprise to us, since we know from our earlier study of $n \times n$ matrices that $(\lambda A)^* = \lambda^* A^*$.

**Exercise 14.3.13** *Let $U$ be a real or complex inner product finite dimensional space with a subspace $V$.*
   *(i) Show that $(V^\perp)^\perp = V$.*
   *(ii) Show that there is a unique $\pi : U \to U$ such that, if $\mathbf{u} \in U$, then $\pi\mathbf{u} \in V$ and $(\iota - \pi)\mathbf{u} \in V^\perp$.*
   *(iii) If $\pi$ is defined as in (ii), show that $\pi \in \mathcal{L}(U, U)$, $\pi^* = \pi$ and $\pi^2 = \pi$.*

The next exercise forms a companion to Exercise 12.1.15.

**Exercise 14.3.14** *Prove that the following conditions on an endomorphism $\alpha$ of a finite dimensional real or complex inner product space $V$ are equivalent.*
   *(i) $\alpha^* = \alpha$ and $\alpha^2 = \alpha$.*
   *(ii) $\alpha V$ and $(\iota - \alpha)V$ are orthogonal complements and $\alpha|_{\alpha V}$ is the identity map on $\alpha V$, $\alpha|_{(\iota - \alpha)V}$ is the zero map on $(\iota - \alpha)V$.*
   *(ii) There is a subspace $U$ of $V$ such that $\alpha|_U$ is the identity mapping of $U$ and $\alpha|_{U^\perp}$ is the zero mapping of $U^\perp$.*
   *(iii) An orthonormal basis of $V$ can be chosen so that all the non-zero elements of the matrix representing $\alpha$ lie on the main diagonal and take the value 1.*
   *[You may find it helpful to use the identity $\iota = \alpha + (\iota - \alpha)$.]*
   *An endomorphism of $V$ satisfying any (and hence all) of the above conditions is called an* orthogonal projection[8] *of $V$. Explain why an orthogonal projection is automatically a projection in the sense of Exercise 12.1.15 and give an example to show that the converse is false.*
   *Suppose that $\alpha$, $\beta$ are both orthogonal projections of $V$. Prove that, if $\alpha\beta = \beta\alpha$, then $\alpha\beta$ is also an orthogonal projection of $V$.*

We can thus call the $\pi$ considered in Exercise 14.3.13 the *orthogonal projection* of $U$ onto $V$.

**Exercise 14.3.15** *Let $U$ be a real or complex inner product finite dimensional space. If $W$ is a subspace of $V$, we write $\pi_W$ for the orthogonal projection onto $W$.*
   *Let $\rho_W = 2\pi_W - \iota$. Show that $\rho_W \rho_W = \iota$ and $\rho_W$ is an isometry. Identify $\rho_W \rho_{W^\perp}$.*

---

[8] Or, when the context is plain, just a *projection*.

## 14.4 Further exercises

**Exercise 14.4.1** Let $V$ be a finite dimensional real vector space and let $\langle\ ,\ \rangle_1$ and $\langle\ ,\ \rangle_2$ be two inner products on $V$. Prove the following results.

(i) There exists a unique endomorphism $\alpha : V \to V$ such that $\langle \mathbf{u}, \mathbf{v}\rangle_2 = \langle \mathbf{u}, \alpha\mathbf{v}\rangle_1$ for all $\mathbf{u}, \mathbf{v} \in V$.

(ii) $\alpha$ is self-adjoint with respect to $\langle\ ,\ \rangle_1$.

(iii) If $\beta : V \to V$ is an endomorphism which is self-adjoint with respect to $\langle\ ,\ \rangle_1$, then $\beta$ is self-adjoint with respect to $\langle\ ,\ \rangle_2$ if and only if $\alpha$ and $\beta$ commute.

**Exercise 14.4.2** Let $V$ be vector space over $\mathbb{R}$ of dimension $n$.

(i) Show that, given $n - 1$ different inner products $\langle\ ,\ \rangle_j$, and $\mathbf{x} \in V$ we can find a non-zero $\mathbf{y} \in V$ with $\langle \mathbf{x}, \mathbf{y}\rangle_j = 0$ for all $1 \leq j \leq n - 1$.

(ii) Give an example with $n = 2$ to show that we cannot replace $n - 1$ by $n$ in (i).

(iii) Show that, if $n \geq 2$, then, given $n$ different inner products $\langle\ ,\ \rangle_j$, we can find non-zero $\mathbf{x}, \mathbf{y} \in V$ with $\langle \mathbf{x}, \mathbf{y}\rangle_j = 0$ for all $1 \leq j \leq n$.

**Exercise 14.4.3** If we write

$$q_n(x) = \frac{d^n}{dx^n}(x^2 - 1)^n,$$

prove, by repeated integration by parts, that

$$\int_{-1}^{1} q_n(x)q_m(x)\,dx = \frac{\delta_{nm}}{2n + 1}$$

for all non-negative integers $n$ and $m$ with $n \neq m$. Compute $q_0$, $q_1$ and $q_2$.

Show that $q_n$ is a polynomial of degree $n$ and find the coefficient $a_n$ of $x^n$. Explain why, if we define $p_n$ as in Exercise 14.1.5, we have $p_n = a_n^{-1}q_n$.

**Exercise 14.4.4 [Tchebychev polynomials]** Recall de Moivre's theorem which tells us that

$$\cos n\theta + i \sin n\theta = (\cos \theta + i \sin \theta)^n.$$

Deduce that

$$\cos n\theta = \sum_{2r \leq n}(-1)^r \binom{n}{2r} \cos^{n-2r} \theta(1 - \cos^2 \theta)^r$$

and so

$$\cos n\theta = T_n(\cos \theta)$$

where $T_n$ is a polynomial of degree $n$. We call $T_n$ the $n$th Tchebychev polynomial. Compute $T_0$, $T_1$, $T_2$ and $T_3$ explicitly.

By considering $(1 + 1)^n - (1 - 1)^n$, or otherwise, show that the coefficient of $t^n$ in $T_n(t)$ is $2^{n-1}$ for all $n \geq 1$. Explain why $|T_n(t)| \leq 1$ for $|t| \leq 1$. Does this inequality hold for all $t \in \mathbb{R}$? Give reasons.

Show that

$$\int_{-1}^{1} T_n(x)T_m(x)\frac{1}{(1-x^2)^{1/2}}\, dx = \delta_{nm}a_n,$$

where $a_0 = \pi$ and $a_n = \pi/2$ otherwise. (Thus, provided we relax our conditions on $r$ a little, the Tchebychev polynomials can be considered as orthogonal polynomials of the type discussed in Exercise 14.1.12.)

Verify the three term recurrence relation

$$T_{n+1}(t) - 2tT_n(t) + T_{n-1}(t) = 0$$

for $|t| \le 1$. Does this equality hold for all $t \in \mathbb{R}$? Give reasons.

Compute $T_4$, $T_5$ and $T_6$ using the recurrence relation.

**Exercise 14.4.5 [Hermite polynomials]** (Do this question formally if you must and rigorously if you can.)

(i) Show that

$$e^{2tx-t^2} = \sum_{n=0}^{\infty} \frac{H_n(x)}{n!} t^n \qquad \qquad \bigstar$$

where $H_n$ is a polynomial. (The $H_n$ are called Hermite polynomials. As usual there are several versions differing only in the scaling chosen.)

(ii) By using Taylor's theorem (or, more easily justified, differentiating both sides of $\bigstar$ $n$ times with respect to $t$ and then setting $t = 0$), show that

$$H_n(x) = (-1)^n e^{x^2} \frac{d^n}{dx^n} e^{-x^2}$$

and deduce, or prove otherwise, that $H_n$ is a polynomial of degree exactly $n$.
[Hint: $e^{2tx-t^2} = e^{x^2}e^{-(t-x)^2}$.]

(iii) By differentiating both sides of $\bigstar$ with respect to $x$ and then equating coefficients, obtain the relation

$$H_n'(x) = 2n H_{n-1}(x).$$

(iv) By integration by parts, or otherwise, show that

$$\int_{\infty}^{\infty} H_m(x)H_n(x)e^{-x^2}\, dx = \delta_{n,m}2^n n! \pi^{1/2}.$$

Observe (without looking too closely at the details) the parallels with Exercise 14.1.12.

(v) By differentiating both sides of $\bigstar$ with respect to $t$ and then equating coefficients, obtain the three term recurrence relation

$$H_{n+1}(x) - 2x H_{n-1}(x) + 2n H_{n+1}(x) = 0.$$

(iv) Show that $H_0(x) = 1$ and $H_1(x) = 2x$. Compute $H_2$, $H_3$ and $H_4$ using (iv).

**Exercise 14.4.6** Suppose that $U$ is a finite dimensional inner product space over $\mathbb{C}$. If $\alpha, \beta : U \to U$ are Hermitian (that is to say, self-adjoint) linear maps, show that $\alpha\beta + \beta\alpha$ is always Hermitian, but that $\alpha\beta$ is Hermitian if and only if $\alpha\beta = \beta\alpha$. Give, with proof, a necessary and sufficient condition for $\alpha\beta - \beta\alpha$ to be Hermitian.

**Exercise 14.4.7** Suppose that $U$ is a finite dimensional inner product space over $\mathbb{C}$. If $\mathbf{a}, \mathbf{b} \in U$, set

$$T_{\mathbf{a},\mathbf{b}}\mathbf{u} = \langle \mathbf{u}, \mathbf{b}\rangle\mathbf{a}$$

for all $\mathbf{u} \in U$. Prove the following results.
  (i) $T_{\mathbf{a},\mathbf{b}}$ is an endomorphism of $U$.
  (ii) $T_{\mathbf{a},\mathbf{b}}^* = T_{\mathbf{b},\mathbf{a}}$.
  (iii) $\operatorname{Tr} T_{\mathbf{a},\mathbf{b}} = \langle \mathbf{a}, \mathbf{b}\rangle$.
  (iv) $T_{\mathbf{a},\mathbf{b}}T_{\mathbf{c},\mathbf{d}} = T_{\mathbf{a},\langle\mathbf{b},\mathbf{c}\rangle\mathbf{d}}$.
Establish a necessary and sufficient condition for $T_{\mathbf{a},\mathbf{b}}$ to be self-adjoint.

**Exercise 14.4.8** (i) Let $V$ be a vector space over $\mathbb{C}$. If $\phi : V^2 \to \mathbb{C}$ is an inner product on $V$ and $\gamma : V \to V$ is an isomorphism, show that

$$\psi(\mathbf{x}, \mathbf{y}) = \phi(\gamma\mathbf{x}, \gamma\mathbf{y})$$

defines an inner product $\psi : V^2 \to \mathbb{C}$.
  (ii) For the rest of this exercise $V$ will be a finite dimensional vector space over $\mathbb{C}$ and $\phi : V^2 \to \mathbb{C}$ an inner product on $V$. If $\psi : V^2 \to \mathbb{C}$ is an inner product on $V$, show that there is an isomorphism $\gamma : V \to V$ such that

$$\psi(\mathbf{x}, \mathbf{y}) = \phi(\gamma\mathbf{x}, \gamma\mathbf{y})$$

for all $(\mathbf{x}, \mathbf{y}) \in V^2$.
  (iii) If $\mathcal{A}$ is a basis $\mathbf{a}_1, \mathbf{a}_2, \ldots, \mathbf{a}_n$ for $V$ and $\mathcal{E}$ is a $\phi$-orthonormal basis $\mathbf{e}_1, \mathbf{e}_2, \ldots, \mathbf{e}_n$ for $V$, we set

$$\mu_\phi(\mathcal{A}, \mathcal{E}) = \det\alpha$$

where $\alpha$ is the linear map with $\alpha\mathbf{a}_j = \mathbf{e}_j$. Show that $\mu_\phi(\mathcal{A}, \mathcal{E})$ is independent of the choice of $\mathcal{E}$, so we may write $\mu_\phi(\mathcal{A}) = \mu_\phi(\mathcal{A}, \mathcal{E})$.
  (iv) If $\gamma$ and $\psi$ are as in (i), find $\mu_\phi(\mathcal{A})/\mu_\psi(\mathcal{A})$ in terms of $\gamma$.

**Exercise 14.4.9** In this question we work in a finite dimensional real or complex vector space $V$ with an inner product.
  (i) Let $\alpha$ be an endomorphism of $V$. Show that $\alpha$ is an orthogonal projection if and only if $\alpha^2 = \alpha$ and $\alpha\alpha^* = \alpha^*\alpha$. (In the language of Section 15.4, $\alpha$ is a *normal* linear map.) [Hint: It may be helpful to prove first that, if $\alpha\alpha^* = \alpha^*\alpha$, then $\alpha$ and $\alpha^*$ have the same kernel.]

(ii) Show that, given a subspace $U$, there is a unique orthogonal projection $\tau_U$ with $\tau_U(V) = U$. If $U$ and $W$ are two subspaces, show that

$$\tau_U \tau_W = 0$$

if and only if $U$ and $W$ are orthogonal. (In other words, $\langle \mathbf{u}, \mathbf{w} \rangle = 0$ for all $\mathbf{u} \in U$, $\mathbf{v} \in V$.)

(iii) Let $\beta$ be a projection, that is to say an endomorphism such that $\beta^2 = \beta$. Show that we can define an inner product on $V$ in such a way that $\beta$ is an orthogonal projection with respect to the new inner product.

**Exercise 14.4.10** Let $V$ be the usual vector space of $n \times n$ matrices. If we give $V$ the basis of matrices $E(i, j)$ with 1 in the $(i, j)$th place and zero elsewhere, identify the dual basis for $V'$.

If $B \in V$, we define $\tau_B : V \to \mathbb{R}$ by $\tau_B(A) = \text{Tr}(AB)$. Show that $\tau_B \in V'$ and that the map $\tau : V \to V'$ defined by $\tau(B) = \tau_B$ is a vector space isomorphism.

If $S$ is the subspace of $V$ consisting of the symmetric matrices and $S^0$ is the annihilator of $S$, identify $\tau^{-1}(S^0)$.

**Exercise 14.4.11** (If you get confused by this exercise just ignore it.) We remarked that the function $\theta$ of Lemma 14.2.8 could be used to set up an identification between $U$ and $U'$ when $U$ was a real finite dimensional inner product space. In this exercise we look at the consequences of making such an identification by setting $\mathbf{u} = \theta \mathbf{u}$ and $U = U'$.

(i) Show that, if $\mathbf{e}_1, \mathbf{e}_2, \ldots, \mathbf{e}_n$ is an orthonormal basis for $U$, then the dual basis $\hat{\mathbf{e}}_j = \mathbf{e}_j$.

(ii) If $\alpha : U \to U$ is linear, then we know that it induces a dual map $\alpha' : U' \to U'$. Since we identify $U$ and $U'$, we have $\alpha' : U \to U$. Show that $\alpha' = \alpha^*$.

**Exercise 14.4.12** The object of this question is to show that infinite dimensional inner product spaces may have unexpected properties. It involves a small amount of analysis.

Consider the space $V$ of all real sequences

$$\mathbf{u} = (u_1, u_2, \ldots)$$

where $n^2 u_n \to 0$ as $n \to \infty$. Show that $V$ is real vector space under the standard coordinatewise operations

$$\mathbf{u} + \mathbf{v} = (u_1 + v_1, u_2 + v_2, \ldots), \quad \lambda \mathbf{u} = (\lambda u_1, \lambda u_2, \ldots)$$

and an inner product space under the inner product

$$\langle \mathbf{u}, \mathbf{v} \rangle = \sum_{j=1}^{\infty} u_j v_j.$$

Show that the set $M$ of all sequences $\mathbf{u}$ with only finitely many $u_j$ non-zero is a subspace of $V$. Find $M^{\perp}$ and show that $M + M^{\perp} \neq V$.

**Exercise 14.4.13** Here is another way in which infinite inner product spaces may exhibit unpleasant properties.

Consider the space $M$ of all real sequences

$$\mathbf{u} = (u_1, u_2, \ldots)$$

with only finitely many of the $u_j$ non-zero. Show that $M$ is a real vector space under the standard coordinatewise operations and an inner product space under the inner product

$$\langle \mathbf{u}, \mathbf{v} \rangle = \sum_{j=1}^{\infty} u_j v_j.$$

Show that $\alpha \mathbf{u} = \sum_{j=1}^{\infty} u_j$ is a well defined linear map from $M$ to $\mathbb{R}$, but that there does not exist an $\mathbf{a} \in M$ with

$$\alpha \mathbf{u} = \langle \mathbf{u}, \mathbf{a} \rangle.$$

Thus the Riesz representation theorem does not hold for $M$.

# 15

# More distances

## 15.1 Distance on $\mathcal{L}(U, U)$

When dealing with matrices, it is natural to say that that the matrix

$$A = \begin{pmatrix} 1 & 10^{-3} \\ 10^{-3} & 1 \end{pmatrix}$$

is close to the $2 \times 2$ identity matrix $I$. On the other hand, if we deal with $10^4 \times 10^4$ matrices, we may suspect that the matrix $B$ given by $b_{ii} = 1$ and $b_{ij} = 10^{-3}$ for $i \neq j$ behaves very differently from the identity matrix. It is natural to seek a notion of distance between $n \times n$ matrices which would measure closeness in some useful way.

What do we wish to mean by saying that we want to 'measure closeness in some useful way'? Surely, we do not wish to say that two matrices are close when they *look* similar, but, rather, to say that two matrices are close when they *act* similarly, in other words, that they represent two linear maps which *act* similarly.

Reflecting our motto

**linear maps for understanding, matrices for computation,**

we look for an appropriate notion of distance between linear maps in $\mathcal{L}(U, U)$. We decide that, in order to have a distance on $\mathcal{L}(U, U)$, we first need a distance on $U$. For the rest of this section, $U$ will be a finite dimensional real or complex inner product space, but the reader will loose nothing if she takes $U$ to be $\mathbb{R}^n$ with the standard inner product $\langle \mathbf{x}, \mathbf{y} \rangle = \sum_{j=1}^{n} x_j y_j$.

Let $\alpha, \beta \in \mathcal{L}(U, U)$ and suppose that we fix an orthonormal basis $\mathbf{e}_1, \mathbf{e}_2, \ldots, \mathbf{e}_n$ for $U$. If $\alpha$ has matrix $A = (a_{ij})$ and $\beta$ matrix $B = (b_{ij})$ with respect to this basis, we could define the distance between $A$ and $B$ by

$$d_\infty(A, B) = \max_{i,j} |a_{ij} - b_{ij}|, \quad d_1(A, B) = \sum_{i,j} |a_{ij} - b_{ij}|,$$

$$d_2(A, B) = \left( \sum_{i,j} |a_{ij} - b_{ij}|^2 \right)^{1/2}.$$

**Exercise 15.1.1** *Check that, for the $d_j$ just defined, we have $d_j(A, B) = \|A - B\|_j$ with*

(i) $\|A\|_j \geq 0$,

(ii) $\|A\|_j = 0 \Rightarrow A = 0$,

(iii) $\|\lambda A\|_j = |\lambda| \|A\|_j$,

(iv) $\|A + B\|_j \leq \|A\|_j + \|B\|_j$.

[*In other words, each $d_j$ is derived from a norm $\| \ \|_j$.*]

However, all these distances depend on the choice of basis. We take a longer and more indirect route which produces a basis independent distance.

**Lemma 15.1.2** *Suppose that $U$ is a finite dimensional inner product space over $\mathbb{F}$ and $\alpha \in \mathcal{L}(U, U)$. Then*

$$\{\|\alpha \mathbf{x}\| \ : \ \|\mathbf{x}\| \leq 1\}$$

*is a non-empty bounded subset of $\mathbb{R}$.*

*Proof* Fix an orthonormal basis $\mathbf{e}_1, \mathbf{e}_2, \ldots, \mathbf{e}_n$ for $U$. If $\alpha$ has matrix $A = (a_{ij})$ with respect to this basis, then, writing $\mathbf{x} = \sum_{j=1}^{n} x_j \mathbf{e}_j$, we have

$$\|\alpha \mathbf{x}\| = \left\| \sum_{i=1}^{n} \left( \sum_{j=1}^{n} a_{ij} x_j \right) \mathbf{e}_i \right\| \leq \sum_{i=1}^{n} \left| \sum_{j=1}^{n} a_{ij} x_j \right|$$

$$\leq \sum_{i=1}^{n} \sum_{j=1}^{n} |a_{ij}| |x_j| \leq \sum_{i=1}^{n} \sum_{j=1}^{n} |a_{ij}|$$

for all $\|\mathbf{x}\| \leq 1$. $\qquad\qquad\qquad\qquad\qquad\qquad\qquad\qquad\qquad\qquad$ $\square$

**Exercise 15.1.3** *We use the notation of the proof of Lemma 15.1.2. Use the Cauchy–Schwarz inequality to obtain*

$$\|\alpha \mathbf{x}\| \leq \left( \sum_{i=1}^{n} \sum_{j=1}^{n} |a_{ij}|^2 \right)^{1/2}$$

*for all $\|\mathbf{x}\| \leq 1$.*

Since every non-empty bounded subset of $\mathbb{R}$ has a least upper bound (that is to say, supremum), Lemma 15.1.2 allows us to make the following definition.

**Definition 15.1.4** *Suppose that $U$ is a finite dimensional inner product space over $\mathbb{F}$ and $\alpha \in \mathcal{L}(U, U)$. We define*

$$\|\alpha\| = \sup\{\|\alpha \mathbf{x}\| \ : \ \|\mathbf{x}\| \leq 1\}.$$

We call $\|\alpha\|$ the *operator norm* of $\alpha$. The following lemma gives a more concrete way of looking at the operator norm.

**Lemma 15.1.5** *Suppose that* $U$ *is a finite dimensional inner product space over* $\mathbb{F}$ *and* $\alpha \in \mathcal{L}(U, U)$.

  *(i)* $\|\alpha\mathbf{x}\| \leq \|\alpha\|\|\mathbf{x}\|$ *for all* $\mathbf{x} \in U$.

  *(ii) If* $\|\alpha\mathbf{x}\| \leq K\|\mathbf{x}\|$ *for all* $\mathbf{x} \in U$, *then* $\|\alpha\| \leq K$.

*Proof* (i) The result is trivial if $\mathbf{x} = \mathbf{0}$. If not, then $\|\mathbf{x}\| \neq 0$ and we may set $\mathbf{y} = \|\mathbf{x}\|^{-1}\mathbf{x}$. Since $\|\mathbf{y}\| = 1$, the definition of the supremum shows that

$$\|\alpha\mathbf{x}\| = \|\alpha(\|\mathbf{x}\|\mathbf{y})\| = \big\|\|\mathbf{x}\|\alpha\mathbf{y}\big\| = \|\mathbf{x}\|\|\alpha\mathbf{y}\| \leq \|\alpha\|\|\mathbf{x}\|$$

as stated.

  (ii) The hypothesis implies

$$\|\alpha\mathbf{x}\| \leq K \quad \text{for } \|\mathbf{x}\| \leq 1$$

and so, by the definition of the supremum, $\|\alpha\| \leq K$.      $\square$

**Theorem 15.1.6** *Suppose that* $U$ *is a finite dimensional inner product space over* $\mathbb{F}$, $\alpha, \beta \in \mathcal{L}(U, U)$ *and* $\lambda \in \mathbb{F}$. *Then the following results hold.*

  *(i)* $\|\alpha\| \geq 0$.

  *(ii)* $\|\alpha\| = 0$ *if and only if* $\alpha = 0$.

  *(iii)* $\|\lambda\alpha\| = |\lambda|\|\alpha\|$.

  *(iv)* $\|\alpha + \beta\| \leq \|\alpha\| + \|\beta\|$.

  *(v)* $\|\alpha\beta\| \leq \|\alpha\|\|\beta\|$.

*Proof* The proof of Theorem 15.1.6 is easy. However, this shows, not that our definition is trivial, but that it is appropriate.

  (i) Observe that $\|\alpha\mathbf{x}\| \geq 0$.

  (ii) If $\alpha \neq 0$ we can find an $\mathbf{x}$ such that $\alpha\mathbf{x} \neq 0$ and so $\|\alpha\mathbf{x}\| > 0$, whence

$$\|\alpha\| \geq \|\mathbf{x}\|^{-1}\|\alpha\mathbf{x}\| > 0.$$

If $\alpha = 0$, then

$$\|\alpha\mathbf{x}\| = \|\mathbf{0}\| = 0 \leq 0\|\mathbf{x}\|$$

so $\|\alpha\| = 0$.

  (iii) Observe that

$$\|(\lambda\alpha)\mathbf{x}\| = \|\lambda(\alpha\mathbf{x})\| = |\lambda|\|\alpha\mathbf{x}\|.$$

  (iv) Observe that

$$\begin{aligned}\big\|(\alpha + \beta)\mathbf{x}\big\| = \|\alpha\mathbf{x} + \beta\mathbf{x}\| &\leq \|\alpha\mathbf{x}\| + \|\beta\mathbf{x}\| \\ &\leq \|\alpha\|\|\mathbf{x}\| + \|\beta\|\|\mathbf{x}\| = \big(\|\alpha\| + \|\beta\|\big)\|\mathbf{x}\|\end{aligned}$$

for all $\mathbf{x} \in U$.

(v) Observe that

$$\|(\alpha\beta)\mathbf{x}\| = \|\alpha(\beta\mathbf{x})\| \le \|\alpha\|\|\beta\mathbf{x}\| \le \|\alpha\|\|\beta\|\|\mathbf{x}\|$$

for all $\mathbf{x} \in U$. □

**Important note** In Exercise 15.5.6 we give a simple argument to show that the operator norm cannot be obtained in an obvious way from an inner product. In fact, the operator norm behaves very differently from any norm derived from an inner product and, when thinking about the operator norm, the reader must be careful not to use results or intuitions developed when studying inner product norms without checking that they do indeed extend.

**Exercise 15.1.7** *Suppose that $\pi$ is a non-zero orthogonal projection (see Exercise 14.3.14) on a finite dimensional real or complex inner product space $V$. Show that $\|\pi\| = 1$.*

*Let $V = \mathbb{R}^2$ with the standard inner product. Show that, given any $K > 0$, we can find a projection $\alpha$ (see Exercise 12.1.15) with $\|\alpha\| \ge K$.*

Having defined the operator norm for linear maps, it is easy to transfer it to matrices.

**Definition 15.1.8** *If $A$ is an $n \times n$ matrix with entries in $\mathbb{F}$ then*

$$\|A\| = \sup\{\|A\mathbf{x}\| \, : \, \|\mathbf{x}\| \le 1\}$$

*where $\|\mathbf{x}\|$ is the usual Euclidean norm of the column vector $\mathbf{x}$.*

**Exercise 15.1.9** *Let $\alpha \in \mathcal{L}(U, U)$ where $U$ is a finite dimensional inner product space over $\mathbb{F}$. If $\alpha$ has matrix $A$ with respect to some orthonormal basis, show that $\|\alpha\| = \|A\|$.*

**Exercise 15.1.10** *Give examples of $2 \times 2$ real matrices $A_j$ and $B_j$ with $\|A_j\| = \|B_j\| = 1$ having the following properties.*
  *(i) $\|A_1 + B_1\| = 0$.*
  *(ii) $\|A_2 + B_2\| = 2$.*
  *(iii) $\|A_3 B_3\| = 0$.*
  *(iv) $\|A_4 B_4\| = 1$.*

To see one reason why the operator norm is an appropriate measure of the 'size of a linear map', look at the the system of $n \times n$ linear equations

$$A\mathbf{x} = \mathbf{y}.$$

If we make a small change in $\mathbf{x}$, replacing it by $\mathbf{x} + \delta\mathbf{x}$, then, taking

$$A(\mathbf{x} + \delta\mathbf{x}) = \mathbf{y} + \delta\mathbf{y},$$

we see that

$$\|\delta\mathbf{y}\| = \|\mathbf{y} - A(\mathbf{x} + \delta\mathbf{x})\| = \|A\delta\mathbf{x}\| \le \|A\|\|\delta\mathbf{x}\|$$

so $\|A\|$ gives us an idea of 'how sensitive $\mathbf{y}$ is to small changes in $\mathbf{x}$'.

If $A$ is invertible, then $\mathbf{x} = A^{-1}\mathbf{y}$. $A^{-1}(\mathbf{y} + \delta\mathbf{y}) = \mathbf{x} + \delta\mathbf{x}$ and

$$\|\delta\mathbf{x}\| \le \|A^{-1}\|\|\delta\mathbf{y}\|$$

so $\|A^{-1}\|$ gives us an idea of 'how sensitive $\mathbf{x}$ is to small changes in $\mathbf{y}$'. Just as dividing by a very small number is both a cause and a symptom of problems, so, if $\|A^{-1}\|$ is large, this is both a cause and a symptom of problems in the numerical solution of simultaneous linear equations.

**Exercise 15.1.11** *(i) Let $\eta$ be a small but non-zero real number, take $\mu = (1 - \eta)^{1/2}$ (use the positive square root) and*

$$A = \begin{pmatrix} 1 & \mu \\ \mu & 1 \end{pmatrix}.$$

*Find the eigenvalues and eigenvectors of $A$. Show that, if we look at the equation,*

$$A\mathbf{x} = \mathbf{y}$$

*'small changes in $\mathbf{x}$ produce small changes in $\mathbf{y}$', but write down explicitly a small $\mathbf{y}$ for which $\mathbf{x}$ is large.*

*(ii) Consider the $4 \times 4$ matrix made up from the $2 \times 2$ matrices $A$, $A^{-1}$ and $0$ (where $A$ is the matrix in part (i)) as follows*

$$B = \begin{pmatrix} A & 0 \\ 0 & A^{-1} \end{pmatrix}.$$

*Show that $\det B = 1$, but, if $\eta$ is very small, both $\|B\|$ and $\|B^{-1}\|$ are very large.*

**Exercise 15.1.12** *Suppose that $A$ is a non-singular square matrix. Numerical analysts often use the* condition number $c(A) = \|A\|\|A^{-1}\|$ *as a miner's canary to warn of troublesome $n \times n$ non-singular matrices.*

*(i) Show that $c(A) \ge 1$.*

*(ii) Show that, if $\lambda \ne 0$, $c(\lambda A) = c(A)$. (This is a good thing, since floating point arithmetic is unaffected by a simple change of scale.)*

*[We shall not make any use of this concept.]*

If it does not seem possible[1] to write down a neat algebraic expression for the norm $\|A\|$ of an $n \times n$ matrix $A$, Exercise 15.5.15 shows how, in principle, it is possible to calculate it. However, the main use of the operator norm is in theoretical work, where we do not need to calculate it, or in practical work, where very crude estimates are often all that is required. (These remarks also apply to the condition number.)

**Exercise 15.1.13** *Suppose that $U$ is an $n$-dimensional inner product space over $\mathbb{F}$ and $\alpha \in \mathcal{L}(U, U)$. If $\alpha$ has matrix $(a_{ij})$ with respect to some orthonormal basis,*

---

[1] Though, as usual, I would encourage you to try.

*show that*

$$n^{-2} \sum_{i,j} |a_{ij}| \leq \max_{r,s} |a_{rs}| \leq \|\alpha\| \leq \sum_{i,j} |a_{ij}| \leq n^2 \max_{r,s} |a_{rs}|.$$

*Use Exercise 15.1.3 to show that*

$$\|\alpha\| \leq \left( \sum_{i,j} a_{ij}^2 \right)^{1/2}.$$

As the reader knows, the differential and partial differential equations of physics are often solved numerically by introducing grid points and replacing the original equation by a system of linear equations. The result is then a system of $n$ equations in $n$ unknowns of the form

$$A\mathbf{x} = \mathbf{b}, \qquad\qquad ★$$

where $n$ may be very large indeed. If we try to solve the system using the best method we have met so far, that is to say, Gaussian elimination, then we need about $Cn^3$ operations. (The constant $C$ depends on how we count operations, but is typically taken as $2/3$.)

However, the $A$ which arise in this manner are certainly not random. Often we expect $\mathbf{x}$ to be close to $\mathbf{b}$ (the weather in a minute's time will look very much like the weather now) and this is reflected by having $A$ very close to $I$.

Suppose that this is the case and $A = I + B$ with $\|B\| < \epsilon$ and $0 < \epsilon < 1$. Let $\mathbf{x}^*$ be the solution of ★ and suppose that we make the initial guess that $\mathbf{x}_0$ is close to $\mathbf{x}^*$ (a natural choice would be $\mathbf{x}_0 = \mathbf{b}$). Since $(I + B)\mathbf{x}^* = \mathbf{b}$, we have

$$\mathbf{x}^* = \mathbf{b} - B\mathbf{x}^* \approx \mathbf{b} - B\mathbf{x}_0$$

so, using an idea that dates back many centuries, we make the new guess $\mathbf{x}_1$, where

$$\mathbf{x}_1 = \mathbf{b} - B\mathbf{x}_0.$$

We can repeat the process as many times as we wish by defining

$$\mathbf{x}_{j+1} = \mathbf{b} - B\mathbf{x}_j.$$

Now

$$\|\mathbf{x}_{j+1} - \mathbf{x}^*\| = \left\| (\mathbf{b} - B\mathbf{x}_j) - (\mathbf{b} - B\mathbf{x}^*) \right\| = \|B\mathbf{x}^* - B\mathbf{x}_j\|$$
$$= \left\| B(\mathbf{x}_j - \mathbf{x}^*) \right\| \leq \|B\| \|\mathbf{x}_j - \mathbf{x}^*\|.$$

Thus, by induction,

$$\|\mathbf{x}_k - \mathbf{x}^*\| \leq \|B\|^k \|\mathbf{x}_0 - \mathbf{x}^*\| \leq \epsilon^k \|\mathbf{x}_0 - \mathbf{x}^*\|.$$

The reader may object that we are 'merely approximating the true answer' but, because computers only work to a certain accuracy, this is true whatever method we use. After only a few iterations, a term bounded by $\epsilon^k \|\mathbf{x}_0 - \mathbf{x}^*\|$ will be 'lost in the noise' and we will have satisfied ★ to the level of accuracy allowed by the computer.

It requires roughly $2n^2$ additions and multiplications to compute $\mathbf{x}_{j+1}$ from $\mathbf{x}_j$ and so about $2kn^2$ operations to compute $\mathbf{x}_k$ from $\mathbf{x}_0$. When $n$ is large, this is much more efficient than Gaussian elimination.

Sometimes things are even more favourable and most of the entries in the matrix $A$ are zero. (The weather at a given point depends only on the weather at points close to it a minute before.) Such matrices are called *sparse matrices*. If $B$ only has $ln$ non-zero entries, then it will only take about $2lkn$ operations[2] to compute $\mathbf{x}_k$.

The King of Brobdingnag (in Swift's *Gulliver's Travels* [29]) was of the opinion 'that whoever could make two ears of corn, or two blades of grass, to grow upon a spot of ground where only one grew before, would deserve better of mankind, and do more essential service to his country, than the whole race of politicians put together.' Those who devise efficient methods for solving systems of linear equations deserve well of mankind.

In practice, we usually know that the system of equations $A\mathbf{x} = \mathbf{b}$ that we wish to solve must have a unique solution $\mathbf{x}^*$. If we undertake theoretical work and do not know this in advance, a sufficient condition for existence and uniqueness is given by $\|I - A\| < 1$. The interested reader should do Exercise 15.5.16 or 15.5.17 and then Exercise 15.5.18. (Alternatively, she could invoke the contraction mapping theorem, if she has met it.)

A very simple modification of the preceding discussion gives the *Gauss–Jacobi method*.

**Lemma 15.1.14 [Gauss–Jacobi]** *Let $A$ be an $n \times n$ matrix with non-zero diagonal entries and $\mathbf{b} \in \mathbb{F}^n$ a column vector. Suppose that the equation*

$$A\mathbf{x} = \mathbf{b}$$

*has the solution $\mathbf{x}^*$. Let us write $D$ for the $n \times n$ diagonal matrix with diagonal entries the same as those of $A$ and set $B = A - D$. If $\mathbf{x}_0 \in \mathbb{R}^n$ and*

$$\mathbf{x}_{j+1} = D^{-1}(\mathbf{b} - B\mathbf{x}_j)$$

*then*

$$\|\mathbf{x}_j - \mathbf{x}^*\| \leq \|D^{-1}B\|^j \|\mathbf{x}_0 - \mathbf{x}^*\|$$

*and $\|\mathbf{x}_j - \mathbf{x}^*\| \to 0$ whenever $\|D^{-1}B\| < 1$.*

*Proof* Proof left to the reader. □

Note that $D^{-1}$ is easy to compute.

Another, slightly more sophisticated, modification gives the *Gauss–Siedel method*.

**Lemma 15.1.15 [Gauss–Siedel]** *Let $A$ be an $n \times n$ matrix with non-zero diagonal entries and $\mathbf{b} \in \mathbb{F}^n$ a column vector. Suppose that the equation*

$$A\mathbf{x} = \mathbf{b}$$

---

[2] This may be over optimistic. In order to exploit sparsity, the non-zero entries must form a simple pattern and we must be able to exploit that pattern.

has the solution $\mathbf{x}^*$. Let us write $A = L + U$ where $L$ is a lower triangular matrix and $U$ a strictly upper triangular matrix (that is to say, an upper triangular matrix with all diagonal terms zero). If $\mathbf{x}_0 \in \mathbb{F}^n$ and

$$L\mathbf{x}_{j+1} = (\mathbf{b} - U\mathbf{x}_j), \qquad \qquad ★$$

then

$$\|\mathbf{x}_j - \mathbf{x}^*\| \le \|L^{-1}U\|^j \|\mathbf{x}_0 - \mathbf{x}^*\|$$

and $\|\mathbf{x}_j - \mathbf{x}^*\| \to 0$ whenever $\|L^{-1}U\| < 1$.

*Proof* Proof left to the reader.                                    □

Note that, because $L$ is lower triangular, the equation ★ is easy to solve. We have stated our theorems for real and complex matrices but, of course, in practice, they are used for real matrices.

We shall look again at conditions for the convergence of these methods when we discuss the spectral radius in Section 15.3. (See Lemma 15.3.9 and Exercise 15.3.11.)

## 15.2 Inner products and triangularisation

If we deal with complex inner product spaces, we have a more precise version of Theorem 12.2.5.

**Theorem 15.2.1** *If $V$ is a finite dimensional complex inner product vector space and $\alpha : V \to V$ is linear, we can find an orthonormal basis for $V$ with respect to which $\alpha$ has an upper triangular matrix $A$ (that is to say, a matrix $A = (a_{ij})$ with $a_{ij} = 0$ for $i > j$).*

*Proof* We use induction on the dimension $m$ of $V$. Since every $1 \times 1$ matrix is upper triangular, the result is true when $m = 1$. Suppose that the result is true when $m = n - 1$ and that $V$ has dimension $n$.

Let $a_{11}$ be a root of the characteristic polynomial of $\alpha$ and let $\mathbf{e}_1$ be a corresponding eigenvector of norm 1. Let $W = \text{span}\{\mathbf{e}_1\}$, let $U = W^\perp$ and let $\pi$ be the orthogonal projection of $V$ onto $U$.

Now $(\pi\alpha)|_U$ is a linear map from $U$ to $U$ and $U$ has dimension $n - 1$, so, by the inductive hypothesis, we can find an orthonormal basis $\mathbf{e}_2, \mathbf{e}_3, \ldots, \mathbf{e}_n$ for $U$ with respect to which $(\pi\alpha)|_U$ has an upper triangular matrix. The statement that $(\pi\alpha)|_U$ has an upper triangular matrix means that

$$\pi\alpha\mathbf{e}_j \in \text{span}\{\mathbf{e}_2, \mathbf{e}_3, \ldots, \mathbf{e}_j\} \qquad \qquad ★$$

for $2 \le j \le n$.

Now $\mathbf{e}_1, \mathbf{e}_2, \ldots, \mathbf{e}_n$ form a basis of $V$. But ★ tells us that

$$\alpha\mathbf{e}_j \in \text{span}\{\mathbf{e}_1, \mathbf{e}_2, \ldots, \mathbf{e}_j\}$$

for $2 \leq j \leq n$ and the statement

$$\alpha \mathbf{e}_1 \in \operatorname{span}\{\mathbf{e}_1\}$$

is automatic. Thus $\alpha$ has upper triangular matrix $(a_{ij})$ with respect to the given basis. $\square$

**Exercise 15.2.2** *By considering roots of the characteristic polynomial or otherwise, show, by example, that the result corresponding to Theorem 15.2.1 is false if $V$ is a finite dimensional vector space of dimension greater than 1 over $\mathbb{R}$. What can we say if $\dim V = 1$?*

*Why is there no contradiction between the example asked for in the previous paragraph and the fact that every square matrix has a $QR$ factorisation?*

We know that it is more difficult to deal with $\alpha \in \mathcal{L}(U, U)$ when the characteristic polynomial has repeated roots. The next result suggests one way round this difficulty.

**Theorem 15.2.3** *If $U$ is a finite dimensional complex inner product vector space and $\alpha \in \mathcal{L}(U, U)$, then we can find $\alpha_n \in \mathcal{L}(U, U)$ with $\|\alpha_n - \alpha\| \to 0$ as $n \to \infty$ such that the characteristic polynomials of the $\alpha_n$ have no repeated roots.*

*Proof of Theorem 15.2.3* By Theorem 15.2.1 we can find an orthonormal basis with respect to which $\alpha$ is represented by upper triangular matrix $A$. We can certainly find $d_{jj}^{(n)} \to 0$ such that the $a_{jj} + d_{jj}^{(n)}$ are distinct for each $n$. Let $D_n$ be the diagonal matrix with diagonal entries $d_{jj}^{(n)}$. If we take $\alpha_n$ to be the linear map represented by $A + D_n$, then the characteristic polynomial of $\alpha_n$ will have the distinct roots $a_{jj} + d_{jj}^{(n)}$. Exercise 15.1.13 tells us that $\|\alpha_n - \alpha\| \to 0$ as $n \to \infty$, so we are done. $\square$

If we use Exercise 15.1.13, we can translate Theorem 15.2.3 into a result on matrices.

**Exercise 15.2.4** *If $A = (a_{ij})$ is an $m \times m$ complex matrix, show, using Theorem 15.2.3, that we can find a sequence $A(n) = (a_{ij}(n))$ of $m \times m$ complex matrices such that*

$$\max_{i,j} |a_{ij}(n) - a_{ij}| \to 0 \text{ as } n \to \infty$$

*and the characteristic polynomials of the $A(n)$ have no repeated roots.*

As an example of the use of Theorem 15.2.3, let us produce yet another proof of the Cayley–Hamilton theorem. We need a preliminary lemma.

**Lemma 15.2.5** *Consider the space of $m \times m$ matrices $M_m(\mathbb{C})$. The multiplication map $M_m(\mathbb{C}) \times M_m(\mathbb{C}) \to M_m(\mathbb{C})$ given by $(A, B) \mapsto AB$, the addition map $M_m(\mathbb{C}) \times M_m(\mathbb{C}) \to M_m(\mathbb{C})$ given by $(A, B) \mapsto A + B$ and the scalar multiplication map $\mathbb{C} \times M_m(\mathbb{C}) \to M_m(\mathbb{C})$ given by $(\lambda, A) \to \lambda A$ are all continuous.*

*Proof* We prove that the multiplication map is continuous and leave the other verifications to the reader.

To prove that multiplication is continuous, we observe that, whenever $\|A_n - A\|$, $\|B_n - B\| \to 0$, we have

$$
\begin{aligned}
\|A_n B_n - AB\| &= \|(A - A_n)B + (B - B_n)A_n\| \\
&\leq \|(A - A_n)B\| + \|(B - B_n)A_n\| \\
&\leq \|A - A_n\|\|B\| + \|B - B_n\|\|A_n\| \\
&\leq \|A - A_n\|\|B\| + \|B - B_n\|(\|A\| + \|A - A_n\|) \\
&\to 0 + 0(\|A\| + 0) = 0
\end{aligned}
$$

as $n \to \infty$. This is the desired result. $\qquad\square$

**Theorem 15.2.6 [Cayley–Hamilton for complex matrices]** *If $A$ is an $m \times m$ matrix and*

$$
P_A(t) = \sum_{k=0}^{m} a_k t^k = \det(tI - A),
$$

*we have $P_A(A) = 0$.*

*Proof* By Theorem 15.2.3, we can find a sequence $A_n$ of matrices whose characteristic polynomials have no repeated roots such that $\|A - A_n\| \to 0$ as $n \to \infty$. Since the Cayley–Hamilton theorem is immediate for *diagonal* and so for *diagonalisable* matrices, we know that, setting

$$
\sum_{k=0}^{m} a_k(n)t^k = \det(tI - A_n),
$$

we have

$$
\sum_{k=0}^{m} a_k(n)A_n^k = 0.
$$

Now $a_k(n)$ is some multinomial[3] in the the entries $a_{ij}(n)$ of $A(n)$. Since Exercise 15.1.13 tells us that $a_{ij}(n) \to a_{ij}$, it follows that $a_k(n) \to a_k$. Lemma 15.2.5 now yields

$$
\left\| \sum_{k=0}^{m} a_k A^k \right\| = \left\| \sum_{k=0}^{m} a_k A^k - \sum_{k=0}^{m} a_k(n)A_n^k \right\| \to 0
$$

as $n \to \infty$, so

$$
\left\| \sum_{k=0}^{m} a_k A^k \right\| = 0
$$

and

$$
\sum_{k=0}^{m} a_k A^k = 0. \qquad\square
$$

Before deciding that Theorem 15.2.3 entitles us to neglect the 'special case' of multiple roots, the reader should consider the following informal argument. We are often interested

---

[3] That is to say, some polynomial in several variables.

in $m \times m$ matrices $A$, all of whose entries are real, and how the behaviour of some system (for example, a set of simultaneous linear differential equations) varies as the entries in $A$ vary. Observe that, as $A$ varies continuously, so do the coefficients in the associated characteristic polynomial

$$t^k + \sum_{j=0}^{k-1} a_k t^k = \det(tI - A).$$

As the coefficients in the polynomial vary continuously so do the roots of the polynomial.[4]

Since the coefficients of the characteristic polynomial are real, the roots are either real or occur in conjugate pairs. As the matrix changes to make the number of non-real roots increase by 2, two real roots must come together to form a repeated real root and then separate as conjugate complex roots. When the number of non-real roots reduces by 2 the situation is reversed. Thus, in the interesting case when the system passes from one regime to another, the characteristic polynomial must have a double root.

Of course, this argument only shows that we need to consider Jordan normal forms of a rather simple type. However, we are often interested not in 'general systems' but in 'particular systems' and their 'particularity' may be reflected in a more complicated Jordan normal form.

## 15.3 The spectral radius

When we looked at iterative procedures like Gauss–Siedel for solving systems of linear equations, we were particularly interested in the question of when $\|A^n \mathbf{x}\| \to 0$. The following result gives an almost complete answer.

**Lemma 15.3.1** *If $A$ is a complex $m \times m$ matrix with $m$ distinct eigenvalues, then $\|A^n \mathbf{x}\| \to 0$ as $n \to \infty$ if and only if all the eigenvalues of $A$ have modulus less than 1.*

*Proof* Take a basis $\mathbf{u}_j$ of eigenvectors with associated eigenvalues $\lambda_j$. If $|\lambda_k| \geq 1$ then

$$\|A^n \mathbf{u}_k\| = |\lambda_k|^n \|\mathbf{u}_k\| \nrightarrow 0.$$

On the other hand, if $|\lambda_j| < 1$ for all $1 \leq j \leq m$, then

$$\left\| A^n \sum_{j=1}^m x_j \mathbf{u}_j \right\| = \left\| \sum_{j=1}^m x_j A^n \mathbf{u}_j \right\|$$

$$= \left\| \sum_{j=1}^m x_j \lambda_j^n \mathbf{u}_j \right\|$$

$$\leq \sum_{j=1}^m x_j |\lambda_j|^n \|\mathbf{u}_j\| \to 0$$

as $n \to \infty$. Thus $\|A^n \mathbf{x}\| \to 0$ as $n \to \infty$ for all $\mathbf{x} \in \mathbb{C}^m$. $\qquad \square$

---

[4] This looks obvious and is, indeed, true, but the proof requires complex variable theory. See Exercise 15.5.14 for a proof.

Note that, as the next exercise shows, although the statement that all the eigenvalues of $A$ are small tells us that $\|A^n\mathbf{x}\|$ tends to zero, it does not tell us about the behaviour of $\|A^n\mathbf{x}\|$ when $n$ is small.

**Exercise 15.3.2** *Let*

$$A = \begin{pmatrix} \frac{1}{2} & K \\ 0 & \frac{1}{4} \end{pmatrix}.$$

*What are the eigenvalues of $A$? Show that, given any integer $N \geq 1$ and any $L > 0$, we can find a $K$ such that, taking $\mathbf{x} = (0, 1)^T$, we have $\|A^N\mathbf{x}\| > L$.*

In numerical work, we are frequently only interested in matrices and vectors with real entries. The next exercise shows that this makes no difference.

**Exercise 15.3.3** *Suppose that $A$ is a real $m \times m$ matrix. Show, by considering real and imaginary parts, or otherwise, that $\|A^n\mathbf{z}\| \to 0$ as $n \to \infty$ for all column vectors $\mathbf{z}$ with entries in $\mathbb{C}$ if and only if $\|A^n\mathbf{x}\| \to 0$ as $n \to \infty$ for all column vectors $\mathbf{x}$ with entries in $\mathbb{R}$.*

Life is too short to stuff an olive and I will not blame readers who mutter something about 'special cases' and ignore the rest of this section which deals with the situation when some of the roots of the characteristic polynomial coincide.

We shall prove the following neat result.

**Theorem 15.3.4** *If $U$ is a finite dimensional complex inner product space and $\alpha \in \mathcal{L}(U, U)$, then*

$$\|\alpha^n\|^{1/n} \to \rho(\alpha) \quad as \ n \to \infty,$$

*where $\rho(\alpha)$ is the largest absolute value of the eigenvalues of $\alpha$.*

Translation gives the equivalent matrix result.

**Lemma 15.3.5** *If $A$ is an $m \times m$ complex matrix, then*

$$\|A^n\|^{1/n} \to \rho(A) \quad as \ n \to \infty,$$

*where $\rho(A)$ is the largest absolute value of the eigenvalues of $A$.*

We call $\rho(\alpha)$ the *spectral radius* of $\alpha$. At this level, we hardly need a special name, but a generalisation of the concept plays an important role in more advanced work. Here is the result of Lemma 15.3.1 without the restriction on the roots of the characteristic equation.

**Theorem 15.3.6** *If $A$ is a complex $m \times m$ matrix, then $\|A^n\mathbf{x}\| \to 0$ as $n \to \infty$ if and only if all its eigenvalues have modulus less than 1.*

*Proof of Theorem 15.3.6 using Theorem 15.3.4* Suppose that $\rho(A) < 1$. Then we can find a $\mu$ with $1 > \mu > \rho(A)$. Since $\|A^n\|^{1/n} \to \rho(A)$, we can find an $N$ such that $\|A^n\|^{1/n} \leq \mu$

for all $n \geq N$. Thus, provided $n \geq N$,

$$\|A^n \mathbf{x}\| \leq \|A^n\| \|\mathbf{x}\| \leq \mu^n \|\mathbf{x}\| \to 0$$

as $n \to \infty$.

If $\rho(A) \geq 1$, then, choosing an eigenvector $\mathbf{u}$ with eigenvalue having absolute value $\rho(A)$, we observe that $\|A^n \mathbf{u}\| \nrightarrow 0$ as $n \to \infty$. $\qquad\square$

In Exercise 15.5.1, we outline an alternative proof of Theorem 15.3.6, using the Jordan canonical form rather than the spectral radius.

Our proof of Theorem 15.3.4 makes use of the following simple results which the reader is invited to check explicitly.

**Exercise 15.3.7** (*i*) *Suppose that $r$ and $s$ are non-negative integers. Let $A = (a_{ij})$ and $B = (b_{ij})$ be two $m \times m$ matrices such that $a_{ij} = 0$ whenever $1 \leq i \leq j + r$, and $b_{ij} = 0$ whenever $1 \leq i \leq j + s$. If $C = (c_{ij})$ is given by $C = AB$, show that $c_{ij} = 0$ whenever $1 \leq i \leq j + r + s + 1$.*

(*ii*) *If $D$ is a diagonal matrix show that $\|D\| = \rho(D)$.*

**Exercise 15.3.8** *By means of a proof or counterexample, establish whether the result of Exercise 15.3.7 remains true if we drop the restriction that $r$ and $s$ should be non-negative.*

*Proof of Theorem 15.3.4* Let $U$ have dimension $m$. By Theorem 15.2.1, we can find an orthonormal basis for $U$ with respect to which $\alpha$ has an upper triangular $m \times m$ matrix $A$ (that is to say, a matrix $A = (a_{ij})$ with $a_{ij} = 0$ for $i > j$). We need to show that $\|A^n\|^{1/n} \to \rho(A)$.

To this end, write $A = B + D$ where $D$ is a diagonal matrix and $B = (b_{ij})$ is strictly upper triangular (that is to say, $b_{ij} = 0$ whenever $i \geq j$). Since the eigenvalues of a triangular matrix are its diagonal entries,

$$\rho(A) = \rho(D) = \|D\|.$$

If $D = 0$, then $\rho(A) = 0$ and $A^n = 0$ for $n \geq m - 1$, so we are done. From now on, we suppose that $D \neq 0$ and so $\|D\| > 0$.

Let $n \geq m$. If $m \leq k \leq n$, it follows from Exercise 15.3.7 that the product of $k$ copies of $B$ and $n - k$ copies of $D$ taken in any order is 0. If $0 \leq k \leq m - 1$ we can multiply $k$ copies of $B$ and $n - k$ copies of $D$ in $\binom{n}{k}$ different orders, but, in each case, the product $C$, say, will satisfy $\|C\| \leq \|B\|^k \|D\|^{n-k}$. Thus

$$\|A^n\| = \|(B + D)^n\| \leq \sum_{k=0}^{m-1} \binom{n}{k} \|B\|^k \|D\|^{n-k}.$$

It follows that

$$\|A^n\| \leq K n^{m-1} \|D\|^n$$

for some $K$ depending on $m$, $B$ and $D$, but not on $n$.

If $\lambda$ is an eigenvalue of $A$ with largest absolute value and $\mathbf{u}$ is an associated eigenvalue, then

$$\|A^n\|\|\mathbf{u}\| \geq \|A^n\mathbf{u}\| = \|\lambda^n\mathbf{u}\| = |\lambda|^n\|\mathbf{u}\|,$$

so $\|A^n\| \geq |\lambda|^n = \rho(A)^n$. We thus have

$$\rho(A)^n \leq \|A^n\| \leq Kn^{m-1}\|D\|^n = Kn^{m-1}\rho(A)^n,$$

whence

$$\rho(A) \leq \|A^n\|^{1/n} \leq K^{1/n}n^{(m-1)/n}\rho(A) \to \rho(A)$$

and $\|A^n\|^{1/n} \to \rho(A)$ as required. (We use the result from analysis which states that $n^{1/n} \to 1$ as $n \to \infty$.) $\qquad\square$

Theorem 15.3.6 gives further information on the Gauss–Jacobi, Gauss–Siedel and similar iterative methods.

**Lemma 15.3.9 [Gauss–Siedel revisited]** *Let $A$ be an $n \times n$ matrix over $\mathbb{F}$ with non-zero diagonal entries and $\mathbf{b} \in \mathbb{F}^n$ a column vector. Suppose that the equation*

$$A\mathbf{x} = \mathbf{b}$$

*has the solution $\mathbf{x}^*$. Let us write $A = L + U$ where $L$ is a lower triangular matrix and $U$ a strictly upper triangular matrix (that is to say, an upper triangular matrix with all diagonal terms zero). If $\mathbf{x}_0 \in \mathbb{F}^n$ and*

$$L\mathbf{x}_{j+1} = \mathbf{b} - U\mathbf{x}_j,$$

*then $\|\mathbf{x}_j - \mathbf{x}^*\| \to 0$ whenever $\rho(L^{-1}U) < 1$. If $\rho(L^{-1}U) \geq 1$ we can choose $\mathbf{x}_0$ so that $\|\mathbf{x}_j - \mathbf{x}^*\| \not\to 0$.*

*Proof* Left to reader. Note that Exercise 15.3.3 tells that, whether we work over $\mathbb{C}$ or $\mathbb{R}$, it is the size of the spectral radius which tells us whether we always have convergence. $\qquad\square$

**Lemma 15.3.10** *Let $A$ be an $n \times n$ matrix with*

$$|a_{rr}| > \sum_{s \neq r}|a_{rs}| \text{ for all } 1 \leq r \leq n.$$

*Then the Gauss–Siedel method described in the previous lemma will converge. (More specifically $\|\mathbf{x}_j - \mathbf{x}^*\| \to 0$ as $j \to \infty$.)*

*Proof* Suppose, if possible, that we can find an eigenvector $\mathbf{y}$ of $L^{-1}U$ with eigenvalue $\lambda$ such that $|\lambda| \geq 1$. Then $L^{-1}U\mathbf{y} = \lambda\mathbf{y}$ and so $U\mathbf{y} = \lambda L\mathbf{y}$. Thus

$$a_{ii}y_i = \lambda^{-1}\sum_{j=1}^{i-1}a_{ij}y_j - \sum_{j=i+1}^{n}a_{ij}y_j$$

and so

$$|a_{ii}||y_i| \le \sum_{j \neq i} |a_{ij}||y_j|$$

for each $i$.

Summing over all $i$ and interchanging the order of summation, we get (bearing in mind that $\mathbf{y} \neq \mathbf{0}$ and so $|y_j| > 0$ for at least one value of $j$)

$$\sum_{i=1}^{n} |a_{ii}||y_i| < \sum_{i=1}^{n} \sum_{j \neq i} |a_{ij}||y_j| = \sum_{j=1}^{n} \sum_{i \neq j} |a_{ij}||y_j|$$

$$= \sum_{j=1}^{n} |y_j| \sum_{i \neq j} |a_{ij}| < \sum_{j=1}^{n} |y_j||a_{jj}| = \sum_{i=1}^{n} |a_{ii}||y_i|$$

which is absurd.

Thus all the eigenvalues of $A$ have absolute value less than 1 and Lemma 15.3.9 applies. $\qquad \square$

**Exercise 15.3.11** *State and prove a result corresponding to Lemma 15.3.9 for the Gauss–Jacobi method (see Lemma 15.1.14) and use it to show that, if $A$ is an $n \times n$ matrix with*

$$|a_{rs}| > \sum_{s \neq r} |a_{rs}| \quad for \ all \ 1 \le r \le n,$$

*then the Gauss–Jacobi method applied to the system $A\mathbf{x} = \mathbf{b}$ will converge.*

## 15.4 Normal maps

In Exercise 8.4.18 the reader was invited to show, in effect, that a Hermitian map is characterised by the fact that it has orthonormal eigenvectors with associated real eigenvalues. Here is an alternative (though, in my view, less instructive) proof using Theorem 15.2.1.

**Theorem 15.4.1** *If $V$ is a finite dimensional complex inner product vector space and $\alpha : V \to V$ is linear, then $\alpha^* = \alpha$ if and only if we can find a an orthonormal basis for $V$ with respect to which $\alpha$ has a diagonal matrix with real diagonal entries.*

*Proof* Sufficiency is obvious. If $\alpha$ is represented with respect to some orthonormal basis by the diagonal matrix $D$ with real entries, then $\alpha^*$ is represented by $D^* = D$ and so $\alpha = \alpha^*$.

To prove necessity, observe that, by Theorem 15.2.1, we can find an orthonormal basis for $V$ with respect to which $\alpha$ has an upper triangular matrix $A$ (that is to say, a matrix $A = (a_{ij})$ with $a_{ij} = 0$ for $i > j$). Now $\alpha^* = \alpha$, so $A^* = A$ and

$$a_{ij} = a_{ji}^* = 0 \quad \text{for } j < i$$

whilst $a_{jj}^* = a_{jj}$ for all $j$. Thus $A$ is, in fact, diagonal with real diagonal entries. $\qquad \square$

It is natural to ask which endomorphisms of a complex inner product space have an associated orthogonal basis of eigenvalues. Although it might take some time and many trial calculations, it is possible to imagine how the answer could have been obtained.

**Definition 15.4.2** *If V is a complex inner product space, we say that $\alpha \in \mathcal{L}(V, V)$ is normal[5] if $\alpha^*\alpha = \alpha\alpha^*$.*

**Exercise 15.4.3** *Let V be a finite dimensional complex inner product space with an orthonormal basis. Show that $\alpha \in \mathcal{L}(V, V)$ is normal if and only if its matrix A relative to the given basis satisfies $A^*A = AA^*$. (Such a matrix A is naturally called a normal matrix.)*

**Theorem 15.4.4** *If V is a finite dimensional complex inner product vector space and $\alpha : V \to V$ is linear, then $\alpha$ is normal if and only if we can find a an orthonormal basis for V with respect to which $\alpha$ has a diagonal matrix.*

*Proof* Sufficiency is obvious. If $\alpha$ is represented with respect to some orthonormal basis by the diagonal matrix $D$, then $\alpha^*$ is represented by $D^* = D$. Since diagonal matrices commute, $DD^* = D^*D$ and so $\alpha\alpha^* = \alpha^*\alpha$.

To prove necessity, observe that, by Theorem 15.2.1, we can find an orthonormal basis for V with respect to which $\alpha$ has an upper triangular matrix $A$ (that is to say, a matrix $A = (a_{ij})$ with $a_{ij} = 0$ for $i > j$). Now $\alpha\alpha^* = \alpha^*\alpha$ so $AA^* = A^*A$ and

$$\sum_{j=1}^{n} a_{rj}a_{sj}^* = \sum_{j=1}^{n} a_{jr}^* a_{js} = \sum_{j=1}^{n} a_{js}a_{jr}^*$$

for all $n \geq r, s \geq 1$. It follows that

$$\sum_{j \geq \max\{r,s\}} a_{rj}a_{sj}^* = \sum_{j \leq \min\{r,s\}} a_{js}a_{jr}^*$$

for all $n \geq r, s \geq 1$. In particular, if we take $r = s = 1$, we get

$$\sum_{j=1}^{n} a_{1j}^* a_{1j} = a_{11}a_{11}^*,$$

so

$$\sum_{j=2}^{n} |a_{1j}|^2 = 0.$$

Thus $a_{1j} = 0$ for $j \geq 2$.

If we now set $r = s = 2$ and note that $a_{12} = 0$, we get

$$\sum_{j=3}^{n} |a_{2j}|^2 = 0,$$

---

[5] The use of the word normal here and elsewhere is a testament to the deeply held human belief that, by declaring something to be normal, we make it normal.

so $a_{2j} = 0$ for $j \geq 3$. Continuing inductively, we obtain $a_{ij} = 0$ for all $j > i$ and so $A$ is diagonal. $\qquad\square$

If our object were only to prove results and not to understand them, we could leave things as they stand. However, I think that it is more natural to seek a proof along the lines of our earlier proofs of the diagonalisation of symmetric and Hermitian maps. (Notice that, if we try to prove part (iii) of Theorem 15.4.5, we are almost automatically led back to part (ii) and, if we try to prove part (ii), we are, after a certain amount of head scratching, led back to part (i).)

**Theorem 15.4.5** *Suppose that $V$ is a finite dimensional complex inner product vector space and $\alpha : V \to V$ is normal.*

*(i) If $\mathbf{e}$ is an eigenvector of $\alpha$ with eigenvalue $0$, then $\mathbf{e}$ is an eigenvalue of $\alpha^*$ with eigenvalue $0$.*

*(ii) If $\mathbf{e}$ is an eigenvector of $\alpha$ with eigenvalue $\lambda$, then $\mathbf{e}$ is an eigenvalue of $\alpha^*$ with eigenvalue $\lambda^*$.*

*(iii) $\alpha$ has an orthonormal basis of eigenvalues.*

*Proof* (i) Observe that

$$\alpha\mathbf{e} = \mathbf{0} \Rightarrow \langle \alpha\mathbf{e}, \alpha\mathbf{e} \rangle = 0 \Rightarrow \langle \mathbf{e}, \alpha^*\alpha\mathbf{e} \rangle = 0$$
$$\Rightarrow \langle \mathbf{e}, \alpha\alpha^*\mathbf{e} \rangle = 0 \Rightarrow \langle \alpha\alpha^*\mathbf{e}, \mathbf{e} \rangle = 0$$
$$\Rightarrow \langle \alpha^*\mathbf{e}, \alpha^*\mathbf{e} \rangle = 0 \Rightarrow \alpha^*\mathbf{e} = \mathbf{0}.$$

(ii) If $\mathbf{e}$ is an eigenvalue of $\alpha$ with eigenvalue $\lambda$, then $\mathbf{e}$ is an eigenvalue of $\beta = \lambda\iota - \alpha$ with associated eigenvalue $0$.

Since $\beta^* = \lambda^*\iota - \alpha^*$, we have

$$\beta\beta^* = (\lambda\iota - \alpha)(\lambda^*\iota - \alpha^*) = |\lambda|^2\iota - \lambda^*\alpha - \lambda\alpha^* + \alpha\alpha^*$$
$$= (\lambda^*\iota - \alpha^*)(\lambda\iota - \alpha) = \beta^*\beta.$$

Thus $\beta$ is normal, so, by (i), $\mathbf{e}$ is an eigenvalue of $\beta^*$ with eigenvalue $0$. It follows that $\mathbf{e}$ is an eigenvalue of $\alpha^*$ with eigenvalue $\lambda^*$.

(iii) We follow the pattern set out in the proof of Theorem 8.2.5 by performing an induction on the dimension $n$ of $V$.

If $n = 1$, then, since every $1 \times 1$ matrix is diagonal, the result is trivial.

Suppose now that the result is true for $n = m$, that $V$ is an $m + 1$ dimensional complex inner product space and that $\alpha \in \mathcal{L}(V, V)$ is normal. We know that the characteristic polynomial must have a root, so we can find an eigenvalue $\lambda_1 \in \mathbb{C}$ and a corresponding eigenvector $\mathbf{e}_1$ of norm 1. Consider the subspace

$$\mathbf{e}_1^\perp = \{\mathbf{u} : \langle \mathbf{e}_1, \mathbf{u} \rangle = 0\}.$$

We observe (and this is the key to the proof) that

$$\mathbf{u} \in \mathbf{e}_1^\perp \Rightarrow \langle \mathbf{e}_1, \alpha\mathbf{u} \rangle = \langle \alpha^*\mathbf{e}_1, \mathbf{u} \rangle = \langle \lambda_1^*\mathbf{e}_1, \mathbf{u} \rangle = \lambda_1^*\langle \mathbf{e}_1, \mathbf{u} \rangle = 0 \Rightarrow \alpha\mathbf{u} \in \mathbf{e}_1^\perp.$$

Thus we can define $\alpha|_{\mathbf{e}_1^\perp} : \mathbf{e}_1^\perp \to \mathbf{e}_1^\perp$ to be the restriction of $\alpha$ to $\mathbf{e}_1^\perp$. We observe that $\alpha|_{\mathbf{e}_1^\perp}$ is normal and $\mathbf{e}_1^\perp$ has dimension $m$ so, by the inductive hypothesis, we can find $m$ orthonormal eigenvectors of $\alpha|_{\mathbf{e}_1^\perp}$ in $\mathbf{e}_1^\perp$. Let us call them $\mathbf{e}_2, \mathbf{e}_3, \ldots, \mathbf{e}_{m+1}$. We observe that $\mathbf{e}_1, \mathbf{e}_2, \ldots, \mathbf{e}_{m+1}$ are orthonormal eigenvectors of $\alpha$ and so $\alpha$ is diagonalisable. The induction is complete. $\qquad\square$

(We give yet another proof of Theorem 15.4.4 in Exercise 15.5.9.)

**Exercise 15.4.6 [A spectral theorem]**[6] *Let $U$ be a finite dimensional inner product space over $\mathbb{C}$ and $\alpha$ an endomorphism. Show that $\alpha$ is normal if and only if there exist distinct non-zero $\lambda_j \in \mathbb{C}$ and orthogonal projections $\pi_j$ such that $\pi_k \pi_j = 0$ when $k \neq j$,*

$$\iota = \pi_1 + \pi_2 + \cdots + \pi_m \quad \text{and} \quad \alpha = \lambda_1 \pi_1 + \lambda_2 \pi_2 + \cdots + \lambda_m \pi_m.$$

*State the appropriate theorem for self-adjoint linear maps.*

The unitary maps (that is to say, the linear maps $\alpha$ with $\alpha^* \alpha = \iota$) are normal.

**Exercise 15.4.7** *Explain why the unitary maps are normal. Let $\alpha$ be an automorphism of a finite dimensional complex inner product space. Show that $\beta = \alpha^{-1} \alpha^*$ is unitary if and only if $\alpha$ is normal.*

**Theorem 15.4.8** *If $V$ is a finite dimensional complex inner product vector space, then $\alpha \in \mathcal{L}(V, V)$ is unitary if and only if we can find an orthonormal basis for $V$ with respect to which $\alpha$ has a diagonal matrix*

$$U = \begin{pmatrix} e^{i\theta_1} & 0 & 0 & \cdots & 0 \\ 0 & e^{i\theta_2} & 0 & \cdots & 0 \\ 0 & 0 & e^{i\theta_3} & \cdots & 0 \\ \vdots & \vdots & \vdots & & \vdots \\ 0 & 0 & 0 & \cdots & e^{i\theta_n} \end{pmatrix}$$

*where $\theta_1, \theta_2, \ldots, \theta_n \in \mathbb{R}$.*

*Proof* Observe that, if $U$ is the matrix written out in the statement of our theorem, we have

$$U^* = \begin{pmatrix} e^{-i\theta_1} & 0 & 0 & \cdots & 0 \\ 0 & e^{-i\theta_2} & 0 & \cdots & 0 \\ 0 & 0 & e^{-i\theta_3} & \cdots & 0 \\ \vdots & \vdots & \vdots & & \vdots \\ 0 & 0 & 0 & \cdots & e^{-i\theta_n} \end{pmatrix}$$

so $UU^* = I$. Thus, if $\alpha$ is represented by $U$ with respect to some orthonormal basis, it follows that $\alpha\alpha^* = \iota$ and $\alpha$ is unitary.

---

[6] Calling this a spectral theorem is rather like referring to Mr and Mrs Smith as royalty on the grounds that Mr Smith is 12th cousin to the Queen.

We now prove the converse. If $\alpha$ is unitary, then $\alpha$ is invertible with $\alpha^{-1} = \alpha^*$. Thus

$$\alpha\alpha^* = \iota = \alpha^*\alpha$$

and $\alpha$ is normal. It follows that $\alpha$ has an orthonormal basis of eigenvectors $\mathbf{e}_j$. If $\lambda_j$ is the eigenvalue associated with $\mathbf{e}_j$, then, since unitary maps preserve norms,

$$|\lambda_j| = |\lambda_j|\|\mathbf{e}_j\| = \|\lambda_j\mathbf{e}_j\| = \|\alpha\mathbf{e}_j\| = \|\mathbf{e}_j\| = 1$$

so $\lambda_j = e^{i\theta_j}$ for some real $\theta_j$ and we are done. $\qquad\square$

The next exercise, which is for amusement only, points out an interesting difference between the group of norm preserving linear maps for $\mathbb{R}^n$ and the group of norm preserving linear maps for $\mathbb{C}^n$.

**Exercise 15.4.9** *(i) Let V be a finite dimensional complex inner product space and consider* $\mathcal{L}(V, V)$ *with the operator norm. By considering diagonal matrices whose entries are of the form* $e^{i\theta_j t}$, *or otherwise, show that, if* $\alpha \in \mathcal{L}(V, V)$ *is unitary, we can find a continuous map*

$$f : [0, 1] \to \mathcal{L}(V, V)$$

*such that* $f(t)$ *is unitary for all* $0 \le t \le 1$, $f(0) = \iota$ *and* $f(1) = \alpha$.
*Show that, if* $\beta, \gamma \in \mathcal{L}(V, V)$ *are unitary, we can find a continuous map*

$$g : [0, 1] \to \mathcal{L}(V, V)$$

*such that* $g(t)$ *is unitary for all* $0 \le t \le 1$, $g(0) = \beta$ *and* $g(1) = \gamma$.
*(ii) Let U be a finite dimensional real inner product space and consider* $\mathcal{L}(U, U)$ *with the operator norm. We take* $\rho$ *to be a reflection*

$$\rho(\mathbf{x}) = \mathbf{x} - 2\langle\mathbf{x}, \mathbf{n}\rangle\mathbf{n}$$

*with* $\mathbf{n}$ *a unit vector.*
*If* $f : [0, 1] \to \mathcal{L}(U, U)$ *is continuous and* $f(0) = \iota$, $f(1) = \rho$, *show, by considering the map* $t \mapsto \det f(t)$, *or otherwise, that there exists an* $s \in [0, 1]$ *with* $f(s)$ *not invertible.*

## 15.5 Further exercises

**Exercise 15.5.1** The object of this question is to give a more algebraic proof of Theorem 15.3.6. This states that, if $U$ is a finite dimensional complex inner product vector space over $\mathbb{C}$ and $\alpha \in \mathcal{L}(U, U)$ is such that all the roots of its characteristic polynomial have modulus strictly less than 1, then $\|\alpha^n\mathbf{x}\| \to 0$ as $n \to \infty$ for all $\mathbf{x} \in U$.

(i) Suppose that $\alpha$ satisfies the hypothesis and $\mathbf{u}_1, \mathbf{u}_2, \ldots, \mathbf{u}_m$ is a basis for $U$ (but not necessarily an orthonormal basis). Show that the required conclusion will follow if we can show that $\|\alpha^n(\mathbf{u}_j)\| \to 0$ as $n \to \infty$ for all $1 \le j \le m$.

(ii) Suppose that $\beta \in \mathcal{L}(U, U)$ and

$$\beta \mathbf{u}_j = \begin{cases} \lambda \mathbf{u}_j + \mathbf{u}_{j+1} & \text{if } 1 \le j \le k - 1, \\ \lambda \mathbf{u}_k & \text{if } j = k. \end{cases}$$

Show that there exists some constant $K$ (depending on the $\mathbf{u}_j$ but not on $n$) such that

$$\|\beta^n \mathbf{u}_1\| \le K n^{k-1} |\lambda|^n$$

and deduce that, if $|\lambda| < 1$,

$$\|\beta^n \mathbf{u}_1\| \to 0$$

as $n \to \infty$.

(iii) Use the Jordan normal form to prove the result stated at the beginning of the question.

**Exercise 15.5.2** Show that $\langle A, B \rangle = \text{Tr}(AB^*)$ defines an inner product on the vector space of $n \times n$ matrices with complex matrices. Deduce that

$$\text{Tr}(AB^*)^2 \le \text{Tr}(AA^*) \text{Tr}(BB^*),$$

giving necessary and sufficient conditions for equality.

If $n = 2$ and

$$C = \begin{pmatrix} 1 & 1 \\ 2 & 0 \end{pmatrix}, \quad D = \begin{pmatrix} 1 & 1 \\ 0 & -2 \end{pmatrix},$$

find an orthonormal basis for $(\text{span}\{C, D\})^\perp$.

Suppose that $A$ is normal (that is to say, $AA^* = A^*A$). By considering $\langle A^*B - B^*A, A^*B - B^*A \rangle$, show that if $B$ commutes with $A$, then $B$ commutes with $A^*$.

**Exercise 15.5.3** Let $U$ be an $n$-dimensional complex vector space and let $\alpha \in \mathcal{L}(U, U)$. Let $\lambda_1, \lambda_2, \ldots, \lambda_n$ be the roots of the characteristic polynomial of $\alpha$ (multiple roots counted multiply) and let $\mu_1, \mu_2, \ldots, \mu_n$ be the roots of the characteristic polynomial of $\alpha\alpha^*$ (multiple roots counted multiply). Explain why all the $\mu_j$ are real and positive.

By using triangularisation, or otherwise, show that

$$|\lambda_1|^2 + |\lambda_2|^2 + \cdots + |\lambda_n|^2 \le \mu_1 + \mu_2 + \cdots + \mu_n$$

with equality if and only if $\alpha$ is normal.

**Exercise 15.5.4** Let $U$ be an $n$-dimensional complex vector space and let $\alpha \in \mathcal{L}(U, U)$. Show that $\alpha$ is normal if and only if we can find a polynomial such that $\alpha^* = P(\alpha)$.

**Exercise 15.5.5** Consider the matrix

$$A = \begin{pmatrix} \frac{1}{\mu} & 1 & 0 & 0 \\ -1 & \frac{1}{\mu} & 1 & 0 \\ 0 & -1 & \frac{1}{\mu} & 1 \\ 0 & 0 & -1 & \frac{1}{\mu} \end{pmatrix}$$

with $\mu$ real and non-zero. Construct the appropriate matrices for the solution of $A\mathbf{x} = \mathbf{b}$ by the Gauss–Jacobi and by the Gauss–Seidel methods.

Determine the range of $\mu$ for which each of the two procedures converges.

**Exercise 15.5.6 [The parallelogram law revisited]** If $\langle \, , \, \rangle$ is an inner product on a vector space $V$ over $\mathbb{F}$ and we define $\|\mathbf{u}\|_2 = \langle \mathbf{u}, \mathbf{u} \rangle^{1/2}$ (taking the positive square root), show that

$$\|\mathbf{a} + \mathbf{b}\|_2^2 + \|\mathbf{a} - \mathbf{b}\|_2^2 = 2(\|\mathbf{a}\|_2^2 + \|\mathbf{b}\|_2^2)$$

for all $\mathbf{a}, \mathbf{b} \in V$.

Show that if $V$ is a finite dimensional space over $\mathbb{F}$ of dimension at least 2 and we use the operator norm $\| \; \|$, there exist $T, S \in \mathcal{L}(V, V)$ such that

$$\|T + S\|^2 + \|T - S\|^2 \neq 2(\|T\|^2 + \|S\|^2).$$

Thus the operator norm does not come from an inner product.

**Exercise 15.5.7** The object of this question is to give a proof of the Riesz representation theorem (Theorem 14.2.4) which has some chance of carrying over to an appropriate infinite dimensional context. Naturally, it requires some analysis. We work in $\mathbb{R}^n$ with the usual inner product.

Suppose that $\alpha : \mathbb{R}^n \to \mathbb{R}$ is linear. Show, by using the operator norm, or otherwise, that

$$\|\mathbf{x}_n - \mathbf{x}\| \to 0 \Rightarrow \alpha \mathbf{x}_n \to \alpha \mathbf{x}$$

as $n \to \infty$. Deduce that, if we write

$$\Pi = \{\mathbf{x} : \alpha \mathbf{x} = 0\},$$

then

$$\mathbf{x}_n \in \Pi, \|\mathbf{x}_n - \mathbf{x}\| \to 0 \Rightarrow \mathbf{x} \in \Pi.$$

Now suppose that $\alpha \neq 0$. Explain why we can find $\mathbf{c} \notin \Pi$ and why

$$\{\|\mathbf{x} - \mathbf{c}\| : \mathbf{x} \in \Pi\}$$

is a non-empty subset of $\mathbb{R}$ bounded below by 0.

Set $M = \inf\{\|\mathbf{x} - \mathbf{c}\| : \mathbf{x} \in \Pi\}$. Explain why we can find $\mathbf{y}_n \in \Pi$ with $\|\mathbf{y}_n - \mathbf{c}\| \to M$. The Bolzano–Weierstrass theorem for $\mathbb{R}^n$ tells us that every bounded sequence has a convergent subsequence. Deduce that we can find $\mathbf{x}_n \in \Pi$ and a $\mathbf{d} \in \Pi$ such that $\|\mathbf{x}_n - \mathbf{c}\| \to M$ and $\|\mathbf{x}_n - \mathbf{d}\| \to 0$.

Show that $\mathbf{d} \in \Pi$ and $\|\mathbf{d} - \mathbf{c}\| = M$. Set $\mathbf{b} = \mathbf{c} - \mathbf{d}$. If $\mathbf{u} \in \Pi$, explain why

$$\|\mathbf{b} + \eta \mathbf{u}\| \geq \|\mathbf{b}\|$$

for all real $\eta$. By squaring both sides and considering what happens when $\eta$ take very small positive and negative values, show that $\langle \mathbf{a}, \mathbf{u} \rangle = 0$ for all $\mathbf{u} \in \Pi$.

Deduce the Riesz representation theorem.

[Exercise 8.5.8 runs through a similar argument.]

**Exercise 15.5.8** In this question we work with an inner product on a finite dimensional complex vector space $U$. However, the results can be extended to more general situations so you should prove the results without using bases.

(i) If $\alpha \in \mathcal{L}(U, U)$, show that $\alpha^*\alpha = 0 \Rightarrow \alpha = 0$.

(ii) If $\alpha$ and $\beta$ are Hermitian, show that $\alpha\beta = 0 \Rightarrow \beta\alpha = 0$.

(iii) Using (i) and (ii), or otherwise, show that, if $\phi$ and $\psi$ are normal, $\phi\psi = 0 \Rightarrow \psi\phi = 0$.

**Exercise 15.5.9 [Simultaneous diagonalisation for Hermitian maps]** Suppose that $U$ is a complex inner product $n$-dimensional vector space and $\alpha$ and $\beta$ are Hermitian endomorphisms. Show that there exists an orthonormal basis $e_1, e_2, \ldots, e_n$ of $U$ such that each $e_j$ is an eigenvector of both $\alpha$ and $\beta$ if and only if $\alpha\beta = \beta\alpha$.

If $\gamma$ is a normal endomorphism, show, by considering $\alpha = 2^{-1}(\gamma + \gamma^*)$ and $\beta = 2^{-1}i(\gamma - \gamma^*)$ that there exists an orthonormal basis $e_1, e_2, \ldots, e_n$ of $U$ consisting of eigenvectors of $\gamma$. (We thus have another proof that normal maps are diagonalisable.)

**Exercise 15.5.10 [Square roots]** In this question and the next we look at 'square roots' of linear maps. We take $U$ to be a vector space of dimension $n$ over $\mathbb{F}$.

(i) Let $U$ be a finite dimensional vector space over $\mathbb{F}$. Suppose that $\alpha$, $\beta : U \to U$ are linear maps with $\beta^2 = \alpha$. By observing that $\alpha\beta = \beta^3$, or otherwise, show that $\alpha\beta = \beta\alpha$.

(ii) Suppose that $\alpha$, $\beta : U \to U$ are linear maps with $\alpha\beta = \beta\alpha$. If $\alpha$ has $n$ distinct eigenvalues, show that every eigenvector of $\alpha$ is an eigenvector of $\beta$. Deduce that there is a basis for $\mathbb{F}^n$ with respect to which the matrices associated with $\alpha$ and $\beta$ are diagonal.

(iii) Let $\mathbb{F} = \mathbb{C}$. If $\alpha : U \to U$ is a linear map with $n$ distinct eigenvalues, show, by considering an appropriate basis, or otherwise, that there exists a linear map $\beta : U \to U$ with $\beta^2 = \alpha$. Show that, if zero is not an eigenvalue of $\alpha$, the equation $\beta^2 = \alpha$ has exactly $2^n$ distinct solutions. (Part (ii) may be useful in showing that there are no more than $2^n$ solutions.) What happens if $\alpha$ has zero as an eigenvalue?

(iv) Let $\mathbb{F} = \mathbb{R}$. Write down a $2 \times 2$ symmetric matrix $A$ with two distinct eigenvalues such that there is no real $2 \times 2$ matrix $B$ with $B^2 = A$. Explain why this is so.

(v) Consider the $2 \times 2$ real matrix

$$R_\theta = \begin{pmatrix} \cos\theta & \sin\theta \\ \sin\theta & -\cos\theta \end{pmatrix}.$$

Show that $R_\theta^2 = I$ for all $\theta$. What geometric fact does this reflect? Why does this result not contradict (iii)?

(vi) Let $\mathbb{F} = \mathbb{C}$. Give, with proof, a $2 \times 2$ matrix $A$ such that there does not exist a matrix $B$ with $B^2 = A$. (Hint: What is our usual choice of a problem $2 \times 2$ matrix?) Why does your result not contradict (iii)?

(vii) Run through this exercise with square roots replaced by cube roots. (Part (iv) will need to be rethought. For an example where there exist infinitely many cube roots, it may be helpful to consider maps in $SO(\mathbb{R}^3)$.)

**Exercise 15.5.11** [**Square roots of positive semi-definite linear maps**] Throughout this question, $U$ is a finite dimensional inner product vector space over $\mathbb{F}$. A self-adjoint map $\alpha : U \to U$ is called *positive semi-definite* if $\langle \alpha \mathbf{u}, \alpha \mathbf{u} \rangle \geq 0$ for all $\mathbf{u} \in U$. (This concept is discussed further in Section 16.3.)

(i) Suppose that $\alpha, \beta : U \to U$ are self-adjoint linear maps with $\alpha\beta = \beta\alpha$. If $\alpha$ has an eigenvalue $\lambda$ and we write

$$E_\lambda = \{\mathbf{u} \in U \,:\, \alpha\mathbf{u} = \lambda\mathbf{u}\}$$

for the associated eigenspace, show that $\beta E_\lambda \subseteq E_\lambda$. Deduce that we can find an orthonormal basis for $E_\lambda$ consisting of eigenvectors of $\beta$. Conclude that we can find an orthonormal basis for $U$ consisting of vectors which are eigenvectors for both $\alpha$ and $\beta$.

(ii) Suppose that $\alpha : U \to U$ is a positive semi-definite linear map. Show that there is a unique positive semi-definite $\beta$ such that $\beta^2 = \alpha$.

(iii) Briefly discuss the differences between the results of this question and Exercise 15.5.10.

**Exercise 15.5.12** (We use the notation and results of Exercise 15.5.11.)

(i) Let us say that a self-adjoint map $\alpha$ is strictly positive definite if $\langle \alpha \mathbf{u}, \alpha \mathbf{u} \rangle > 0$ for all non-zero $\mathbf{u} \in U$. Show that a positive semi-definite symmetric linear map is strictly positive definite if and only if it is invertible. Show that the positive square root of a strictly positive definite linear map is strictly positive definite.

(ii) If $\alpha : U \to U$ is an invertible linear map, show that $\alpha^*\alpha$ is a strictly positive self-adjoint map. Hence, or otherwise, show that there is a unique unitary map $\gamma$ such that $\alpha = \gamma\beta$ where $\beta$ is strictly positive definite.

(iii) State the results of part (ii) in terms of $n \times n$ matrices. If $n = 1$ and we work over $\mathbb{C}$, to what elementary result on the representation of complex numbers does the result correspond?

**Exercise 15.5.13** (i) By using the Bolzano–Weierstrass theorem, or otherwise, show that, if $A(m) = \big(a_{ij}(m)\big)$ is a sequence of $n \times n$ complex matrices with $|a_{ij}(m)| \leq K$ for all $1 \leq i, j \leq n$ and all $m$, we can find a sequence $m(k) \to \infty$ and a matrix $A = (a_{ij})$ such that $a_{ij}\big(m(k)\big) \to a_{ij}$ as $k \to \infty$.

(ii) Deduce that if $\alpha_m$ is a sequence of endomorphisms of a complex finite dimensional inner product space with $\|\alpha_m\|$ bounded, we can find a sequence $m(k) \to \infty$ and an endomorphism $\alpha$ such that $\|\alpha_{m(k)} - \alpha\| \to 0$ as $k \to \infty$.

(iii) If $\gamma_m$ is a sequence of unitary endomorphisms of a complex finite dimensional inner product space and $\gamma$ is an endomorphism such that $\|\gamma_m - \gamma\| \to 0$ as $m \to \infty$, show that $\gamma$ is unitary.

(iv) Show that, even if we drop the condition $\alpha$ invertible in Exercise 15.5.12, there exists a factorisation $\alpha = \beta\gamma$ with $\gamma$ unitary and $\beta$ positive definite.

(v) Can we take $\beta$ strictly positive definite in (iv)? Is the factorisation in (iv) always unique? Give reasons.

**Exercise 15.5.14** (Requires the theory of complex variables.) We work in $\mathbb{C}$. State Rouché's theorem. Show that, if $a_n \neq 0$ and the equation

$$\sum_{j=0}^{n} a_j z^j = 0$$

has roots $\lambda_1, \lambda_2, \ldots, \lambda_n$ (multiple roots represented multiply), then, given $\epsilon > 0$, we can find a $\delta > 0$ such that, whenever $|b_j - a_j| < \delta$ for $0 \leq j \leq n$, we can find roots $\mu_1, \mu_2, \ldots, \mu_n$ (multiple roots represented multiply) of the equation

$$\sum_{j=0}^{n} b_j z^j = 0$$

with $|\mu_k - \lambda_k| < \epsilon$ $[1 \leq k \leq n]$.

**Exercise 15.5.15** [**Finding the operator norm for** $\mathcal{L}(U, U)$] We work on a finite dimensional real inner product vector space $U$ and consider an $\alpha \in \mathcal{L}(U, U)$.

(i) Show that $\langle \mathbf{x}, \alpha^* \alpha \mathbf{x} \rangle = \|\alpha(\mathbf{x})\|^2$ and deduce that

$$\|\alpha^* \alpha\| \geq \|\alpha\|^2.$$

Conclude that $\|\alpha^*\| \geq \|\alpha\|$.

(ii) Use the results of (i) to show that $\|\alpha\| = \|\alpha^*\|$ and that

$$\|\alpha^* \alpha\| = \|\alpha\|^2.$$

(iii) By considering an appropriate basis, show that, if $\beta \in \mathcal{L}(U, U)$ is symmetric, then $\|\beta\|$ is the largest absolute value of its eigenvalues, i.e.

$$\|\beta\| = \max\{|\lambda| : \lambda \text{ an eigenvalue of } \beta\}.$$

(iv) Deduce that $\|\alpha\|$ is the positive square root of the largest absolute value of the eigenvalues of $\alpha^* \alpha$.

(v) Show that all the eigenvalues of $\alpha^* \alpha$ are positive. Thus $\|\alpha\|$ is the positive square root of the largest eigenvalues of $\alpha^* \alpha$.

(vi) State and prove the corresponding result for a complex inner product vector space $U$.

[Note that a demonstration that a particular quantity can be computed shows neither that it is desirable to compute that quantity nor that the best way of computing the quantity is the one given.]

**Exercise 15.5.16** [**Completeness of the operator norm**] Let $U$ be an $n$-dimensional inner product space over $\mathbb{F}$. The object of this exercise is to show that the operator norm is complete, that is to say, every Cauchy sequence converges. (If the last sentence means nothing to you, go no further.)

(i) Suppose that $\alpha(r) \in \mathcal{L}(U, U)$ and $\|\alpha(r) - \alpha(s)\| \to 0$ as $r, s \to \infty$. If we fix an orthonormal basis for $U$ and $\alpha(r)$ has matrix $A(r) = (a_{ij}(r))$ with respect to that basis, show that

$$|a_{ij}(r) - a_{ij}(s)| \to 0 \quad \text{as } r, s \to \infty$$

for all $1 \le i, j \le n$. Explain why this implies the existence of $a_{ij} \in \mathbb{F}$ with

$$|a_{ij}(r) - a_{ij}| \to 0 \quad \text{as } r \to \infty$$

as $n \to \infty$.

(ii) Let $\alpha \in \mathcal{L}(U, U)$ be the linear map with matrix $A = (a_{ij})$. Show that

$$\|\alpha(r) - \alpha\| \to 0 \quad \text{as } r \to \infty.$$

**Exercise 15.5.17** The proof outlined in Exercise 15.5.16 is a very natural one, but goes via bases. Here is a basis free proof of completeness.

(i) Suppose that $\alpha_r \in \mathcal{L}(U, U)$ and $\|\alpha_r - \alpha_s\| \to 0$ as $r, s \to \infty$. Show that, if $\mathbf{x} \in U$,

$$\|\alpha_r \mathbf{x} - \alpha_s \mathbf{x}\| \to 0 \quad \text{as } r, s \to \infty.$$

Explain why this implies the existence of an $\alpha \mathbf{x} \in \mathbb{F}$ with

$$\|\alpha_r \mathbf{x} - \alpha \mathbf{x}\| \to 0 \quad \text{as } r \to \infty$$

as $n \to \infty$.

(ii) We have obtained a map $\alpha : U \to U$. Show that $\alpha$ is linear.

(iii) Explain why

$$\|\alpha_r \mathbf{x} - \alpha \mathbf{x}\| \le \|\alpha_s \mathbf{x} - \alpha \mathbf{x}\| + \|\alpha_r - \alpha_s\|\|\mathbf{x}\|.$$

Deduce that, given any $\epsilon > 0$, there exists an $N(\epsilon)$, depending on $\epsilon$, but not on $\mathbf{x}$, such that

$$\|\alpha_r \mathbf{x} - \alpha \mathbf{x}\| \le \|\alpha_s \mathbf{x} - \alpha \mathbf{x}\| + \epsilon\|\mathbf{x}\|$$

for all $r, s \ge N(\epsilon)$.

Deduce that

$$\|\alpha_r \mathbf{x} - \alpha \mathbf{x}\| \le \epsilon\|\mathbf{x}\|$$

for all $r \ge N(\epsilon)$ and all $\mathbf{x} \in U$. Conclude that

$$\|\alpha_r - \alpha\| \to 0 \quad \text{as } r \to \infty.$$

**Exercise 15.5.18** Once we know that the operator norm is complete (see Exercise 15.5.16 or Exercise 15.5.17), we can apply analysis to the study of the existence of the inverse. As before, $U$ is an $n$-dimensional inner product space over $\mathbb{F}$ and $\alpha$, $\beta$, $\gamma \in \mathcal{L}(U, U)$.

(i) Let us write

$$S_n(\alpha) = \iota + \alpha + \cdots + \alpha^n.$$

If $\|\alpha\| < 1$, show that $\|S_n(\alpha) - S_m(\alpha)\| \to 0$ as $n, m \to \infty$. Deduce that there is an $S(\alpha) \in \mathcal{L}(U, U)$ such that $\|S_n(\alpha) - S(\alpha)\| \to 0$ as $n \to \infty$.

(ii) Compute $(\iota - \alpha)S_n(\alpha)$ and deduce carefully that $(\iota - \alpha)S(\alpha) = \iota$. Conclude that $\beta$ is invertible whenever $\|\iota - \beta\| < 1$.

(iii) Show that, if $U$ is non-trivial and $c \geq 1$, there exists a $\beta$ which is not invertible with $\|\iota - \beta\| = c$ and an invertible $\gamma$ with $\|\iota - \gamma\| = c$.

(iv) Suppose that $\alpha$ is invertible. By writing

$$\beta = \alpha\big(\iota - \alpha^{-1}(\alpha - \beta)\big),$$

or otherwise, show that there is a $\delta > 0$, depending only on the value of $\|\alpha\|^{-1}$, such that $\beta$ is invertible whenever $\|\alpha - \beta\| < \delta$.

(v) (If you know the meaning of the word open.) Check that we have shown that the collection of invertible elements in $\mathcal{L}(U, U)$ is open. Why does this result also follow from the continuity of the map $\alpha \mapsto \det \alpha$?

[However, the proof via determinants does not generalise, whereas the proof of this exercise has echos throughout mathematics.]

(vi) If $\alpha_n \in \mathcal{L}(U, U)$, $\|\alpha_n\| < 1$ and $\|\alpha_n\| \to 0$, show, by the methods of this question, that

$$\|(\iota - \alpha_n)^{-1} - \iota\| \to 0$$

as $n \to \infty$. (That is to say, the map $\alpha \mapsto \alpha^{-1}$ is continuous at $\iota$.)

(iv) If $\beta_n \in \mathcal{L}(U, U)$, $\beta_n$ is invertible, $\beta$ is invertible and $\|\beta_n - \beta\| \to 0$, show that

$$\|\beta_n^{-1} - \beta^{-1}\| \to 0$$

as $n \to \infty$. (That is to say, the map $\alpha \mapsto \alpha^{-1}$ is continuous on the set of invertible elements of $\mathcal{L}(U, U)$.)

[We continue with some of these ideas in Questions 15.5.19 and 15.5.22.]

**Exercise 15.5.19** We continue with the hypotheses and notation of Exercise 15.5.18.

(i) Show that there is a $\gamma \in \mathcal{L}(U, U)$ such that

$$\left\| \sum_{j=0}^{n} \frac{1}{j!}\alpha^j - \gamma \right\| \to 0 \quad \text{as } n \to \infty.$$

We write $\exp \alpha = \gamma$.

(ii) Suppose that $\alpha$ and $\beta$ commute. Show that

$$\left\| \sum_{u=0}^{n} \frac{1}{u!}\alpha^u \sum_{v=0}^{n} \frac{1}{v!}\beta^v - \sum_{k=0}^{n} \frac{1}{k!}(\alpha + \beta)^k \right\|$$

$$\leq \sum_{u=0}^{n} \frac{1}{u!}\|\alpha\|^u \sum_{v=0}^{n} \frac{1}{v!}\|\beta\|^v - \sum_{k=0}^{n} \frac{1}{k!}(\|\alpha\| + \|\beta\|)^k$$

(note the minus sign) and deduce that

$$\exp \alpha \exp \beta = \exp(\alpha + \beta).$$

(iii) Show that $\exp \alpha$ is invertible with inverse $\exp(-\alpha)$.

(iv) Let $t \in \mathbb{F}$. Show that

$$|t|^{-2} \left\| \exp(t\alpha) - \iota - t\alpha - \frac{t^2}{2}\alpha^2 \right\| \to 0 \quad \text{as } t \to 0.$$

(v) We now drop the assumption that $\alpha$ and $\beta$ commute. Show that

$$t^{-2} \left\| \exp(t\alpha)\exp(t\beta) - \exp\big(t(\alpha + \beta)\big) \right\| \to \frac{1}{2}\|\alpha\beta - \beta\alpha\|$$

as $t \to 0$.

Deduce that, if $\alpha$ and $\beta$ do not commute and $t$ is sufficiently small, but non-zero, then $\exp(t\alpha)\exp(t\beta) \neq \exp\big(t(\alpha + \beta)\big)$.

Show also that, if $\alpha$ and $\beta$ do not commute and $t$ is sufficiently small, but non-zero, then $\exp(t\alpha)\exp(t\beta) \neq \exp(t\beta)\exp(t\alpha)$.

(vi) Show that

$$\left\| \sum_{k=0}^{n} \frac{1}{k!}(\alpha + \beta)^k - \sum_{k=0}^{n} \frac{1}{k!}\alpha^k \right\| \leq \sum_{k=0}^{n} \frac{1}{k!}(\|\alpha\| + \|\beta\|)^k - \sum_{k=0}^{n} \frac{1}{k!}\|\alpha\|^k$$

and deduce that

$$\| \exp(\alpha + \beta) - \exp\alpha \| \leq e^{\|\alpha\| + \|\beta\|} - e^{\|\alpha\|}.$$

Hence show that the map $\alpha \mapsto \exp\alpha$ is continuous.

(vii) If $\alpha$ has matrix $A$ and $\exp\alpha$ has matrix $C$ with respect to some basis of $U$, show that, writing

$$S(n) = \sum_{r=0}^{n} \frac{A^r}{r!},$$

we have $s_{ij}(n) \to c_{ij}$ as $n \to \infty$ for each $1 \leq i, j \leq n$.

**Exercise 15.5.20** By considering Jordan normal forms, or otherwise, show that, if $A$ is an $n \times n$ complex matrix,

$$\det\left( \sum_{r=0}^{n} \frac{A^r}{r!} \right) \to \exp(\operatorname{Tr} A).$$

If you have done the previous question, conclude that

$$\det \exp\alpha = \exp \operatorname{Tr}\alpha.$$

**Exercise 15.5.21** Let $A$ and $B$ be $m \times m$ complex matrices and let $\mu \in \mathbb{C}$. Which, if any, of the following statements about the spectral radius $\rho$ are true and which are false? Give proofs or counterexamples as appropriate.

(i) $\rho(\mu A) = |\mu|\rho(A)$.

(ii) $\rho(A) = 0 \Rightarrow A = 0$.

(iii) If $\rho(A) = 0$, then $A$ is nilpotent.

(iv) If $A$ is nilpotent, then $\rho(A) = 0$.

(v) $\rho(A + B) \leq \rho(A) + \rho(B)$.

(vi) $\rho(AB) \leq \rho(A)\rho(B)$.

(vii) $\rho(AB) \geq \rho(A)\rho(B)$.

(viii) If $A$ and $B$ commute, then $\rho(A + B) = \rho(A) + \rho(B)$.

(ix) If $A$ and $B$ commute, then $\rho(AB) = \rho(A)\rho(B)$.

(x) If $A$ and $B$ commute, then $\rho(AB) \leq \rho(A)\rho(B)$.

(xi) $\det A \leq \rho(A)^m$.

(xii) $\det A = 0 \Rightarrow \rho(A) = 0$.

(xiii) Given $K \geq 1$ we can find an $m \times m$ matrix $C$ with $\rho(C) = 0$, $\|C\| \geq K$ and $\|C^{r+1}\| \geq K\|C^r\|$ for all $1 \leq r \leq m - 2$.

**Exercise 15.5.22** In this question we look at the spectral radius using as much analysis and as little algebra as possible. As usual, $U$ is a finite dimensional complex inner product vector space over $\mathbb{C}$ and $\alpha$, $\beta \in \mathcal{L}(U, U)$. Naturally you must not use the result of Theorem 15.3.4.

(i) Write $\rho_B(\alpha) = \liminf_{n\to\infty} \|\alpha^n\|^{1/n}$. If $\epsilon > 0$, then, by definition, we can find an $N$ such that $\|\alpha^N\|^{1/N} \leq \rho_B(\alpha) + \epsilon$. Show that, if $r$ is fixed,

$$\|\alpha^{kN+r}\| \leq C_r(\rho_B(\alpha) + \epsilon)^{Nk+r},$$

where $C_r$ is independent of $N$. Deduce that

$$\limsup_{n\to\infty} \|\alpha^n\|^{1/n} \leq \rho_B(\alpha) + \epsilon$$

and conclude that $\|\alpha^n\|^{1/n} \to \rho_B(\alpha)$.

(ii) Let $\epsilon > 0$. Show that we can we choose $K$ so that

$$\|\alpha^r\| \leq K(\rho_B(\alpha) + \epsilon)^r \text{ and } \|\beta^r\| \leq K(\rho_B(\beta) + \epsilon)^r$$

for all $r \geq 0$. If we choose such a $K$, show that

$$\|(\alpha + \beta)^n\| \leq K(\rho_B(\alpha) + \rho_B(\beta) + 2\epsilon)^n$$

for all $n \geq 0$. Deduce that

$$\rho_B(\alpha + \beta) \leq \rho_B(\alpha) + \rho_B(\beta).$$

(iii) If $\beta^m = 0$ for some $m$, show that

$$\rho_B(\alpha + \beta) \leq \rho_B(\alpha).$$

(iv) Show, in the manner of Exercise 15.5.18, that, if $\rho_B(\alpha) < 1$, then $\iota - \alpha$ is invertible and

$$\left\| \sum_{j=1}^{n} \alpha^j - (\iota - \alpha)^{-1} \right\| \to 0.$$

(v) Using (iv), show that $t\iota - \alpha$ is invertible whenever $t > \rho_B(\alpha)$. Deduce that $\rho_B(\alpha) \geq \lambda$ whenever $\lambda$ is an eigenvalue of $\alpha$.

(vi) If $\beta$ is invertible, show that $\rho_B(\beta^{-1}\alpha\beta) = \rho_B(\alpha)$.

(vii) Explain why Theorem 12.2.5 tells us that we can find $\beta$, $\gamma$, $\delta \in \mathcal{L}(U, U)$ such that $\beta^m = 0$ for some $m$, $\gamma$ is invertible, $\delta$ has the same eigenvalues as $\alpha$,

$$\rho_B(\delta) = \max\{|\lambda| \; : \; \lambda \text{ an eigenvalue of } \delta\} = \rho(\delta)$$

and

$$\alpha = \gamma^{-1}(\delta + \beta)\gamma.$$

Hence show that

$$\rho_B(\alpha) = \max\{|\lambda| \; : \; \lambda \text{ an eigenvalue of } \delta\} = \rho(\alpha).$$

(ix) Show that, if $B$ is an $m \times m$ matrix satisfying the condition

$$b_{rr} = 1 > \sum_{s \neq r} |b_{rs}| \text{ for all } 1 \leq r \leq m,$$

then $B$ is invertible.

By considering $DB$, where $D$ is a diagonal matrix, or otherwise, show that, if $A$ is an $m \times m$ matrix with

$$|a_{rr}| > \sum_{s \neq r} |a_{rs}| \text{ for all } 1 \leq r \leq m$$

then $A$ is invertible. (This is relevant to Lemma 15.3.9 since it shows that the system $A\mathbf{x} = \mathbf{b}$ considered there is always soluble. A short proof, together with a long list of independent discoverers in given in '*A recurring theorem on determinants*' by Olga Taussky [30].)

**Exercise 15.5.23 [Over-relaxation]** The following is a modification of the Gauss–Siedel method for finding the solution $\mathbf{x}^*$ of the system

$$A\mathbf{x} = \mathbf{b},$$

where $A$ is a non-singular $m \times m$ matrix with non-zero diagonal entries. Write $A$ as the sum of $m \times m$ matrices $A = L + D + U$ where $L$ is strictly lower triangular (so has zero diagonal entries), $D$ is diagonal and $U$ is strictly upper triangular. Choose an initial $\mathbf{x}_0$ and take

$$(D + \omega L)\mathbf{x}_{j+1} = \left(-\omega U + (1 - \omega)D\right)\mathbf{x}_j + \omega\mathbf{b}.$$

with $\omega \neq 0$.

Check that, if $\omega = 1$, this is the Gauss–Siedel method. Check also that, if $\|\mathbf{x}_j - \mathbf{x}\| \to 0$, as $j \to \infty$, then $\mathbf{x} = \mathbf{x}^*$.

Show that the method will certainly fail to converge for some choices of $\mathbf{x}_0$ unless $0 < \omega < 2$. (However, in appropriate circumstances, taking $\omega$ close to 2 can be very effective.) [Hint: If $|\det B| \geq 1$ what can you say about $\rho(B)$?]

**Exercise 15.5.24** (This requires elementary group theory.) Let $G$ be a finite Abelian group and write $C(G)$ for the set of functions $f : G \to \mathbb{C}$.

(i) Show that

$$\langle f, g \rangle = \sum_{x \in G} f(x)g(x)^*$$

is an inner product on $C(G)$. We use this inner product for the rest of this question.

(ii) If $y \in G$, we define $\alpha_y f(x) = f(x + y)$. Show that $\alpha_y : C(G) \to C(G)$ is a unitary linear map and so there exists an orthonormal basis of eigenvectors of $\alpha_y$.

(iii) Show that $\alpha_y \alpha_w = \alpha_w \alpha_y$ for all $y, w \in G$. Use this result to show that there exists an orthonormal basis $B_0$ each of whose elements is an eigenvector for $\alpha_y$ for all $y \in G$.

(iv) If $\phi \in B_0$, explain why $\phi(x + y) = \lambda_y \phi(x)$ for all $x \in G$ and some $\lambda_y$ with $|\lambda_y| = 1$. Deduce that $|\phi(x)|$ is a non-zero constant independent of $x$. We now write

$$B = \{\phi(0)^{-1}\phi \ : \ \phi \in B_0\}$$

so that $B$ is an orthogonal basis of eigenvectors of each $\alpha_y$ with $\chi \in B \Rightarrow \chi(0) = 1$.

(v) Use the relation $\chi(x + y) = \lambda_y \chi(x)$ to deduce that $\chi(x + y) = \chi(x)\chi(y)$ for all $x, y \in G$.

(vi) Explain why

$$f = |G|^{-1} \sum_{\chi \in B} \langle f, \chi \rangle \chi,$$

where $|G|$ is the number of elements in $G$.

[Observe that we have produced a kind of Fourier analysis on $G$.]

# 16

# Quadratic forms and their relatives

## 16.1 Bilinear forms

In Section 8.3 we discussed functions of the form

$$(x, y) \mapsto ux^2 + 2vxy + wy^2.$$

These are special cases of the idea of a bilinear function.

**Definition 16.1.1** *If $U$, $V$ and $W$ are vector spaces over $\mathbb{F}$, then $\alpha : U \times V \to W$ is a bilinear function if the map*

$$\alpha_{,\mathbf{v}} : U \to W \quad \text{given by } \alpha_{,\mathbf{v}}(\mathbf{u}) = \alpha(\mathbf{u}, \mathbf{v})$$

*is linear for each fixed $\mathbf{v} \in V$ and the map*

$$\alpha_{\mathbf{u},} : V \to W \quad \text{given by } \alpha_{\mathbf{u},}(\mathbf{v}) = \alpha(\mathbf{u}, \mathbf{v})$$

*is linear for each fixed $\mathbf{u} \in U$.*

Exercise 16.5.1 discusses a result on general bilinear functions which is important in multivariable analysis, but, apart from this one exercise, we shall only discuss the special case when we take $U = V$ and $W = \mathbb{F}$. Although much of the algebra involved applies to both $\mathbb{R}$ and $\mathbb{C}$ (and indeed to more general fields), it is often more natural to look at *sesquilinear functions* (see Exercise 16.1.26) rather than bilinear functions (see Exercise 16.2.10) when considering $\mathbb{C}$. For this reason, we will make a further restriction and, initially, only consider the case when $\mathbb{F} = \mathbb{R}$.

**Definition 16.1.2** *If $U$ is a vector space over $\mathbb{R}$, then a bilinear function $\alpha : U \times U \to \mathbb{R}$ is called a bilinear form.*

Before discussing what a bilinear form looks like, we note a link with dual spaces. The following exercise is a small paper tiger.

**Lemma 16.1.3** *Suppose that $U$ is vector space over $\mathbb{R}$ and $\alpha : U \times U \to \mathbb{R}$ is a bilinear form. If we set*

$$\theta_L(\mathbf{x})\mathbf{y} = \alpha(\mathbf{x}, \mathbf{y}) \quad \text{and} \quad \theta_R(\mathbf{y})\mathbf{x} = \alpha(\mathbf{x}, \mathbf{y})$$

*for* $\mathbf{x}, \mathbf{y} \in U$, *we have* $\theta_L(\mathbf{x}), \theta_R(\mathbf{y}) \in U'$. *Further* $\theta_L$ *and* $\theta_R$ *are linear maps from* $U$ *to* $U'$.

*Proof* Left to the reader as an exercise in disentangling notation.                    □

**Lemma 16.1.4** *We use the notation of Lemma 16.1.3. If* $U$ *is finite dimensional and*

$$\alpha(\mathbf{x}, \mathbf{y}) = 0 \quad for\ all\ \mathbf{y} \in U \Rightarrow \mathbf{x} = 0,$$

*then* $\theta_L : U \to U'$ *is an isomorphism.*

*Proof* We observe that the stated condition tells us that $\theta_L$ is injective since

$$\theta_L(\mathbf{x}) = 0 \Rightarrow \theta_L(\mathbf{x})(\mathbf{y}) = 0 \quad \text{for all } \mathbf{y} \in U$$
$$\Rightarrow \alpha(\mathbf{x}, \mathbf{y}) = 0 \quad \text{for all } \mathbf{y} \in U$$
$$\Rightarrow \mathbf{x} = 0.$$

Since $\dim U = \dim U'$, it follows that $\theta_L$ is an isomorphism.                    □

**Exercise 16.1.5** *State and prove the corresponding result for* $\theta_R$.
*[In lemma 16.1.7 we shall see that things are simpler than they look at the moment.]*

Lemma 16.1.4 is a generalisation of Lemma 14.2.8. Our proof of Lemma 14.2.8 started with the geometric observation that (for an $n$ dimensional inner product space) the null-space of a non-zero linear functional is an $n - 1$ dimensional subspace, so we could use an appropriate vector perpendicular to that subspace to obtain the required map. Our proof of Lemma 16.1.4 is much less constructive. We observe that a certain linear map between two vector spaces of the same dimension is injective and conclude that is must be bijective. (However, the next lemma shows that, if we use a coordinate system, it is easy to write down $\theta_L$ and $\theta_R$ explicitly.)

We now introduce a specific basis.

**Lemma 16.1.6** *Let* $U$ *be a finite dimensional vector space over* $\mathbb{R}$ *with basis* $\mathbf{e}_1, \mathbf{e}_2, \ldots,$ $\mathbf{e}_n$.

(i) *If* $A = (a_{ij})$ *is an* $n \times n$ *matrix with entries in* $\mathbb{R}$, *then*

$$\alpha\left(\sum_{i=1}^{n} x_i \mathbf{e}_i, \sum_{j=1}^{n} y_j \mathbf{e}_j\right) = \sum_{i=1}^{n}\sum_{j=1}^{n} x_i a_{ij} y_j$$

*(for* $x_i, y_j \in \mathbb{R}$*) defines a bilinear form.*

(ii) *If* $\alpha : U \times U$ *is a bilinear form, then there is a unique* $n \times n$ *real matrix* $A = (a_{ij})$ *with*

$$\alpha\left(\sum_{i=1}^{n} x_i \mathbf{e}_i, \sum_{j=1}^{n} y_j \mathbf{e}_j\right) = \sum_{i=1}^{n}\sum_{j=1}^{n} x_i a_{ij} y_j$$

*for all* $x_i, y_j \in \mathbb{R}$.

(iii) *We use the notation of Lemma 16.1.3 and part (ii) of this lemma. If we give $U'$ the dual basis (see Lemma 11.4.1) $\hat{\mathbf{e}}_1, \hat{\mathbf{e}}_2, \ldots, \hat{\mathbf{e}}_n$, then $\theta_L$ has matrix $A^T$ with respect to the bases we have chosen for $U$ and $U'$ and $\theta_R$ has matrix $A$.*

*Proof* Left to the reader. Note that $a_{ij} = \alpha(\mathbf{e}_i, \mathbf{e}_j)$. $\qquad\qquad\qquad\qquad\qquad\square$

We know that $A$ and $A^T$ have the same rank and that an $n \times n$ matrix is invertible if and only if it has full rank. Thus part (iii) of Lemma 16.1.6 yields the following improvement on Lemma 16.1.4.

**Lemma 16.1.7** *We use the notation of Lemma 16.1.3. If $U$ is finite dimensional, the following conditions are equivalent.*

(i) *$\alpha(\mathbf{x}, \mathbf{y}) = 0$ for all $\mathbf{y} \in U \Rightarrow \mathbf{x} = 0$.*

(ii) *$\alpha(\mathbf{x}, \mathbf{y}) = 0$ for all $\mathbf{x} \in U \Rightarrow \mathbf{y} = 0$.*

(iii) *If $\mathbf{e}_1, \mathbf{e}_2, \ldots, \mathbf{e}_n$ is a basis for $U$, then the $n \times n$ matrix with entries $\alpha(\mathbf{e}_i, \mathbf{e}_j)$ is invertible.*

(iv) *$\theta_L : U \to U'$ is an isomorphism.*

(v) *$\theta_R : U \to U'$ is an isomorphism.*

**Definition 16.1.8** *If $U$ is finite dimensional and $\alpha : U \times U \to \mathbb{R}$ is bilinear, we say that $\alpha$ is degenerate (or singular) if there exists a non-zero $\mathbf{x} \in U$ such that $\alpha(\mathbf{x}, \mathbf{y}) = 0$ for all $\mathbf{y} \in U$.*

**Exercise 16.1.9** (i) *If $\beta : \mathbb{R}^2 \times \mathbb{R}^2 \to \mathbb{R}$ is given by*

$$\beta\big((x_1, x_2)^T, (y_1, y_2)^T\big) = x_1 y_2 - x_2 y_1,$$

*show that $\beta$ is a non-degenerate bilinear form, but $\beta(\mathbf{x}, \mathbf{x}) = 0$ for all $\mathbf{x} \in U$.*

*Is it possible to find a degenerate bilinear form $\alpha$ associated with a vector space $U$ of non-zero dimension with $\alpha(\mathbf{x}, \mathbf{x}) \neq 0$ for all $\mathbf{x} \neq \mathbf{0}$? Give reasons.*

Exercise 16.1.17 shows that, from one point of view, the description 'degenerate' is not inappropriate. However, there are several parts of mathematics where degenerate bilinear forms are neither rare nor useless, so we shall consider general bilinear forms whenever we can.

Bilinear forms can be decomposed in a rather natural way.

**Definition 16.1.10** *Let $U$ be a vector space over $\mathbb{R}$ and let $\alpha : U \times U \to \mathbb{R}$ be a bilinear form.*

*If $\alpha(\mathbf{u}, \mathbf{v}) = \alpha(\mathbf{v}, \mathbf{u})$ for all $\mathbf{u}, \mathbf{v} \in U$, we say that $\alpha$ is symmetric.*

*If $\alpha(\mathbf{u}, \mathbf{v}) = -\alpha(\mathbf{v}, \mathbf{u})$ for all $\mathbf{u}, \mathbf{v} \in U$, we say that $\alpha$ is antisymmetric (or skew-symmetric).*

**Lemma 16.1.11** *Let $U$ be a vector space over $\mathbb{R}$. Every bilinear form $\alpha : U \times U \to \mathbb{R}$ can be written in a unique way as the sum of a symmetric and an antisymmetric bilinear form.*

*Proof* This should come as no surprise to the reader. If $\alpha_1 : U \times U \to \mathbb{R}$ is a symmetric and $\alpha_2 : U \times U \to \mathbb{R}$ an antisymmetric form with $\alpha = \alpha_1 + \alpha_2$, then

$$\alpha(\mathbf{u}, \mathbf{v}) = \alpha_1(\mathbf{u}, \mathbf{v}) + \alpha_2(\mathbf{u}, \mathbf{v})$$
$$\alpha(\mathbf{v}, \mathbf{u}) = \alpha_1(\mathbf{u}, \mathbf{v}) - \alpha_2(\mathbf{u}, \mathbf{v})$$

and so

$$\alpha_1(\mathbf{u}, \mathbf{v}) = \frac{1}{2}\big(\alpha(\mathbf{u}, \mathbf{v}) + \alpha(\mathbf{v}, \mathbf{u})\big)$$
$$\alpha_2(\mathbf{u}, \mathbf{v}) = \frac{1}{2}\big(\alpha(\mathbf{u}, \mathbf{v}) - \alpha(\mathbf{v}, \mathbf{u})\big).$$

Thus the decomposition, if it exists, is unique.

We leave it to the reader to check that, conversely, if $\alpha_1$, $\alpha_2$ are defined using the second set of formulae in the previous paragraph, then $\alpha_1$ is a symmetric linear form, $\alpha_2$ an antisymmetric linear form and $\alpha = \alpha_1 + \alpha_2$. $\square$

Symmetric forms are closely connected with *quadratic forms*.

**Definition 16.1.12** *Let $U$ be a vector space over $\mathbb{R}$. If $\alpha : U \times U \to \mathbb{R}$ is a symmetric form, then $q : U \to \mathbb{R}$, defined by*

$$q(\mathbf{u}) = \alpha(\mathbf{u}, \mathbf{u}),$$

*is called a* quadratic form.

The next lemma recalls the link between inner product and norm.

**Lemma 16.1.13** *With the notation of Definition 16.1.12,*

$$\alpha(\mathbf{u}, \mathbf{v}) = \frac{1}{4}\big(q(\mathbf{u} + \mathbf{v}) - q(\mathbf{u} - \mathbf{v})\big)$$

*for all $\mathbf{u}$, $\mathbf{v} \in U$.*

*Proof* The computation is left to the reader. $\square$

**Exercise 16.1.14** [**A 'parallelogram' law**] *We use the notation of Definition 16.1.12. If $\mathbf{u}$, $\mathbf{v} \in U$, show that*

$$q(\mathbf{u} + \mathbf{v}) + q(\mathbf{u} - \mathbf{v}) = 2\big(q(\mathbf{u}) + q(\mathbf{v})\big).$$

*Remark 1* Although all we have said applies when we replace $\mathbb{R}$ by $\mathbb{C}$, it is more natural to use *Hermitian* and *skew-Hermitian* forms when working over $\mathbb{C}$. We leave this natural development to Exercise 16.1.26 at the end of this section.

*Remark 2* As usual, our results can be extended to vector spaces over more general fields (see Section 13.2), but we will run into difficulties when we work with a field in which $2 = 0$ (see Exercise 13.2.8) since both Lemmas 16.1.11 and 16.1.13 depend on division by 2.

The reason behind the usage 'symmetric form' and 'quadratic form' is given in the next lemma.

**Lemma 16.1.15** *Let $U$ be a finite dimensional vector space over $\mathbb{R}$ with basis $\mathbf{e}_1, \mathbf{e}_2, \ldots, \mathbf{e}_n$.*

*(i) If $A = (a_{ij})$ is an $n \times n$ real symmetric matrix, then*

$$\alpha \left( \sum_{i=1}^n x_i \mathbf{e}_i, \sum_{j=1}^n y_j \mathbf{e}_j \right) = \sum_{i=1}^n \sum_{j=1}^n x_i a_{ij} y_j$$

*(for $x_i$, $y_j \in \mathbb{R}$) defines a symmetric form.*

*(ii) If $\alpha : U \times U$ is a symmetric form, then there is a unique $n \times n$ real symmetric matrix $A = (a_{ij})$ with*

$$\alpha \left( \sum_{i=1}^n x_i \mathbf{e}_i, \sum_{j=1}^n y_j \mathbf{e}_j \right) = \sum_{i=1}^n \sum_{j=1}^n x_i a_{ij} y_j$$

*for all $x_i$, $y_j \in \mathbb{R}$.*

*(iii) If $\alpha$ and $A$ are as in (ii) and $q(\mathbf{u}) = \alpha(\mathbf{u}, \mathbf{u})$, then*

$$q \left( \sum_{i=1}^n x_i \mathbf{e}_i \right) = \sum_{i=1}^n \sum_{j=1}^n x_i a_{ij} x_j$$

*for all $x_i \in \mathbb{R}$.*

*Proof* Left to the reader. Note that $a_{ij} = \alpha(\mathbf{e}_i, \mathbf{e}_j)$. □

**Exercise 16.1.16** *If $B = (b_{ij})$ is an $n \times n$ real matrix, show that*

$$q \left( \sum_{i=1}^n x_i \mathbf{e}_i \right) = \sum_{i=1}^n \sum_{j=1}^n x_i b_{ij} x_j$$

*(for $x_i \in \mathbb{R}$) defines a quadratic form. Find the associated symmetric matrix $A$ in terms of $B$.*

**Exercise 16.1.17** *This exercise may strike the reader as fairly tedious, but I think that it is quite helpful in understanding some of the ideas of the chapter. Describe, or sketch, the sets in $\mathbb{R}^3$ given by*

$$x_1^2 + x_2^2 + x_3^2 = k, \quad x_1^2 + x_2^2 - x_3^2 = k,$$
$$x_1^2 + x_2^2 = k, \quad x_1^2 - x_2^2 = k, \quad x_1^2 = k, 0 = k$$

*for $k = 1$, $k = -1$ and $k = 0$.*

We have a nice change of basis formula.

**Lemma 16.1.18** *Let $U$ be a finite dimensional vector space over $\mathbb{R}$. Suppose that $\mathbf{e}_1$, $\mathbf{e}_2, \ldots, \mathbf{e}_n$ and $\mathbf{f}_1, \mathbf{f}_2, \ldots, \mathbf{f}_n$ are bases with*

$$\mathbf{f}_j = \sum_{i=1}^n m_{ij} \mathbf{e}_i$$

*for $1 \leq i \leq n$. Then, if*

$$\alpha \left( \sum_{i=1}^n x_i \mathbf{e}_i, \sum_{j=1}^n y_j \mathbf{e}_j \right) = \sum_{i=1}^n \sum_{j=1}^n x_i a_{ij} y_j,$$

$$\alpha \left( \sum_{i=1}^n x_i \mathbf{f}_i, \sum_{j=1}^n y_j \mathbf{f}_j \right) = \sum_{i=1}^n \sum_{j=1}^n x_i b_{ij} y_j,$$

*and we write $A = (a_{ij})$, $B = (b_{ij})$, $M = (m_{ij})$, we have $B = M^T A M$.*

*Proof* Observe that

$$\alpha(\mathbf{f}_r, \mathbf{f}_s) = \alpha \left( \sum_{i=1}^n m_{ir} \mathbf{e}_i, \sum_{j=1}^n m_{js} \mathbf{e}_j \right)$$

$$= \sum_{i=1}^n \sum_{j=1}^n m_{ir} m_{js} \alpha(\mathbf{e}_i, \mathbf{e}_j) = \sum_{i=1}^n \sum_{j=1}^n m_{ir} a_{ij} m_{js}$$

as required. $\square$

Although we have adopted a slightly different point of view, the reader should be aware of the following definition. (See, for example, Exercise 16.5.3.)

**Definition 16.1.19** *If $A$ and $B$ are symmetric $n \times n$ matrices, we say that the quadratic forms $\mathbf{x} \mapsto \mathbf{x}^T A \mathbf{x}$ and $\mathbf{x} \mapsto \mathbf{x}^T B \mathbf{x}$ are equivalent (or congruent) if there exists a non-singular $n \times n$ matrix $M$ with $B = M^T A M$.*

It is often more natural to consider 'change of coordinates' than 'change of basis'.

**Lemma 16.1.20** *Let $A$ be an $n \times n$ real symmetric matrix, let $M$ be an invertible $n \times n$ real matrix and take*

$$q(\mathbf{x}) = \mathbf{x}^T A \mathbf{x}$$

*for all column vectors $\mathbf{x}$. If we set $\mathbf{X} = M \mathbf{x}$, then*

$$q(\mathbf{X}) = \mathbf{x}^T M^T A M \mathbf{x}.$$

*Proof* Immediate. $\square$

*Remark* Note that, although we still deal with matrices, they now represent *quadratic forms* and not *linear maps*. If $q_1$ and $q_2$ are quadratic forms with corresponding matrices $A_1$ and

$A_2$, then $A_1 + A_2$ corresponds to $q_1 + q_2$, but it is hard to see what meaning to give to $A_1 A_2$. The difference in nature is confirmed by the very different change of basis formulae. Taking one step back and looking at *bilinear forms*, the reader should note that, though these have a very useful matrix representation, the moment we look at *trilinear forms* (the definition of which is left to the reader) the natural associated array does not form a matrix. Notwithstanding this note of warning, we can make use of our knowledge of symmetric matrices.

**Theorem 16.1.21** *Let $U$ be a real $n$-dimensional inner product space. If $\alpha : U \times U \to \mathbb{R}$ is a symmetric form, then we can find an orthonormal basis $\mathbf{e}_1, \mathbf{e}_2, \ldots, \mathbf{e}_n$ and real numbers $d_1, d_2, \ldots, d_n$ such that*

$$\alpha\left(\sum_{i=1}^{n} x_i \mathbf{e}_i, \sum_{j=1}^{n} y_j \mathbf{e}_j\right) = \sum_{k=1}^{n} d_k x_k y_k$$

*for all $x_i, y_j \in \mathbb{R}$. The $d_k$ are the roots of the characteristic polynomial of $\alpha$ with multiple roots appearing multiply.*

*Proof* Choose any orthonormal basis $\mathbf{f}_1, \mathbf{f}_2, \ldots, \mathbf{f}_n$ for $U$. We know, by Lemma 16.1.15, that there exists a symmetric matrix $A = (a_{ij})$ such that

$$\alpha\left(\sum_{i=1}^{n} x_i \mathbf{f}_i, \sum_{j=1}^{n} y_j \mathbf{f}_j\right) = \sum_{i=1}^{n} \sum_{j=1}^{n} x_i a_{ij} y_j$$

for all $x_i, y_j \in \mathbb{R}$. By Theorem 8.2.5, we can find an orthogonal matrix $M$ such that $M^T A M = D$, where $D$ is a diagonal matrix whose entries are the eigenvalues of $A$ appearing with the appropriate multiplicities. The change of basis formula now gives the required result. $\square$

Recalling Exercise 8.1.7, we get the following a result on quadratic forms.

**Lemma 16.1.22** *Suppose that $q : \mathbb{R}^n \to \mathbb{R}$ is given by*

$$q(\mathbf{x}) = \sum_{i=1}^{n} \sum_{j=1}^{n} x_i a_{ij} x_j$$

*where $\mathbf{x} = (x_1, x_2, \ldots, x_n)^T$ with respect to some orthogonal coordinate system $S$. Then there exists another coordinate system $S'$ obtained from the first by rotation of axes such that*

$$q(\mathbf{y}) = \sum_{i=1}^{n} d_i y_i^2$$

*where $\mathbf{y} = (y_1, y_2, \ldots, y_n)^T$ with respect to $S'$.*

**Exercise 16.1.23** *Explain, using informal arguments (you are not asked to prove anything, or, indeed, to make the statements here precise), why the volume of an ellipsoid given by*

$\sum_{j=1}^{n} x_i a_{ij} x_j \leq L$ (where $L > 0$ and the matrix $(a_{ij})$ is symmetric with all its eigenvalues strictly positive) in ordinary $n$-dimensional space is $(\det A)^{-1/2} L^{n/2} V_n$, where $V_n$ is the volume of the unit sphere.

Here is an important application used the study of multivariate (that is to say, multidimensional) normal random variables in probability.

**Example 16.1.24** *Suppose that the random variable* $\mathbf{X} = (X_1, X_2, \ldots, X_n)^T$ *has density function*

$$f_{\mathbf{X}}(\mathbf{x}) = K \exp\left(-\frac{1}{2}\sum_{i=1}^{n}\sum_{j=1}^{n} x_i a_{ij} x_j\right) = K \exp(-\tfrac{1}{2}\mathbf{x}^T A \mathbf{x})$$

*for some real symmetric matrix* $A$. *Then all the eigenvalues of* $A$ *are strictly positive and we can find an orthogonal matrix* $M$ *such that*

$$\mathbf{X} = M\mathbf{Y}$$

*where* $\mathbf{Y} = (Y_1, Y_2, \ldots, Y_n)^T$, *the* $Y_j$ *are independent, each* $Y_j$ *is normal with mean 0 and variance* $d_j^{-1}$ *and the* $d_j$ *are the roots of the characteristic polynomial for* $A$ *with multiple roots counted multiply.*

*Sketch proof* (We freely use results on probability and integration which are not part of this book.) We know, by Exercise 8.1.7 (ii), that there exists a special orthogonal matrix $P$ such that

$$P^T A P = D,$$

where $D$ is a diagonal matrix with entries the roots of the characteristic polynomial of $A$. By the change of variable theorem for integrals (we leave it to the reader to fill in or ignore the details), $\mathbf{Y} = P^{-1}\mathbf{X}$ has density function

$$f_{\mathbf{Y}}(\mathbf{y}) = K \exp\left(-\frac{1}{2}\sum_{j=1}^{n} d_j y_j^2\right)$$

(we know that rotation preserves volume). We note that, in order that the integrals converge, we must must have $d_j > 0$. Interpreting our result in the standard manner, we see that the $Y_j$ are independent and each $Y_j$ is normal with mean 0 and variance $d_j^{-1}$.

Setting $M = P^T$ gives the result. $\qquad\qquad\square$

**Exercise 16.1.25** (*Requires some knowledge of multidimensional calculus.*) *We use the notation of Example 16.1.24. What is the value of* $K$ *in terms of* $\det A$?

We conclude this section with two exercises in which we consider appropriate analogues for the complex case.

**Exercise 16.1.26** *If* $U$ *is a vector space over* $\mathbb{C}$, *we say that* $\alpha : U \times U \to \mathbb{C}$ *is a sesquilinear form if*

$$\alpha_{,\mathbf{v}} : U \to \mathbb{C} \text{ given by } \alpha_{,\mathbf{v}}(\mathbf{u}) = \alpha(\mathbf{u}, \mathbf{v})$$

*is linear for each fixed* $\mathbf{v} \in U$ *and the map*

$$\alpha_{\mathbf{u},} : U \to \mathbb{C} \text{ given by } \alpha_{\mathbf{u},}(\mathbf{v}) = \alpha(\mathbf{u}, \mathbf{v})^*$$

*is linear for each fixed* $\mathbf{u} \in U$. *We say that a sesquilinear form* $\alpha : U \times U \to \mathbb{C}$ *is* Hermitian *if* $\alpha(\mathbf{u}, \mathbf{v}) = \alpha(\mathbf{v}, \mathbf{u})^*$ *for all* $\mathbf{u}, \mathbf{v} \in U$. *We say that* $\alpha$ *is* skew-Hermitian *if* $\alpha(\mathbf{v}, \mathbf{u}) = -\alpha(\mathbf{u}, \mathbf{v})^*$ *for all* $\mathbf{u}, \mathbf{v} \in U$.

(*i*) *Show that every sesquilinear form can be expressed uniquely as the sum of a Hermitian and a skew-Hermitian form.*

(*ii*) *If* $\alpha$ *is a Hermitian form, show that* $\alpha(\mathbf{u}, \mathbf{u})$ *is real for all* $\mathbf{u} \in U$.

(*iii*) *If* $\alpha$ *is skew-Hermitian, show that we can write* $\alpha = i\beta$ *where* $\beta$ *is Hermitian.*

(*iv*) *If* $\alpha$ *is a Hermitian form show that*

$$4\alpha(\mathbf{u}, \mathbf{v}) = \alpha(\mathbf{u}+\mathbf{v}, \mathbf{u}+\mathbf{v}) - \alpha(\mathbf{u}-\mathbf{v}, \mathbf{u}-\mathbf{v}) + i\alpha(\mathbf{u}+i\mathbf{v}, \mathbf{u}+i\mathbf{v}) - i\alpha(\mathbf{u}-i\mathbf{v}, \mathbf{u}-i\mathbf{v})$$

*for all* $\mathbf{u}, \mathbf{v} \in U$.

(*v*) *Suppose now that* $U$ *is an inner product space of dimension* $n$ *and* $\alpha$ *is Hermitian. Show that we can find an orthonormal basis* $\mathbf{e}_1, \mathbf{e}_2, \ldots, \mathbf{e}_n$ *and real numbers* $d_1, d_2, \ldots, d_n$ *such that*

$$\alpha\left(\sum_{r=1}^{n} z_r \mathbf{e}_r, \sum_{s=1}^{n} w_s \mathbf{e}_s\right) = \sum_{t=1}^{n} d_t z_t w_t^*$$

*for all* $z_r, w_s \in \mathbb{C}$.

**Exercise 16.1.27** *Consider Hermitian forms for a vector space* $U$ *over* $\mathbb{C}$. *Find and prove analogues for those parts of Lemmas 16.1.3 and 16.1.4 which deal with* $\theta_R$. *You will need the following definition of an anti-linear map. If* $U$ *and* $V$ *are vector spaces over* $\mathbb{C}$, *then a function* $\phi : U \to V$ *is an anti-linear map if* $\phi(\lambda\mathbf{x} + \mu\mathbf{y}) = \lambda^*\phi\mathbf{x} + \mu^*\phi\mathbf{y}$ *for all* $\mathbf{x}, \mathbf{y} \in U$ *and all* $\lambda, \mu \in \mathbb{C}$. *A bijective anti-linear map is called an anti-isomorphism.*

[*The treatment of* $\theta_L$ *would follow similar lines if we were prepared to develop the theme of anti-linearity a little further.*]

## 16.2 Rank and signature

Often we we need to deal with vector spaces which have no natural inner product or with circumstances when the inner product is irrelevant. Theorem 16.1.21 is then replaced by the Theorem 16.2.1.

**Theorem 16.2.1** *Let* $U$ *be a real* $n$-*dimensional vector space. If* $\alpha : U \times U \to \mathbb{R}$ *is a symmetric form, then we can find a basis* $\mathbf{e}_1, \mathbf{e}_2, \ldots, \mathbf{e}_n$ *and positive integers* $p$ *and* $m$ *with* $p + m \leq n$ *such that*

$$\alpha\left(\sum_{i=1}^{n} x_i \mathbf{e}_i, \sum_{j=1}^{n} y_j \mathbf{e}_j\right) = \sum_{k=1}^{p} x_k y_k - \sum_{k=p+1}^{p+m} x_k y_k$$

*for all* $x_i, y_j \in \mathbb{R}$.

By the change of basis formula of Lemma 16.1.18, Theorem 16.2.1 is equivalent to the following result on matrices.

**Lemma 16.2.2** *If $A$ is an $n \times n$ symmetric real matrix, we can find positive integers $p$ and $m$ with $p + m \leq n$ and an invertible real matrix $B$ such that $B^T A B = E$ where $E$ is a diagonal matrix whose first $p$ diagonal terms are 1, whose next $m$ diagonal terms are $-1$ and whose remaining diagonal terms are 0.*

*Proof* By Theorem 8.2.5, we can find an orthogonal matrix $M$ such that $M^T A M = D$ where $D$ is a diagonal matrix whose first $p$ diagonal entries are strictly positive, whose next $m$ diagonal terms are strictly negative and whose remaining diagonal terms are 0. (To obtain the appropriate order, just interchange the numbering of the associated eigenvectors.) We write $d_j$ for the $j$th diagonal term.

Now let $\Delta$ be the diagonal matrix whose $j$th entry is $\delta_j = |d_j|^{-1/2}$ for $1 \leq j \leq p + m$ and $\delta_j = 1$ otherwise. If we set $B = M\Delta$, then $B$ is invertible (since $M$ and $\Delta$ are) and

$$B^T A B = \Delta^T M^T A M \Delta = \Delta D \Delta = E$$

where $E$ is as stated in the lemma. $\qquad\square$

We automatically get a result on quadratic forms.

**Lemma 16.2.3** *Suppose that $q : \mathbb{R}^n \to \mathbb{R}$ is given by*

$$q(\mathbf{x}) = \sum_{i=1}^{n} \sum_{j=1}^{n} x_i a_{ij} x_j$$

*where $\mathbf{x} = (x_1, x_2, \ldots, x_n)^T$ with respect to some (not necessarily orthogonal) coordinate system $S$. Then there exists another (not necessarily orthogonal) coordinate system $S'$ such that*

$$q(\mathbf{y}) = \sum_{i=1}^{p} y_i^2 - \sum_{i=p+1}^{p+m} y_i^2$$

*where $\mathbf{y} = (y_1, y_2, \ldots, y_n)^T$ with respect to $S'$.*

**Exercise 16.2.4** *Obtain Lemma 16.2.3 directly from Lemma 16.1.22.*

If we are only interested in obtaining the largest number of results in the shortest time, it makes sense to obtain results which do not involve inner products from previous results involving inner products. However, there is a much older way of obtaining Lemma 16.2.3 which involves nothing more complicated than completing the square.

**Lemma 16.2.5** *(i) Suppose that $A = (a_{ij})_{1 \leq i, j \leq n}$ is a real symmetric $n \times n$ matrix with $a_{11} \neq 0$, and we set $b_{ij} = |a_{11}|^{-1}(|a_{11}|a_{ij} - a_{1i}a_{1j})$. Then $B = (b_{ij})_{2 \leq i, j \leq n}$ is a real symmetric matrix. Further, if $\mathbf{x} \in \mathbb{R}^n$ and we set*

$$y_1 = |a_{11}|^{1/2} \left( x_1 + \sum_{j=2}^{n} a_{1j} |a_{11}|^{-1/2} x_j \right)$$

*and* $y_j = x_j$ *otherwise, we have*

$$\sum_{i=1}^{n}\sum_{j=1}^{n} x_i a_{ij} x_j = \epsilon y_1^2 + \sum_{i=2}^{n}\sum_{j=2}^{n} y_i b_{ij} y_j$$

*where* $\epsilon = 1$ *if* $a_{11} > 0$ *and* $\epsilon = -1$ *if* $a_{11} < 0$.

(ii) *Suppose that* $A = (a_{ij})_{1\le i,j\le n}$ *is a real symmetric* $n \times n$ *matrix. Suppose further that* $\sigma : \{1, 2, \dots, n\} \to : \{1, 2, \dots, n\}$ *is a permutation (that is to say,* $\sigma$ *is a bijection) and we set* $c_{ij} = a_{\sigma i \sigma j}$. *Then* $C = (c_{ij})_{1\le i,j\le n}$ *is a real symmetric matrix. Further, if* $\mathbf{x} \in \mathbb{R}^n$ *and we set* $y_j = x_{\sigma(j)}$, *we have*

$$\sum_{i=1}^{n}\sum_{j=1}^{n} x_i a_{ij} x_j = \sum_{i=1}^{n}\sum_{j=1}^{n} y_i c_{ij} y_j.$$

(iii) *Suppose that* $n \ge 2$, $A = (a_{ij})_{1\le i,j\le n}$ *is a real symmetric* $n \times n$ *matrix and* $a_{11} = a_{22} = 0$, *but* $a_{12} \ne 0$. *Then there exists a real symmetric* $n \times n$ *matrix* $C$ *with* $c_{11} \ne 0$ *such that, if* $\mathbf{x} \in \mathbb{R}^n$ *and we set* $y_1 = (x_1 + x_2)/2$, $y_2 = (x_1 - x_2)/2$, $y_j = x_j$ *for* $j \ge 3$, *we have*

$$\sum_{i=1}^{n}\sum_{j=1}^{n} x_i a_{ij} x_j = \sum_{i=1}^{n}\sum_{j=1}^{n} y_i c_{ij} y_j.$$

(iv) *Suppose that* $A = (a_{ij})_{1\le i,j\le n}$ *is a non-zero real symmetric* $n \times n$ *matrix. Then we can find an* $n \times n$ *invertible matrix* $M = (m_{ij})_{1\le i,j\le n}$ *and a real symmetric* $(n - 1) \times (n - 1)$ *matrix* $B = (b_{ij})_{2\le i,j\le n}$ *such that, if* $\mathbf{x} \in \mathbb{R}^n$ *and we set* $y_i = \sum_{j=1}^{n} m_{ij} x_j$, *then*

$$\sum_{i=1}^{n}\sum_{j=1}^{n} x_i a_{ij} x_j = \epsilon y_1^2 + \sum_{i=2}^{n}\sum_{j=2}^{n} y_i b_{ij} y_j$$

*where* $\epsilon = \pm 1$.

*Proof* The first three parts are direct computation which is left to the reader, who should not be satisfied until she feels that all three parts are 'obvious'.[1] Part (iv) follows by using (ii), if necessary, and then either part (i) or part (iii) followed by part (i). □

Repeated use of Lemma 16.2.5 (iv) (and possibly Lemma 16.2.5 (ii) to reorder the diagonal) gives a constructive proof of Lemma 16.2.3. We shall run through the process in a particular case in Example 16.2.11 (second method).

It is clear that there are many different ways of reducing a quadratic form to our standard form $\sum_{j=1}^{p} x_j^2 - \sum_{j=p+1}^{p+m} x_j^2$, but it is not clear that each way will give the *same* value of $p$ and $m$. The matter is settled by the next theorem.

**Theorem 16.2.6 [Sylvester's law of inertia]** *Suppose that* $U$ *is a vector space of dimension* $n$ *over* $\mathbb{R}$ *and* $q : U \to \mathbb{R}^n$ *is a quadratic form. If* $\mathbf{e}_1, \mathbf{e}_2, \dots, \mathbf{e}_n$ *is a basis for* $U$ *such that*

$$q\left(\sum_{j=1}^{n} x_j \mathbf{e}_j\right) = \sum_{j=1}^{p} x_j^2 - \sum_{j=p+1}^{p+m} x_j^2$$

---

[1] This may take some time.

*and* $\mathbf{f}_1, \mathbf{f}_2, \ldots, \mathbf{f}_n$ *is a basis for* $U$ *such that*

$$q\left(\sum_{j=1}^n y_j \mathbf{f}_j\right) = \sum_{j=1}^{p'} y_j^2 - \sum_{j=p'+1}^{p'+m'} y_j^2,$$

*then* $p = p'$ *and* $m = m'$.

**Proof** The proof is short and neat, but requires thought to absorb fully.

Let $E$ be the subspace spanned by $\mathbf{e}_1, \mathbf{e}_2, \ldots, \mathbf{e}_p$ and $F$ the subspace spanned by $\mathbf{f}_{p'+1}$, $\mathbf{f}_{p'+2}, \ldots, \mathbf{f}_n$. If $\mathbf{x} \in E$ and $\mathbf{x} \neq \mathbf{0}$, then $\mathbf{x} = \sum_{j=1}^p x_j \mathbf{e}_j$ with not all the $x_j$ zero and so

$$q(\mathbf{x}) = \sum_{j=1}^p x_j^2 > 0.$$

If $\mathbf{x} \in F$, then a similar argument shows that $q(\mathbf{x}) \leq 0$. Thus

$$E \cap F = \{\mathbf{0}\}$$

and, by Lemma 5.4.10,

$$\dim(E + F) = \dim E + \dim F - \dim(E \cap F) = \dim E + \dim F.$$

But $E + F$ is a subspace of $U$, so $n \geq \dim(E + F)$ and

$$n \geq \dim(E + F) = \dim E + \dim F = p + (n - p').$$

Thus $p' \geq p$. Symmetry (or a similar argument) shows that $p \geq p'$, so $p = p'$. A similar argument (or replacing $q$ by $-q$) shows that $m = m'$.                    $\square$

*Remark 1.* In the next section we look at positive definite forms. Once you have looked at that section, the argument above can be thought of as follows. Suppose that $p \geq p'$. Let us ask 'What is the maximum dimension $\tilde{p}$ of a subspace $W$ for which the restriction of $q$ is positive definite?' Looking at $E$ we see that $\tilde{p} \geq p$, but this only gives a lower bound and we cannot be sure that $W \supseteq U$. If we want an upper bound we need to find 'the largest obstruction' to making $W$ large and it is natural to look for a large subspace on which $q$ is negative semi-definite. The subspace $F$ is a natural candidate.

*Remark 2.* Sylvester considered his law of inertia to be *obvious*. By looking at Exercise 16.1.17, convince yourself that it is, indeed, obvious if the dimension of $U$ is 3 or less (and you think sufficiently long and sufficiently geometrically). However, as the dimension of $U$ increases, it becomes harder to convince yourself (and very much harder to convince other people) that the result is obvious. The result was rediscovered and proved by Jacobi.

**Definition 16.2.7** *Suppose that $U$ is vector space of dimension $n$ over $\mathbb{R}$ and $q : U \to \mathbb{R}$ is a quadratic form. If $\mathbf{e}_1$, $\mathbf{e}_2$, ..., $\mathbf{e}_n$ is a basis for $U$ such that*

$$q\left(\sum_{j=1}^{n} x_j \mathbf{e}_j\right) = \sum_{j=1}^{p} x_j^2 - \sum_{j=p+1}^{p+m} x_j^2,$$

*we say that $q$ has rank $p + m$ and signature[2] $p - m$.*

Naturally, the rank and signature of a symmetric bilinear form or a symmetric matrix is defined to be the rank and signature of the associated quadratic form.

**Exercise 16.2.8** *Let $q : \mathbb{R}^2 \to \mathbb{R}$ be given by $q(x, y) = x^2 - y^2$. Sketch*

$$A = \{(x, y) \in \mathbb{R}^2 : q(x, y) \geq 0\} \text{ and } B = \{(x, y) \in \mathbb{R}^2 : q(x, y) = 0\}.$$

*Is either of $A$ or $B$ a subspace of $\mathbb{R}^2$? Give reasons.*

**Exercise 16.2.9** *(i) If $A = (a_{ij})$ is an $n \times n$ real symmetric matrix and we define $q : \mathbb{R}^n \to \mathbb{R}$ by $q(\mathbf{x}) = \sum_{i=1}^{n} \sum_{j=1}^{n} x_i a_{ij} x_j$, show that the 'signature rank' of $q$ is the 'matrix rank' (that is to say, the 'row rank') of $A$.*

*(ii) Given the roots of the characteristic polynomial of $A$ with multiple roots counted multiply, explain how to compute the rank and signature of $q$.*

**Exercise 16.2.10** *Although we have dealt with real quadratic forms, the same techniques work in the complex case. Note, however, that we must distinguish between complex quadratic forms (as discussed in this exercise) which do not mesh well with complex inner products and Hermitian forms (see the remark at the end of this exercise) which do.*

*(i) Consider the complex quadratic form $\gamma : \mathbb{C}^n \to \mathbb{C}$ given by*

$$\gamma(\mathbf{z}) = \sum_{u=1}^{n} \sum_{v=1}^{n} a_{uv} z_u z_v = \mathbf{z}^T A \mathbf{z}$$

*where $A$ is a symmetric complex matrix (that is to say, $A^T = A$). Show that, if $P$ is an invertible $n \times n$ matrix,*

$$\gamma(P\mathbf{z}) = \mathbf{z}^T (P^T A P)\mathbf{z}.$$

*(ii) We continue with the notation of (i). Show that we can choose $P$ so that*

$$\gamma(P\mathbf{z}) = \sum_{u=1}^{r} z_u^2$$

*for some $r$.*

---

[2] This is the definition of signature used in the Fenlands. Unfortunately there are several definitions of signature in use. The reader should always make clear which one she is using and always check which one anyone else is using. In my view, the convention that the triple $(p, m, n)$ forms the signature of $q$ is the best, but it is not worth making a fuss.

(iii) *Show, by considering the rank of appropriate matrices, or otherwise, that, if Q is an invertible n × n matrix such that*

$$\gamma(Q\mathbf{z}) = \sum_{u=1}^{r'} z_u^2$$

*for some r', we must have r = r'.*

(iv) *Show that there does not exist a non-zero subspace U of $\mathbb{C}^n$ such that*

$$\mathbf{z} \in U, \ \mathbf{z} \neq 0 \Rightarrow \gamma(\mathbf{z}) \quad \text{real and strictly positive.}$$

(v) *Show that, if $n \geq 2m$, there exists a subspace E of $\mathbb{C}^n$ of dimension m such that $\gamma(\mathbf{z}) = 0$ for all $\mathbf{z} \in E$.*

(vi) *Show that if $\gamma_1, \gamma_2, \ldots, \gamma_k$ are quadratic forms on $\mathbb{C}^n$ and $n \geq 2^k$, there exists a non-zero $\mathbf{z} \in \mathbb{C}^n$ such that*

$$\gamma_1(\mathbf{z}) = \gamma_2(\mathbf{z}) = \ldots = \gamma_k(\mathbf{z}) = 0.$$

*[Of course, it is often more interesting to look at* Hermitian *forms $\sum_{u=1}^{n} \sum_{v=1}^{n} a_{uv} z_u z_v^*$ with $a_{vu} = a_{uv}^*$. We do this in Exercise 16.2.13.]*

We now turn to actual calculation of the rank and signature.

**Example 16.2.11**  *Find the rank and signature of the quadratic form $q : \mathbb{R}^3 \to \mathbb{R}$ given by*

$$q(x_1, x_2, x_3) = x_1 x_2 + x_2 x_3 + x_3 x_1.$$

*Solution*  We give three methods. Most readers are only likely to meet this sort of problem in an artificial setting where any of the methods might be appropriate, but, in the absence of special features, the third method is not appropriate and I would expect the arithmetic for the first method to be rather horrifying. Fortunately the second method is easy to apply and will always work. (See also Theorem 16.3.10 for the case when you are only interested in whether the rank and signature are both $n$.)

*First method*  Since $q$ and $2q$ have the same signature, we need only look at $2q$. The quadratic form $2q$ is associated with the symmetric matrix

$$A = \begin{pmatrix} 0 & 1 & 1 \\ 1 & 0 & 1 \\ 1 & 1 & 0 \end{pmatrix}.$$

We now seek the eigenvalues of $A$. Observe that

$$\det(tI - A) = \det \begin{pmatrix} t & -1 & -1 \\ -1 & t & -1 \\ -1 & -1 & t \end{pmatrix}$$

$$= t \det \begin{pmatrix} t & -1 \\ -1 & t \end{pmatrix} + \det \begin{pmatrix} -1 & -1 \\ -1 & t \end{pmatrix} - \det \begin{pmatrix} -1 & t \\ -1 & -1 \end{pmatrix}$$

$$= t(t^2 - 1) - (t + 1) - (t + 1) = (t + 1)\big(t(t - 1) - 2\big)$$

$$= (t + 1)(t^2 - t - 2) = (t + 1)^2(t - 2).$$

Thus the characteristic polynomial of $A$ has one strictly positive root and two strictly negative roots (multiple roots being counted multiply). We conclude that $q$ has rank $1 + 2 = 3$ and signature $1 - 2 = -1$.

*Second method* We use the ideas of Lemma 16.2.5. The substitutions $x_1 = y_1 + y_2$, $x_2 = y_1 - y_2$, $x_3 = y_3$ give

$$q(x_1, x_2, x_3) = x_1 x_2 + x_2 x_3 + x_3 x_1$$

$$= (y_1 + y_2)(y_1 - y_2) + (y_1 - y_2)y_3 + y_3(y_1 + y_2)$$

$$= y_1^2 - y_2^2 + 2y_1 y_3$$

$$= (y_1 + y_3)^2 - y_2^2 - y_3^2,$$

so $q$ has rank $1 + 2 = 3$ and signature $1 - 2 = -1$.

The substitutions $w_1 = y_1 + y_3$, $w_2 = y_2$, $w_3 = y_3$ give

$$q(x_1, x_2, x_3) = w_1^2 - w_2^2 - w_3^2,$$

but we do not need this step to determine the rank and signature.

*Third method* This method will only work if we have some additional insight into where our quadratic form comes from or if we are doing an examination and the examiner gives us a hint. Observe that, if we take

$$E_1 = \{(x_1, x_2, x_3) : x_1 + x_2 + 2x_3 = 0\},$$
$$E_2 = \{(x_1, x_2, x_3) : x_1 = x_2, x_3 = 0\},$$

then $q(\mathbf{e}) < 0$ for $\mathbf{e} \in E_1 \setminus \{\mathbf{0}\}$ and $q(\mathbf{e}) > 0$ for $\mathbf{e} \in E_2 \setminus \{\mathbf{0}\}$. Since $\dim E_1 = 2$ and $\dim E_2 = 1$ it follows, using the notation of Definition 16.2.7, that $p \geq 1$ and $m \geq 2$. Since $\mathbb{R}^3$ has dimension 3, $p + m \leq 3$ so $p = 1$, $m = 2$ and $q$ must have rank 3 and signature $-1$. $\qquad\square$

**Exercise 16.2.12** *Find the rank and signature of the real quadratic form*

$$q(x_1, x_2, x_3) = x_1 x_2 + x_2 x_3 + x_3 x_1 + a x_1^2$$

*for all values of a.*

**Exercise 16.2.13** *Suppose that we work over* $\mathbb{C}$ *and consider* Hermitian *rather than* symmetric *forms and matrices.*

(i) *Show that, given any Hermitian matrix A, we can find an invertible matrix P such that* $P^*AP$ *is diagonal with diagonal entries taking the values* $1, -1$ *or* $0$.

(ii) *If* $A = (a_{rs})$ *is an* $n \times n$ *Hermitian matrix, show that* $\sum_{r=1}^{n} \sum_{s=1}^{n} z_r a_{rs} z_s^*$ *is real for all* $z_r \in \mathbb{C}$.

(iii) *Suppose that A is a Hermitian matrix and* $P_1$, $P_2$ *are invertible matrices such that* $P_1^*AP_1$ *and* $P_2^*AP_2$ *are diagonal with diagonal entries taking the values* $1, -1$ *or* $0$. *Prove that the number of entries of each type is the same for both diagonal matrices.*

## 16.3 Positive definiteness

For many mathematicians the most important quadratic forms are the *positive definite* quadratic forms.

**Definition 16.3.1** *Let U be a vector space over* $\mathbb{R}$. *A quadratic form* $q : U \rightarrow \mathbb{R}$ *is said to be* positive semi-definite *if*

$$q(\mathbf{u}) \geq 0 \quad \text{for all } \mathbf{u} \in U$$

*and* strictly positive definite *if*

$$q(\mathbf{u}) > 0 \quad \text{for all } \mathbf{u} \in U \text{ with } \mathbf{u} \neq \mathbf{0}.$$

As might be expected, mathematicians sometimes use the words 'positive definite' to mean 'strictly positive definite' and sometimes to mean 'strictly positive definite or positive semi-definite'.[3]

Naturally, a symmetric bilinear form or a symmetric matrix is said to be positive semi-definite or strictly positive definite if the associated quadratic form is.

**Exercise 16.3.2** (i) *Write down definitions of negative semi-definite and strictly negative definite quadratic forms in the style of Definition 16.3.1 so that the following result holds. A quadratic form* $q : U \rightarrow \mathbb{R}$ *is strictly negative definite (respectively negative semi-definite) if and only if* $-q$ *is strictly positive definite (respectively positive semi-definite).*

(ii) *Show that every quadratic form over a real finite dimensional vector space is the sum of a positive semi-definite and a negative semi-definite quadratic form. Is the decomposition unique? Give a proof or a counterexample.*

**Exercise 16.3.3** (*Requires a smidgen of probability theory.*) *Let* $X_1, X_2, \ldots, X_n$ *be bounded real valued random variables. Show that the matrix* $E = (\mathbb{E}X_i X_j)$ *is symmetric and positive semi-definite. Show that E is not strictly positive definite if and only if we can find* $c_i \in \mathbb{R}$ *not all zero such that*

$$\Pr(c_1 X_1 + c_2 X_2 + \cdots + c_n X_n = 0) = 1.$$

---

[3] Compare the use of 'positive' sometimes to mean 'strictly positive' and sometimes to mean 'non-negative'.

**Exercise 16.3.4** *Show that a quadratic form q over an n-dimensional real vector space is strictly positive definite if and only if it has rank n and signature n. State and prove conditions for q to be positive semi-definite. State and prove conditions for q to be negative semi-definite.*

The next exercise indicates one reason for the importance of positive definiteness.

**Exercise 16.3.5** *Let U be a real vector space. Show that $\alpha : U^2 \to \mathbb{R}$ is an inner product if and only if $\alpha$ is a symmetric form which gives rise to a strictly positive definite quadratic form.*

An important example of a quadratic form which is neither positive nor negative semi-definite appears in Special Relativity[4] as

$$q(x_1, x_2, x_3, x_4) = x_1^2 + x_2^2 + x_3^2 - x_4^2$$

or, in a form more familiar in elementary texts,

$$q_c(x, y, z, t) = x^2 + y^2 + z^2 - c^2 t^2.$$

The reader who has done multidimensional calculus will be aware of another important application.

**Exercise 16.3.6** *(To be answered according to the reader's background.) Suppose that $f : \mathbb{R}^n \to \mathbb{R}$ is a smooth function.*
*(i) If f has a minimum at $\mathbf{a}$, show that $(\partial f / \partial x_i)(\mathbf{a}) = 0$ for all i and the Hessian matrix*

$$\left( \frac{\partial^2 f}{\partial x_i \partial x_j}(\mathbf{a}) \right)_{1 \le i, j \le n}$$

*is positive semi-definite.*
*(ii) If $(\partial f / \partial x_i)(\mathbf{a}) = 0$ for all i and the Hessian matrix*

$$\left( \frac{\partial^2 f}{\partial x_i \partial x_j}(\mathbf{a}) \right)_{1 \le i, j \le n}$$

*is strictly positive definite, show that f has a strict minimum at $\mathbf{a}$.*
*(iii) Give an example in which f has a strict minimum but the Hessian is not strictly positive definite. (Note that there are examples with $n = 1$.)*
[*We gave an unsophisticated account of these matters in Section 8.3, but the reader may well be able to give a deeper treatment.*]

**Exercise 16.3.7** *Locate the maxima, minima and saddle points of the function $f : \mathbb{R}^2 \to \mathbb{R}$ given by $f(x, y) = \sin x \sin y$.*

---

[4] Because of this, non-degenerate symmetric forms are sometimes called inner products. Although non-degenerate symmetric forms play the role of inner products in Relativity theory, this nomenclature is *not consistent* with normal usage. The analyst will note that, since conditional convergence is hard to handle, the theory of non-degenerate symmetric forms will not generalise to infinite dimensional spaces in the same way as the theory of strictly positive definite symmetric forms.

If the reader has ever wondered how to check whether a Hessian matrix is, indeed, strictly positive definite when $n$ is large, she should observe that the matter can be settled by finding the rank and signature of the associated quadratic form by the methods of the previous section. The following discussion gives a particularly clean version of the 'completion of squares' method when we are interested in positive definiteness.

**Lemma 16.3.8**  *If $A = (a_{ij})_{1 \leq i, j \leq n}$ is a strictly positive definite matrix, then $a_{11} > 0$.*

*Proof* If $x_1 = 1$ and $x_j = 0$ for $2 \leq j \leq n$, then $\mathbf{x} \neq \mathbf{0}$ and, since $A$ is strictly positive definite,

$$a_{11} = \sum_{i=1}^{n} \sum_{j=1}^{n} x_i a_{ij} x_j > 0.$$
                                                                                        $\square$

**Lemma 16.3.9**  *(i) If $A = (a_{ij})_{1 \leq i, j \leq n}$ is a real symmetric matrix with $a_{11} > 0$, then there exists a unique column vector $\mathbf{l} = (l_{i1})_{1 \leq i \leq n}$ with $l_{11} > 0$ such that $A - \mathbf{l}\mathbf{l}^T$ has all entries in its first column and first row zero.*

*(ii) Let $A$ and $\mathbf{l}$ be as in (i) and let*

$$b_{ij} = a_{ij} - l_i l_j \quad \text{for } 2 \leq i, j \leq n.$$

*Then $B = (b_{ij})_{2 \leq i, j \leq n}$ is strictly positive definite if and only if $A$ is.*

(The matrix $B$ is called the *Schur complement* of $a_{11}$ in $A$, but we shall not make use of this name.)

*Proof* (i) If $A - \mathbf{l}\mathbf{l}^T$ has all entries in its first column and first row zero, then

$$l_1^2 = a_{11},$$
$$l_1 l_i = a_{i1} \quad \text{for } 2 \leq i \leq n.$$

Thus, if $l_{11} > 0$, we have $l_{ii} = a_{11}^{1/2}$, the positive square root of $a_{11}$, and $l_i = a_{i1} a_{11}^{-1/2}$ for $2 \leq i \leq n$.

Conversely, if $l_{11} = a_{11}^{1/2}$, the positive square root of $a_{11}$, and $l_i = a_{i1} a_{11}^{1/2}$ for $2 \leq i \leq n$ then, by inspection, $A - \mathbf{l}\mathbf{l}^T$ has all entries in its first column and first row zero.

(ii) If $B$ is strictly positive definite,

$$\sum_{i=1}^{n} \sum_{j=1}^{n} x_i a_{ij} x_j = a_{11} \left( x_1 + \sum_{i=2}^{n} a_{i1} a_{11}^{-1/2} x_i \right)^2 + \sum_{i=2}^{n} \sum_{j=2}^{n} x_i b_{ij} x_j \geq 0$$

with equality if and only if $x_i = 0$ for $2 \leq i \leq n$ and

$$x_1 + \sum_{i=2}^{n} a_{i1} a_{11}^{-1/2} x_i = 0,$$

that is to say, if and only if $x_i = 0$ for $1 \leq i \leq n$. Thus $A$ is strictly positive definite.

If $A$ is strictly positive definite, then setting $x_1 = -\sum_{i=2}^{n} a_{i1} a_{11}^{-1/2} y_i$ and $x_i = y_i$ for $2 \le i \le n$, we have

$$\sum_{i=2}^{n} \sum_{j=2}^{n} y_i b_{ij} y_j = a_{11} \left( x_1 + \sum_{i=2}^{n} a_{i1} a_{11}^{-1/2} x_i \right)^2 + \sum_{i=2}^{n} \sum_{j=2}^{n} x_i b_{ij} x_j$$

$$= \sum_{i=1}^{n} \sum_{j=1}^{n} x_i a_{ij} x_j \ge 0$$

with equality if and only if $x_i = 0$ for $1 \le i \le n$, that is to say, if and only if $y_i = 0$ for $2 \le i \le n$. Thus $B$ is strictly positive definite. $\square$

The following theorem is a simple consequence.

**Theorem 16.3.10 [The Cholesky factorisation]** *An $n \times n$ real symmetric matrix $A$ is strictly positive definite if and only if there exists a lower triangular matrix $L$ with all diagonal entries strictly positive such that $LL^T = A$. If the matrix $L$ exits, it is unique.*

*Proof* If $A$ is strictly positive definite, the existence and uniqueness of $L$ follow by induction, using Lemmas 16.3.8 and 16.3.9. If $A = LL^T$, then

$$\mathbf{x}^T A \mathbf{x} = \mathbf{x}^T L L^T \mathbf{x} = \|L^T \mathbf{x}\|^2 \ge 0$$

with equality if and only if $L^T \mathbf{x} = \mathbf{0}$ and so (since $L^T$ is triangular with non-zero diagonal entries) if and only if $\mathbf{x} = \mathbf{0}$. $\square$

**Exercise 16.3.11** *If $L$ is a lower triangular matrix with all diagonal entries non-zero, show that $A = LL^T$ is a symmetric strictly positive definite matrix.*

*If $\tilde{L}$ is a lower triangular matrix, what can you say about $\tilde{A} = \tilde{L}\tilde{L}^T$? (See also Exercise 16.5.33.)*

The proof of Theorem 16.3.10 gives an easy computational method of obtaining the factorisation $A = LL^T$ when $A$ is strictly positive definite. If $A$ is not positive definite, then the method will reveal the fact.

**Exercise 16.3.12** *How will the method reveal that $A$ is not strictly positive definite and why?*

**Example 16.3.13** *Find a lower triangular matrix $L$ such that $LL^T = A$ where*

$$A = \begin{pmatrix} 4 & -6 & 2 \\ -6 & 10 & -5 \\ 2 & -5 & 14 \end{pmatrix},$$

*or show that no such matrix $L$ exists.*

*Solution* If $\mathbf{l}_1 = (2, -3, 1)^T$, we have

$$A - \mathbf{l}_1\mathbf{l}_1^T = \begin{pmatrix} 4 & -6 & 2 \\ -6 & 10 & -5 \\ 2 & -5 & 14 \end{pmatrix} - \begin{pmatrix} 4 & -6 & 2 \\ -6 & 9 & -3 \\ 2 & -3 & 1 \end{pmatrix} = \begin{pmatrix} 0 & 0 & 0 \\ 0 & 1 & -2 \\ 0 & -2 & 13 \end{pmatrix}.$$

If $\mathbf{l}_2 = (1, -2)^T$, we have

$$\begin{pmatrix} 1 & -2 \\ -2 & 13 \end{pmatrix} - \mathbf{l}_2\mathbf{l}_2^T = \begin{pmatrix} 1 & -2 \\ -2 & 13 \end{pmatrix} - \begin{pmatrix} 1 & -2 \\ -2 & 4 \end{pmatrix} = \begin{pmatrix} 0 & 0 \\ 0 & 9 \end{pmatrix}.$$

Thus $A = LL^T$ with

$$L = \begin{pmatrix} 2 & 0 & 0 \\ -3 & 1 & 0 \\ 1 & -2 & 3 \end{pmatrix}.$$

$\square$

**Exercise 16.3.14** *Show, by attempting the factorisation procedure just described, that*

$$A = \begin{pmatrix} 4 & -6 & 2 \\ -6 & 8 & -5 \\ 2 & -5 & 14 \end{pmatrix},$$

*is not positive semi-definite.*

**Exercise 16.3.15** *(i) Check that you understand why the method of factorisation given by the proof of Theorem 16.3.10 is essentially just a sequence of 'completing the squares'.*

*(ii) Show that the method will either give a factorisation $A = LL^T$ for an $n \times n$ symmetric matrix or reveal that $A$ is not strictly positive definite in less than $Kn^3$ operations. You may choose the value of $K$.*
*[We give a related criterion for strictly positive definiteness in Exercise 16.5.27.]*

**Exercise 16.3.16** *Apply the method just given to*

$$H_3 = \begin{pmatrix} 1 & \frac{1}{2} & \frac{1}{3} \\ \frac{1}{2} & \frac{1}{3} & \frac{1}{4} \\ \frac{1}{3} & \frac{1}{4} & \frac{1}{5} \end{pmatrix} \quad \text{and} \quad H_4(\lambda) = \begin{pmatrix} 1 & \frac{1}{2} & \frac{1}{3} & \frac{1}{4} \\ \frac{1}{2} & \frac{1}{3} & \frac{1}{4} & \frac{1}{5} \\ \frac{1}{3} & \frac{1}{4} & \frac{1}{5} & \frac{1}{6} \\ \frac{1}{4} & \frac{1}{5} & \frac{1}{6} & \lambda \end{pmatrix}.$$

*Find the smallest $\lambda_0$ such that $\lambda > \lambda_0$ implies $H_4(\lambda)$ strictly positive definite. Compute $\frac{1}{7} - \lambda_0$.*

**Exercise 16.3.17** *(This is just a slightly more general version of Exercise 10.5.18.)*
*(i) By induction on the degree of $P$ and considering the zeros of $P'$, or otherwise, show that, if $a_j > 0$ for each $j$, the real roots of*

$$P(t) = t^n + \sum_{j=1}^n (-1)^{n-j} a_j t^j$$

*(if any) are all strictly positive.*

*(ii) Let A be an n × n real matrix with n real eigenvalues $\lambda_j$ (repeated eigenvalues being counted multiply). Show that the eigenvalues of A are strictly positive if and only if*

$$\sum_{j=1}^{n} \lambda_j > 0, \quad \sum_{j \neq i} \lambda_i \lambda_j > 0, \quad \sum_{i,j,k \text{ distinct}} \lambda_i \lambda_j \lambda_k > 0, \ \dots$$

*(iii) Find a real 2 × 2 matrix $A = (a_{ij})$ such that*

$$\det(tI - A) = t^2 - b_1 t + b_0$$

*with $b_1$, $b_0 > 0$, but A has no real eigenvalues.*

*(iv) Let $A = (a_{ij})$ be a diagonalisable 3 × 3 real matrix with 3 real eigenvalues. Show that the eigenvalues of A are strictly positive if and only if*

$$\operatorname{Tr} A = a_{11} + a_{22} + a_{33} > 0$$

$$a_{11}a_{22} + a_{22}a_{33} + a_{33}a_{22} - a_{12}a_{21} - a_{23}a_{32} - a_{31}a_{13} > 0$$

$$\det A > 0.$$

*If the reader reflects, she will see that using a 'nice explicit formula' along the lines of (iii) for an n × n matrix A amounts to using row expansion to evaluate $\det(tI - A)$ which is computationally a* very bad idea *when n is large.*

The ideas of this section come in very useful in mechanics. It will take us some time to come to the point of the discussion, so the reader may wish to skim through what follows and then reread it more slowly. The next paragraph is not supposed to be rigorous, but it may be helpful.

Suppose that we have two strictly positive definite forms $p_1$ and $p_2$ in $\mathbb{R}^3$ with the usual coordinate axes. The equations $p_1(x, y, z) = 1$ and $p_2(x, y, z) = 1$ define two ellipsoids $\Gamma_1$ and $\Gamma_2$. Our algebraic theorems tell us that, by rotating the coordinate axes, we can ensure that the axes of symmetry of $\Gamma_1$ lie along the new axes. By rescaling along each of the new axes, we can convert $\Gamma_1$ to a sphere $\Gamma_1'$. The rescaling converts $\Gamma_2$ to a new ellipsoid $\Gamma_2'$. A further rotation allows us to ensure that axes of symmetry of $\Gamma_2'$ lie along the resulting coordinate axes. Such a rotation leaves the sphere $\Gamma_1'$ unaltered.

Replacing our geometry by algebra and strengthening our results slightly, we obtain the following theorem.

**Theorem 16.3.18[5]** *If A is an n × n strictly positive definite real symmetric matrix and B is an n × n real symmetric matrix, we can find an invertible matrix M such that $M^T A M = I$ and $M^T B M$ is a diagonal matrix D. The diagonal entries of D are the roots of*

$$\det(tA - B) = 0$$

*multiple roots being counted multiply.*

---

[5] This is a famous result of Weierstrass. If the reader reflects, she will see that it is not surprising that mathematicians were interested in the diagonalisation of quadratic forms long before the ideas involved were used to study the diagonalisation of matrices.

*Proof* Since $A$ is positive definite, we can find an invertible matrix $P$ such that $P^T A P = I$. Since $P^T B P$ is symmetric, we can find an orthogonal matrix $Q$ such that $Q^T P^T B P Q = D$ a diagonal matrix. If we set $M = PQ$, it follows that $M$ is invertible and

$$M^T A M = Q^T P^T A P Q = Q^T I Q = Q^T Q = I, \quad M^T B M = Q^T P^T B P Q = D.$$

Since

$$\det(tI - D) = \det M^T (tA - B) M$$
$$= \det M \det M^T \det(tA - B) = (\det M)^2 \det(tA - B),$$

we know that $\det(tI - D) = 0$ if and only if $\det(tA - B) = 0$, so the final sentence of the theorem follows. $\qquad\square$

**Exercise 16.3.19** *We work with the real numbers.*

*(i) Suppose that*

$$A = \begin{pmatrix} 1 & 0 \\ 0 & -1 \end{pmatrix} \quad \text{and} \quad B = \begin{pmatrix} 0 & 1 \\ 1 & 0 \end{pmatrix}.$$

*Show that, if there exists an invertible matrix $P$ such that $P^T A P$ and $P^T B P$ are diagonal, then there exists an invertible matrix $M$ such that $M^T A M = A$ and $M^T B M$ is diagonal.*

*By writing out the corresponding matrices when*

$$M = \begin{pmatrix} a & b \\ c & d \end{pmatrix}$$

*show that no such $M$ exists and so $A$ and $B$ cannot be simultaneously diagonalisable. (See also Exercise 16.5.21.)*

*(ii) Show that the matrices $A$ and $B$ in (i) have rank 2 and signature 0. Sketch*

$$\{(x, y) : x^2 - y^2 \geq 0\}, \ \{(x, y) : 2xy \geq 0\} \quad \text{and} \quad \{(x, y) : ux^2 - vy^2 \geq 0\}$$

*for $u > 0 > v$ and $v > 0 > u$. Explain geometrically, as best you can, why no $M$ of the type required in (i) can exist.*

*(iii) Suppose that*

$$C = \begin{pmatrix} -1 & 0 \\ 0 & -1 \end{pmatrix}.$$

*Without doing any calculation, decide whether $C$ and $B$ are simultaneously diagonalisable and give your reason.*

We now introduce some mechanics. Suppose that we have a mechanical system whose behaviour is described in terms of coordinates $q_1, q_2, \ldots, q_n$. Suppose further that the system rests in equilibrium when $\mathbf{q} = \mathbf{0}$. Often the system will have a *kinetic energy* $E(\dot{\mathbf{q}}) = E(\dot{q}_1, \dot{q}_2, \ldots, \dot{q}_n)$ and a *potential energy* $V(\mathbf{q})$. We expect $E$ to be a strictly positive definite quadratic form with associated symmetric matrix $A$. There is no loss in generality in taking $V(\mathbf{0}) = 0$. Since the system is in equilibrium, $V$ is stationary at $\mathbf{0}$ and (at least to the second order of approximation) $V$ will be a quadratic form. We assume that (at least initially) higher order terms may be neglected and we may take $V$ to be a quadratic form with associated matrix $B$.

We observe that

$$M\dot{\mathbf{q}} = \frac{d}{dt} M\mathbf{q}$$

and so, by Theorem 16.3.18, we can find new coordinates $Q_1, Q_2, \ldots, Q_n$ such that the kinetic energy of the system with respect to the new coordinates is

$$\tilde{E}(\dot{\mathbf{Q}}) = \dot{Q}_1^2 + \dot{Q}_2^2 + \cdots + \dot{Q}_n^2$$

and the potential energy is

$$\tilde{V}(\mathbf{Q}) = \lambda_1 Q_1^2 + \lambda_2 Q_2^2 + \cdots + \lambda_n Q_n^2$$

where the $\lambda_j$ are the roots of $\det(tA - B) = 0$.

If we take $Q_i = 0$ for $i \neq j$, then our system reduces to one in which the kinetic energy is $\dot{Q}_j^2$ and the potential energy is $\lambda_j Q_j^2$. If $\lambda_j < 0$, this system is unstable. If $\lambda_j > 0$, we have a harmonic oscillator frequency $\lambda_j^{1/2}$. Clearly the system is unstable if any of the roots of $\det(tA - B) = 0$ are strictly negative. If all the roots $\lambda_j$ are strictly positive it seems plausible that the *general solution* for our system is

$$Q_j(t) = K_j \cos(\lambda_j^{1/2} t + \phi_j) \, [1 \le j \le n]$$

for some constants $K_j$ and $\phi_j$. More sophisticated analysis, using Lagrangian mechanics, enables us to make precise the notion of a 'mechanical system given by coordinates $\mathbf{q}$' and confirms our conclusions.

The reader may wish to look at Exercise 8.5.4 if she has not already done so.

## 16.4 Antisymmetric bilinear forms

In view of Lemma 16.1.11 it is natural to seek a reasonable way of looking at antisymmetric bilinear forms.

As we might expect, there is a strong link with *antisymmetric matrices* (that is to say, square matrices $A$ with $A^T = -A$).

**Lemma 16.4.1** *Let $U$ be a finite dimensional vector space over $\mathbb{R}$ with basis $\mathbf{e}_1, \mathbf{e}_2, \ldots,$ $\mathbf{e}_n$.*

*(i) If $A = (a_{ij})$ is an $n \times n$ real antisymmetric matrix, then*

$$\alpha \left( \sum_{i=1}^{n} x_i \mathbf{e}_i, \sum_{j=1}^{n} y_j \mathbf{e}_j \right) = \sum_{i=1}^{n} \sum_{j=1}^{n} x_i a_{ij} y_j$$

*(for $x_i, y_j \in \mathbb{R}$) defines an antisymmetric form.*

*(ii) If $\alpha : U \times U \to \mathbb{R}$ is an antisymmetric form, then there is a unique $n \times n$ real antisymmetric matrix $A = (a_{ij})$ with*

$$\alpha\left(\sum_{i=1}^{n} x_i \mathbf{e}_i, \sum_{j=1}^{n} y_j \mathbf{e}_j\right) = \sum_{i=1}^{n}\sum_{j=1}^{n} x_i a_{ij} y_j$$

*for all $x_i$, $y_j \in \mathbb{R}$.*

The proof is left as an exercise for the reader.

**Exercise 16.4.2** *If $\alpha$ and $A$ are as in Lemma 16.4.1 (ii), find $q(\mathbf{u}) = \alpha(\mathbf{u}, \mathbf{u})$.*

**Exercise 16.4.3** *Suppose that $A$ is an $n \times n$ matrix with real entries such that $A^T = -A$. Show that, if we work in $\mathbb{C}$, $iA$ is a Hermitian matrix. Deduce that the eigenvalues of $A$ have the form $\lambda i$ with $\lambda \in \mathbb{R}$.*

*If $B$ is an antisymmetric matrix with real entries and $M$ is an invertible matrix with real entries such that $M^T B M$ is diagonal, what can we say about $B$ and why?*

Clearly, if we want to work over $\mathbb{R}$, we cannot hope to 'diagonalise' a general antisymmetric form. However, we can find another 'canonical reduction' along the lines of our first proof that a symmetric matrix can be diagonalised (see Theorem 8.2.5). We first prove a lemma which contains the essence of the matter.

**Lemma 16.4.4** *Let $U$ be a vector space over $\mathbb{R}$ and let $\alpha$ be a non-zero antisymmetric form.*

*(i) We can find $\mathbf{e}_1$, $\mathbf{e}_2 \in U$ such that $\alpha(\mathbf{e}_1, \mathbf{e}_2) = 1$.*

*(ii) Let $\mathbf{e}_1$, $\mathbf{e}_2 \in U$ obey the conclusions of (i). Then the two vectors are linearly independent. Further, if we write*

$$E = \{\mathbf{u} \in U \ : \ \alpha(\mathbf{e}_1, \mathbf{u}) = \alpha(\mathbf{e}_2, \mathbf{u}) = 0\},$$

*then $E$ is a subspace of $U$ and*

$$U = \text{span}\{\mathbf{e}_1, \mathbf{e}_2\} \oplus E.$$

*Proof* (i) If $\alpha$ is non-zero, there must exist $\mathbf{u}_1$, $\mathbf{u}_2 \in U$ such that $\alpha(\mathbf{u}_1, \mathbf{u}_2) \neq 0$. Set

$$\mathbf{e}_1 = \frac{1}{\alpha(\mathbf{u}_1, \mathbf{u}_2)}\mathbf{u}_1 \quad \text{and} \quad \mathbf{e}_2 = \mathbf{u}_2.$$

(ii) We leave it to the reader to check that $E$ is a subspace. If

$$\mathbf{u} = \lambda_1 \mathbf{e}_1 + \lambda_2 \mathbf{e}_2 + \mathbf{e}$$

with $\lambda_1, \lambda_2 \in \mathbb{R}$ and $\mathbf{e} \in E$, then

$$\alpha(\mathbf{u}, \mathbf{e}_1) = \lambda_1 \alpha(\mathbf{e}_1, \mathbf{e}_1) + \lambda_2 \alpha(\mathbf{e}_2, \mathbf{e}_1) + \alpha(\mathbf{e}, \mathbf{e}_1) = 0 - \lambda_2 + 0 = -\lambda_2,$$

so $\lambda_2 = -\alpha(\mathbf{u}, \mathbf{e}_1)$. Similarly $\lambda_1 = \alpha(\mathbf{u}, \mathbf{e}_2)$ and so

$$\mathbf{e} = \mathbf{u} - \alpha(\mathbf{u}, \mathbf{e}_2)\mathbf{e}_1 + \alpha(\mathbf{u}, \mathbf{e}_1)\mathbf{e}_2.$$

Conversely, if

$$\mathbf{v} = \alpha(\mathbf{u}, \mathbf{e}_2)\mathbf{e}_1 - \alpha(\mathbf{u}, \mathbf{e}_1)\mathbf{e}_2 \quad \text{and} \quad \mathbf{e} = \mathbf{u} - \mathbf{v},$$

then

$$\alpha(\mathbf{e}, \mathbf{e}_1) = \alpha(\mathbf{u}, \mathbf{e}_1) - \alpha(\mathbf{u}, \mathbf{e}_2)\alpha(\mathbf{e}_1, \mathbf{e}_1) + \alpha(\mathbf{u}, \mathbf{e}_1)\alpha(\mathbf{e}_2, \mathbf{e}_1)$$
$$= \alpha(\mathbf{u}, \mathbf{e}_1) - 0 - \alpha(\mathbf{u}, \mathbf{e}_1) = 0$$

and, similarly, $\alpha(\mathbf{e}, \mathbf{e}_2) = 0$, so $\mathbf{e} \in E$.

Thus any $\mathbf{u} \in U$ can be written in one and only one way as

$$\mathbf{u} = \lambda_1 \mathbf{e}_1 + \lambda_2 \mathbf{e}_2 + \mathbf{e}$$

with $\mathbf{e} \in E$. In other words,

$$U = \text{span}\{\mathbf{e}_1, \mathbf{e}_2\} \oplus E. \qquad \square$$

**Theorem 16.4.5** *Let $U$ be a vector space of dimension $n$ over $\mathbb{R}$ and let $\alpha : U^2 \to \mathbb{R}$ be an antisymmetric form. Then we can find an integer $m$ with $0 \le 2m \le n$ and a basis $\mathbf{e}_1$, $\mathbf{e}_2, \ldots, \mathbf{e}_n$ such that*

$$\alpha(\mathbf{e}_i, \mathbf{e}_j) = \begin{cases} 1 & \text{if } i = 2r - 1, \ j = 2r \text{ and } 1 \le r \le m, \\ -1 & \text{if } i = 2r, \ j = 2r - 1 \text{ and } 1 \le r \le m, \\ 0 & \text{otherwise.} \end{cases}$$

*Proof* We use induction. If $n = 0$ the result is trivial. If $n = 1$, then $\alpha = 0$ since

$$\alpha(x\mathbf{e}, y\mathbf{e}) = xy\alpha(\mathbf{e}, \mathbf{e}) = 0,$$

and the result is again trivial. Now suppose that the result is true for all $0 \le n \le N$ where $N \ge 1$. We wish to prove the result when $n = N + 1$.

If $\alpha = 0$, any choice of basis will do. If not, the previous lemma tells us that we can find $\mathbf{e}_1$, $\mathbf{e}_2$ linearly independent vectors and a subspace $E$ such that

$$U = \text{span}\{\mathbf{e}_1\mathbf{e}_2\} \oplus E,$$

$\alpha(\mathbf{e}_1, \mathbf{e}_2) = 1$ and $\alpha(\mathbf{e}_1, \mathbf{e}) = \alpha(\mathbf{e}_2, \mathbf{e}) = 0$ for all $\mathbf{e} \in E$. Automatically, $E$ has dimension $N - 1$ and the restriction $\alpha|_{E^2}$ of $\alpha$ to $E^2$ is an antisymmetric form. Thus we can find an $m$ with $2m \le N + 1$ and a basis $\mathbf{e}_3, \mathbf{e}_4, \ldots, \mathbf{e}_{N+1}$ such that

$$\alpha(\mathbf{e}_i, \mathbf{e}_j) = \begin{cases} 1 & \text{if } i = 2r - 1, \ j = 2r \text{ and } 2 \le r \le m, \\ -1 & \text{if } i = 2r, \ j = 2r - 1 \text{ and } 2 \le r \le m, \\ 0 & \text{otherwise.} \end{cases}$$

The vectors $\mathbf{e}_1, \mathbf{e}_2, \ldots, \mathbf{e}_{N+1}$ form a basis such that

$$\alpha(\mathbf{e}_i, \mathbf{e}_j) = \begin{cases} 1 & \text{if } i = 2r - 1, j = 2r \text{ and } 2 \leq r \leq m, \\ -1 & \text{if } i = 2r, j = 2r - 1 \text{ and } 2 \leq r \leq m, \\ 0 & \text{otherwise,} \end{cases}$$

so the induction is complete. $\square$

Part (ii) of the next exercise completes our discussion.

**Exercise 16.4.6** (i) *If A is an $n \times n$ real antisymmetric matrix, show that there is an $n \times n$ real invertible matrix M, such that $M^T A M$ is a matrix with matrices taking the form*

$$(0) \quad or \quad \begin{pmatrix} 0 & 1 \\ -1 & 0 \end{pmatrix}$$

*laid out along the diagonal and all other entries 0.*

(ii) *Show that if $\alpha$ is fixed in the statement of Theorem 16.4.5, then m is unique.*

**Exercise 16.4.7** *If U is a finite dimensional vector space over $\mathbb{R}$ and there exists a non-singular antisymmetric form $\alpha : U^2 \to \mathbb{R}$, show that the dimension of U is even.*

**Exercise 16.4.8** *Which of the following statements about an $n \times n$ real matrix A are true for all appropriate A and all n and which are false? Give reasons.*

(i) *If A is antisymmetric, then the rank of A is even.*

(ii) *If A is antisymmetric, then its characteristic polynomial has the form $P_A(t) = t^{n-2m}(t^2 + 1)^m$.*

(iii) *If A is antisymmetric, then its characteristic polynomial has the form $P_A(t) = t^{n-2m} \prod_{j=1}^{m}(t^2 + d_j^2)$ with $d_j$ real.*

(iv) *If the characteristic polynomial of A takes the form*

$$P_A(t) = t^{n-2m} \prod_{j=1}^{m}(t^2 + d_j^2)$$

*with $d_j$ real, then A is antisymmetric.*

(v) *Given m with $0 \leq 2m \leq n$ and $d_j$ real, there exists a real antisymmetric matrix A with characteristic polynomial $P_A(t) = t^{n-2m} \prod_{j=1}^{m}(t^2 + d_j^2)$.*

**Exercise 16.4.9** *If we work over $\mathbb{C}$, we have results corresponding to Exercise 16.1.26 (iii) which make it very easy to discuss skew-Hermitian forms and matrices.*

(i) *Show that, if A is a skew-Hermitian matrix, then there is a unitary matrix P such that $P^* A P$ is diagonal and its diagonal entries are purely imaginary.*

(ii) *Show that, if A is a skew-Hermitian matrix, then there is an invertible matrix P such that $P^* A P$ is diagonal and its diagonal entries take the values $i, -i$ or 0.*

(iii) *State and prove an appropriate form of Sylvester's law of inertia for skew-Hermitian matrices.*

## 16.5 Further exercises

**Exercise 16.5.1** The result of this exercise is important for the study of Taylor's theorem in many dimensions. However, the reader should treat it as just another exercise involving functions of functions of the type we have referred to as paper tigers.

Let $E$, $F$ and $G$ be finite dimensional vector spaces over $\mathbb{R}$. We write $\mathcal{B}(E, F; G)$ for the space of bilinear maps $\alpha : E \times F \to G$. Define

$$(\Theta(\alpha)(\mathbf{u}))(\mathbf{v}) = \alpha(\mathbf{u}, \mathbf{v})$$

for all $\alpha \in \mathcal{B}(E, F; G)$, $\mathbf{u} \in E$ and $\mathbf{v} \in F$.

(i) Show that $\Theta(\alpha)(\mathbf{u}) \in \mathcal{L}(F, G)$.

(ii) Show that, if $\mathbf{v}$ is fixed,

$$\big(\Theta(\alpha)(\lambda_1\mathbf{u}_1 + \lambda_2\mathbf{u}_2)\big)(\mathbf{v}) = \big(\lambda_1\Theta(\alpha)(\mathbf{u}_1) + \lambda_2\Theta(\alpha)(\mathbf{u}_2)\big)(\mathbf{v})$$

and deduce that

$$\Theta(\alpha)(\lambda_1\mathbf{u}_1 + \lambda_2\mathbf{u}_2) = \lambda_1\Theta(\alpha)(\mathbf{u}_1) + \lambda_2\Theta(\alpha)(\mathbf{u}_2)$$

for all $\lambda_1$, $\lambda_2 \in \mathbb{R}$ and $\mathbf{u}_1$, $\mathbf{u}_2 \in E$. Conclude that $\Theta(\alpha) \in \mathcal{L}(E, \mathcal{L}(F, G))$.

(iii) By arguments similar in spirit to those of (ii), show that $\Theta : \mathcal{B}(E, F; G) \to \mathcal{L}(E, \mathcal{L}(F, G))$ is linear.

(iv) Show that if $(\Theta(\alpha)(\mathbf{u}))(\mathbf{v}) = \mathbf{0}$ for all $\mathbf{u} \in E$, $\mathbf{v} \in F$, then $\alpha = 0$. Deduce that $\Theta$ is injective.

(v) By computing the dimensions of $\mathcal{B}(E, F; G)$ and $\mathcal{L}(E, \mathcal{L}(F, G))$, show that $\Theta$ is an isomorphism.

**Exercise 16.5.2** Let $\beta$ be a bilinear form on a finite dimensional vector space $V$ over $\mathbb{R}$ such that

$$\beta(\mathbf{x}, \mathbf{x}) = 0 \Rightarrow \mathbf{x} = \mathbf{0}.$$

Show that we can find a basis $\mathbf{e}_1, \mathbf{e}_2, \ldots, \mathbf{e}_n$ for $V$ such that $\beta(\mathbf{e}_j, \mathbf{e}_k) = 0$ for $n \geq k > j \geq 1$. What can you say in addition if $\beta$ is symmetric? What result do we recover if $\beta$ is an inner product?

**Exercise 16.5.3** What does it mean to say that two quadratic forms $\mathbf{x}^T A\mathbf{x}$ and $\mathbf{x}^T B\mathbf{x}$ are equivalent? (See Definition 16.1.19 if you have forgotten.) Show, using matrix algebra, that equivalence is, indeed, an equivalence relation on the space of quadratic forms in $n$ variables.

Show that, if $A$ and $B$ are symmetric matrices with $\mathbf{x}^T A\mathbf{x}$ and $\mathbf{x}^T B\mathbf{x}$ equivalent, then the determinants of $A$ and $B$ are both strictly positive, both strictly negative or both zero.

Are the following statements true? Give a proof or counterexample.

(i) If $A$ and $B$ are symmetric matrices with strictly positive determinant, then $\mathbf{x}^T A\mathbf{x}$ and $\mathbf{x}^T B\mathbf{x}$ are equivalent quadratic forms.

(ii) If $A$ and $B$ are symmetric matrices with $\det A = \det B > 0$, then $\mathbf{x}^T A \mathbf{x}$ and $\mathbf{x}^T B \mathbf{x}$ are equivalent quadratic forms.

**Exercise 16.5.4** Consider the real quadratic form

$$c_1 x_1 x_2 + c_2 x_2 x_3 + \cdots + c_{n-1} x_{n-1} x_n,$$

where $n \geq 2$ and all the $c_i$ are non-zero. Explain, without using long calculations, why both the rank and signature are independent of the values of the $c_i$.

Now find the rank and signature.

**Exercise 16.5.5** (i) Let $f_1, f_2, \ldots, f_t, f_{t+1}, f_{t+2}, \ldots, f_{t+u}$ be linear functionals on the finite dimensional real vector space $U$. Let

$$q(\mathbf{x}) = f_1(\mathbf{x})^2 + \cdots + f_t(\mathbf{x})^2 - f_{t+1}(\mathbf{x})^2 - \cdots - f_{t+u}(\mathbf{x})^2.$$

Show that $q$ is a quadratic form of rank $p + q$ and signature $p - q$ where $p \leq t$ and $q \leq u$.

Give an example with all the $f_j$ non-zero for which $p = t = 2$, $q = u = 2$ and an example with all the $f_j$ non-zero for which $p = 1, t = 2, q = u = 2$.

(ii) Consider a quadratic form $Q$ on a finite dimensional real vector space $U$. Show that $Q$ has rank 2 and signature 0 if and only if we can find linearly independent linear functionals $f$ and $g$ such that $Q(\mathbf{x}) = f(\mathbf{x})g(\mathbf{x})$.

**Exercise 16.5.6** Let $q$ be a quadratic form over a real vector space $U$ of dimension $n$. If $V$ is a subspace of $U$ having dimension $n - m$, show that the restriction $q|_V$ of $q$ to $V$ has signature $l$ differing from $k$ by at most $m$.

Show that, given $l$, $k$, $m$ and $n$ with $0 \leq m \leq n$, $|k| \leq n$, $|l| \leq m$ and $|l - k| \leq m$ we can find a quadratic form $q$ on $\mathbb{R}^n$ and a subspace $V$ of dimension $n - m$ such that $q$ has signature $k$ and $q|_V$ has signature $l$.

**Exercise 16.5.7** Suppose that $V$ is a subspace of a real finite dimensional space $U$. Let $V$ have dimension $m$ and $U$ have dimension $2n$. Find a necessary and sufficient condition relating $m$ and $n$ such that any antisymmetric bilinear form $\beta : V^2 \to \mathbb{R}$ can be written as $\beta = \alpha|_{V^2}$ where $\alpha$ is a non-singular antisymmetric form on $U$.

**Exercise 16.5.8** Let $V$ be a real finite dimensional vector space and let $\alpha : V^2 \to \mathbb{R}$ be a symmetric bilinear form. Call a subspace $U$ of $V$ strictly positive definite if $\alpha|_{U^2}$ is a strictly positive definite form on $U$. If $W$ is a subspace of $V$, write

$$W^\perp = \{\mathbf{v} \in V \ : \ \alpha(\mathbf{v}, \mathbf{w}) = 0 \text{ for all } \mathbf{w} \in W\}.$$

(i) Show that $W^\perp$ is a subspace of $V$.

(ii) If $U$ is strictly positive definite, show that $V = U \oplus U^\perp$.

(iii) Give an example of $V$, $\alpha$ and a subspace $W$ such that $V$ is not the direct sum of $W$ and $W^\perp$.

(iv) Give an example of $V$, $\alpha$ and a subspace $W$ such that $V = W \oplus W^\perp$, but $W$ is not strictly positive definite.

(v) If $p$ is the largest number such that $V$ has a strictly positive definite subspace of dimension $p$, establish that every strictly positive definite subspace $U$ is a subspace of a strictly positive definite subspace $X$ of dimension $p$. Is $X$ necessarily uniquely determined by $U$? Give a proof or counterexample.

(vi) Let $V = M_n$, the vector space of $n \times n$ real matrices. If we set $\alpha(A, B) = \operatorname{Tr} AB$, show that $\alpha$ is a symmetric bilinear form. Find a strictly positive definite subspace of $V$ of maximum dimension, justifying your answer.

**Exercise 16.5.9 [Hadamard's inequality revisited]** We work over $\mathbb{C}$. Let $M$ be an $n \times n$ matrix with columns $\mathbf{m}_i$ satisfying $\|\mathbf{m}_i\| \le 1$ for the Euclidean norm. If we set $P = MM^*$, show that $P$ is a positive semi-definite symmetric matrix all of whose entries $p_{ij}$ satisfy $|p_{ij}| \le 1$.

By applying the arithmetic-geometric inequality to the eigenvalues $\lambda_1, \lambda_2, \ldots, \lambda_n$ of $P$, show that

$$(\det P)^{1/n} \le n^{-1} \operatorname{Tr} P \le 1.$$

Deduce that $|\det M| \le 1$. Show that we have equality (that is to say, $|\det M| = 1$) if and only if $M$ is unitary.

Deduce that if $A$ is a complex $n \times n$ matrix with columns $\mathbf{a}_j$

$$|\det A| \le \prod_{j=1}^{n} \|\mathbf{a}_j\|$$

with equality if and only if either one of the columns is the zero vector or the column vectors are orthogonal.

**Exercise 16.5.10** Let $\mathcal{P}_n$ be the set of strictly positive definite $n \times n$ symmetric matrices. Show that

$$A \in \mathcal{P}_n, \ t > 0 \Rightarrow tA \in \mathcal{P}_n$$

and

$$A, B \in \mathcal{P}_n, \ 1 \ge t \ge 0 \Rightarrow tA + (1 - t)B \in \mathcal{P}_n.$$

(We say that $\mathcal{P}_n$ is a *convex cone*.)

Show that, if $A \in \mathcal{P}_n$, then

$$\log \det A - \operatorname{Tr} A + n \le 0$$

with equality if and only if $A = I$.

If $A \in \mathcal{P}_n$, let us write $\phi(A) = -\log \det A$. Show that

$$A, B \in \mathcal{P}_n, \ 1 \ge t \ge 0 \Rightarrow \phi\big(tA + (1 - t)B\big) \le t\phi(A) + (1 - t)\phi(B).$$

(We say that $\phi$ is a *convex function*.)

[Hint: Recall that, under appropriate circumstances, we can find a non-singular $P$ such that $P^T A P$ and $P^T B P$ are both diagonal.]

**Exercise 16.5.11 [Drawing a straight line]** Suppose that we have a quantity $y$ that we believe satisfies the equation

$$y = a + bt \qquad\qquad \bigstar$$

with $a$ and $b$ unknown. We make observations of $y$ at $n$ distinct times $t_1, t_2, \ldots, t_n$ (with $n \geq 2$), obtaining $y_1, y_2, \ldots, y_n$, but, because there will be errors in our observations, we do not expect to have $y_j = a + bt_j$.

The discussion of the method of least squares in Section 7.5 suggests that we should estimate $a$ and $b$ as $\hat{a}$ and $\hat{b}$, where $\hat{a}, \hat{b}$ are the values of $a$ and $b$ which minimise

$$\sum_{j=1}^{n} \left(y_j - (a + bt_j)\right)^2.$$

Find $\hat{a}$ and $\hat{b}$.

Show that we can find $\alpha$ and $\beta$ such that, writing $s_j = \beta t_j + \alpha$, we have

$$\sum_{j=1}^{n} s_j = 0 \quad \text{and} \quad \sum_{j=1}^{n} s_j^2 = 1.$$

Find $u$ and $v$ so that, writing $s = \beta t + \alpha$, $\bigstar$ takes the form

$$y = u + vs.$$

Find $\hat{u}$ and $\hat{v}$ so that $\sum_{j=1}^{n}(y_j - (u + vs_j))^2$ is minimised when $u = \hat{u}$, $v = \hat{v}$ and write $\hat{a}$ and $\hat{b}$ in terms of $\hat{u}$ and $\hat{v}$.

**Exercise 16.5.12** (Requires a small amount of probability theory.) Suppose that we make observations $Y_j$ at time $t_j$ which we believe are governed by the equation

$$Y_j = a + bt_j + \sigma Z_j$$

where $Z_1, Z_2, \ldots, Z_n$ are independent normal random variables each with mean 0 and variance 1. As usual, $\sigma > 0$. (So, whereas in the previous question we just talked about errors, we now make strong assumptions about the way the errors arise.) The discussion in the previous question shows that there is no loss in generality and some gain in computational simplicity if we suppose that

$$\sum_{j=1}^{n} t_j = 0 \quad \text{and} \quad \sum_{j=1}^{n} t_j^2 = 1.$$

We shall suppose that $n \geq 3$ since we can always fit a line through two points.

We are interested in estimating $a$, $b$ and $\sigma^2$. Check, from the results of the last question, that the least squares method gives the estimates

$$\hat{a} = \bar{Y} = n^{-1} \sum_{j=1}^{n} Y_j \quad \text{and} \quad \hat{b} = \sum_{j=1}^{n} t_j Y_j.$$

We can get some feeling[6] for the size of $\sigma$ by looking at

$$\hat{\sigma}^2 = (n-2)^{-1} \sum_{j=1}^{n} \left(Y_j - (\hat{a} + \hat{b}t_j)\right)^2.$$

Show that, if we write $\tilde{a} = \hat{a} - a$ and $\tilde{b} = \hat{b} - b$, we have

$$\tilde{a} = \sigma \bar{Z} = \sigma n^{-1} \sum_{j=1}^{n} Z_j, \quad \hat{b} = \sigma \sum_{j=1}^{n} t_j Z_j, \quad \hat{\sigma}^2 = \sigma^2 (n-2)^{-1} \sum_{j=1}^{n} \left(Z_j - (\tilde{a} + \tilde{b}t_j)\right)^2.$$

In this question we shall derive the joint distribution of $\tilde{a}$, $\tilde{b}$ and $\hat{\sigma}^2$ (and so of $\hat{a}$, $\hat{b}$ and $\hat{\sigma}^2$) by elementary matrix algebra.

(i) Show that $\mathbf{e}_1 = (n^{-1/2}, n^{-1/2}, \ldots, n^{-1/2})^T$ and $\mathbf{e}_2 = (t_1, t_2, \ldots, t_n)^T$ are orthogonal column vectors of norm 1 with respect to the standard inner product on $\mathbb{R}^n$. Deduce that there exists an orthonormal basis $\mathbf{e}_1, \mathbf{e}_2, \ldots, \mathbf{e}_n$ for $\mathbb{R}^n$. If we let $M$ be the $n \times n$ matrix whose $r$th row is $\mathbf{e}_r^T$, explain why $M \in O(\mathbb{R}^n)$ and so preserves inner products.

(ii) Let $W_1, W_2, \ldots, W_n$ be the set of random variables defined by

$$\mathbf{W} = M\mathbf{Z}$$

that is to say, by $W_i = \sum_{j=1}^{n} e_{ij} Z_j$. Show that $Z_1, Z_2, \ldots, Z_n$ have joint density function

$$f(z_1, z_2, \ldots, z_n) = (2\pi)^{-n/2} \exp(-\|\mathbf{z}\|^2/2)$$

and use the fact that $M \in O(\mathbb{R}^n)$ to show that $W_1, W_2, \ldots, W_n$ have joint density function

$$g(w_1, w_2, \ldots, w_n) = (2\pi)^{-n/2} \exp(-\|\mathbf{w}\|^2/2).$$

Conclude that $W_1, W_2, \ldots, W_n$ are independent normal random variables each with mean 0 and variance 1.

(iii) Explain why

$$W_1^2 + W_2^2 + \cdots + W_n^2 = Z_1^2 + Z_2^2 + \cdots + Z_n^2.$$

Show that $\tilde{a} = n^{-1/2} \sigma W_1$, $\tilde{b} = \sigma W_2$ and

$$\sum_{j=1}^{n} \left(Z_j - (\tilde{a} + \tilde{b}t_j)\right)^2 = W_3^2 + W_4^2 + \cdots + W_n^2.$$

---

[6] This sentence is deliberately vague. The reader should simply observe that $\hat{\sigma}^2$, or something like it, is likely to be interesting. In particular, it does not really matter whether we divide by $n-2$, $n-1$ or $n$.

(iv) Deduce the following results. The random variables $\hat{a}$, $\hat{b}$ and $\hat{\sigma}^2$ are independent.[7] The random variable $\hat{a}$ is normally distributed with mean $a$ and variance $\sigma^2/n$. The random variable $\hat{b}$ is normally distributed with mean $b$ and variance $\sigma^2$. The random variable $(n-2)\sigma^{-2}\hat{\sigma}^2$ is distributed like the sum of the squares of $n-2$ independent normally distributed random variables with mean 0 and variance 1.

**Exercise 16.5.13** Suppose that we make observations $X_1, X_2 \ldots, X_n$ which we believe to be independent normal random variables each with mean $\mu$ and variance $\sigma^2$.

(i) Suppose that $\mu = 0$ and $\sigma = 1$. By imitating the arguments of the previous question, show that

$$\bar{X} = n^{-1}\sum_{j=1}^{n} X_j \quad \text{and} \quad \hat{\sigma}^2 = (n-1)^{-1}\sum_{j=1}^{n}(X_j - \bar{X})^2$$

are independent and find their distributions.

(ii) Now let $\mu$ and $\sigma^2$ take general values. Show that $\bar{X}$ and $\hat{\sigma}^2$ remain independent and find their distributions.

**Exercise 16.5.14** Let $V$ be the vector space of $n \times n$ matrices over $\mathbb{R}$. Show that

$$q(A) = \mathrm{Tr}(A^2) - (\mathrm{Tr}\, A)^2$$

is a quadratic form on $V$. By considering the subspaces $V_1$ consisting of matrices $\lambda I$, $V_2$ consisting of matrices of trace 0, and $V_3$ consisting of antisymmetric matrices, or otherwise, find the signature of $q$.

**Exercise 16.5.15** Two quadratic forms $h$ and $k$ on $\mathbb{R}^3$ are defined by

$$h(x, y, z) = 2x^2 + 5y^2 + 4z^2 + 6xy + 14yz + 8zx$$

and

$$k(x, y, z) = 2x^2 + 14y^2 + 3z^2 + 10xy - 4yz.$$

Show that one of these forms is strictly positive definite and one of them is not. Determine $\lambda$, $\mu$ and $\nu$ so that, in an appropriate coordinate system, the form of the strictly positive definite one becomes $X^2 + Y^2 + Z^2$ and the form of the other becomes $\lambda X^2 + \mu Y^2 + \nu Z^2$.

**Exercise 16.5.16** Find a linear transformation that simultaneously reduces the quadratic forms

$$2x^2 + y^2 + 2z^2 + 2yz - 2zx$$
$$x^2 + 2y^2 + 2z^2 + 4yz$$

to the forms $X^2 + Y^2 + Z^2$ and $\lambda X^2 + \mu Y^2 + \nu Z^2$ where $\lambda$, $\mu$ and $\nu$ are to be found.

---

[7] Observe that, if random variables are independent, their joint distribution is known once their individual distributions are known.

**Exercise 16.5.17** Let $n \geq 2$. Find the eigenvalues and corresponding eigenvectors of the $n \times n$ matrix

$$A = \begin{pmatrix} 1 & 1 & \cdots & 1 \\ 1 & 1 & \cdots & 1 \\ \vdots & \vdots & \ddots & \vdots \\ 1 & 1 & \cdots & 1 \end{pmatrix}.$$

If the quadratic form $F$ is given by

$$c \sum_{r=1}^{n} x_r^2 + \sum_{r=1}^{n} \sum_{s=r+1}^{n} x_r x_s,$$

explain why there exists an orthonormal change of coordinates such that $F$ takes the form $\sum_{r=1}^{n} \mu_r y_y^2$ with respect to the new coordinate system. Find the $\mu_r$ explicitly.

**Exercise 16.5.18** (A result required for Exercise 16.5.19.) Suppose that $n$ is a strictly positive integer and $x$ is a positive rational number with $x^2 = n$. Let $x = p/q$ with $p$ and $q$ positive integers with no common factor. By considering the equation

$$p^2 \equiv nq^2 \pmod{q},$$

show that $x$ is, in fact, an integer.

Suppose that $a_j \in \mathbb{Z}$. Show that any rational root of the monic polynomial $t^n + a_{n-1}t^{n-1} + \cdots + a_0$ is, in fact, an integer.

**Exercise 16.5.19** Let $A$ be the Hermitian matrix

$$\begin{pmatrix} 1 & i & 2i \\ -i & 3 & -i \\ -2i & i & 5 \end{pmatrix}.$$

Show, using Exercise 16.5.18, or otherwise, that there does not exist a unitary matrix $U$ and a diagonal matrix $E$ with rational entries such that $U^*AU = E$.

Setting out your method carefully, find an invertible matrix $B$ (with entries in $\mathbb{C}$) and a diagonal matrix $D$ with rational entries such that $B^*AB = D$.

**Exercise 16.5.20** Let $f(x_1, x_2, \ldots, x_n) = \sum_{1 \leq i,j \leq n} a_{ij} x_i x_j$ where $A = (a_{ij})$ is a real symmetric matrix and $g(x_1, x_2, \ldots, x_n) = \sum_{1 \leq i \leq n} x_i^2$. Show that the stationary points and associated values of $f$, subject to the restriction $g(\mathbf{x}) = 1$, are given by the eigenvectors of norm 1 and associated eigenvalues of $A$. (Recall that $\mathbf{x}$ is such a stationary point if $g(\mathbf{x}) = 1$ and

$$\|\mathbf{h}\|^{-1}|f(\mathbf{x}+\mathbf{h}) - f(\mathbf{x})| \to 0$$

as $\|\mathbf{h}\| \to 0$ with $g(\mathbf{x}+\mathbf{h}) = 1$.)

Deduce that the eigenvectors of $A$ give the stationary points of $f(\mathbf{x})/g(\mathbf{x})$ for $\mathbf{x} \neq \mathbf{0}$. What can you say about the eigenvalues?

**Exercise 16.5.21** (This continues Exercise 16.3.19 (i) and harks back to Theorem 7.3.1 which the reader should reread.) We work in $\mathbb{R}^2$ considered as a space of column vectors.

(i) Take $\alpha$ to be the quadratic form given by $\alpha(\mathbf{x}) = x_1^2 - x_2^2$ and set

$$A = \begin{pmatrix} 1 & 0 \\ 0 & -1 \end{pmatrix}.$$

Let $M$ be a $2 \times 2$ real matrix. Show that the following statements are equivalent.

(a) $\alpha(M\mathbf{x}) = \alpha\mathbf{x}$ for all $\mathbf{x} \in \mathbb{R}^2$.

(b) $M^T A M = A$.

(c) There exists a real $s$ such that

$$M = \pm \begin{pmatrix} \cosh s & \sinh s \\ \sinh s & \cosh s \end{pmatrix} \quad \text{or} \quad M = \pm \begin{pmatrix} \cosh s & \sinh s \\ -\sinh s & -\cosh s \end{pmatrix}.$$

(ii) Deduce the result of Exercise 16.3.19 (i).

(iii) Let $\beta(\mathbf{x}) = x_1^2 + x_2^2$ and set $B = I$. Write down results parallelling those of part (i).

**Exercise 16.5.22** (i) Let $\phi_1$ and $\phi_2$ be non-degenerate bilinear forms on a finite dimensional real vector space $V$. Show that there exists an isomorphism $\alpha : V \to V$ such that

$$\phi_2(\mathbf{u}, \mathbf{v}) = \phi_1(\mathbf{u}, \alpha\mathbf{v})$$

for all $\mathbf{u}, \mathbf{v} \in V$.

(ii) Show, conversely, that, if $\psi_1$ is a non-degenerate bilinear form on a finite dimensional real vector space $V$ and $\beta : V \to V$ is an isomorphism, then the equation

$$\psi_2(\mathbf{u}, \mathbf{v}) = \psi_1(\mathbf{u}, \beta\mathbf{v})$$

for all $\mathbf{u}, \mathbf{v} \in V$ defines a non-degenerate bilinear form $\psi_2$.

(iii) If, in part (i), both $\phi_1$ and $\phi_2$ are symmetric, show that $\alpha$ is self-adjoint with respect to $\phi_1$ in the sense that

$$\phi_1(\alpha\mathbf{u}, \mathbf{v}) = \phi_1(\mathbf{u}, \alpha\mathbf{v})$$

for all $\mathbf{u}, \mathbf{v} \in V$. Is $\alpha$ necessarily self-adjoint with respect to $\phi_2$? Give reasons.

(iv) If, in part (i), both $\phi_1$ and $\phi_2$ are inner products (that is to say, symmetric and strictly positive definite), show that, in the language of Exercise 15.5.11, $\alpha$ is strictly positive definite. Deduce that we can find an isomorphism $\gamma : V \to V$ such that $\gamma$ is strictly positive definite and

$$\phi_2(\mathbf{u}, \mathbf{v}) = \phi_1(\gamma\mathbf{u}, \gamma\mathbf{v})$$

for all $\mathbf{u}, \mathbf{v} \in V$.

(v) If $\psi_1$ is an inner product on $V$ and $\gamma : V \to V$ is strictly positive definite, show that the equation

$$\psi_2(\mathbf{u}, \mathbf{v}) = \psi_1(\gamma\mathbf{u}, \gamma\mathbf{v})$$

for all $\mathbf{u}, \mathbf{v} \in V$ defines an inner product.

**Exercise 16.5.23** Let $U$ be a real vector space of dimension $n$ and $\phi$ a non-degenerate symmetric bilinear form on $U$. If $M$ is a subspace of $U$ show that

$$M^\perp = \{\mathbf{y} \in U : \phi(\mathbf{x}, \mathbf{y}) = 0 \quad \text{for all } \mathbf{x} \in M\}$$

is a subspace of $M$.

By recalling Lemma 11.4.13 and Lemma 16.1.7, or otherwise, show that

$$\dim M + \dim M^\perp = n.$$

If $U = \mathbb{R}^2$ and

$$\phi\big((x_1, x_2)^T, (y_1, y_2)^T\big) = x_1 y_1 - x_2 y_2,$$

find one dimensional subspaces $M_1$ and $M_2$ such that $M_1 \cap M_1^\perp = \{\mathbf{0}\}$ and $M_2 = M_2^\perp$. If $M$ is a subspace of $U$ with $M^\perp = M$ show that $\dim M \le n/2$.

**Exercise 16.5.24** (A short supplement to Exercise 16.5.23.) Let $U$ be a real vector space of dimension $n$ and $\phi$ a symmetric bilinear form on $U$. If $M$ is a subspace of $U$ show that

$$M^\perp = \{\mathbf{y} \in U : \phi(\mathbf{x}, \mathbf{y}) = 0 \quad \text{for all } \mathbf{x} \in M\}$$

is a subspace of $M$. By using Exercise 16.5.23, or otherwise, show that

$$\dim M + \dim M^\perp \ge n.$$

**Exercise 16.5.25** We continue with the notation and hypotheses of Exercise 16.5.23. An automorphism $\gamma$ of $U$ is said to be an $\phi$-*automorphism* if it preserves $\phi$ (that is to say

$$\phi(\gamma\mathbf{x}, \gamma\mathbf{y}) = \phi(\mathbf{x}, \mathbf{y})$$

for all $\mathbf{x}, \mathbf{y} \in U$). A vector $\mathbf{x} \in U$ is said to be *isotropic* (with respect to $\phi$) if $\phi(\mathbf{x}, \mathbf{x}) = 0$. We shall call a subspace $M$ of $U$ *non-isotropic* if $M \cap M^\perp = \{\mathbf{0}\}$. Prove the following results.

(i) There exist vectors which are not isotropic.
(ii) If $\gamma$ is a $\phi$-automorphism, then $\gamma M = M \Leftrightarrow \gamma M^\perp = M^\perp$.
(iii) If $M$ is a non-isotropic subspace, there exists a unique $\phi$-automorphism $\theta_M$ (the *symmetry with respect to M*) such that

$$\theta_M \mathbf{x} = \begin{cases} \mathbf{x} & \text{if } \mathbf{x} \in M, \\ -\mathbf{x} & \text{if } \mathbf{x} \in M^\perp. \end{cases}$$

Let $\alpha$ be a $\phi$-automorphism and $\mathbf{e}_1$ a non-isotropic vector. Show that the two vectors $\alpha\mathbf{e}_1 \pm \mathbf{e}_1$ cannot both be isotropic and deduce that there exists a non-isotropic subspace $M_1$ of dimension $1$ or $n - 1$ such that $\theta_{M_1}\alpha\mathbf{e}_1 = \mathbf{e}_1$. By using induction on $n$, or otherwise, show that every $\phi$-automorphism is the product of at most $n$ symmetries with respect to non-isotropic subspaces.

To what earlier result on distance preserving maps does this result correspond?

**Exercise 16.5.26** Let $A$ be an $n \times n$ matrix with *real* entries such that $AA^T = I$. Explain why, if we work over the *complex numbers*, $A$ is a unitary matrix. If $\lambda$ is a real eigenvalue, show that we can find a basis of orthonormal vectors with *real* entries for the space

$$\{\mathbf{z} \in \mathbb{C}^n : A\mathbf{z} = \lambda\mathbf{z}\}.$$

Now suppose that $\lambda$ is an eigenvalue which is not real. Explain why we can find a $0 < \theta < 2\pi$ such that $\lambda = e^{i\theta}$. If $\mathbf{z} = \mathbf{x} + i\mathbf{y}$ is a corresponding eigenvector such that the entries of $\mathbf{x}$ and $\mathbf{y}$ are real, compute $A\mathbf{x}$ and $A\mathbf{y}$. Show that the subspace

$$\{\mathbf{z} \in \mathbb{C}^n : A\mathbf{z} = \lambda\mathbf{z}\}$$

has even dimension $2m$, say, and has a basis of orthonormal vectors with *real* entries $\mathbf{e}_1$, $\mathbf{e}_2, \ldots, \mathbf{e}_{2m}$ such that

$$A\mathbf{e}_{2r-1} = \cos\theta\mathbf{e}_{2r-1} + i\sin\theta\mathbf{e}_{2r}$$
$$A\mathbf{e}_{2r} = -\sin\theta\mathbf{e}_{2r-1} + i\cos\theta\mathbf{e}_{2r}.$$

Hence show that, if we now work over the *real numbers*, there is an $n \times n$ orthogonal matrix $M$ such that $MAM^T$ is a matrix with matrices $K_j$ taking the form

$$(1), \ (-1) \text{ or } \begin{pmatrix} \cos\theta_j & -\sin\theta_j \\ \sin\theta_j & \cos\theta_j \end{pmatrix}$$

laid out along the diagonal and all other entries 0.

[The more elementary approach of Exercise 7.6.18 is probably better, but this method brings out the connection between the results for unitary and orthogonal matrices.]

**Exercise 16.5.27 [Routh's rule]**[8] Suppose that $A = (a_{ij})_{1 \le i,j \le n}$ is a real symmetric matrix. If $a_{11} > 0$ and $B = (b_{ij})_{2 \le i,j \le n}$ is the Schur complement given (as in Lemma 16.3.9) by

$$b_{ij} = a_{ij} - l_i l_j \quad \text{for } 2 \le i, j \le n,$$

show that

$$\det A = a_{11} \det B.$$

Show, more generally, that

$$\det(a_{ij})_{1 \le i,j \le r} = a_{11}\det(b_{ij})_{2 \le i,j \le r}.$$

Deduce, by induction, or otherwise, that $A$ is strictly positive definite if and only if

$$\det(a_{ij})_{1 \le i,j \le r} > 0$$

for all $1 \le r \le n$ (in traditional language 'all the leading minors of $A$ are strictly positive').

---

[8] Routh beat Maxwell in the Cambridge undergraduate mathematics exams, but is chiefly famous as a great teacher. Rayleigh, who had been one of his pupils, recalled an 'undergraduate [whose] primary difficulty lay in conceiving how anything could float. This was so completely removed by Dr Routh's lucid explanation that he went away sorely perplexed as to how anything could sink!' Routh was an excellent mathematician and this is only one of several results named after him.

**Exercise 16.5.28** Use the method of the previous question to show that, if $A$ is a real symmetric matrix with non-zero leading minors

$$A_r = \det(a_{ij})_{1 \le i, j \le r},$$

the real quadratic form $\sum_{1 \le i, j \le n}^n x_i a_{ij} x_j$ can reduced by a real non-singular change of coordinates to $\sum_{i=1}^n (A_i/A_{i-1}) y_i^2$ (where we set $A_0 = 1$).

**Exercise 16.5.29** (A slightly, but not very, different proof of Sylvester's law of inertia). Let $\phi_1, \phi_2, \ldots, \phi_k$ be linear forms (that is to say, linear functionals) on $\mathbb{R}^n$ and let $W$ be a subspace of $\mathbb{R}^n$. If $k < \dim W$, show that there exists a non-zero $\mathbf{y} \in W$ such that

$$\phi_1(\mathbf{y}) = \phi_2(\mathbf{y}) = \ldots = \phi_k(\mathbf{y}) = 0.$$

Now let $f$ be the quadratic form on $\mathbb{R}^n$ given by

$$f(x_1, x_2, \ldots, x_n) = x_1^2 + \cdots + x_p^2 - x_{p+1}^2 - \cdots - x_{p+q}^2$$

with $p, q \ge 0$ and $p + q \le n$. Suppose that $\psi_1, \psi_2, \ldots, \psi_{r+s}$ are linear forms on $\mathbb{R}^n$ such that

$$f(\mathbf{y}) = \psi_1(\mathbf{y})^2 + \cdots + \psi_r(\mathbf{y})^2 - \psi_{r+1}(\mathbf{y})^2 - \psi_{r+s}(\mathbf{y})^2.$$

Show that $p \le r$.

Deduce Sylvester's law of inertia.

**Exercise 16.5.30 [An analytic proof of Sylvester's law of inertia]** In this question we work in the space $M_n(\mathbb{R})$ of $n \times n$ matrices with distances given by the operator norm.

(i) Use the result of Exercise 7.6.18 (or Exercise 16.5.26) to show that, given any special orthogonal $n \times n$ matrix $P_1$, we can find a continuous map

$$P : [0, 1] \to M_n(\mathbb{R})$$

such that $P(0) = I$, $P(1) = P_1$ and $P(s)$ is special orthogonal for all $s \in [0, 1]$.

(ii) Deduce that, given any special orthogonal $n \times n$ matrices $P_0$ and $P_1$, we can find a continuous map

$$P : [0, 1] \to M_n(\mathbb{R})$$

such that $P(0) = P_0$, $P(1) = P_1$ and $P(s)$ is special orthogonal for all $s \in [0, 1]$.

(iii) Suppose that $A$ is a non-singular $n \times n$ matrix. If $P$ is as in (ii), show that $P(s)^T A P(s)$ is non-singular and so the eigenvalues of $P(s)^T A P(s)$ are non-zero for all $s \in [0, 1]$. Assuming that, as the coefficients in a polynomial vary continuously, so do the roots of the polynomial (see Exercise 15.5.14), deduce that the number of strictly positive roots of the characteristic polynomial of $P(s)^T A P(s)$ remains constant as $s$ varies.

(iv) Continuing with the notation and hypotheses of (ii), show that, if $P(0)^T A P(0) = D_0$ and $P(1)^T A P(1) = D_1$ with $D_0$ and $D_1$ diagonal, then $D_0$ and $D_1$ have the same number of strictly positive and strictly negative terms on their diagonals.

(v) Deduce that, if $A$ is a non-singular symmetric matrix and $P_0$, $P_1$ are special orthogonal with $P_0^T A P_0 = D_0$ and $P_1^T A P_1 = D_1$ diagonal, then $D_0$ and $D_1$ have the same number of strictly positive and strictly negative terms on their diagonals. Thus we have proved Sylvester's law of inertia for non-singular symmetric matrices.

(vi) To obtain the full result, explain why, if $A$ is a symmetric matrix, we can find $\epsilon_n \to 0$ such that $A + \epsilon_n I$ and $A - \epsilon_n I$ are non-singular. By applying part (v) to these matrices, show that, if $P_0$ and $P_1$ are special orthogonal with $P_0^T A P_0 = D_0$ and $P_1^T A P_1 = D_1$ diagonal, then $D_0$ and $D_1$ have the same number of strictly positive, strictly negative terms and zero terms on their diagonals.

**Exercise 16.5.31** Consider the quadratic forms $a_n$, $b_n$, $c_n : \mathbb{R}^n \to \mathbb{R}$ given by

$$a_n(x_1, x_2, \ldots, x_n) = \sum_{i=1}^{n} \sum_{j=1}^{n} x_i x_j,$$

$$b_n(x_1, x_2, \ldots, x_n) = \sum_{i=1}^{n} \sum_{j=1}^{n} x_i \min\{i, j\} x_j,$$

$$c_n(x_1, x_2, \ldots, x_n) = \sum_{i=1}^{n} \sum_{j=1}^{n} x_i \max\{i, j\} x_j.$$

By completing the square, find a simple expression for $a_n$. Deduce the rank and signature of $a_n$.

By considering

$$b_n(x_1, x_2, \ldots, x_n) - b_{n-1}(x_2, \ldots, x_n),$$

diagonalise $b_n$. Deduce the rank and signature of $b_n$.

Find a simple expression for

$$c_n(x_1, x_2, \ldots, x_n) + b_n(x_n, x_{n-1}, \ldots, x_1)$$

in terms of $a_n(x_1, x_2, \ldots, x_n)$ and use it to diagonalise $c_n$. Deduce the rank and signature of $c_n$.

**Exercise 16.5.32** The real non-degenerate quadratic form $q(x_1, x_2, \ldots, x_n)$ vanishes if $x_{k+1} = x_{k+2} = \ldots = x_n = 0$. Show, by induction on $t = n - k$, or otherwise, that the form $q$ is equivalent to

$$y_1 y_{k+1} + y_2 y_{k+2} + \cdots + y_k y_{2k} + p(y_{2k+1}, y_{2k+2}, \ldots, y_n),$$

where $p$ is a non-degenerate quadratic form. Deduce that, if $q$ is reduced to diagonal form

$$w_1^2 + \cdots + w_s^2 - w_{s+1}^2 - \cdots - w_n^2$$

we have $k \leq s \leq n - k$.

**Exercise 16.5.33** (i) If $A = (a_{ij})$ is an $n \times n$ positive semi-definite symmetric matrix, show that either $a_{11} > 0$ or $a_{i1} = 0$ for all $i$.

(ii) Let $A$ be an $n \times n$ real matrix. Show that $A$ is a positive semi-definite symmetric matrix if and only if $A = L^T L$, where $L$ is a real lower triangular matrix.

**Exercise 16.5.34** Use the 'Cholesky method' to determine the values of $\lambda \in \mathbb{R}$ for which

$$\begin{pmatrix} 2 & -4 & 2 \\ -4 & 10 + \lambda & 2 + 3\lambda \\ 2 & 2 + 3\lambda & 23 + 9\lambda \end{pmatrix}$$

is strictly positive definite and find the Cholesky factorisation for those values.

**Exercise 16.5.35** (i) Suppose that $n \geq 1$. Let $A$ be an $n \times n$ real matrix of rank $n$. By considering the Cholesky factorisation of $B = A^T A$, prove the existence and uniqueness of the $QR$ factorisation

$$A = QR$$

where $Q$ is an $n \times n$ orthogonal matrix and $R$ is an $n \times n$ upper triangular matrix with strictly positive diagonal entries.

(ii) Suppose that we drop the condition that $A$ has rank $n$. Does the $QR$ factorisation always exist? If the factorisation exists it is unique? Give proofs or counterexamples.

(iii) In parts (i) and (ii) we considered an $n \times n$ matrix. Use the method of (i) to prove that if $n \geq m \geq 1$ and $A$ is an $n \times m$ real matrix of rank $m$ then we can find an $n \times n$ orthogonal matrix $Q$ and an $n \times m$ thin upper triangular matrix $R$ such that $A = QR$.

# References

[1] A. A. Albert and Benjamin Muckenhoupt. On matrices of trace zero. *Michigan Mathematical Journal*, **4** (1957) 1–3.

[2] Aristotle. *Nicomachean Ethics*, translated by Martin Ostwald. Macmillan, New York, 1962.

[3] S. Axler. Down with determinants. *American Mathematical Monthly*, **102** (1995) 139–154.

[4] S. Axler. *Linear Algebra Done Right*, 2nd edn. Springer, New York, 1997.

[5] George Boole. *A Treatise on Differential Equations*. Macmillan, Cambridge, 1859.

[6] George Boole. *A Treatise on the Calculus of Finite Differences*. Macmillan, London, 1860.

[7] Robert Burton. *The Anatomy of Melancholy*. Henry Cripps, Oxford, 1621.

[8] I. Calvino. *Invisible Cities*, translated by W. Weaver of *Le Città Invisibili*. Harcourt, Brace, Jovanovich, New York, 1972.

[9] Lewis Carroll. *Through the Looking-Glass, and What Alice Found There*. Macmillan, London, 1872.

[10] Lewis Carroll. *A Tangled Tale*. Macmillan, London, 1885. Available on Gutenberg.

[11] G. Chrystal. *Algebra*. Adam and Charles Black, Edinburgh, 1889. There is a Dover reprint.

[12] Arthur Conan Doyle. *The Sign of Four*. Spencer Blackett, London, 1890.

[13] Charles Darwin. Letter to Asa Gray of 20 April 1863. Available at www.darwinproject.ac.uk/entry-4110.

[14] Patrice Dubré. *Louis Pasteur*. Johns Hopkins University Press, Baltimore, MD, 1994.

[15] Freeman Dyson. Turning points: a meeting with Enrico Fermi. *Nature*, **427** (22 January 2004) 297.

[16] Albert Einstein. Considerations concerning the fundamentals of theoretical physics. *Science*, **91** (1940) 487–492.

[17] J. H. Ewing and F. W. Gehring, editors. *Paul Halmos, Celebrating Fifty Years of Mathematics*. Springer, New York, 1991.

[18] P. R. Halmos. *Finite-Dimensional Vector Spaces*, 2nd edn. Van Nostrand, Princeton, NJ, 1958.

[19] Y. Katznelson and Y. R. Katznelson. *A (Terse) Introduction to Linear Algebra*. American Mathematical Society, Providence, RI, 2008.

[20] D. E. Knuth. *The Art of Computer Programming*, 3rd edn. Addison-Wesley, Reading, MA, 1997.

[21] J. C. Maxwell. *A Treatise on Electricity and Magnetism*. Oxford University Press, Oxford, 1873. There is a Cambridge University Press reprint.

[22] J. C. Maxwell. *Matter and Motion.* Society for Promoting Christian Knowledge, London, 1876. There is a Cambridge University Press reprint.

[23] L. Mirsky. *An Introduction to Linear Algebra.* Oxford University Press, Oxford, 1955. There is a Dover reprint.

[24] R. E. Moritz. *Memorabilia Mathematica.* Macmillan, New York, 1914. Reissued by Dover as *On Mathematics and Mathematicians.*

[25] T. Muir. *The Theory of Determinants in the Historical Order of Development.* Macmillan, London, 1890. There is a Dover reprint.

[26] A. Pais. *Subtle is the Lord: The Science and the Life of Albert Einstein.* Oxford University Press, New York, 1982.

[27] A. Pais. *Inward Bound.* Oxford University Press, Oxford, 1986.

[28] G. Strang. *Linear Algebra and Its Applications*, 2nd edn. Harcourt Brace Jovanovich, Orlando, FL, 1980.

[29] Jonathan Swift. *Gulliver's Travels.* Benjamin Motte, London, 1726.

[30] Olga Taussky. A recurring theorem on determinants. *American Mathematical Monthly*, **56** (1949) 672–676.

[31] Giorgio Vasari. *The Lives of the Artists*, translated by J. C. and P. Bondanella. Oxford World's Classics, Oxford, 1991.

[32] James E. Ward III. Vector spaces of magic squares. *Mathematics Magazine*, **53** (2) (March 1980) 108–111.

[33] H. Weyl. *Symmetry.* Princeton University Press, Princeton, NJ, 1952.

# Index